Huaqun Guo · Ian McLoughlin ·
Umayal Lakshmanan · Xiaoxiao Miao ·
Eyasu Getahun Chekole · Weizhi Meng ·
Peng Cheng Wang · Jiqiang Lu ·
Nicholas Heng Loong Wong
Editors

Proceedings of the 10th IRC Conference on Science, Engineering and Technology

IRC-SET 2024, 17 August, Singapore

Springer

Editors
Huaqun Guo
Infocomm Technology Cluster
Singapore Institute of Technology
Singapore, Singapore

Umayal Lakshmanan
International Researchers Club
Singapore, Singapore

Eyasu Getahun Chekole
Department of Information Systems
Technology and Design
Singapore University of Technology
and Design
Singapore, Singapore

Peng Cheng Wang
Engineering Cluster
Singapore Institute of Technology
Singapore, Singapore

Nicholas Heng Loong Wong
Infocomm Technology Cluster
Singapore Institute of Technology
Singapore, Singapore

Ian McLoughlin
Infocomm Technology Cluster
Singapore Institute of Technology
Singapore, Singapore

Xiaoxiao Miao
Infocomm Technology Cluster
Singapore Institute of Technology
Singapore, Singapore

Weizhi Meng
Department of Applied Mathematics
and Computer Science
Technical University of Denmark (DTU)
Kongens Lyngby, Denmark

Jiqiang Lu
School of Cyber Science and Technology
Beihang University
Beijing, China

ISSN 2948-2321 ISSN 2948-233X (electronic)
Proceedings in Technology Transfer
ISBN 978-981-96-3769-0 ISBN 978-981-96-3770-6 (eBook)
https://doi.org/10.1007/978-981-96-3770-6

This Springer imprint is published by the registered company Springer Nature Singapore Pte Ltd.
The registered company address is: 152 Beach Road, #21-01/04 Gateway East, Singapore 189721, Singapore

Proceedings in Technology Transfer

The book series "Proceedings in Technology Transfer" provides a valuable platform for researchers, practitioners, and policymakers to share their experiences and insights related to technology transfer. Our aim is to facilitate knowledge exchange between academic researchers and industry professionals, accelerating the introduction of new technologies and driving economic growth.

Our series publishes books and conference proceedings that focus on original research findings, best practice recommendations, emerging trends, and technology evaluations. The series covers all topics which explore the transfer of technology in various industry sectors that require innovations and economic growth.

Join us in advancing the technology transfer and its impact on innovation. Publish your proceedings with us and offer insights and expertise to professionals in your field.

Preface

The International Researchers Club (IRC) (www.irc.org.sg) was set up in 2001. The vision of IRC is to create a vibrant and innovative research community for Singapore and beyond, through contributions of technical specialties and occupational experiences from its members, and fostering strong networking and social interactions of expatriates and new citizens with the local community.

With the vision of IRC, it is our great pleasure to organize the IRC conference on Science, Engineering and Technology (IRC-SET, www.ircset.org) for the younger talents and researchers. IRC-SET 2015 was the inaugural conference of IRC and IRC-SET 2024 is now in its tenth run. The IRC-SET conference aims to provide a platform for young and talented researchers to share their research findings, to obtain comments and feedback from experts in the field, and to exchange innovative ideas of leading-edge research in multi-disciplinary areas. Researchers from universities, junior colleges, polytechnics and secondary schools are warmly welcome to participate in this conference to showcase and present their research projects, results and findings. Unlike other academic conferences, this conference focuses specifically on Education and Youth development, and has officially been given technical sponsorship from six universities, namely National University of Singapore (NUS), Nanyang Technological University (NTU), Singapore Management University (SMU), Singapore University of Technology and Design (SUTD), Singapore Institute of Technology (SIT), and Singapore University of Social Sciences (SUSS). The IRC-SET 2024 conference is also supported by the IEEE Intelligent Transportation Systems Society (ITSS) Singapore Chapter, the IEEE Broadcast Technology Society (BTS) Singapore Chapter, and the IEEE Singapore Section Women in Engineering (WIE) Affinity Group.

The program of IRC-SET 2024 advocates the importance of innovative technology backed by the strong foundation of science and engineering education. Exposing our young participants to the key technology enablers will encourage more interest into the fields of science, engineering and technology. To select good quality papers, the IRC-SET 2024 Call for Papers was broadcast to universities, junior colleges, polytechnics and secondary schools. The researchers then submitted their technical papers to the conference online system. To meet the criteria of the conference, the submitted papers (which had to follow the required template) were peer reviewed by the conference technical program committee, which comprised of a few reviewers comprising the IRC researchers, professors, lecturers and teachers. Comments and recommendations based on novelty of the work, scientific, engineering and technology relevance, technical treatment plausible, and clarity in writing, tables, graphs and illustrations were then provided back to the authors. Through this rigorous review, the technical program committee has selected these papers to be presented in the IRC-SET 2024 conference, which are now published in this proceeding.

The IRC-SET 2024 Conference was held online and consisted of the Opening Speech by General Chair Prof. Ian McLoughlin (Professor in Singapore Institute of Technology), and nine parallel presentation sessions. The nine presentation sessions covered the key focus areas which included ITSS Session, BTS Session, Biomedical Science, WIE session, Computer Science, Data Science and AI, Life Science, Cyber Security, and Chemical Engineering. In the closing ceremony, A/Prof. Huaqun Guo (President of IRC and Associate Professor in Singapore Institute of Technology), announced the winners of the best paper awards and best presenter awards, and concluded the conference with an introduction to the International Researchers Club and a closing speech.

This proceeding is dedicated to the International Researchers Club and its members.

September 2024

Huaqun Guo
Ian McLoughlin
Umayal Lakshmanan
Xiaoxiao Miao
Eyasu Getahun Chekole
Weizhi Meng
Peng Cheng Wang
Jiqiang Lu
Nicholas Heng Loong Wong

Acknowledgments

We would like to extend our utmost gratitude to the people who have, in one way or other, inspired, aided and contributed to the successful completion of this book.

First of all, we would like to thank all members of the conference organizing committee who have contributed their time and professional knowledge to make the conference a big success.

Special thanks to all reviewers for their expertise, time, effort and timely response throughout the peer evaluation process.

Special thanks to all Session Chairs for their expertise, time and effort to make the presentations smooth and have the fruitful knowledge sharing at the conference day.

Last but not least, our greatest appreciation to all IRC members for their unity.

Organizing Committee

General Chair

Ian McLoughlin	Singapore Institute of Technology, Singapore

Technical Program Chairs

Umayal Lakshmanan	SenzeHub Social Enterprise, Singapore
Xiaoxiao Miao	Singapore Institute of Technology, Singapore
Eyasu Getahun Chekole	Singapore University of Technology and Design, Singapore

Publicity Chairs

Weizhi Meng	Technical University of Denmark, Denmark
Noori Kim	Purdue University, United States

Publication Chairs

Peng Cheng Wang	Singapore Institute of Technology, Singapore
Jiqiang Lu	Beihang University, China
Nicholas Heng Loong Wong	Singapore Institute of Technology, Singapore

ePoster Chairs

Rong Li	Shanghai University of Sport, China
Elena Lisova	Malardalen University, Sweden and Volvo Construction Equipment, Sweden

Local Arrangement Chair

Kasthurirengan Sampath	JOil (s) Pte. Ltd, Singapore

Communication Chairs

Yongqing Zhu	Singapore University of Social Sciences, Singapore
Yu Wang	A*STAR Research Entities, Singapore

Steering Committee

Huaqun Guo	Singapore Institute of Technology, Singapore
Anand Bhojan	National University of Singapore, Singapore

International Advisory Panel

Chau Yuen	Nanyang Technological University, Singapore

TPC Members

Aldy Gunawan	Singapore Management University, Singapore
Boon Seng Chew	Singapore Polytechnic, Singapore
Casey Kiam Cheng How	Singapore University of Social Sciences, Singapore
Elena Lisova	Malardalen University, Sweden and Volvo Construction Equipment, Sweden
Eyasu Getahun Chekole	Singapore University of Technology and Design, Singapore
Homin Park	Singapore University of Social Sciences, Singapore
Huaqun Guo	Singapore Institute of Technology, Singapore
Jiqiang Lu	Beihang University, China
Judice Koh	National University of Singapore, Singapore
Nicholas Heng Loong Wong	Singapore Institute of Technology, Singapore
Noori Kim	Purdue University, United States
Umayal Lakshmanan	SenzeHub Social Enterprise, Singapore
Xiaoxiao Miao	Singapore Institute of Technology, Singapore
Yongqing Zhu	Singapore University of Social Sciences, Singapore

Contents

The Evaluation of Music's Effect on Pain Tolerance and Perception Using EEG

Zhuangyi Tao[1], Faith Ong[1(✉)], Lexi Liu[1], and Aung Aung Phyo Wai[2]

[1] Nanyang Girls' High School, Singapore, Singapore
faithongmm@gmail.com
[2] Nanyang Technological University, Singapore, Singapore

Abstract. Studies have demonstrated that music profoundly influences human emotions and physiological states, serving as an effective non-pharmacological intervention in pain management. This study investigates whether pain perception and modulation can be influenced by music. We hypothesised that music would reduce pain perception regardless of the music type. In our experiment, we adopted cold-induced pain to simulate pain conditions to collect EEG data from six healthy students. We analysed EEG data to assess the effect of music on pain perception. Our preliminary analysis revealed disparities in EEG patterns between control and music conditions. Significant changes in brain waves were observed in association with pain perception, especially delta, theta, and alpha band power, which are linked to the emotional and cognitive dimensions of pain. Additionally, the presence of music significantly increased alpha and beta band powers, leading to more relaxed and alert states. Our findings also showed variability in pain tolerance among individuals under different music and experimental conditions. Specific musical elements (e.g., rhythm, melody, harmony) also play critical roles in modulating pain perception, along with varying cognitive, social, and behavioural parameters of the subjects.

1 Introduction

Pain and emotion are closely linked, as they can be measured by brain wave activities reflecting different emotional states and pain perceptions (Bushnell et al., 2013). The pain conditions can be classified into acute and chronic pain (Craig et al., 2021). Acute pain usually occurs suddenly and is caused by something specific such as headaches, injury-induced pain, dental pain and post- surgical pain. Chronic pain is usually ongoing and lasts more than six months including arthritis, cancer-related pain and irritable bowel syndrome. EEG is highly relevant to study pain perception due to its non-invasive measurement with high temporal resolution of brain activity associated with pain (Schulz et al., 2012). Our research uses EEG modality to investigate the impact of music on pain perception. EEG is instrumental in exploring how factors such as attention and emotion modulate pain, and it reveals distinct neural patterns in chronic pain conditions (Lin et al., 2005). EEG also aids in developing pain management strategies, including neurofeedback and brain- computer interfaces. In addition, EEG supports personalised pain medicine due to its ability to capture individual variability, enhancing the effectiveness of treatments for pain conditions (Huang et al., 2013).

H. Guo et al. (Eds.): IRC-SET 2024, PTT, pp. 1–7, 2025.
https://doi.org/10.1007/978-981-96-3770-6_1

Numerous scientific studies have already recognised music for its therapeutic and soothing effect, and potential effectiveness in pain modulation (Lu et al., 2019, Lu et al., 2023). However, our research compares the differences in EEG activities between two music genres on pain perception, an area which has not been widely researched. We hypothesised that listing to classical and pop music genres will cause varying level of pain perception though listening to both music reduces pain perception.

2 Methodology

We recruited six subjects, all female secondary school students, for our experiment. We obtained the parental consents of those subjects before the experiment. The following Fig. 1 depicts the overview of experiment tasks, timing, and with two major components: baseline and cold-induced pain tasks. Participants will undergo controlled pain stimuli exposure across three conditions: listening to music from two distinct genres and a condition of silence (control) while recording neural activity on pain-related neural patterns and the potential of modulatory effects induced by music. Our study employed a controlled experimental design using the muse 2 headband (Interaxon Inc.) to record brain activity from four channels (TP9, TP10, AF3, AF4). We chose classical music after reviewing literature which demonstrate classical music's many benefits to health and mental state (Lu et al., 2019). Pop music was the other genre that we selected as it is popular. Baseline tasks were conducted prior to the main experiment as a framework for each participant's varying mental states. The Attention Task establishes a standard measure of the subject's baseline level of attention, serving as a control to compare against performance in pain tolerance tasks. The emotion task measures the subject's initial emotional state, enabling us to assess changes in emotion during the pain tolerance tasks. The memory task, which utilizes the 1 back and 2 back test assesses the subject's baseline memory performance. Lastly, the pain tolerance test measures the subject's initial pain tolerance level for establishing a reference point for comparing the effects of different music, as well as a control for individual differences in pain tolerance among individuals through their subjective ratings (Nahman-Averbuch et al., 2023).

We exposed cold-induced pain to subjects by dipping hands into a water bucket, with and without the accompaniment of music of two different genres as shown in Fig. 2. For the cold induced pain task, subjects undergo 2 blocks of testing. In 1 block, the subject is first exposed to cold induced pain without music, then exposed to cold induced pain with the accompaniment of music (music genre is randomly selected). This procedure is repeated for the other genre of music. Subjects were first exposed to pain without music in both blocks to ensure consistency, hence any change in pain tolerance observed can be more confidently attributed to the effects of music rather than other factors. Introducing music first could potentially influence the subject's perception and response in the subsequent no-music condition, introducing bias or carryover effects.

We collected data from 6 participants using the experiment design as shown in Fig. 1. Before conducting the experiments, we ensured that subjects wore the MUSE 2 headband as per correct usage guidelines. During the cold induced pain tasks, the subjects were blindfolded, and their hand was first spontaneously dipped into buckets of water at around 27 degrees Celsius for 2 min. Afterwards, their hand was dipped into

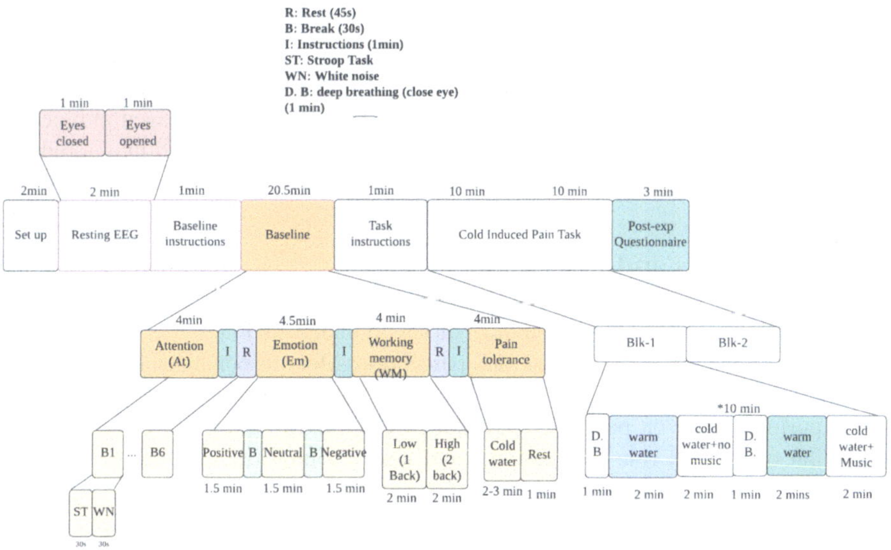

Fig. 1. Experiment Design and Tasks Sequence

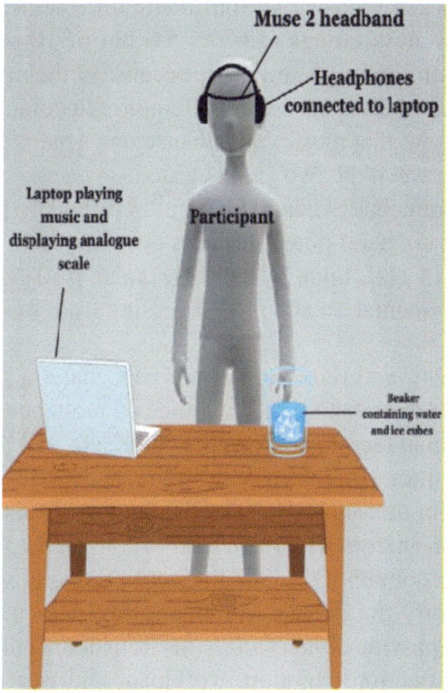

Fig. 2. Experiment Setup with Test Subject

buckets of ice-water mixture at around 15 degrees Celsius for 2 min without any music played. Subjects were given a 1-min break where they engage in deep breathing activity.

After 1 min of deep breathing, block 1 testing continued and subjects were exposed to 27 degrees Celsius water for 2 min, followed by exposing them to ice-water mixture at around 15 degrees Celsius for 2 min with either music 1 or music 2 played (random sequence). This is repeated for block 2 with the music that was not played in block 1. We recorded the initial temperature of water mixture in the bucket and monitored the temperature change of the mixture using a thermometer.

3 Data Analysis and Discussion

We also conducted online public opinion survey to understand how users perceive the influences of music on emotion and pain perception. Most respondents answer beliefs in music's efficacy in aiding stress relief, in line with our hypothesis that music could have a positive effect on pain tolerance and perception. So, there is a positive public perception of music's role in emotional and possibly physical well-being. The pre-experiment survey data highlighted 75% of participants rated their pain tolerance level as medium whereas 25% rated as higher pain tolerance. So, there is not a high contrast in perceived pain tolerance level among our subjects resulting in less discrepancy on user's pain perception responses in our experiment.

From post-experiment survey data, we saw more diversified perceived pain level ratings among test subjects. Out of two music stimuli, no clear winner identified as similar levels of effectiveness ratings (5.67 vs 5.0 out of 10 scales between pop and classical music). These similar results might be because of their personal preference and likeness on music we selected as per research literature. In comparison between pre and post survey ratings, it shows that music can enhance one's pain tolerance, however there is no clear distinction between the two genres tested.

We collected four channels EEG data (Fp1, Fp2, Tp9 and Tp10) with a sampling rate of 256 Hz. We extracted six band-power features per channel in δ(1–4 Hz), Θ(4–8 Hz), α(8–12 Hz), low β(12–18 Hz), high β(18–30 Hz), and γ(30–45 Hz) frequency bands after performing data segmentation and pre-processing from raw EEG data as shown in Fig. 3.

To ensure the integrity and relevance of the EEG data, a preprocessing step was implemented to remove unwanted frequency ranges. Frequencies below 0.5 Hz, which are often caused by external factors or movement artefacts, and those above 45 Hz were filtered using a bypass filter. Therefore, only frequencies between 0.5 Hz and 45 Hz were deemed desirable for this analysis. Next, the feature extraction process utilised the short-time Fast Fourier Transform (FFT) method to convert the EEG data from the time domain to the frequency domain. From the frequency domain, six band-power features were computed to quantify the energy within specific frequency bands. Furthermore, mental states detection provides binary classifier for identifying the accuracy of four distinct mental states: attention, emotion, workload, and pain tolerance as shown in Fig. 4.

A repeated measures ANOVA reveal that participants report lower pain scores when music is present ($M = 2.5$, $SD = 0.8$) compared to silence ($M = 3.5$, $SD = 0.7$); $F(1,29) = 10.32$, $p < 0.01$, indicating a significant effect of music on increasing pain tolerance. During pain stimuli without music, increased power in delta (1–4 Hz) and theta (4–8 Hz)

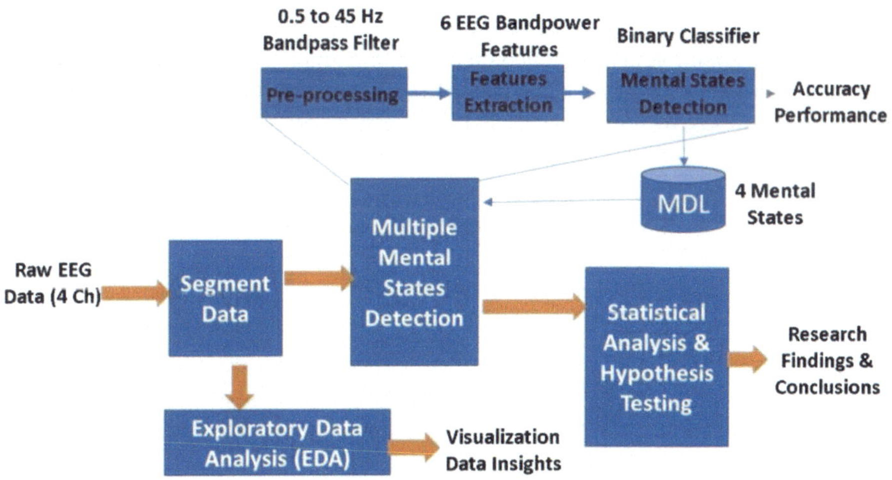

Fig. 3. Overview of EEG Data Analysis Steps

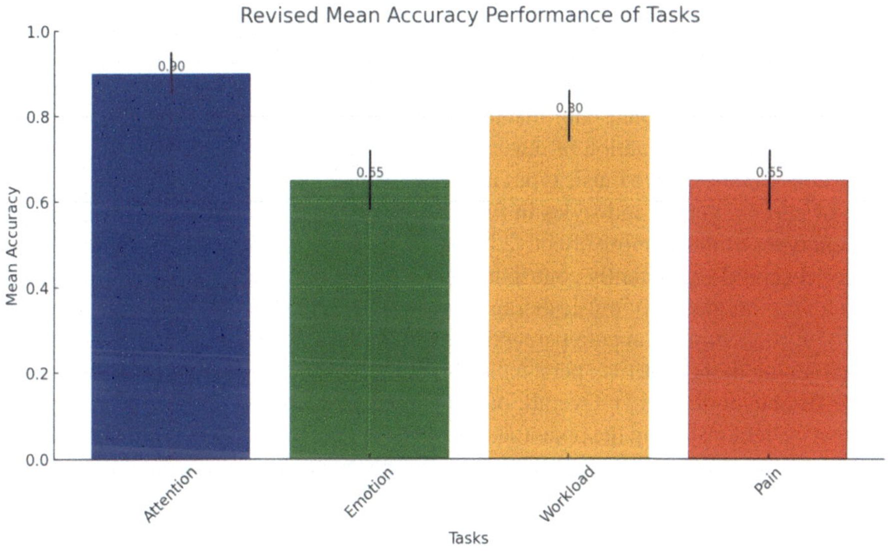

Fig. 4. Accuracy Results of User's Baseline Task Performance using EEG data

waves can be seen as indicators of pain processing and emotional distress (derived from post experiment survey). In contrast, decrease significantly in both frequencies, with a reduction in perceived pain intensity. A paired t- test showed a significant reduction in delta bandpower during music exposure ($t(29) = 4.25, p < 0.01$), indicating a statistically significant effect of music on pain perception as shown in Table 1. With music, there's an increase in alpha (8–13 Hz) and beta (13–30 Hz) wave activity, reflecting increased relaxed states, respectively. An ANOVA results showed a significant condition (music vs.

no music) x wave type (alpha, beta) interaction ($F(1,29) = 5.67$, $p < 0.05$), underscoring music's role in attention diversion and emotional modulation during pain. Participants reported lower pain scores in the presence of music compared to silence, supported by repeated measures ANOVA that revealed a significant effect of music on increasing pain tolerance.

Table 1. Statistical Test Results Between Different Experiment Conditions

Experiment Conditions	Significant EEG Bandpower features	Significant Among Attention, Emotion, Workload Scores
Control VS Music1 (classical)	δ: $t(29) = 2.10$, $p < 0.05$; θ: $t(29) = 2.15$, $p < 0.05$	$F(1, 29) = 5.33$, $p < 0.05$
Control Vs Music2 (pop)	α: $t(29) = 3.25$, $p < 0.01$; β: $t(29) = 3.40$, $p < 0.01$	$F(1, 29) = 6.27$, $p < 0.05$
Music1 (classical) Vs Music2 (pop)	β: $F(1, 58) = 4.05$, $p < 0.05$	$\mathbf{F(1, 29) = 3.22, p = 0.08}$

With small sample size, these findings offer preliminary insights that could be explored further in a larger study. The data underscores the importance of psychological factors and personal pain management strategies, suggesting areas for deeper investigation on the evaluation of music's effect on pain tolerance and perception. The suggestions for alternative music types underscore the importance of considering a wide range of musical genres and styles in future research, as different individuals may find different types of music more effective in managing pain. The anticipated findings from this project could significantly contribute to the field of pain management, offering evidence for the integration of music therapy into clinical settings. By elucidating the neural basis of music's impact on pain perception and tolerance, the study could pave the way for developing more effective, personalised non-pharmacological interventions for pain relief (Trimble et al., 2017). Overall, our research contributes to the field of pain management by demonstrating the potential of EEG neurofeedback interventions and paving the way for personalized music therapy as a non-pharmacological intervention for pain relief (Holden et al., 2013). These findings underscore the significance of integrating music therapy into clinical settings to develop more effective and personalized pain management strategies. These differences in EEG patterns could be more pronounced with certain types of music beyond the two options we considered. Future research should explore a wider range of musical genres and styles, employ larger sample sizes for greater statistical power, and utilise machine learning techniques to identify patterns within EEG data that correlate with pain perception and cognitive performance.

4 Conclusion

In conclusion, this study explores the influence of music on pain perception and modulation through an experimental design utilising cold-induced pain and EEG measurements. Our analysis focused on EEG band power features and cognitive states derived from EEG to assess the effect of music on pain perception. Our results indicate significant disparities in EEG patterns between control and music conditions. Notably, brain waves associated with pain perception, particularly delta, theta, and alpha band power, exhibited significant changes. The presence of music significantly increased alpha and beta band powers, correlating with more relaxed and alert states. These findings suggest that music effectively enhances pain tolerance and modulates pain perception. Increased power in delta and theta waves during pain stimuli without music indicated pain processing and emotional distress, while a significant reduction in these frequencies during music exposure was observed. This reduction corresponds with a decrease in perceived pain intensity and an increase in alpha and beta wave activity, reflecting a state of relaxation and alertness. Despite the small sample size, the study provides preliminary insights into the role of music in pain management, highlighting the importance of psychological factors and personal pain management strategies.

References

Bushnell, M.C., Čeko, M., Low, L.A.: Cognitive and emotional control of pain and its disruption in chronic pain. Nat. Rev. Neurosci. **14**(7), 502–511 (2013). https://doi.org/10.1038/nrn3516

Craig, K.D., MacKenzie, N.E.: What is pain: Are cognitive and social features core components? Paediatric and Neonatal Pain **3**(3), 106–118 (2021). https://doi.org/10.1002/pne2.12046

Schulz, E., Zherdin, A., Tiemann, L., Plant, C., Ploner, M.: Decoding an individual's sensitivity to pain from the multivariate analysis of EEG data. Cereb. Cortex **22**(5), 1118–1123 (2012). https://doi.org/10.1093/cercor/bhr186

Huang, G., Xiao, P., Hu, L., Hung, Y.S., Zhang, Z.: Single-trial laser-evoked potentials feature extraction for prediction of pain perception. In: 2013 35th Annual International Conference of the IEEE Engineering in Medicine and Biology Society (EMBC), pp. 4207–4210 (2013). https://doi.org/10.1109/EMBC.2013.6610473

Lu, X., Thompson, W.F., Zhang, L., Hu, L.: Music reduces pain unpleasantness: Evidence from an EEG study. J. Pain Res. **12**, 3331–3342 (2019). https://doi.org/10.2147/jpr.s212080

Lu, X., et al.: The effect of background liked music on acute pain perception and its neural correlates. Hum. Brain Mapp. **44**(9), 3493–3505 (2023). https://doi.org/10.1002/hbm.26293

Lin, W.C., Chiu, H.W., Hsu, C.Y.: Discovering EEG Signals Response to Musical Signal Stimuli by Time-frequency analysis and Independent Component Analysis. Conference proceedings of IEEE Engineering in Medicine and Biology Society **2005**, 2765–2768 (2005). https://doi.org/10.1109/IEMBS.2005.1617045

Trimble, M., Hesdorffer, D.: Music and the brain: the neuroscience of music and musical appreciation. BJPsych. International **14**(2), 28–31 (2017). https://doi.org/10.1192/s2056474000001720

Nahman-Averbuch, H., et al.: Communication of pain intensity and unpleasantness through magnitude ratings: Influence of scale type, but not gender of the participant. Eur. J. Pain **27**(10), 1161–1176 (2023). https://doi.org/10.1002/ejp.2147

Holden, R., Holden, J.: Music: a better alternative than pain? Br. J. Gen. Pract. **63**(615), 536 (2013). https://doi.org/10.3399/bjgp13x673748

MElodious MEmory (MEME): Syncing Music and Electroencephalogram (EEG) Neurofeedback to Enhance Short-Term Memory Retention and Recall

Chng Yu Ye Kelly[1], Lim Ke Jia Khloe[1], Song An An Shermaine[1](✉), and Aung Aung Phyo Wai[2]

[1] Nanyang Girls High School, Singapore, Singapore
{2021_chng_yu_ye_kelly,2021_lim_ke_jia_khloe, 2021_song_an_an_shermaine}@nygh.edu.sg

[2] Nanyang Technological University, Singapore, Singapore
apwaung@ntu.edu.sg

Abstract. Music's omnipresence in our daily routines is facilitated by a variety of channels such as streaming services, digital platforms, and traditional radio. This widespread integration of music into activities like studying and exercising highlights its significant role in our lives. Given this backdrop, extensive research has explored music's influence on various performance metrics across numerous disciplines. Our project aims to delve deeper into this realm by examining how different music genres, the familiarity of music, and rhythmic variations can augment short-term memory capabilities or not. We conducted an experiment with 14 healthy students using consumer EEG headbands to investigate the influence of music tempo and familiarity on short-term memory performance. Our EEG data analysis revealed significant increases in alpha and theta band power during memory tasks with 50 BPM and familiar music, because of an enhanced state of cognitive processing. Post-hoc analyses confirm that the 50 BPM and familiar music significantly outperformed the control and 150 bpm ($p < 0.05$). This finding was consistent with subjective reports aligned with the objective EEG and task performance outcomes. In conclusion, our study supports that music with slower tempos and familiar melodies can improve short-term memory retention and recall. The correlation between EEG band powers and task performance offers applying music and EEG neurofeedback in memory training and therapeutic interventions.

Keywords: music · EEG · short-term memory performance · tempo · familiarity

1 Introduction

A lot of research has shown the effects of music and sounds on performance in many areas of study. According to Fassbender et al. (2012), music influences memory. However, there are mixed results about what kind of effects music can influence. Fassbender et al. (2012) found that music negatively affected memory during a study or learning phase, but increased mood and sports performance.

H. Guo et al. (Eds.): IRC-SET 2024, PTT, pp. 8–16, 2025.
https://doi.org/10.1007/978-981-96-3770-6_2

However, the positive impacts of music on memory are also substantially supported. The study by Wallace (1994) demonstrated that "the melody of a song can facilitate learning and recall". Moreover, Wallace also proved that the actual melody of the song contributes to learning of the words more than just the rhythm of the music. When text is sung, people chunk words and verses together and use other memory aids "to assist in learning, in retrieving, and if necessary, in reconstructing a text" (Wallace, 1994). On the other hand, Samson and Zatorre (1991) suggest that it may be because music employs different encoding mechanisms and stores music in a different place in the brain and therefore hearing the melody again can facilitate retrieval. The music may make more connections in one's brain and therefore makes it easier and faster to retrieve (Wallace, 1994). Regarding music and memory, Kilgour et al. (2000) implied, adding any structure to students' study time is preferable than none at all to increase retention; thus it is possible that music can be used as that structure; music implements a tempo, add rhythm and melody, and create repetition, presenting information to be more easily learned and retrieved from memory because it provides the listener other cues in addition to verbal information (Hyman and Rubin, 1990).

Further benefits of music are positive emotions and mood regulation (Sloboda and O'neill, 2001; Saarikallio and Erkkilä, 2007). Studies showed that listening to music, which individuals found pleasurable, yielded a significant increase in dopamine, the hormone of happiness (Nadler et al., 2010). Moreover, some studies imply that the Mozart Effect can boost cognitive performance (Rauscher et al., 1993). However, it is still unknown if there are causal effects between these factors. Therefore, further subsequent studies have revealed that Mozart's compositions do not directly affect cognitive performance, but rather affect mood and exploit positive emotions (Nantais and Schellenberg, 1999; Thompson et al., 2001). In the same vein, Carpenter (2012) concluded that the effect of feeling good can have an effect on short-term memory and in the decision making process. Based on similar studies, the Chinese University of Hong Kong used music as a training method for memory. They found that pupils who undergo musical training demonstrated better verbal but not visual memory than did their counterparts without such training.

Thus, the causality of the effect of music on short-term memory is still unknown.

Given the rise of online music platforms, more students are listening to music while studying, with 58.8% of our survey respondents (total respondents of 68) indicating that they listen to music multiple times a day for reasons such as relaxation (80.9%) and concentration (45.6%) across genres such as pop (94.1%), classical (51.5%) and R&B (22.1%). Thus, we have decided to investigate the direct relationship between the beats of music as well as the familiarity of the music tunes to enhance short term memory retention and recall. This is especially relevant to students' daily life as they are required to memorize large amounts of content for school work, and can possibly allow them to study and memorize more effectively.

2 Hypothesis

In this study, we have made 2 hypotheses: the first being music with slower beats helps improve short term memory and the second being familiar music improving short term memory.

Music with slower beats improves short term memory: One study conducted by Janata and colleagues (2012) investigated the impact of slow-tempo music on memory recall. They found that participants who listened to music with a slower beat had significantly better memory performance compared to those who listened to fast-tempo or no music at all. The slower beats seemed to enhance the consolidation and retrieval processes involved in short-term memory. Another study by Thaut and colleagues (2014) explored the effects of rhythmic auditory stimulation (RAS) on working memory. RAS involves the synchronization of movements or cognitive processes with an external rhythm. The researchers found that synchronized tapping to a slower beat improved working memory accuracy and response times compared to tapping at a faster tempo. This suggests that slower beats can facilitate the encoding and retrieval of information in short-term memory. However, it is believed that slower rhythms may promote a relaxed state, reducing anxiety and stress, which can impede memory performance. Additionally, slower beats may facilitate focused attention and enhance the encoding and consolidation of information in short-term memory.

Familiar music has been shown to improve short-term memory performance. When individuals are exposed to music that is familiar to them, it can enhance their ability to remember and recall information more effectively. Familiar music often elicits emotional responses and can create a state of emotional engagement. This emotional connection enhances memory encoding and retrieval processes, leading to better short-term memory performance. Familiar music can activate rich associative networks in the brain. The neural connections formed between the music and associated memories or experiences can facilitate the encoding and MElodious MEmory (MEME): Syncing Music and Electroencephalogram (EEG) Neurofeedback to Enhance Short-Term Memory Retention and Recallretrieval of information, leading to improved short-term memory. Familiar music can capture and sustain attention, aiding in the concentration and focus required for memory tasks. It can help individuals maintain a heightened level of alertness, leading to improved encoding and retention of information in short-term memory. Familiar music acts as a contextual cue, providing a familiar and structured environment for memory processes. The presence of familiar music can serve as a retrieval cue, facilitating the recall of associated information stored in short-term memory. Familiar music engages various cognitive processes, such as pattern recognition, prediction, and anticipation. These cognitive mechanisms can enhance the encoding, storage, and retrieval of information in short-term memory. Overall, the research suggests that familiar music can improve short-term memory performance by enhancing emotional engagement, activating associative networks, providing attentional focus, and serving as a contextual cue. Incorporating familiar music into learning or memory tasks may be a beneficial strategy for optimizing memory performance.

3 Methodology

The following Fig. 1 depicts the overview of tasks, timing and duration of experiment with two major components: baseline and test tasks. We recruited 14 subjects, all female secondary school students, through an online survey and obtained parental consents of potential subjects before the experiment. Prior to data collection, we collected user's music preferences and cognitive profiles via pre-experiment survey questions (Table 1).

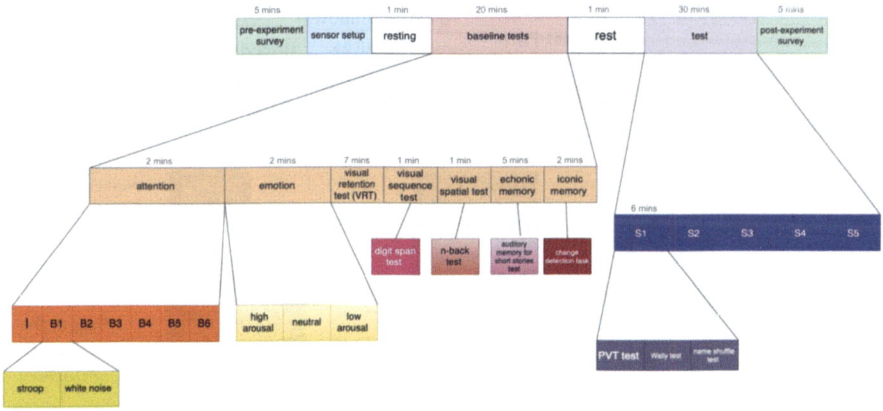

Fig. 1. Overview of Experiment Design and Sequence of Tasks

Table 1. Types of Music Employed in Experiment Blocks

Block 1	Block 2 (wally test)	Block 3 (wally test)	Block 4 (w ally test)	Block 5 (name shuffle test)	Block 6 (name shuffle test)
No music (Control)	Instrumental music (50 beats per minute)	Instrumental music (100 beats per minute)	Instrumental music (150 beats per minute)	Familiar Music (Personalized)	Unfamiliar Music (Personalized)

The baseline session can further breakdown into cognitive modeling (attention and arousal states) and 5 different memory assessment tasks as described above. The test session has 5 blocks, each consisting of Psychomotor Vigilance Task, Wally Test and the Name Shuffle Test. In each block, we play music with different bpms to compare and evaluate the influence of different music beats on the performance of the above three tests. After all experiment tasks are completed, subjects will go through post-experiment surveys providing their feedback, experience and tasks completed in the experiment (Fig. 2).

Muse2 EEG headband (InteraXon Inc.) were used to measure brain's electrical activities from frontal and temporal-parietal areas (Fp1, Fp2, Tp9 and Tp10 channels) during the experiment. Data was then segmented based on timestamps for each task and music condition block. Afterwhich, six band-power features per channel for δ(1–4 Hz), Θ(4–8 Hz), α(8–12 Hz), low β(12–18 Hz), high β(18–30 Hz), and γ(30–45 Hz) frequency bands were extracted. Lastly, data-driven machine learning, namely the support vector machine classifier, was used to determine the accuracy of the participant's performance, reflecting their attention level. This data is then inputted into the SVM binary classifier model. This data allowed it to learn to match new incoming EEG data from the main tests with the user's attention state and memory capabilities.

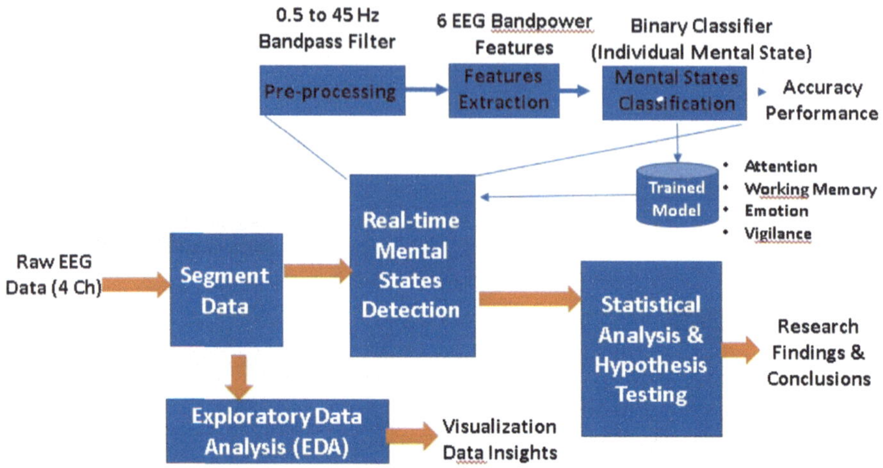

Fig. 2. Data Analysis Steps and Processes using EEG (electroencephalogram)

4 Results and Discussion

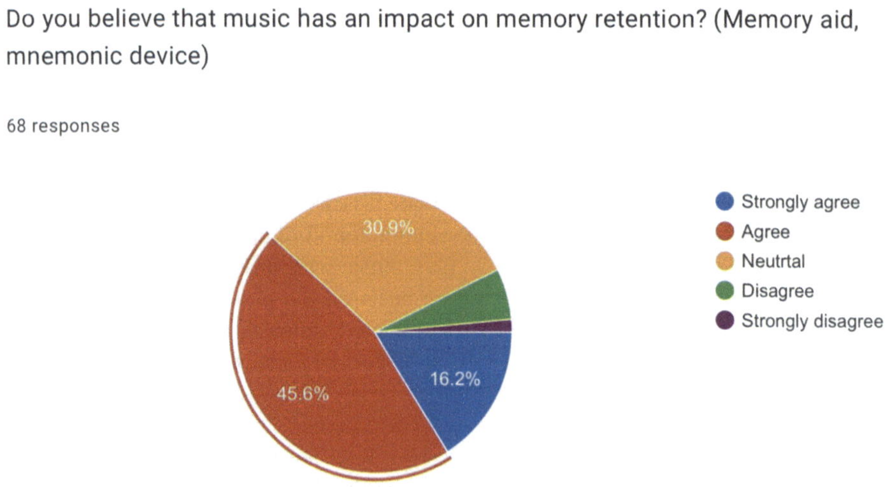

Fig. 3. Pie Chart of the Public Opinion on the Impact of Music on Memory

Firstly, survey results are discussed. General demographic survey among high-school students reflected that the student population generally believed there to be a connection between music and memory retention, supporting evidence of our research objective. The overall positive reception and engagement ratings from post-survey did not affect negatively though diverse subject profiles identified in pre-survey, highlighting the effectiveness of our proposed experiment design across multiple subjects. We also noticed the consistent interest and engagement ratings from pre to post-experiment phases (Fig. 3 and Table 2).

Table 2. Memory Test Metrics and Outcomes

Memory test	Metric	Outcome (Mean ± SD)
Visual Retention	Reaction Time (Correct)	250 ms ± 15 ms
	Reaction Time (Incorrect)	300 ms ± 20 ms
	Accuracy	80% ± 5%
Sequence Memory	Accuracy (Correct)	80% ± 6%
	Accuracy (Incorrect)	70% ± 5%
Spatial Memory	Accuracy (Correct)	85% ± 5%
	Accuracy (Incorrect)	75% ± 6%
Echoic Memory	Reaction Time (Correct)	320 ms ± 18 ms
	Reaction Time (Incorrect)	370 ms ± 22 ms
	Correctness	90% ± 4%
Iconic Memory	Reaction Time (Correct)	310 ms ± 17 ms
	Reaction Time (Incorrect)	360 ms ± 25 ms
	Correctness	88% ± 3%

From the baseline sub-task of mental states modeling, we achieved binary classification accuracy of 76.21 ± 3.10% and 75.41 ± 4.21% using subject-dependent ten-fold cross-validation for attention and vigilance states respectively. From behavior performance observation, correctness in identifying attention and alertness tasks showed 90 ± 2% and 87 ± 3% respectively. These findings highlight that most of the subjects follow the calibration baseline tasks properly, which serve as reliable models to compute the respective mental state scores with different blocks of the main tests (Fig. 4).

From computed EEG band power features, we observe a significant increase in theta band power (4–7 Hz) while listening to music with slower tempos and familiar melodies. This would suggest an increase in deep relaxation states conducive to memory encoding that requires deep absorbance and focus into the task at hand. The 40% increase in theta band power indicates such an enhanced state. Similarly, 35% elevation in alpha band power (8–13 Hz) with similar music conditions reflect conducive conditions for increased cognitive processing and memory consolidation. Considering the fact that the alpha band is dominant in the inhibition of irrelevant stimuli for better concentration, it can be concluded that familiar music would enable better focus and memory recall (Table 3).

As seen above, recall accuracy percentage is highest for block 2 and 5 respectively, corresponding to subjects listening to 50 bpm music and familiar music. This is in relation to exposure to music with higher bpm (Block 1; no music, Block 3; 100 bpm, Block 4; 150 bpm) and in comparison with unfamiliar music (Block 6).

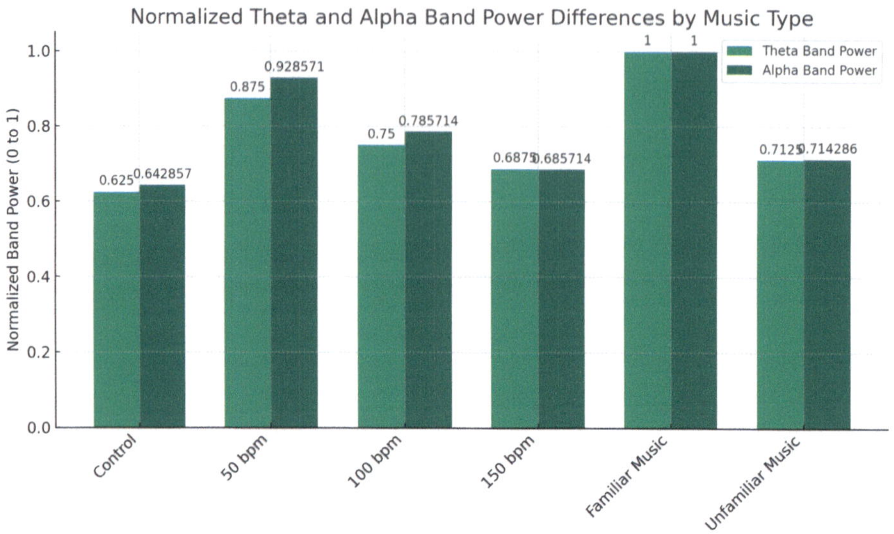

Fig. 4. Normalized Theta and Alpha Band Power Differences by Music Type

Table 3. Performance Metrics of Different Music Blocks

Block	Recall Accuracy	Response Time	Error Rate	Correct Rejections	False
1	70%	900 ms	5%	90%	10%
2	75%	850 ms	4%	92%	8%
3	73%	870 ms	4.5%	91%	9%
4	71%	880 ms	5.2%	89%	11%
5	80%	820 ms	3.5%	94%	6%
6	68%	890 ms	5.8%	88%	12%

Thus it can be concluded that music with slower beats and lesser familiarity to participants result in better attention. This can be explained by the previous findings showing that music with slower beats tend to result in greater theta and alpha band power, which are essential for deep relaxation and focus conducive for memory work. Moreover, familiar music tends to evoke emotions and/ or stimulate cognitive linkages in the participants' brain, enabling the formation of new memories in different regions of the brain. This is supported by secondary data from online literature as well.

The inter-relationships between cognitive states (attention and vigilance), EEG brainwave activity (alpha and beta power), and memory task performance reveal a complex and multidisciplinary research. While attention and relaxed vigilance (high alpha power) appear to enhance memory task performance, excessive vigilance (high beta power) might yield different benefits. These observations highlight the importance of optimizing cognitive arousal and focus for memory tasks, suggesting a delicate balance between being alert and relaxed for optimal cognitive performance. Further research could explore how to develop targeted interventions for desirable memory retention and enhancement.

5 Limitations

1. Experiment was conducted on female students of age 13–16, which is only one specific age group. Short-term memory capacity might differ for different age groups and genders, thus experiment is affected by potential biases introduced
   a. Participants were all female due to environment allowing this to be the only accessible subject population
      i. On the other hand, differing age groups and genders could also mean mental capacity and memorizing abilities present a greater difference, affect results
2. Human reaction time error when noting down the recall time for each participant and the start and end time of each experiment to sync with the electroencephalogram (EEG) data
   a. Possibilities to reduce limitations: conduct experiment a few more times, on more participants

6 Conclusion

The intersection of music and cognitive functions (attention, memory, workload, etc.) garners growing interests in neuroadaptive means for exploring the enhancement of short-term memory retention and recall. Our research goal is to investigate the effects of synchronizing music with EEG neurofeedback on enhancing short-term memory retention and recall. Our current research investigations focus on detecting brain activity changes whilst subjects undergo memory tasks, listening to a variety of music types. The use of various statistical tests ensures the robustness of the findings, while the machine learning model's high classification accuracy adds a layer of predictive capability to the study. Overall, statistical results would support the hypothesis that music with slower tempos and familiar melodies enhances short-term memory retention and recall, as indicated by both objective EEG data and subjective survey responses.

References

Anne, F.: Music can boost memory and mood. Harvard Health Publishing (2015). https://www.health.harvard.edu/mind-and-mood/music-can-boost-memory-and-mood

Darragh, B.: The effect of music genre on a memory task. Modern Psychological Studies (2012). https://scholar.utc.edu/cgi/viewcontent.cgi?article=1214&context=mps

Daniela, P.: The Cultural Impact of Music on Society with a Special Emphasis on Consumerism. Bulgarian Comparative Education Society, Sofia (2021). https://www.google.com.sg/url?sa=t&rct=j&q=&esrc=s&source=web&cd=&ved=2ahUKEwjKy-mXmNKEAxW--DgGHeNNDjwQFnoECEYQAQ&url=https%3A%2F%2Ffiles.eric.ed.gov%2Ffulltext%2FED614071.pdf&usg=AOvVaw3K-5mXBLSOJb1jriTlLg7c&opi=89978449

Jancke, L.: Music, memory and emotion. National Library of Medicine (2007). https://www.levinemusic.org/about/news/how-music-affects-memory-and-concentration/#:~:text=Studies%20have%20shown%20that%20%E2%80%9Cmusic,application%20in%20helping%20dementia%20patients

Janina, A.: The Influence of Background Music on Learning in the Light of Different Theoretical Perspectives and the Role of Working Memory Capacity. Frontiers (2017). https://doi.org/10.3389/fpsyg.2017.01902/full

Madyson, C.: Music's Effects on Memory. Psychology Research Methods Journal (2017). https://digitalcommons.lindenwood.edu/cgi/viewcontent.cgi?article=1023&context=psych_journals

Marlene, C.: Why music causes memories to flood back. The Washington Post (2023). https://www.washingtonpost.com/wellness/2023/02/26/songs-music-memory-connection/

Ming, L.: Slower tempo makes worse performance? The effect of musical tempo on cognitive processing speed. Frontiers (2023). https://www.google.com.sg/url?sa=t&rct=j&q=&esrc=s&source=web&cd=&ved=2ahUKEwi02efWi9KEAxVcSWwGHVWBACcQFnoECBcQAQ&url=https%3A%2F%2Fwww.frontiersin.org%2Fjournals%2Fpsychology%2Farticles%2F10.3389%2Ffpsyg.2023.998460%2Ffull&usg=AOvVaw1WVAWTox5nj9MbUyl7CUq7&opi=89978449

Musliu, A.: The Impact of Music on Memory. Research Gate (2017). https://www.researchgate.net/publication/318539845_The_Impact_of_Music_on_Memory'

Nadia, J.: Differential effect of music on memory depends on emotional valence: an experimental study about listening to music and music training. Taylor and Francis Online (2023). https://doi.org/10.1080/23311908.2023.2234692

Farnsworth, B., iMotions, Seernani, D., Heslinga, O.: 5 basics of EEG 101: Data Collection, Processing & Analysis. iMotions (2024). https://imotions.com/blog/learning/best-practice/5-basics-eeg-data-processing

Sweet, L.H.: N-Back paradigm. In Springer eBooks, pp. 1718–1719 (2011). https://doi.org/10.1007/978-0-387-79948-3_1315

Halcyon: An AI-Powered Forum with Automated Hate Speech Detection

Aneesha Rajesh Babu[1] and Xiaorong Li[2(✉)]

[1] NCS Pte Ltd, 5 Ang Mo Kio Street 62 NCS Hub, Singapore 569141, Singapore
aneesha.rajeshbabu@ncs.com.sg

[2] Singapore Institute of Technology, ICT, 13 New Punggol Rd., Singapore 829979, Singapore
xiaorong.li@singaporetech.edu.sg

Abstract. The growth of hate speech and online harassment on social media platforms makes the creation of efficient solutions to address this persistent problem necessary. A promising method for locating and monitoring hate speech in social media is sentiment analysis, a computational tool that auto detects and assesses the emotional tone of text. To empower platforms to take preventative action against the spread of hate speech, this work addresses the use of sentiment analysis as a tool to identify and analyse hate speech. In this paper, we focus on hate detection in textual content and use machine learning algorithms to categorize and measure hateful content. This approach can assist in content moderation, enhance user safety, and foster a more positive online environment. To improve the precision and dependability of sentiment analysis algorithms, issues including context sensitivity and changing language patterns must be addressed. The AI-enabled discussion forum incorporates an innovative design that addresses hate speech discussions through machine learning algorithms and active tagging to improve accuracy.

Keywords: Sentiment analysis · AI · social media texts · auto detection

1 Introduction

Many platforms claim that they are committed to provide their users freedom of speech. People are able to post what is on their minds without being identified or having any kind consequence. Freedom of speech is not always a negative thing especially in countries where criticising the ways of a corrupt government can put them or their family in danger. However, it also gives a platform for people to bully, harass, threaten, or stalk others while remaining anonymous [1].

Whether it is a social media platform such as Twitter or a publishing platform such as Medium, the amount of likes and comments each published post harnesses can be viewed by all. Most platforms do not display the number of dislikes the post receives as there are many users who dislike posts just to spread hate and negativity. Hate speech has especially increased over the years especially since the lockdown for COVID-19 [2] pandemic as work and studies has been shifted online.

H. Guo et al. (Eds.): IRC-SET 2024, PTT, pp. 17–26, 2025.
https://doi.org/10.1007/978-981-96-3770-6_3

Twitter, to counter this surge of hate, has set some guidelines on dealing with abuse and in aiding victims of abuse [3] in 2022. However, since there is an enormous number of users and posts, it is impossible for these platforms to keep all their content under control. There are some platforms which use content-curation algorithms that scan and lock content. However, there are chances for free and legal content to end up getting blocked [1].

Recently, YouTube had disabled the feature for users to view the number of dislikes of a video. YouTube believed that by removing this feature, it could help better protect their creators from harassment and to put a stop to dislike attacks which is when people work to drive up the number of dislikes on a creator's videos [4]. However, many creators can still get targeted unfairly by other users in the comment section. Since users are still able to post whatever comments that they have typed out, with limited restrictions, the negativity can be channeled using the comments with ease.

2 Problem Formulation

Though the platform should encourage users to post their thoughts freely. It should not enable them to spread hate and abuse other users. Most platforms do not remove negative or spam comments until it gets reported by many users. This process usually takes a long time as many users would have to report the comment and they would have to wait for the report to be processed before action is taken against the comment and the user. Though users have the option of restricting their comment section for the post they are publishing, they would not be able to receive any negative comments, they also would not be able to receive the positive comments as well. Therefore, users in social media platforms are not able to differentiate and block out hateful comments and users entirely without repercussions such as not being able to view all comments.

Most platforms do not remove negative or spam comments until it gets reported by many users. This process usually takes a long time as many users would have to report the comment and wait for the report to be processed before action is taken against the comment and the user. Though users have the option of restricting their comment section for the post they are publishing, they would not be able to receive any negative feedback, they also would not be able to receive the positive feedback as well. Therefore, users in social media platforms are not able to differentiate and block out hateful comments and users entirely without repercussions such as not being able to view all comments.

View Problematic Users: By having an overview of problematic users, which would be determined by the sentiment analysis, users would be able to view how problematic various users are without having to read through their posts. This would make it easier for the admin to take action and deactivate accounts as well. They would not need to wait for reports before having to take action.

Block Out Only Negative Comments: When users get harassed they would have to turn off their comment section in order not to see hate. This also blocks out all the positive comments. By having the option to only block out negative comments, they would still be able to read the positive ones.

Visualisation of Analytics: Users would be able to view which topics receive the most positive or negative sentiments. This enables users to view the average sentiment on that topic, user or category. This would enable the user to know what kind of sentiments their content has created.

3 Literature Review

There has been numerous research on using various models on the study of emotions of the users in social media platforms or forums. Though there have been many projects on sentiment analysis using data scraped from social media sites, most of these projects do not have an actual cause that they are aiding or have sentiment analysis integrated with a "social media" application. They focus on the sentiment analysis portion and try to tune their models to have better accuracy and precision.

3.1 Sentiment Analysis to Detect Hate Speech

To identify misogynistic conversations in social media, Hajarian et al. [5] proposed a method to identify incels in social media based on sentiment analysis on users' comments. NLP and machine learning methods were not used to identify incels in social media before, although they were implemented for detecting hateful and misogyny content. However, this proposed method which has used text mining, profanity-check and TextBlob for sentiment analysis can be used to identify incels in social media texts.

As reviewed literature in this section shows, NLP and machine learning methods can be considered the primary solution for identifying hateful posts and comments in social media. However, most studies in these fields focus on the sentiment analysis part but not actually using it to find hateful comments immediately once it has been posted. Therefore, in this project, the proposed solution aims to fill this research gap. They managed to identify 78.8% of the users correctly while 21.2% were not identified correctly.

3.2 Sentiment Analysis in Social Media

There has been a variety of studies on sentiment analysis of user feedback on websites. In [7], the authors discuss supervised learning strategies that include naive bayes and support vector machines, which are conventional learning methods. Support vector machines offer improved accuracy when compared to other classifiers.

The authors of [8] offer a solution to the categorization of sentiment polarity issues. Online customer reviews of products sold by Amazon are gathered as data for the suggested task. It was suggested to use a process for categorising sentiment polarity. For result estimation, categorization at the sentence and review level was done.

The study [9] presents a method for extracting textual content that is lexicon-based. Using dictionaries of words with their semantic orientation (power and polarity) documented and including negation and intensification, the Semantic Orientation Calculator is used in this.

4 Design of AI-Enabled Platform

4.1 Data Analytics Platform

The proposed application would be an AI-enabled data analytics platform which can analyse the sentiments of the posts containing various types of media such as text, picture, and documents. Users can post their comments to the content that the creators have published. When the comment is posted, users and creators will be able to view the sentiment analysis of the comment. They would be able to view the percentage of negativity or positivity the comment is expressing. The user profile page would be created to display how many negative or positive comments the user has. The purpose of this application is so that the experience for the user could be both pleasant and convenient and they are able to share their content to everyone as well.

AI Models for Sentiment Analysis

Data can be collected from using web scrapers on social media platforms such as twitter and various forum sites or datasets which can be found online. The data would then be cleaned and processed before being used to train the models. Few of the models that might be explored include Naïve Bayes, Logistic Regression and Decision Tree. The model with the best performance measure which includes precision, recall and F1 measure would be integrated with the platform.

Active Tagging

Active learning allows the model to adapt and update its learning strategy based on the evolving needs of the task. As the model learns from labelled data and improves its performance, it can dynamically select new samples for annotation, focusing on areas where it performs poorly or has higher uncertainty. This adaptability helps the model continually refine its knowledge and adapt to changing data distributions. This can be done through the forum where the admin will be able to identify if the sentiments have been tagged accurately and update it where appropriate.

Dashboard

The platform would contain a dashboard which displays an overview of sentiments on the posts, users and categories. Details such as average sentiments of posts by various users and categories would be displayed. An overview on the number of posts, users and categories would also be displayed (Fig. 1).

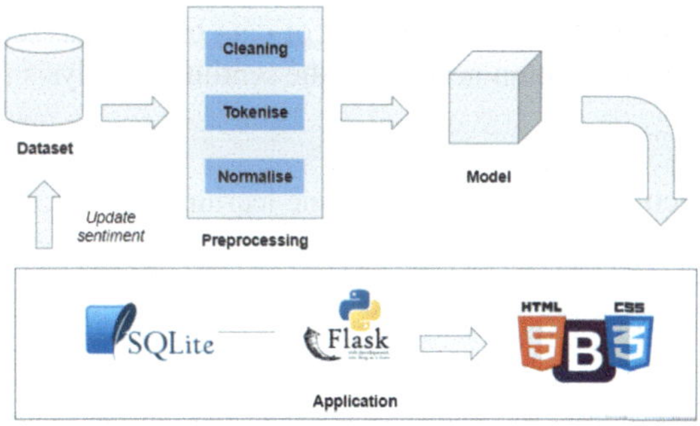

Fig. 1. Current workflow of the AI-powered forum

4.2 Data Pre-processing

Data pre-processing plays a crucial role in Natural Language Processing (NLP) tasks, including sentiment analysis. The essential steps comprise:

- Text cleaning: handling contractions and stopwords, eliminating HTML tags, URLs, special characters, and converting to lowercase.
- Tokenisation: Text is tokenized when it is divided up into individual words, phrases, or sentences. Contextual data can also be gathered using n-grams.
- Normalisation: In this stage, words are reduced to their most fundamental or root form. Lemmatization translates words to their base form based on vocabulary and morphological analysis, whereas stemming lowers words to their base form.
- By standardising and purifying the text data, these preprocessing methods help to prepare it for later analysis or modelling in NLP activities.

4.3 Sentiment Analysis Models

We have developed a few popular models for sentiment analysis:

NLTK VADER
The rule-based sentiment analysis tool NLTK VADER (Valence Aware Dictionary and sEntiment Reasoner) [10] was created specifically for social media text. To analyze sentiment at the sentence level, a combination of lexicons and grammatical rules are used. VADER is renowned for its capacity to manage sentiment expressions that are context-specific, negated, and of varying intensities of emotion.

SKLEARN Naive Bayes
The Naive Bayes algorithm [11] uses the concept of feature independence and is based on the Bayes theorem. It is especially helpful for problems involving text classification in which each feature stands in for a word or term. Naive Bayes models are quick, easy to use, and effective at handling big datasets.

Both NLTK VADER and SKLEARN Naive Bayes have their strengths and can be effective for sentiment analysis. The choice between them depends on factors such as the specific requirements of the task, the type of data being analyzed, and the level of interpretability and customization needed for the sentiment analysis model.

BERT

After trying out lexicon/rule based and machine learning models, the next step would be to train and test using a transformer such as BERT [12]. The fine tuning of the model and transformer-based architecture make it suitable and efficient for sentiment analysis of larger dataset and find the results.

First off, because BERT is bidirectional, it can comprehend word meaning and relationships by considering both left and right context. In contrast to models like the GPT that just consider left context, this offers a more thorough grasp of language.

Second, pre-training on a sizable volume of unlabeled text data aids BERT by enabling it to develop rich representations and capture complex linguistic patterns and semantics.

Thirdly, BERT's adaptability is a result of its capacity to be tailored for particular NLP tasks including sentiment analysis, named entity identification, and question answering. BERT performs better when its pre-trained knowledge is fine-tuned to fit task-specific labelled data.

4.4 A Forum with Dashboard

A forum application was developed using Python and the Flask framework. It utilised SQLite as the database management system. For the backend, SQLAlchemy, a Python SQL toolkit and Object-Relational Mapping (ORM) library, was employed. The dashboard component of the application, as shown in Fig. 2, was implemented using Plotly, a Python library for creating interactive and visually appealing data visualisations.

Fig. 2. Screenshot of an analytics dashboard

5 Performance Study

The first data set used for performance study was the transportation feedback data, which contains the textual feedback of users on transportation. The dataset includes the important information in columns such as title, description, main category and subcategories, etc. 1500 rows were randomly chosen and labelled from 15,000 data entries for each iteration. 0 for a negative sentiment and 1 for not negative.

Iteration 1

Model	Recall	F1 Score	Precision
NLTK VADER	58.58%	54.14%	51.33%
SKLEARN Naive Bayes	35%	33.55%	34.48%

Iteration 2

Model	Recall	F1 Score	Precision
NLTK VADER	56.19%	54.35%	54.29%
SKLEARN Naive Bayes	55.55%	47.50%	48.48%

Iteration 3

Model	Recall	F1 Score	Precision
NLTK VADER	55.42%	54.81%	56.19%
SKLEARN Naive Bayes	75%	61.11%	69.23%

Fig. 3. Results of NLTK VADER and SKLEARN Naive Bayes after each iteration

After each iteration, 10% of the dataset is manually updated. Based on the results shown in the figure above, the Naive Bayes model from SKLEARN outperforms other models. However, in future comparisons with BERT, the performance of the model will be evaluated to determine its integration into the application.

To ensure continuous improvement, an admin will have the ability to identify inaccuracies in the sentiment displayed by the model. When such inaccuracies are noticed, the admin can press a button to update the dataset accordingly and regenerate the results. This iterative process allows the model to be constantly trained with new data, enabling it to improve over time. Figure 3 shows the improvement of the results accuracy after only a few iterations with active tagging.

Figure 4 shows the comparison of the results of different models. The machine learning and deep learning algorithms outperform the rule-based algorithms as shown in Fig. 4 (a)-(c). The best performing machine learning model is Stochastic Gradient Descent (SGD) using CountVectorizer. The tag generation is powered by KeyBERT [13] extracting keywords. It uses BERT embeddings to produce keyphrases and keywords that are most related to a given content with high accurate results.

We also conducted the performance comparison based on another dataset [12] which contains approximately 47000 tweets taken from X (formerly Twitter) which are labelled according to the following classes: Age, Ethnicity, Gender, Religion, Other type of cyberbullying and Not cyberbullying. The data has been balanced in order to contain around 8000 rows of each class. The two columns of the dataset were "tweet_text" and "cyberbullying_type" which contained the tweets and classes respectively.

	Accuracy	Recall	F1	Precision
VADER	80.04%	65.09%	67.12%	73.05%
TextBlob	76.74%	65.80%	66.42%	67.24%
AFINN	80.48%	69.91%	71.13%	72.94%
Pattern	76.74%	65.80%	66.42%	67.24%

(a) Rule-based algorithms

	Accuracy	Recall	F1	Precision		Accuracy	Recall	F1	Precision
TF-IDF MB	87.55%	75.11%	79.33%	90.10%	CV MB	86.50%	90.22%	83.91%	81.67%
TF-IDF CB	95.36%	94.83%	93.76%	92.81%	CV CB	81.43%	87.20%	78.91%	77.81%
TF-IDF BB	85.86%	75.83%	78.37%	82.77%	CV BB	85.23%	73.59%	76.59%	82.87%
TF-IDF SGDC	94.30%	90.79%	91.93%	93.24%	CV SGDC	93.46%	90.54%	90.91%	91.30%
TF-IDF LSVC	94.73%	91.37%	92.53%	93.86%	CV LSVC	94.30%	90.48%	91.88%	93.52%
TF-IDF LR	88.19%	76.74%	80.80%	90.00%	CV LR	92.83%	87.69%	89.60%	92.05%
TF-IDF RF	89.45%	83.35%	84.81%	86.64%	CV RF	90.72%	84.18%	86.36%	89.35%
TF-IDF DT	89.24%	84.12%	84.86%	85.68%	CV DT	87.55%	82.10%	82.59%	83.13%

(b) Machine Learning algorithms

	Accuracy	Recall	F1	Precision
Sequential	95.99%	94.63%	94.50%	94.36%
CNN	93.25%	89.79%	90.53%	91.34%
LSTM	93.88%	90.21%	91.33%	92.62%
BERT	94.30%	91.40%	92.03%	92.72%

(c) Deep Learning algorithms

Fig. 4. (a)-(c). Results of Rule-based, Machine Learning and Deep Learning algorithms for the Transportation Feedback

The machine learning algorithms including CV LR, LSVC and TF-IDF LSVC and deep learning algorithms such as Sequential, CNN, LSTM, and BERT outperform the rule-based algorithms to achieve higher accuracy of hate detection in social media text, as shown in Fig. 5 (a)-(c).

	Accuracy	Recall	F1	Precision
VADER	80.04%	65.09%	67.12%	73.05%
TextBlob	76.74%	65.80%	66.42%	67.24%
AFINN	80.48%	69.91%	71.13%	72.94%
Pattern	76.74%	65.80%	66.42%	67.24%

(a) Rule-based algorithms

	Accuracy	Recall	F1	Precision
Sequential	95.99%	94.63%	94.50%	94.36%
CNN	93.25%	89.79%	90.53%	91.34%
LSTM	93.88%	90.21%	91.33%	92.62%
BERT	94.30%	91.40%	92.03%	92.72%

(b) Machine Learning algorithms

	Accuracy	Recall	F1	Precision		Accuracy	Recall	F1	Precision
TF-IDF MB	87.55%	75.11%	79.33%	90.10%	CV MB	86.50%	90.22%	83.91%	81.67%
TF-IDF CB	95.36%	94.83%	93.76%	92.81%	CV CB	81.43%	87.20%	78.91%	77.81%
TF-IDF BB	85.86%	75.83%	78.37%	82.77%	CV BB	85.23%	73.59%	76.59%	82.87%
TF-IDF SGDC	94.30%	90.79%	91.93%	93.24%	CV SGDC	93.46%	90.54%	90.91%	91.30%
TF-IDF LSVC	94.73%	91.37%	92.53%	93.86%	CV LSVC	94.30%	90.48%	91.88%	93.52%
TF-IDF LR	88.19%	76.74%	80.80%	90.00%	CV LR	92.83%	87.69%	89.60%	92.05%
TF-IDF RF	89.45%	83.35%	84.81%	86.64%	CV RF	90.72%	84.18%	86.36%	89.35%
TF-IDF DT	89.24%	84.12%	84.86%	85.68%	CV DT	87.55%	82.10%	82.59%	83.13%

(c) Deep Learning algorithms

Fig. 5. (a)-(c). Results of Rule-based, Machine Learning and Deep Learning algorithms for the tweets dataset

6 Conclusion

Social media platforms generate vast amounts of data daily, making manual monitoring for hate speech impractical due to the sheer volume of content. This study focuses on development of auto hate detection for social media networks using sentiment analysis methods, such as NLTK and the BERT model, to monitor and evaluate user comments on forums. The proposed Automated sentiment analysis framework can provide valuable insights into the prevalence, types, and trends of hate speech. The analysis offers perspectives of users, and the auto tagging mechanism can direct changes to better serve users' requirements. Thereby, the systems can operate continuously, identifying and mitigating hate speech as soon as it appears, which is essential for maintaining healthy online environments and preventing the escalation of harmful sentiments.

References

1. Liberties: Free speech on social media: Rights and filtering (n.d.). https://www.liberties.eu/en/stories/free-speech-social-media/43773. Accessed: 14 February 2023
2. BBC News: Online hate speech rose 20% during pandemic: 'we've Normalised it', BBC News. BBC2021. https://www.bbc.com/news/newsbeat-59292509. Accessed: 10 November 2022

3. Twitter: Twitter's policy on hateful conduct | twitter help (n.d.). https://help.twitter.com/en/rules-and-policies/hateful-conduct-policy. Accessed: 10 November 2022
4. The US Sun: Why did YouTube remove the dislike count? 2021. https://www.the-sun.com/tech/4268641/why-did-youtube-remove-dislike-count/. Accessed: 10 November 2022
5. Hajarian, M., Khanbabaloo, Z.: Toward Stopping Incel Rebellion: Detecting Incels in Social Media Using Sentiment Analysis. In: 2021 7th International Conference on Web Research (ICWR), pp. 169–174 (2021). https://doi.org/10.1109/ICWR51868.2021.9443027
6. Hajarian, M., Bastanfard, A., Mohammadzadeh, J., Khalilian, M.: SNEFL: Social network explicit fuzzy like dataset and its application for Incel detection. Multimedia Tools and Applications **78**(23), 33457–33486 (2019)
7. Bhavitha, B.K., Rodrigues, A.P., Chiplunkar, N.N.: Comparative study of machine learning techniques in sentiment analysis. In: Proceedings of the International Conference on Inventive Communication and Computational Technologies, ICICCT 2017, pp. 216–221 (2017). https://doi.org/10.1109/ICICCT.2017.7975191
8. Fang, X., Zhan, J.: Sentiment analysis using product review data. J Big Data **2**(1) (2015). https://doi.org/10.1186/s40537-015-0015-2
9. Taboada, M., Brooke, J., Tofiloski, M., Voll, K., Stede, M.: Lexicon-based methods for sentiment analysis. Computational Linguistics **37**(2) (2013)
10. Hutto, C., Gilbert, E.: VADER: A Parsimonious Rule-Based Model for Sentiment Analysis of Social Media Text. Proceedings of the International AAAI Conference on Web and Social Media **8**(1), 216–225 (2014)
11. Wongkar, M., Angdresey, A.: Sentiment Analysis Using Naive Bayes Algorithm Of The Data Crawler: Twitter. 2019 Fourth International Conference on Informatics and Computing (ICIC). Semarang, Indonesia, pp. 1–5 (2019)
12. Devlin, J., Chang, M.W., Lee, K., Toutanova, K.: BERT: Pre-training of Deep Bidirectional Transformers for Language Understanding. In: Proceedings of the 2019 Conference on Empirical Methods in Natural Language Processing (EMNLP), pp. 4171–4186 (2019)
13. Github: KeyBERT (n.d.) https://maartengr.github.io/KeyBERT/

Assessment of the Umami Flavour of Commercial Meat and Plant-Based Meat Analogues

Galen Gay[1(✉)] and Aaron Thong[2]

[1] National Junior College, 37 Hillcrest Road, Singapore 288913, Singapore
galengay433@gmail.com
[2] Singapore Institute of Food and Biotechnology Innovation, 31 Biopolis Way, Nanos 138669, Singapore

Abstract. In view of the environmental repercussions of industrial livestock farming, alternative meat (AM) products have been gaining traction in the global market. However, progress remains slow due to a mismatch of the umami taste profile with conventional meat products. Thus, the development of an analysis platform capable of assessing the umami of meat analogues is critical to consumer palatability and acceptance. Therefore, in this study, an extraction and analysis method were developed to quantify the umami free amino acid and free 5'-nucleotide content in conventional and alternative meat samples. An extraction method using water was developed to mimic the release of umami compounds in the oral environment, while hydrochloric and perchloric acid were used in separate methods to obtain the full profile of the compounds. Separation, identification and quantification of the umami taste compounds using reversed-phase HPLC was achieved and successfully applied in combination with analysis via electronic-tongue to determine the Equivalent Umami Concentration (EUC) of beef and three AM samples. The taste profiles of all the four samples were found to be significantly different and varied significantly across the different extraction procedures, with the oral simulation extraction method achieving 75% extraction efficiency of the full umami compound profile. The umami rankings determined by the e-tongue agreed with that of chemosensory tests: mycoprotein < beef < plant-based meat alternatives (PBMA). The numerical EUCs determined by e-tongue and HPLC were also found to be similar for the mycoprotein and PBMA samples.

Keywords: Equivalent Umami Concentration · High Performance Liquid Chromatography · Electronic Tongue

1 Introduction

Livestock methane emissions account for roughly 32% of anthropogenic methane emissions (UNEP, 2021). Methane has been linked to an estimated 79–121 million tonnes of global crop production losses (CCAC, 2018). This problem is likely to be exacerbated by increasing protein demand in line with projections for population increase (Caputo

H. Guo et al. (Eds.): IRC-SET 2024, PTT, pp. 27–36, 2025.
https://doi.org/10.1007/978-981-96-3770-6_4

et al., 2022). Hence, there has a marked shift away from traditional meat (TM) to alternative meat (AM) products in line with the main aim of reducing methane emissions (Henchion et al., 2017).

To boost consumer acceptance of AM, it is important that they mimic the flavour profile of TM (Eckl et al., 2021). Therefore, it is imperative that a platform that is capable of accurately assessing the flavour profile of AM, is developed. So that their flavour profiles can be further refined to become compatible with consumers' palettes. It has been recognised that free amino acids (FAAs) are a key component of the flavour profile of meats and that 5'-monophosphate nucleotides (5'-NMPs) also act as flavour enhancers (Khan et al., 2015). Umami, which is synonymous with the "meaty flavour" of TM products, is imparted by the FAAs L-glutamate (L-Glu) and L-aspartate (L-Asp), as well as the umami-enhancing 5'-NMPs adenosine monophosphate, guanosine monophosphate and inosine monophosphate (Yamaguchi et al., 1971). In this study, umami flavour compounds were extracted from TM and AM products and were analysed and quantified by means of reversed phase high performance liquid chromatography (HPLC) and electronic tongue (e-tongue). To assess the synergistic effect of the FAAs and 5'-NMPs present in the samples their equivalent umami concentrations (EUC) were empirically determined and analysed as a numerical approximation of their umami flavour profile.

2 Materials and Methods

2.1 Chemicals and Reagents

Ultrapure water (Milli-Q) was provided by an ELGA PURELAB Chorus I Complete Water Purification System. Methanol (>99%) was obtained from VWR Chemicals (Radnor, PA, U.S.A.). Adenosine (>99%, HPLC, A), Cytidine (>99%, HPLC, C) and Guanosine (>99%, HPLC, G) were purchased from Fluka BioChemika. Inosine was obtained from Sigma-Aldrich (St. Louis, MO, U.S.A.), adenosine 5'-monophosphate monohydrate (≥97%, AMP), uridine 5'-monophosphate disodium (≥99%, UMP), thymidine 5'-monophosphate disodium salt hydrate (≥99%, TMP), cytidine 5'-monophosphate disodium salt (≥99%, CMP), guanosine 5'-monophosphate disodium salt hydrate (≥99%, GMP), inosine 5'-monophosphate disodium salt hydrate (≥99.0%, IMP), L-Glutamic acid monosodium salt hydrate (≥99%, MSG), 37% hydrochloric acid, perchloric acid (≥70%), amino acid standard analytical solution (AAS18) and disodium hydrogen phosphate were obtained from the same supplier. Dipotassium hydrogen phosphate (≥99.0%) and potassium dihydrogen phosphate (≥99.5%) were purchased from 1st BASE (Singapore). Acetonitrile was purchased from Fisher Chemicals and Borate buffer, o-phthaldialdehyde (OPA) and 9-fluorenylmethyl chloroformate (FMOC) from Agilent Technologies (Santa Clara, CA, U.S.A.).

2.2 Sampling

TM sample was obtained from Master Grocers' Minced Beef (MM). AM samples were obtained from Beyond Meat's Beyond Beef (BM), Impossible Foods' Impossible Beef (IM) and Quorn's Meat Free Mince (QM). Prior to analysis, raw samples were blended into a fine powder under liquid nitrogen. Samples were stored at −20 °C before cooking *sous-vide* in an 80 °C-water bath (Grant JB2 Series) for 12 minutes prior to analysis as put forth by Rotola-Pukkila et al. (2015).

2.3 Extraction of FAAs and 5'-NMPs

Extraction of Umami Taste Compounds in a Simulated Oral Environment. To simulate the conditions during mastication, an extraction method was adapted from Moerdijk-Poortvliet et al. (2022), involving the addition of 2.5 mL of Milli-Q water to 250 mg of sample at 37 °C, followed by inversion at 90 rpm for 1 minute. This was then followed by centrifugation at 3900 rpm for 10 minutes at 25 °C (Eppendorf Centrifuge 5810R). The supernatant was then filtered using a syringe filter before transferring to a HPLC vial for analysis.

Acid Extraction of Umami Taste Compounds. To obtain the full 5'-NMP profile, the following extraction procedure was adapted from Hwang et al. (2020): To 250 mg of sample, 1 mL of cold 0.5 M HCl was added. This was followed by homogenisation for 5 mins before centrifugation at 3900 rpm for 10 minutes at 4°C. The supernatant was then retrieved before repeating the previous extraction. The supernatants were then microcentrifuged for 5 minutes (Eppendorf 5424R) before filtering using a syringe filter. Samples were then evaporated to dryness using a vacuum centrifuge (Mi-Vac Quattro Concentrator) before reconstitution with 1 mL of mobile phase A, as will be discussed later in section 2.5. Another set of matrix spike duplicates were prepared by spiking 0.4 mL of 1 mM umami 5'-NMP standard into the sample to achieve a final spike concentration of 200 μM, from which the recovery rates could be determined and used to calculate the original concentration via the following formula, where cj represents the concentration of 5'-NMPs (in μM) present in the sample:

$$\%_{\text{recovery}} = \frac{c_{j(\text{spike})} - c_{j(\text{uncorrected})}}{200} \tag{1}$$

This extraction procedure was also repeated using cold 0.5 M perchloric acid as per Hwang et al. (2020), with a neutralisation step using NaOH after the initial centrifugation, followed by microcentrifugation. Since perchloric acid (pKa = −15.2) is a stronger acid than hydrochloric acid (pKa = −5.9) (Trummal et al., 2016), it was theorised that perchloric acid would be able to yield greater extraction of nucleotides.

The following extraction procedure of FAAs was adapted from Palacios et al. (2015) involving the addition of 5 mL of hexane to 500 mg of sample before inversion for 15 min and centrifugation at 3000 rpm for 5 min. The hexane was then decanted before repeating. Extraction was performed using 4950 μL of 0.1 M HCl with addition of 50 μL Norvaline (internal standard). Inversion and centrifugation were repeated as per the above procedure. 1 mL of supernatant was microcentrifuged for 5 min at 13 000 rpm before filtering using a syringe filter for analysis. All samples for each extraction method were prepared and analysed in triplicate.

2.4 Analysis of FAAs via HPLC

Amino acids were quantified based on the *Agilent Amino Acid Analysis: 'How-To'* Guide (2020) using an Agilent Infinity II 1290 HPLC equipped with a diode array detector (G4212A DAD). The amino acids were separated on a reversed-phase column (Agilent AdvanceBio AAA, C18, 4.6 x 100 mm, 2.7 μm) with a guard column (Agilent AdvanceBio AAA, 4.6 x 5 mm, 2.7 μm) at 40°C with a flow rate of 1.5 mL/ min. The amino acids were derivatized on-line before injection using a mixture of OPA and FMOC. The derivatization reagents were mixed with the buffer (borate buffer, 0.4 M, pH 10.2) before adjusting to the initial mobile phase pH (8.2) for injection. Mobile phase A was composed of 10 mM disodium hydrogen phosphate and 10 mM sodium tetraborate buffer with the pH adjusted to 8.2 using concentrated hydrochloric acid (37 %). Mobile phase B was composed of 45% Acetonitrile, 45% methanol and 10% water (v:v:v). Quantification of amino acids was achieved using a 6-point external calibration curve (5–500 μM). Amino acids were identified based on their elution timings with reference to their standard compounds.

2.5 Analysis of 5'-NMPs via HPLC

Nucleotide analysis was adapted from Ranogajec et al. (2010), involving the separation of analytes on a Synergy Hydro reversed phase column (150 mm x 4.6 mm, Phenomenex, USA). The HPLC conditions were as follows: column temperature, 20 °C; flow rate, 0.9 mL/ min; detection, UV 270 nm. The injection volume was 5 μL and elution was performed with a gradient. Mobile phase A consisted of a 50 mM phosphate buffer (pH 5.8) adjusted using 5 M NaOH, while mobile phase B consisted of methanol. Quantification of nucleotides was achieved using a 7-point external calibration curve (5–1000 μM). Nucleotides were identified based on their elution timings with reference to their standard compounds.

2.6 Assessment of Umami Flavour Profile

The equivalent umami concentration (EUC) was derived to assess the synergistic effect of the compounds on the umami taste and to approximate the umami taste sensation, as put forth by Yamaguchi et al. (1971) using the following equation:

$$EUC = \sum a_i b_i + 1.218\left(\sum a_i b_i\right)\left(\sum a_j b_j\right) \tag{2}$$

where the EUC is expressed in mg of MSG/ 100 g and ai and aj denote the concentrations (in mg/ 100 g) of the umami amino acids and nucleotides respectively. bi and bj also denote the synergy factors of the umami amino acids (L-Asp, 0.077; L-Glu, 1) and nucleotides (AMP, 0.18; GMP, 2.3; IMP, 1) respectively.

2.7 Umami Determination by Electronic Tongue

However, given that the EUC is a formula-based approximation of the perceived umami taste attribute, the αAstree electronic tongue (AlphaMOS, Toulouse, France), a potentiometric based system, was also used to determine the true umami flavour profile of the samples. The sensor array comprised seven cross-selective chemically modified field effect transistor technology sensors: AHS, CTS and NMS for sourness, saltiness and umami respectively, as well as general-purpose sensors (Wang et al., 2021). The system also included an autosampler, an Ag/ AgCl reference electrode and a stirrer. This cross-selective system allowed for a more accurate mimicry of gustatory perception *in vivo* and a more comprehensive characterisation of sample flavour profiles by incorporating multiple sensors with different sensitivities.

For the preparation of test samples, 0.3 g of cooked sample was resuspended at 1% w/v in 30 mL of Milli-Q water. This was 0.5% w/v for the control samples, where 0.15 g of mixed sample (1:1:1:1, w:w:w:w) was used instead. All samples were homogenised and incubated using a rotary shaker for 30 min. Samples were then centrifuged for 30 min at 3800 rpm, after which the supernatant was decanted for analysis (25 mL for test samples, 22.5 mL for control). Control extracts were spiked with 2.5 mL MSG solution with varying concentrations (0.002441 to 10 g/100 ml) for the seven-point calibration curve. Samples were prepared and analysed in triplicate.

3 Results and Discussion

3.1 Umami FAA and 5'-NMP Content

Using the peak areas of individual compounds, the 5'-NMP and FAA content was successfully quantified (Fig. 1). A two-way ANOVA was performed to determine the effect of sample type and extraction procedure on the compound concentration. The interaction between the sample and the extraction procedure was found to be significant ($F6,24 = 24.6$, $p < 0.001$). The effect of the extraction method on the AMP and GMP content, as well as their recoveries was found to be significant for all samples ($p< 0.001$ for all) except for IMP ($p$content = 0.421, $p$recovery = 0.259). Additionally, the FAA and 5'-NMP content, as well as their recoveries varied significantly among the four samples ($p < 0.001$ for all). The concentration of IMP in BM was also a magnitude higher than the other samples (Fig. 1), which alludes to the addition of inosinic acid salts as flavouring to mask beany off-flavours (Mittermeier-Kleifiinger et al., 2021). Extraction using water was able to achieve consistent recoveries of >80%. However, the acid extractions were characterised by low recovery rates. Additionally, GMP and IMP were not detected in the MM sample across all extraction procedures, which was inconsistent with the findings of Hwang et al. (2020). This hinted at the possibility of hydrolysis of the nucleotides analytes into their respective nucleobases, as inosine and guanosine peaks were observed for the acid-hydrolysed MM samples but not for the water extraction, therefore, the values were corrected based on the recovery rates from (1) using the formula:

$$a_{j(corrected)} = \%_{recovery} \times a_{j(uncorrected)} \tag{3}$$

This was used to account for loss of analyte during the extraction process and suppression of compound peaks during UV detection due to matrix interference (Fig. 2). With the corrected values, it was observed that the variance in content of the samples across the extraction procedures was indeed reduced for AMP ($p = 0.0395$), however conversely, there was a larger variance for GMP ($p < 0.001$) and IMP ($p = 0.0312$).

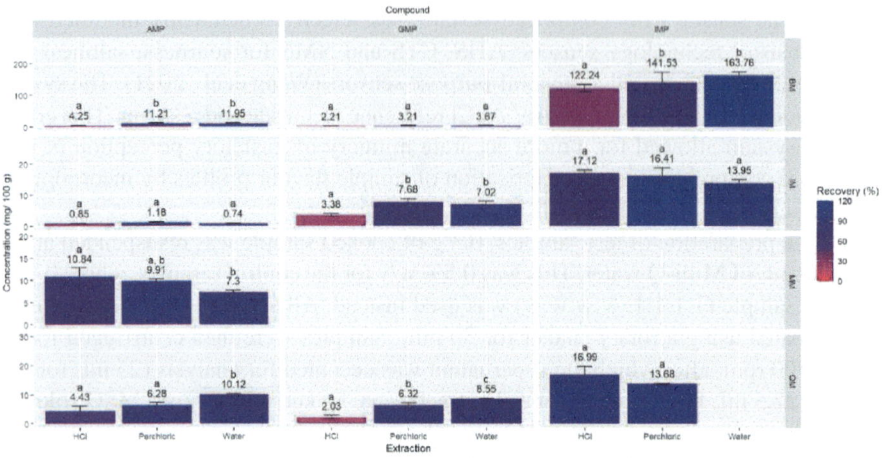

Fig. 1. Umami 5'-NMP content of the four samples and their recoveries, $^{a-c}$ different superscript letters within the same row indicate statistically significant differences in content between extraction methods ($p < 0.05$)

Additionally, for compounds hydrolysed using hydrochloric acid with low *%recovery* , their *aj*(*corrected*) were generally higher than those which underwent perchloric acid hydrolysis, contradictory to the notion that perchloric acid would be capable of performing more rigorous hydrolysis. This suggests that the use of a target analyte as part of the recovery spike, was limited in situations where recoveries were low. This could be caused by differences in the interactions between the spiked and unspiked samples with the matrix due to the increased concentration of target analytes in the spikes, which would result in varying degrees of matrix interference and inaccurate estimation of the loss of target analytes during extraction or matrix effect during detection. However, apart from QM, water extraction averaged a $75.2 \pm 13.1\%$ extraction efficiency as compared to perchloric acid, indicating that perchloric acid was effective in obtaining the full 5'-NMP profile of these samples.

While the difference in umami FAA content between MM and QM was statistically insignificant, IM and BM saw significantly higher L-Asp and L-Glu content (Table 1). This can be attributed to the inclusion of soy, rice and pea protein concentrates rich in such amino acids (Mohsen et al., 2009; Babault et al., 2015; Shih et al., 2000) as well as flavourings in the plant protein formula.

3.2 Principal Component Analysis (PCA) and Principal Component Regression (PCR)

PCA was performed to visualise the seven sensor readings of the e-tongue (Fig. 3). Strong correlation between CPS, CTS and NMS sensors was observed while the MSG calibration curve was well-distributed for concentrations upwards of 0.015625 g/ 100 mL. PCR was also performed to create an EUC estimation model based on the sensor readings that demonstrated good linear correlation with a coefficient of 0.970.

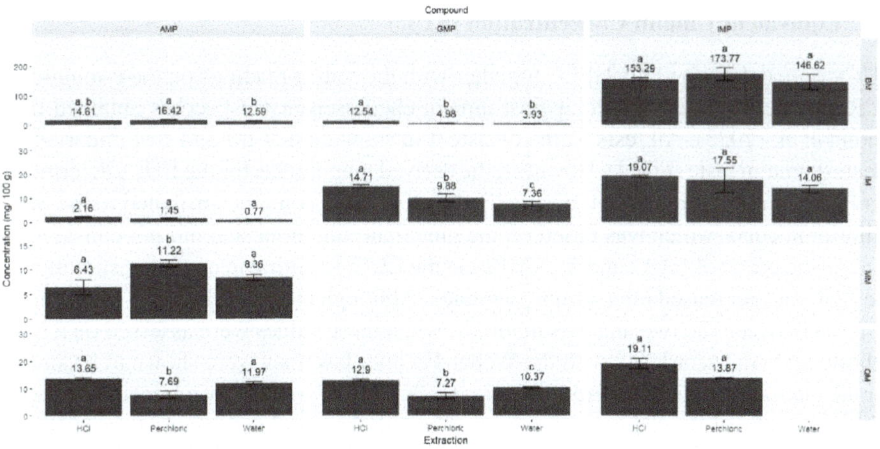

Fig. 2. Corrected umami 5'-NMP content of the four samples, $^{a-c}$ different superscript letters within the same row indicate statistically significant differences in content between extraction methods ($p < 0.05$)

Table 1. Umami FAA content (mg/ 100 g) of TM and PM samples, $^{a-c}$ different superscript letters within the same row indicate statistically significant differences in content

	Content (mg/ 100 g)			
Compound	IM	MM	BM	QM
L-Asp	9.50 ± 0.184^{b}	1.86 ± 0.0508^{a}	6.91 ± 0.231^{c}	1.49 ± 0.0645^{a}
L-Glu	1050 ± 174^{b}	19.9 ± 0.182^{a}	214 ± 4.52^{c}	22.1 ± 0.174^{a}

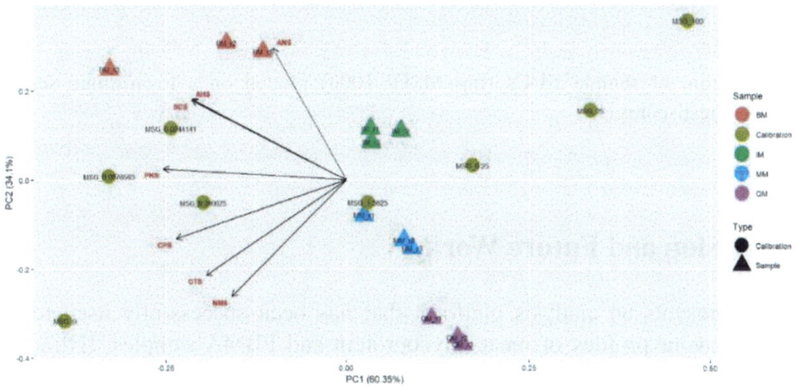

Fig. 3. PCA biplots of the seven sensor responses

3.3 Equivalent Umami Concentration (EUC)

EUCs were determined via HPLC together with the sensor readings on the e-tongue via PCR and they were compared against umami chemosensory test scores obtained from Thong et al. (2023). All tests were consistent in showing that IM and BM imparted the greatest umami taste (Fig. 4). However, the derived magnitudes for the EUCs determined via HPLC and PCR remained drastically different. This could be attributed to the MSG calibration curve which was based on the glutamate salt alone and did not consider the synergistic effect of FAAs and 5'-NMPs via the EUC, resulting in an underestimation of the EUC and decreased inter-sample variance. Although the numerical EUC evaluations differed between the two analysis methods, when these values were assessed on a logarithmic scale in accordance with the Weber-Fechner law, they were shown to be similar for all samples except MM. Additionally the PCR-EUC trend showed good correlation with the results of the chemosensory tests. Overall, the simulated umami taste profile of the samples in the mouth was comparable to its full potential.

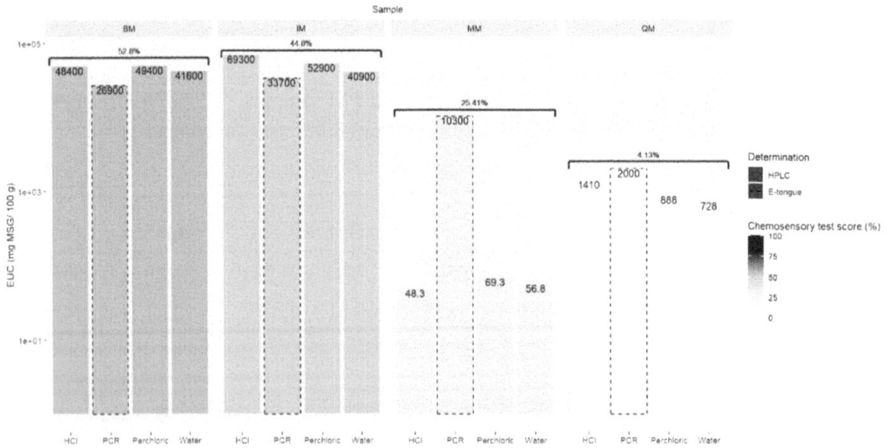

Fig. 4. Histogram of sample EUCs (mg MSG/ 100g) plotted on a logarithmic scale with chemosensory test scores (%)

4 Conclusion and Future Works

This study presents an analysis platform that has been successfully used to assess the umami flavour profiles of meat, mycoprotein and PBMA samples. Extraction of umami taste compounds was achieved using water, hydrochloric acid and perchloric acid, through which it was determined that hydrochloric acid was unsuitable for said extraction due to low and varied recovery rates that could not be rectified using matrix spike duplicates. Through the oral simulation extraction procedure using water, it was determined that the amount of umami compounds released in the mouth during mastication was approximately 75% of the total FAA and free 5'-NMP content of the samples.

The concentration of umami compounds present in the samples was determined and their synergistic effect was evaluated using the EUCs determined both by HPLC and e-tongue in conjunction with chemosensory tests from literature. Although there were discrepancies in the exact placings, it was concluded that the PBMAs (IM and BM) contained the greatest amount of umami. The rankings determined by e-tongue were also found to agree with the results of chemosensory tests, that being QM<MM<BM<IM. Additionally, the predicted EUC values determined by PCR were also found to be similar to those determined by HPLC for IM, BM and QM.

To further this study, extensive method validation of the developed HPLC method could be carried out to determine its intra-and interday precisions, limits of detection and limits of quantification to ensure the stability, sensitivity and accuracy of the HPLC analysis method. Additionally, non-target analyte spikes (e.g. 5′-N-methylcarboxamidoadenosine) can be used for the 5'-NMP analysis method to determine the recoveries of target analytes. This is to obtain a more accurate assessment of the matrix effect since they are less susceptible to it as they are not naturally occurring. Alternatively, matrix minimisation via dilution of the sample can be explored as an alternative to obtain better recoveries by reducing the extent of matrix interference. The EUC prediction model could also be further developed via the establishment of an EUC calibration curve to model the synergistic effects of FAAs and 5'-NMPs. Different unsupervised or supervised linear regression models (e.g. partial least squares) could also be brought in to establish the calibration curves to achieve better sensitivity to lower EUCs. This umami taste analysis platform involving the combined use of HPLC, EUC in conjunction with chemosensory tests could also be further applied to assess other tastes to obtain the full taste profiles of AM, other food samples or pharmaceutical drugs for the development of their taste profiles.

References

Agilent Technologies, Inc. Agilent Biocolumns Amino Acid Analysis "How-To" Guide (2020). https://www.agilent.com/cs/library/brochures/5991-7694EN_AdvanceBio%20AAA_How-To%20Guide_LR.pdf

Babault, N., et al.: Pea proteins oral supplementation promotes muscle thickness gains during resistance training: a double-blind, randomized, Placebo-controlled clinical trial vs. Whey protein. Journal of the International Society of Sports Nutrition **12**(1), 3 (2015). Taylor & Francis Online. https://doi.org/10.1186/s12970-014-0064-5

Caputo, V., Sogari, G., Van Loo, E.J.: Do plant-based and blend meat alternatives taste like meat? A combined sensory and choice experiment study. Applied Economic Perspectives and Policy **45**(1), 86–105 (2022). Wiley Online Library. https://doi.org/10.1002/aepp.13247

Climate & Clean Air Coalition: Tropospheric ozone. Climate & Clean Air Coalition (2018). https://www.ccacoalition.org/en/slcps/tropospheric-ozone

Eckl, M.R., Biesbroek, S., van't Veer, P., Geleijnse, J.M.: Replacement of Meat with Non-Meat Protein Sources: A Review of the Drivers and Inhibitors in Developed Countries. Nutrients **13**(10), 3602 (2021). MDPI. https://doi.org/10.3390/nu13103602

Henchion, M., Hayes, M., Mullen, A., Fenelon, M., Tiwari, B.: Future Protein Supply and Demand: Strategies and Factors Influencing a Sustainable Equilibrium. Foods **6**(7), 53 (2017). MDPI. https://doi.org/10.3390/foods6070053

Hwang, Y.-H., Ismail, I., Joo, S.-T.: Identification of Umami Taste in Sous-Vide Beef by Chemical Analyses, Equivalent Umami Concentration, and Electronic Tongue System. Foods **9**(3), 251 (2020). MDPI. https://doi.org/10.3390/foods9030251

Khan, M.I., Jo, C., Tariq, M.R.: Meat flavor precursors and factors influencing flavor precursors—A systematic review. Meat Science **110**, 278–284 (2015). ScienceDirect. https://doi.org/10.1016/j.meatsci.2015.08.002

Mittermeier-Kleßinger, V.K., Hofmann, T., Dawid, C.: Mitigating Off-Flavors of Plant-Based Proteins. Journal of Agricultural and Food Chemistry **69**(32), 9202–9207 (2021). ACS Publications. https://doi.org/10.1021/acs.jafc.1c03398

Moerdijk-Poortvliet, T.C.W., et al.: Extraction and analysis of free amino acids and 5′-nucleotides, the key contributors to the umami taste of seaweed. Food Chemistry **370**, 131352 (2022). ScienceDirect. https://doi.org/10.1016/j.foodchem.2021.131352

Mohsen, S.M., et al.: Effect of substitution of soy protein isolate on aroma volatiles, chemical composition and sensory quality of wheat cookies. Int. J. Food Sci. Technol. **44**(9), 1705–1712 (2009). Wiley Online Library. https://doi.org/10.1111/j.1365-2621.2009.01978.x

Perez-Palacios, T., Barroso, M.A., Ruiz, J., Antequera, T.: A Rapid and Accurate Extraction Procedure for Analysing Free Amino Acids in Meat Samples by GC-MS. International Journal of Analytical Chemistry, 2015. Hindawi Publishing Corporation (2015). https://doi.org/10.1155/2015/209214

Ranogajec, A., Beluhan, S., Šmit, Z.: Analysis of nucleosides and monophosphate nucleotides from mushrooms with reversed-phase HPLC. J. Separat. Sci. **33**(8), 1024–1033 (2010). Wiley Analytical Science. https://doi.org/10.1002/jssc.200900516

Rotola-Pukkila, M.K., Pihlajaviita, S.T., Kaimainen, M.T., Hopia, A.I.: Concentration of Umami Compounds in Pork Meat and Cooking Juice with Different Cooking Times and Temperatures. Journal of Food Science **80**(12), C2711–C2716 (2015). IFT. https://doi.org/10.1111/1750-3841.13127

Shih, F.F., Daigle, K.W.: Preparation and characterization of rice protein isolates. Journal of the American Oil Chemists' Society **77**(8), 885–889 (2000). Springer Link. https://doi.org/10.1007/s11746-000-0141-2

Thong, A., Tan, V.W.K., Chan, G., Choy, M.J.Y., Forde, C.G.: Comparison of differences in Sensory, Volatile Odour-Activity and Volatile Profile of Commercial Plantbased meats. Food Research International. ScienceDirect (2023). https://doi.org/10.1016/j.foodres.2023.113848

Trummal, A., Lipping, L., Kaljurand, I., Koppel, I.A., Leito, I.: Acidity of Strong Acids in Water and Dimethyl Sulfoxide. The Journal of Physical Chemistry A **120**(20), 3663–3669 (2016). ACS Publications. https://doi.org/10.1021/acs.jpca.6b02253

United Nations Environment Programme: Methane emissions are driving climate change. Here's how to reduce them. United Nations Environment Programme (2021). https://www.unep.org/news-and-stories/story/methane-emissions-are-driving-climate-change-heres-how-reduce-them

Wang, Z., et al.: Taste Masking Study Based on an Electronic Tongue: the Formulation Design of 3D Printed Levetiracetam Instant-Dissolving Tablets. Pharmaceutical Research **38**(5), 831–842 (2021). Springer Link. https://doi.org/10.1007/s11095-021-03041-9

Yamaguchi, S., Yoshikawa, T., Ikeda, S., Ninomiya, T.: Measurement of the relative taste intensity of some L-α-amino acids and 5'-nucleotides. Journal of Food Science **36**(6), 846–849 (1971). IFT. https://doi.org/10.1111/j.1365-2621.1971.tb15541.x

Comparative Study of Tying a Surgical Knot Using Conventional Instrumentation and Haptic Feedback Enabled Forceps

Gillian Ng[1(✉)] and Yoke Rung Wong[1,2]

[1] Biomedical Engineering, School of Science and Technology (SST), Singapore University of Social Sciences, Singapore, Singapore
gillian.ng.w.x@nccs.com.sg, yrwong001@suss.edu.sg

[2] Biomechanics Laboratory, Singapore General Hospital, Singapore, Singapore

Abstract. Laparoscopy is a form of minimally invasive surgery that has demonstrated medical advantages to patients undergone surgery. However, the altered depth perception and counterintuitive movements in the cavity might cause inefficient intrabdominal surgical knot-tying, potentially lengthening the surgical procedure and increasing complication risks to patients. Developments have been made in laparoscopic surgeries using robotics and smart feedback systems to form a more comprehensive cognitive depiction of the intended subject. In this study, we aim to conduct a comparative analysis between two groups - conventional forceps versus haptic feedback-enabled (termed as 'smart' in this study) forceps. By embedding a strain gauge sensor in the smart forceps and conjugating it with a real-time data acquisition system and input/output interface, it allowed the surgeon to gain haptic feedback without jeopardizing the surgical process. Static test was carried out on all the Square knots to determine if there was a statistically significant difference between the two groups. The two key findings shown in this study were the outcome of the comparative analysis and the observation of the failure mechanism presented in the two groups. The synergy of our findings suggests the potentiality of the haptic feedback system (HFS) in becoming a standard component of commercially available surgical robotic systems, a field that is gaining traction.

Keywords: Laparoscopic Surgery · Haptic Feedback System (HFS) · Surgical Instrumentation · Suture Knots

1 Introduction

1.1 Overview of Laparoscopic Surgery - Advantages for the Patient

In the surgical sphere, the types of operative procedures can be split into two major classifications, namely open surgery and laparoscopic surgery. Open surgery involves a large single incision that dissects through the abdominal region, allowing direct access to the surgical site. Laparoscopic surgery also known as keyhole surgery, diagnostic

H. Guo et al. (Eds.): IRC-SET 2024, PTT, pp. 37–46, 2025.
https://doi.org/10.1007/978-981-96-3770-6_5

laparoscopy and/or minimally invasive surgery (MIS), is a surgery that is initiated via a small incisor [1]. Previous research has established laparoscopic surgery as the preferred choice over open surgery, which is indicative of the shift towards the former [2–4].

From year 2018 to 2020, general surgery residents performed an astounding 462% more laparoscopic procedures [5]. With the function of laparoscopy being well-defined over the past few decades, its increase in clinical practice is therefore expected and unsurprising. From the patient's perspective, the benefits of laparoscopic surgery include avoiding large open incisions and thus generally reducing the likelihood of major blood loss, less trauma and mitigating unwanted effects from analgesia because the required amount is significantly lower [6].

The expected rate of postoperative complications related to wound infection and dehiscence was also generally lower [7]. Through the conductance of the operation restricted to within the body cavity, the risk of cooling, drying, excessive handling and retraction of internal organs were lowered as compared to open surgeries. This further reduces the chances of postoperative peritoneal adhesions due to subsequent bowel obstruction [6].

The benefits mentioned help to reduce hospitalisation and promote early mobilisation, ensuring a faster recovery. The risks of osteopenia, muscle wasting, and urinary retention were reported to be lower when they have generally been associated with lengthy inpatient stays and inactivity [6, 8]. Lastly, laparoscopic surgery is less likely to cause postoperative concern linked to self-image, as the scars created by laparoscopic surgery tend to be smaller than those of open surgery.

1.2 Practical Aspects Relating to Laparoscopic Surgery (Suturing)

Given the less direct contact between the surgeon and the patient, laparoscopic surgeries involve the use of a special laparoscope tool that aids in the visualisation of the body cavity by transmitting the video feed captured by the small camera attached to the laparoscope, to the monitor screen usually placed in front of the surgeon. However, there are disadvantages, even amongst experienced surgeons. These include an unstable video camera platform, restricted degree of freedom, two-dimensional imaging, and poor ergonomics [9].

Studies have, to an extent, content the difficulties related to perception. Trocar, a metal tube used in laparoscopic surgeries, serves as a steadying point for the robotic instruments. However, a small movement at the proximal end will manifest into a large movement at the distal end. With the inversion of the normal axis, the trocar must be maneuvered in the opposite direction of the intended direction [10]. Essentially, surgeons are effectively operating in a mirror. Furthermore, haptic feedbacks are lost as three-dimensional structures are transmitted to reflect into a two-dimensional screen, causing difficulties in perceiving depth and cues for identifying anatomical structures [11, 12].

In laparoscopic dissection, the restricted range of motion; where the degree of freedom drops from six to four degrees, makes it more difficult to work with tools and tissues (particularly for suturing). In a survey conducted in 2013 by the Fellowship Council (FC) program directors, more than half of admitting fellows ($\geq$56%) who had completed their training in residency surgery were unable to perform laparoscopic suturing [13]. Additionally, the mastering of laparoscopic intracorporeal suturing has a steep learning curve [14]. A statistically significant variation was established in the average intracorporeal knot-tying times depending on the competence level of the surgeon, including experienced, less experienced, and surgical chief residents (97.3, 237.2, and 265.3 s respectively, P values < 0.01) [14]. Suturing is an advanced laparoscopic skill as they are ever present in gastrointestinal procedures. Despite the provision of simulation-based training, the steep learning curve and lack of standardised training make most residents underprepared for laparoscopic sutures [15]. This can increase intra-operative injury, where excessive clamping force can lead to tissue trauma.

1.3 Surgical Knot and Its Associated Problems

A high-quality suturing requires precise skills as it is considered a delicate task. Despite suture knots being the weakest link, they are used as ligatures during suturing and are thus central to surgical procedures. If the knots fail, disastrous consequences may occur. One common complication related to a failed knot includes wound dehiscence, the presence of which will affect proper wound healing. Likewise, the proportion of incisional hernia was reported to be as high as 20% within the same year following a midline laparotomy [16].

While surgeons vary greatly in their knot-tying methods, it has been demonstrated that adding more throws to a knot consistently lowers the rates of spontaneous untying [17]. However, the tying pre-tension administered by the surgeon when tightening each throw are dependent on individual approximation and intuition rather than experience [18]. This observation was presented by the study, whereby the sequential suture tension varied even amongst experienced surgeons. Although the importance of mechanical analysis of knots has been communicated throughout the medical profession, the existing guidelines for best clinical practices are still primarily empirical and historical, rather than relying on physics-based structural analyses [19]. It is arduous to standardise how tight the knot should be tied in both laboratory and clinical setting. Some studies have shown evidence that within the range of 80–100 N are where knots unravelled at high tension, particularly for Square and surgeons' knot [20].

One of the problems with suturing is that despite the knowledge of the maximum tensile force as reported by other studies, the need for high tensional strength may not be translatable for all laparoscopic surgeries [21]. For cases such as tendon or hernia repair, they require high tensional strength, while routine wound closure procedures require a lower tensile strain [20]. As a result, it may be probable that by building the feedback system into such instruments will be able to bring a higher degree of awareness. Bringing this into context where the force applied onto these surgical knots through conventional instruments (forceps) are based on intuition, there is a higher risk of tissue damage, especially when the limitations presented by inadequate tactile and force perception.

In this study, we aim to conduct a comparative analysis between two groups - conventional forceps versus haptic feedback-enabled (termed as 'smart' in this study) forceps. We hypothesize that the smart forceps allows the surgeon to gain haptic feedback without jeopardizing the surgical process and thus ensure good surgical outcome.

2 Materials and Methods

2.1 Design of the Smart Forceps

Conventional surgical forceps were retrofitted with a strain gauge sensor (Fig. 1). The retrofitted forceps were mounted with a miniature holder designed to accommodate the strain gauge sensor.

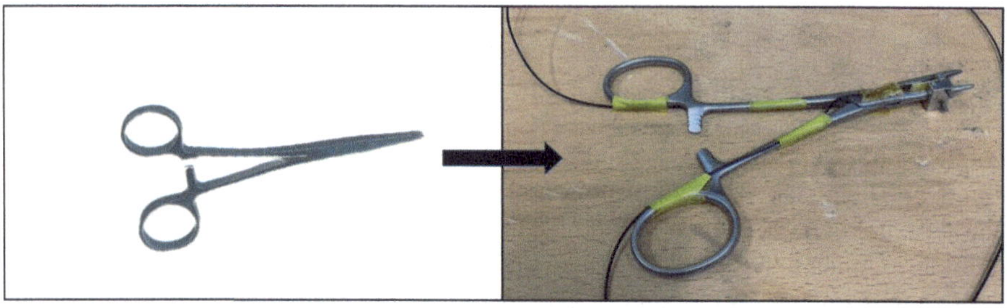

Fig. 1. Conventional Kelly Forceps retrofitted with a strain gauge sensor.

The side of the forceps was fitted with a cable that connects to the P-3500 strain gauge indicator box (Vishay Measurements Group, Raleigh, North Carolina). The commercial indicator box displayed the change in electrical resistance when the strain gauge sensor deformed due to the application of force on the tip of the forceps. This indicator box is based on the Wheatstone bridge circuit, where it facilitates quarter-, half- and full-bridge strain gauge circuits. The indicator box also allows the user to select the preferred gain factor, balancing of the circuit and the output gain, where we ultimately set the gain factor at "5 to 2.0 k", and a balance of 0.

2.2 Strain Gauge Sensor and Wheatstone Bridge Configuration

The measurement of the strain was based on the principles of a Wheatstone bridge circuit. The strain gauge acts as the fourth arm of the bridge where it functions as the variable resistance whilst R1, R2 and R3 have a fixed resistance magnitude. The differentiated voltage across the two junctions was further measured using the direct current (DC) power supply supplied to the circuitry. The electrical resistance of a strain gauge varies in proportion to the applied strain acting on the structure. The strain gauge deformed upon the application of the force to the structure, wherein the change in electrical resistances then measured the change in differential voltage. In the Wheatstone bridge, there were two parallel voltage divider circuits, R1 and R3, and R2 and Rg. These two sets of voltage divider circuits composed the electrical equivalent of the Wheatstone bridge. The output of the Wheatstone bridge, Vmeas, was then measured between the center nodes of the two voltage dividers.

2.3 Force Calibration – Strain Gauge Sensor

The retrofitted Kelly forceps were mounted on a specially designed jig, where the jig was then fixated on the base of the Instron machine (ElectroPuls E1000, Instron Corp.,

Canton MA, USA). The loading process was controlled by the user's command into the proprietary software of Instron. The loading process was manually set to have a fixed increment load of 10 N, starting from 0 N to 50 N, where it subsequently held each amount of force for a few seconds before the user allowed the 10 N increment. To obtain the correct force magnitude, we attempted several balance settings ('4.3–9.9 Gauge Factor (GF)', '2.3–4.5 GF', ''1.7–2.5 GF', '5–2.5 GF' and '5–2.0 GF') within the indicator box and plotted the change in strain with each corresponding incremental force. Figure 2 shows the measured data, the regression line equation and coefficient of determination calculated by linear regression analysis using Excel, and after selecting the gauge factor of '5–2.0 GF'.

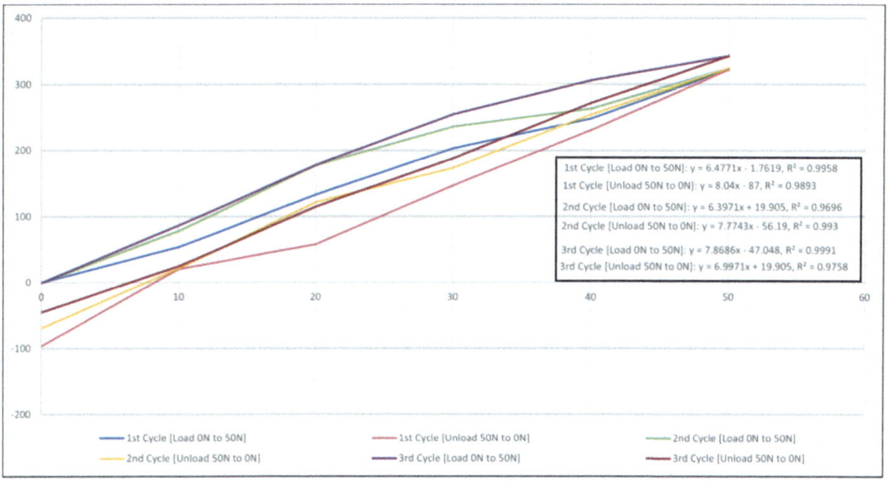

Fig. 2. Strain-Force Plot.

Based on the equation, the sensor has a mean sensitivity of 7.259. In addition, the coefficient of determination of each curve was near 1, which demonstrated that the strain gauge sensor has a strong linear behaviour and is suitable for sensing the applied load. Upon the confirmation of the indicator box's settings, the clamping force was then measured by the Instron load cell, where the deformation of the plate (that holds the strain gauge) caused the change in strain, leading to the changes in the differential voltage as measured by the National Instrument Data Acquisition System (NI DAQ).

2.4 Setting Up the Graphical User Interface (GUI)

The Graphical User Interface (GUI) was developed using LabView 8.6, Full Professional Developer software (National Instruments, Austin, TX). It is a graphical programming language that establishes a dataflow paradigm where the codes were drawn or presented graphically, instead of written, as shown in Fig. 3. The execution of the program was directed by the connector wires that linked the processing nodes together (Elliott et al. 2007). Figure 4 shows the front panel that contained the inputs and controls which displayed the real-time observations. The interface was developed to represent three distinct trigger points by programming specific Boolean to reflect the colour yellow, green and red. This resulted in an effective display of discrete data when either one of the following conditions were instituted; when the clamping force is inadequate (<25 N), excessive (>50 N), or optimal (25 N ≤ X ≤ 50 N).

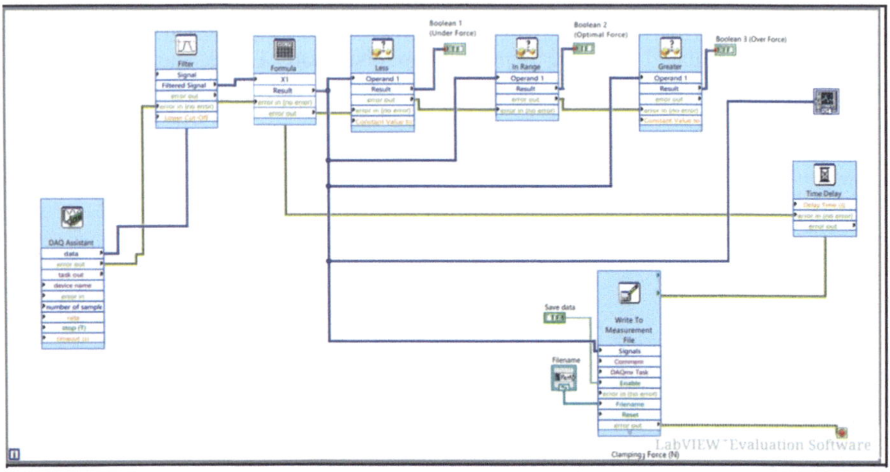

Fig. 3. Block Diagram for the Virtual Instrument (VI).

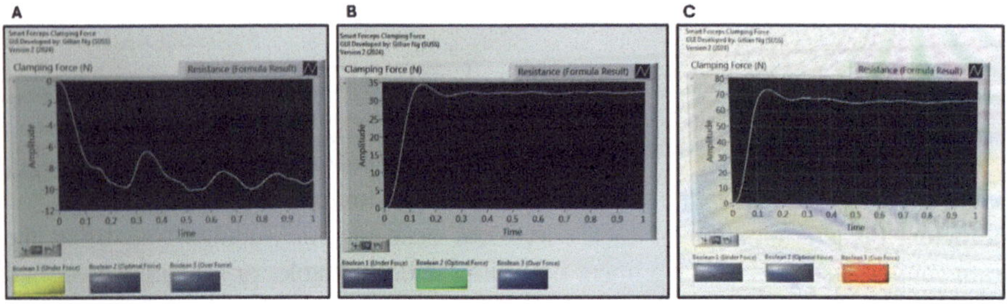

Fig. 4. Front panel of the VI. (A) shows the real-time observation when the clamping force is below 25 N (inadequate force), thereby indicated by the yellow Boolean. (B) shows the real-time observation when the clamping force is between the range of 25 N to 50 N (optimal force), thereby indicated by the green Boolean. (C) shows the real-time observation when the clamping force is above 50 N (excessive force), thereby indicated by the red Boolean.

2.5 Suturing of Porcine Skin (Conventional vs Smart Forceps)

To validate the comparison between the conventional and smart forceps, we conducted the surgical repair procedure (suturing) on 5 pairs of 60 by 30 mm porcine skin for each group. There were no differences in terms of the experimental set-up for both the conventional and smart forceps. The suture material used was polypropylene suture, in which we manipulated the suture needle with both conventional and smart forceps to puncture the porcine skin. The first anchor point was performed at least 1 cm from the laceration, with the second anchor point penetrating the other piece of the porcine skin at the same distance. A basic knot, also known as a Square knot, that consists of two "throws" was used for comparing the surgical knot security for the two test groups. After rejoining the two pieces of porcine skins, we subjected the samples to a strain test.

2.6 Static Test

To conduct a comparative analysis between the two experimental groups (conventional and smart forceps), we obtained the maximum tensile force by conducting static test using the Instron 3400 Series force transducer. This force transducer was operated through the Bluehill® Universal software. Based on its recommendation, we ran pretest before the actual measurement of the data. The pre-test involves the following section - preload and precycling. Preload removed slack from the load string before the test began, where it was set at a displacement rate of 20 mm/min, whereas precycling was set for 5 cycles, at a rate of 120 mm/min. After running the pre-test, the top pneumatic gripper pulled the specimen at a 20 mm/min displacement rate. Measurement was concluded upon the breakage of the suture material.

2.7 Statistical Analysis

Sample size calculation was first conducted to establish if the number of samples per group were adequate to draw an accurate conclusion, where the 80% power calculation = 4.39. A normality test was subsequently conducted to determine whether the samples were normally distributed. Normality distribution was established when the significant value was more than 0.05. Statistical analysis between the two groups was performed using an independent t-test. Statistical analyses were performed using IBM SPSS Statistics (version 25.0). The statistical significance level was set at $P < 0.05$.

3 Results and Discussion

Table 1. Average Maximum Tensile Force (N) and Standard Deviation.

Group	Average (N)	Std. Deviation	Std. Error Mean
Smart	10.11	3.29	1.47
Conventional	9.27	2.85	1.28

Table 1 represents the average maximum tensile force (knot's failure load) obtained from each group, where the average force required to tear the suture material was higher in the smart forceps group (10.11 N) compared to the conventional forceps group (9.27 N).

Table 2 represents the maximum tensile force in Newton from the respective groups (smart and conventional forceps), where the individual measurements observed for the smart forceps group were higher than those in the conventional forceps group. However, the differences observed were not statistically significant ($p = 0.678$), indicating the absence of comparative differences between the two groups. However, it was interesting to note the failure mechanisms for the two groups.

Table 2. Consolidated Data of Smart (A) and Conventional (B) forceps.

A

Sample ID	Maximum Tensile Force (N)	Failure Mechanism
smart_9	8.97	Knot unravelled
smart_2	8.2	Suture break
smart_3	13.83	Suture break
smart_6	13.26	Suture break
smart_11	6.28	Suture break

B

Sample ID	Maximum Tensile Force (N)	Failure Mechanism
conv_1	4.56	Knot unravelled
conv_2	9.64	Knot unravelled
conv_3	11.02	Knot unravelled
conv_4	11.94	Knot unravelled
conv_5	9.18	Suture break

In the smart forceps group (shown in Table 2A), 80% of them obtained a suture break phenomenon while the same proportion obtained knot unravelled in the conventional forceps group instead. Suture break – ideal scenario with regards to suture knot security - in this study was defined as the point at which the material breaks under loading, and knot unravelled was defined as a knot that slips easily along the cord at which it was made. Despite the reported p-value, the type of failure mechanism presented suggests a potential role for smart forceps, nonetheless. While it has been the norm in research to support the study based on the significance of the results, we need to consider the case where separate inferences may be made by observing the individual phenomenon [22].

The mechanism of suture knots is dependent on the static contact between the two strands of suture material which may be composed differently. The strands may exist as either a separate or united entity, but they are essentially viewed as friction between the strands. However, the ability to maintain a tight knot is bounded by several factors, including the number of 'throws', and the force applied when tying a knot.

Square knots have generally been purported in several studies to be rather secure in clinical practice, but this type of knot still permits the unravelling of the knot at high tensile loading [20]. It may be argued that this failure mode could be circumvented simply by increasing the number of knot throws. However, increasing the number of throws increases its bulk, which regardless of the location of the suture, may pose problems with certain repairs, particularly when the sutures need to navigate narrow fibroosseous pulley [23]. Likewise, large knots may potentially interfere with the recovery process and reduce the repair strength [24, 25].

Although majority of the newer suture materials have a low coefficient of friction and are more likely to slip when tied, it is still paramount to be aware of the proper knot holding strength as modern high-strength suture materials cannot be fully utilised without proper strength awareness.

4 Conclusion

Although the comparative analysis showed no significant difference, the failure mode observed between the two groups showed promising results. The success of surgical procedures may even be correlated to the biomechanics properties of the suture, whereby the knot should be tight enough such that it does not unravel while not being too tight to the point it renders the tissue ischemic, leading to wound breakdown. Smart forceps therefore provide the user with a degree of precision and confidence in terms of surgical knot tying, where its principles lie upon approximation instead of strangulation. Although this study design is still in the exploratory stage, it does hold promising potential in clinical applications, and a certain degree of focus should be directed to developing improved prototypes of similar smart forceps. With the technological advancements currently being paved out in terms of medical instruments, equipping users with a feedback-enabled system may enhance the provisional quality of healthcare services.

References

1. Richardson, W.S., et al.: Minimally invasive abdominal surgery. Ochsner J. **2**(3), 153–157 (2000)
2. Gallagher, A.G., O'Sullivan, G.C.: Fundamentals of Surgical Simulation: Principles and Practice. Springer, London (2012)
3. Eckert, M., et al.: The changing face of the general surgeon: national and local trends in resident operative experience. Am. J. Surg. **199**(5), 652–656 (2010)
4. Alkhoury, F., et al.: The impact of laparoscopy on the volume of open cases in general surgery training. J. Surg. Educ. **67**(5), 316–319 (2010)
5. Bingmer, K., et al.: Decline of open surgical experience for general surgery residents. Surg. Endosc. **34**(2), 967–972 (2019)
6. Agha, R., Muir, G.: Does laparoscopic surgery spell the end of the open surgeon? JRSM **96**(11), 544–546 (2003)
7. Huang, A.: Clinical surgery. J. R. Soc. Med. **94**(12), 651–652 (2001)

8. Buia, A., Stockhausen, F., Hanisch, E.: Laparoscopic surgery: a qualified systematic review. World J. Methodol. **5**(4), 238 (2015)
9. Ballantyne, G.H.: The pitfalls of laparoscopic surgery: challenges for robotics and telerobotic surgery. Surg. Laparosc. Endosc. Percutaneous Tech. **12**(1), 1–5 (2002)
10. Xin, H., Zelek. J.S., Carnahan, H.: Laparoscopic surgery, perceptual limitations and force: a review. In: First Canadian Student Conference on Biomedical Computing (2006)
11. Lin, C.J., Chen, H.-J.: The investigation of laparoscopic instrument movement control and learning effect. Biomed. Res. Int. **2013**, 1–16 (2013)
12. Moorthy, K., et al.: Bimodal assessment of laparoscopic suturing skills: construct and concurrent validity. Surg. Endosc. **18**(11), 1608–1612 (2004)
13. Mattar, S.G., et al.: General surgery residency inadequately prepares trainees for fellowship. Ann. Surg. **258**(3), 440–449 (2013)
14. Bermas, H., et al.: Laparoscopic suturing and knot tying: a comparison of standard techniques to a mechanical assist device. JSLS: J. Soc. Laparoendosc. Surg. **8**(2), 187–189 (2004)
15. Wang, X., et al.: Laparoscopic suture training curricula and techniques. Ann. Transl. Med. **6**(11), 215 (2018)
16. Slater, N.J., Bleichrodt, R.P., van Goor, H.: Wound dehiscence and incisional hernia. Surg. Infect. (Larchmt.) **30**(6), 282–289 (2012)
17. Fischer, L., et al.: Variability of surgical knot tying techniques: Do we need to standardize? Langenbeck's Arch. Surg. **395**(4), 445–450 (2009)
18. von Trotha, K.-T., et al.: Surgical sutures: Coincidence or experience? Hernia **21**(4), 505–508 (2017)
19. Trimbos, J.B., Van Rijssel, E.J., Klopper, P.J.: Performance of sliding knots in monofilament and multifilament suture material. Obstet. Gynecol. **68**(3), 425–430 (1986)
20. Wong, Y.-R., McGrouther, D.A.: Biomechanics of surgical knot security: a systematic review. Int. J. Surg. **109**(3), 481–490 (2023)
21. Khiste, S.V., Ranganath, V., Nichani, A.S.: Evaluation of tensile strength of surgical synthetic absorbable suture materials: an in vitro study. J. Periodontal Implant Sci. **43**(3), 130 (2013)
22. Visentin, D.C., Cleary, M., Hunt, G.E.: The earnestness of being important: reporting non-significant statistical results. J. Adv. Nurs. **76**(4), 917–919 (2019)
23. Zhao, C., et al.: Beyond the square knot: a novel knotting technique for surgical use. J. Bone Joint Surg. **95**(11), 1020–1027 (2013)
24. Papandrea, R., et al.: Biomechanical and clinical evaluation of the EPITENON-first technique of flexor tendon repair. J. Hand Surg. **20**(2), 261–266 (1995)
25. Komatsu, F., Mori, R., Uchio, Y.: Optimum surgical suture material and methods to obtain high tensile strength at knots: Problems of conventional knots and the reinforcement effect of adhesive agent. J. Orthop. Sci.Orthop. Sci. **11**(1), 70–74 (2006)

Leveraging Large Language Models for Developing Digital Smart Companions for the Elderly

Casey Kiam Cheng How[1](✉), Wan Qing Ang[1], Yongqing Zhu[1], and Maode Ma[2]

[1] Singapore University of Social Sciences, Singapore, Singapore
caseyhowkc@suss.edu.sg
[2] Qatar University, Doha, Qatar

Abstract. With the anticipated increase in the elderly population and the insufficient resources available to cater to their care needs, there is a surge of interest in companion robots to alleviate loneliness among seniors. This paper presents the development of a digital smart companion aimed at enhancing the quality of life for the elderly, leveraging large language models and Artificial Intelligence (AI) to provide companionship and assistance.

Keywords: large language models (LLM) · digital smart companion · Artificial Intelligence (AI)

1 Introduction

Demographic shifts worldwide have led to the phenomenon of an aging population. This demographic transformation is characterized by a notable surge in the number of individuals aged 65 and above, commonly referred to as seniors or the elderly. According to an article in Straits Times news, Singapore's population is aging rapidly, with citizens aged 65 and above rising to 18.4% in 2022, up from 11.1% in 2012, and projected to reach 23.8% by 2030 [1].

In response to these challenges, the aging population presents both complex challenges and opportunities. There is a growing demand for technology-driven solutions that can enhance the quality of life for seniors, promote their independence, and address social and emotional well-being.

Consequently, the objective of this project is to develop a digital smart companion system leveraging a large language model (LLM) and other advancements in Artificial Intelligence (AI). This system is designed with the explicit purpose of providing companionship and assistance to the elderly population, thereby fostering a supportive environment that enables them to thrive in their later years.

The remainder of this paper includes the literature review of related work, followed by the design and implementation of the digital smart companion system, and the conclusion and future work.

H. Guo et al. (Eds.): IRC-SET 2024, PTT, pp. 47–55, 2025.
https://doi.org/10.1007/978-981-96-3770-6_6

2 Literature Review

Many studies have highlighted the desired features of digital companions or companion robots that can perform a wide range of functions. The features can generally be categorized into three categories: physical assistance and safety, cognitive support and stimulation, and social interaction and entertainment.

Nursing robots are an example in the category of physical assistance and safety, providing crucial support such as monitoring vital signs, assisting with mobility, and ensuring timely responses to emergencies. They play a significant role in enhancing the safety and well-being of patients by automating tasks like medication administration and routine checks, thereby reducing the workload on healthcare staff and improving efficiency in caregiving settings.

Florence [2], the robotic nursing assistant, utilizes AI-based image recognition software to identify patients either by scanning their wrist tag or visually. It then gathers their vital signs using its camera, sensors, and pulse oximeter, with the data seamlessly integrated into the National Electronic Health Record. Additionally, Florence can deliver medication, snacks, or other items directly to patients. Upon approaching a patient's bedside, it autonomously detects the patient's proximity and adjusts its position accordingly to ensure efficient service delivery.

Companions provide cognitive support and stimulation aid in memory recall, stimulate cognitive functions through interactive exercises, and offer personalized reminders for daily activities. They contribute to enhancing mental acuity and emotional well-being, fostering engagement and independence among users. Mabu is a healthcare companion robot that interacts with patients to provide medication reminders, answer health-related questions, and engage in conversations aimed at cognitive stimulation and emotional support [3].

Finally, companions for social interaction and entertainment engage users through activities like dancing, singing, storytelling, playing board games, and providing news updates. They foster emotional well-being and social connectivity by facilitating interactions, entertainment, and leisure activities that enhance quality of life and promote companionship among users. LOVOT [4] is a companion robot from Japan that focuses on providing emotional support and social interaction. It features expressive eyes and a cuddly design, encouraging users to interact through touch, play, and companionship. LOVOT responds to voice commands, recognizes its owner, and seeks out interaction to provide comfort and entertainment.

With the advancement of AI, these companions are becoming increasingly capable of adapting to user needs, learning from interactions, and providing personalized assistance and companionship tailored to individual preferences and health conditions. AI companions have experienced four major shifts [5], as illustrated in Fig. 1.

One example of AI pets is the Tamagotchi, a small virtual pet that users could care for on a digital screen. This toy was designed to simulate the experience of caring for a pet, leading users to develop a sense of attachment and sparking interest in having a digital companion.

The release of KASPAR [6], a child-shaped doll-like robot in 2005 marked the next technological evolution of AI companions. Designed to respond to human touch

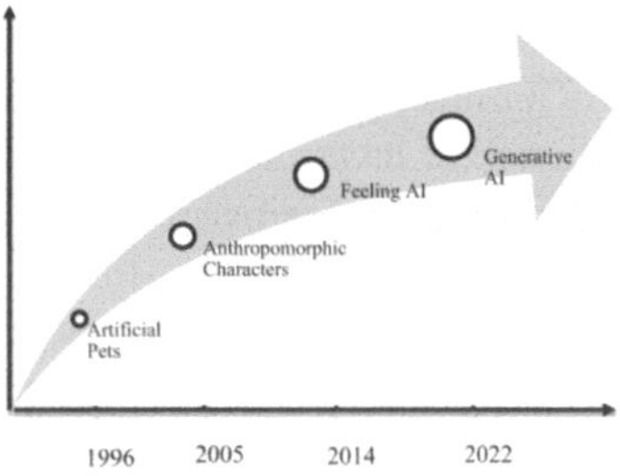

Fig. 1. TechnologyShifts of the Evolution of AI Companions

with facial expressions, KASPAR allows users to care for it. However, its limited conversational capabilities restrictes the depth of interaction.

The third shift in AI companions came with the integration of affective computing, which allows systems to recognize and interpret human emotions. In 2014, Microsoft Asia released XiaoIce [7], an empathetic chatbot capable of understanding user emotions and providing support through sentiment analysis. Utilizing advanced architectures like Convolutional Neural Networks (CNNs) and Long Short-Term Memory networks (LSTM), XiaoIce can engage in prolonged interactions and form human-like bonds with users.

Generative AI marks the fourth and most recent shift. The launch of ChatGPT [8], powered by an LLM using a generative pre-trained transformer model and self-supervised learning, enables it to respond to a wide range of prompts and inquiries, engaging in extended discussions on virtually any topic. We will adopt this advanced capability, which also allows for the recall of past interactions with users, thereby enhancing the personalized experience of our proposed digital smart companion.

3 Design and Implementation

Given the widespread availability and affordability of mobile phones among elders, we have chosen to develop the digital smart companion as a mobile app. Mobile phones are readily accessible to many elders, offering a familiar and convenient platform for interacting with technology. This approach ensures that our solution can reach a broader audience and integrate seamlessly into daily life, enhancing accessibility and usability. Following the feasibility study phase, we have identified the key features for development: companion chat for interaction, task management for organization, and recent activity monitoring for health and safety tracking.

3.1 Companion Chat

The investigation on existing literature reveals significant ethical implications and data privacy concerns related to companion chatbots. These discussions frequently address several key issues. Firstly, there is a concern over the potential impact on interpersonal relationships, with worries that individuals may develop excessive attachments to these robots, potentially harming their social interactions [9].

Secondly, there is apprehension about the deceptive nature of companion robots, which can create a false sense of intimacy despite lacking genuine emotional connections. Additionally, there is a notable risk of invasion of privacy as companion robots collect personal data, posing threats of exploiting sensitive information without explicit consent.

Moreover, inadequate data security measures further exacerbate risks of data breaches and exploitation, potentially compromising users' personal information for malicious purposes. These concerns underscore the critical need for robust ethical frameworks and stringent data privacy protocols to govern the development and deployment of companion chatbot technologies.

To mitigate the above issues, several measures can be implemented. Firstly, addressing concerns over interpersonal relationships can involve designing chatbots with clear boundaries and guidelines to encourage healthy user interactions while promoting real-life social engagement. Educating users about the limitations of robotic companionship and encouraging balanced use can also mitigate excessive attachment issues.

Secondly, to address the deceptive nature concern, transparency in the design and communication of companion robots is essential. Developers should clearly communicate that these interactions are simulated and artificial, emphasizing that they do not replace genuine human relationships. Implementing design principles to prioritize authenticity and ethical communication can help manage expectations and prevent misconceptions.

Regarding privacy concerns, robust data protection measures are crucial. Companion chatbots should adhere to stringent data privacy laws and regulations, ensuring that personal data collection is minimized, anonymized where possible, and obtained with explicit user consent. Implementing encryption protocols and secure data storage practices can safeguard against unauthorized access and data breaches.

To mitigate these issues, we store interaction data that are encrypted using Firebase Cloud, a scalable and reliable cloud database. The backend utilizes ChatGPT, astate-of-the-art language model developed by OpenAI to respond, the app also includes a text-to-speech functionality for converting the text output into spoken audio for ease of accessibility (Fig. 2).

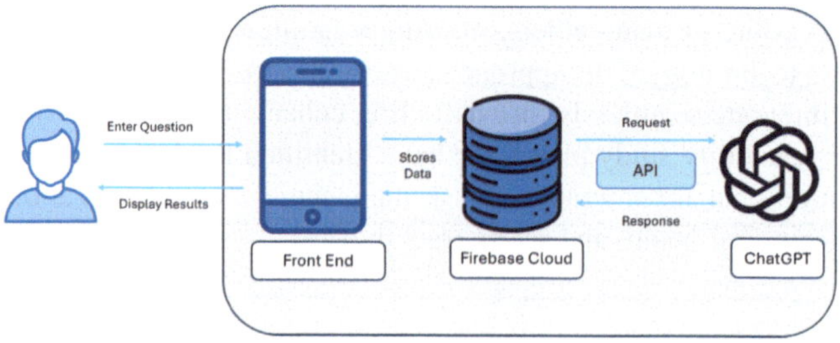

Fig. 2. Architecture Design of the Chat Companion Feature

3.2 Sentimental Analysis on Chat Conversation

Elderly individuals may face various emotional challenges, including loneliness, anxiety, or depression. Utilizing sentiment analysis enables the tracking of their emotional well-being. By examining interactions within chat messages, it can flag potential concerns early on, allowing for timely intervention and support from caregivers or healthcare professionals. A typical flow of sentiment analysis is shown in Fig. 3.

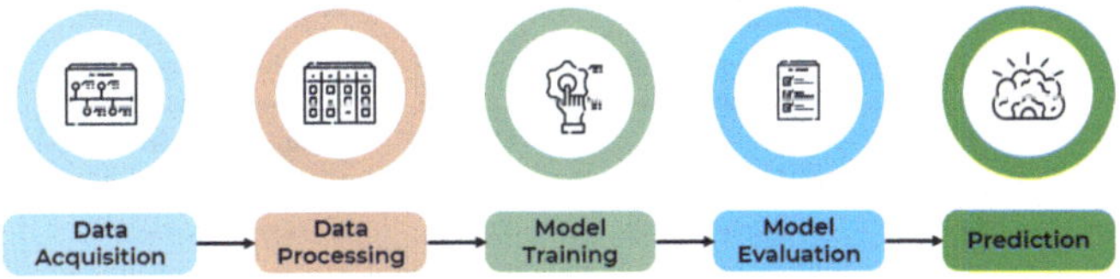

Fig. 3. Sentiment Analysis Flow

Acquisition:
To construct the sentiment analysis model, we start from selecting and training a suitable dataset. While finding an exact match for elderly conversations isn't feasible, we have identified a highly relevant dataset—the Chat Sentiment Dataset obtained from Kaggle. This dataset includes a column featuring three distinct sentiment labels: Positive, Neutral, and Negative. Figure 4 shows the count of each sentiment in the Chat dataset.

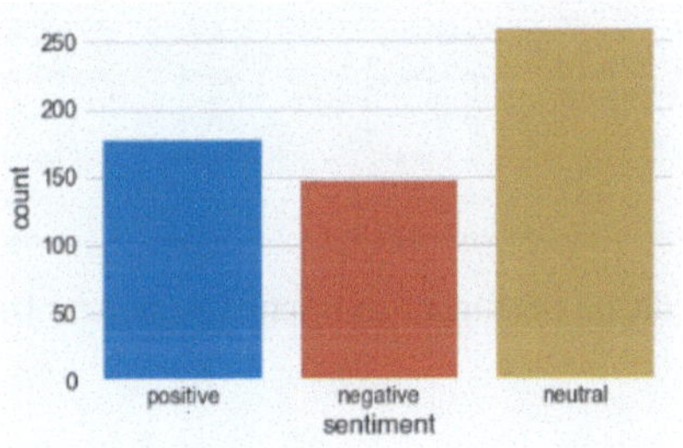

Fig. 4. Count of Sentiment in Chat Sentiment Dataset

Processing
Before performing sentiment analysis, we have preprocessed the text data to ensure that it is suitable for analysis and model training, whichincludes converting all text to lowercase to avoid duplication of vocabulary, removing URLs and special characters, stripping punctuation, eliminating stopwords to reduce noise, and applying lemmatization to reduce words to their base forms.
Training
For model training, we have initially used Valence Aware Dictionary and sEntiment Reasoner (VADER), a prebuilt sentiment analysis model tailored for social media text. VADER operates with a pre-built lexicon where words are rated based on their sentiment polarity and intensity, assigning sentiment scores to words and calculating an overall

sentiment score for the text. Subsequently, we have developed our model using a Support Vector Machine (SVM). Figure 5 illustrates the predicted and actual sentiment using VADER.

	preprocessed_text	sentiment_score	predicted_sentiment	sentiment
210	im particularly interested	0.4576	positive	neutral
452	finished project ive working month 💪	0.0000	neutral	positive
59	regret wasting money product	-0.6705	negative	negative
129	software frustrating use	-0.4404	negative	negative
10	im feeling neutral	0.1280	positive	neutral
72	company terrible	-0.4767	negative	negative
324	service salon topnotch	0.0000	neutral	positive
215	food average	0.0000	neutral	neutral
522	got car accident car totaled 😢 🚗	-0.4767	negative	negative
299	concert incredible	0.0000	neutral	positive

Fig. 5. Predicted Sentiment and Actual Sentiment Using VADER

The SVMis a supervised machine learning algorithm used for classification tasks. It works by finding the optimal hyperplane that separates different classes in the feature space. Figure 6 shows the predicted sentiment and actual sentiment using SVM.

	preporcessed_text	predicted_sentiment	actual_sentiment
1	buying product regret	negative	positive
43	overpriced product	negative	negative
98	im loving lucky partner supportive	positive	positive
80	feeling im neutral	neutral	neutral
96	opinion strong	neutral	neutral
84	company im impressed really	positive	positive
74	buy need orange recipe today	neutral	neutral
86	im neutral	neutral	neutral
46	customer excellent service	positive	positive
25	feeling im productive today	neutral	neutral

Fig. 6. Predicted Sentiment and Actual Sentiment Using SVM

Evaluation

When evaluating the model, besides using accuracy as a metric, we have also aimed to minimize false negatives and false positives, which sometimes come as a trade-off. Therefore, if the need arises, we consider having a high recall more important, as any failure to identify negative emotions can have profound consequences for users' safety and well-being. Table 1 shows a comparison between both VADER and SVM models. Considering the results, we have concluded that the SVM model is better in all aspects.

3.3 To-Do Task Feature

As individuals age, it can become increasingly challenging to manage daily tasks and stay organized. This difficulty often arises due to the factors such as memory lapses or forgetfulness. Integrating a to-do app feature can serve as a reliable memory aid, helping elderly individuals stay on top of their responsibilities and commitments.

Table 1. Model Performance Metrics Comparison

	VADER	SVM
Confusion Matrix	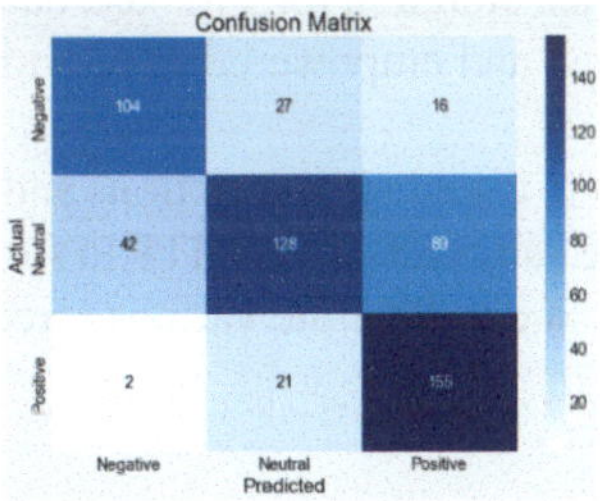	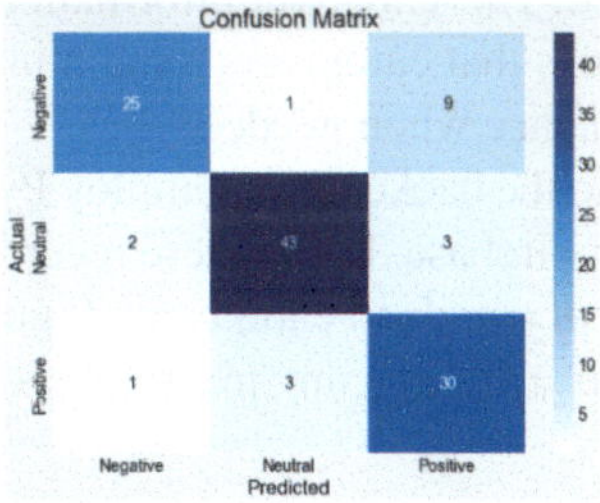
Accuracy	66.27%	83.76%
Precision	68.11%	85.00%
Recall	66.27%	83.76%

We have incorporated Firebase database integration into the to-do task feature, with the intention of leveraging its capabilities for Activity Monitoring Feature. Figure 7 shows the screen of the To-do application as well as the corresponding data storage in Firebase.

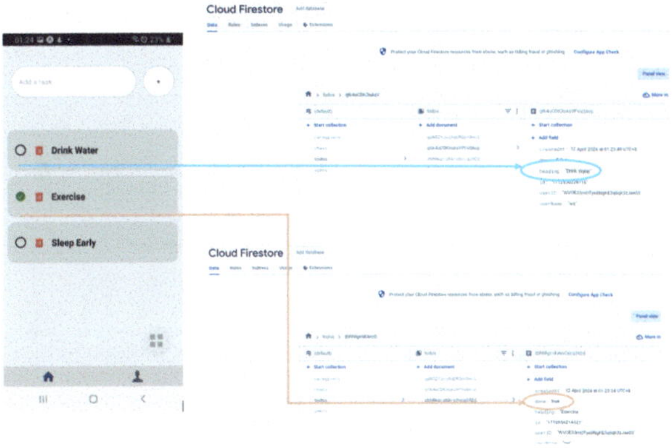

Fig. 7. To-Do Screen and Firebase Integration

3.4 Activity Monitoring

As the app currently stores Chat and To-Do information in Firebase, it can access records of the most recent activity within each feature and pinpoint the last activity to effectively monitor user engagement. In instances of prolonged user inactivity, an automated email notification system could be triggered to alert caregivers.

To seamlessly implement this process, the application needs to maintain records of caregivers associated with each elderly individual. Therefore, we have created a bottom navigation to the profile tab and introduced a screen where caregivers can be conveniently added. This caregiver information is then stored in the Firebase database. This approach ensures that caregivers remain informed and empowered to provide timely support and assistance when needed.

In the backend, we employ Python to establish connections and retrieve data records from Firebase. Using these records, we can connect to SMTP to streamline the transmission of emails to caregivers. Figures 8 and 9 illustrate various screens and the reference email sent upon any inactivity notice.

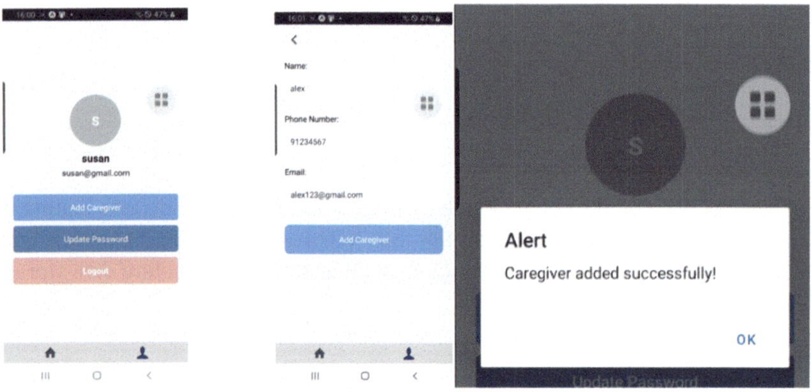

Fig. 8. Activity Monitoring

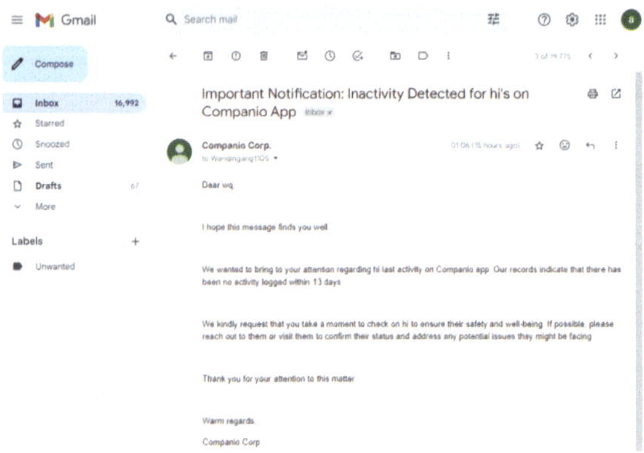

Fig. 9. Email Notification for Inactivity

4 Conclusion and Future Work

In conclusion, through the development of this mobile app, we have established a comprehensive platform designed to function as a companion system facilitating the monitoring and management of elderly individuals' daily lives. By this endeavor, our aspiration is for AI technology to enrich the lives of elderly individuals by promoting independence and enhancing overall well-being. Simultaneously, we aim to alleviate the burdens experienced by caregivers.

As the app expands and incorporates additional features, relying solely on database records of the user's last activity may become inefficient. Therefore, in future stages of development, we propose integrating other analytics applications that can track user activity more effectively by capturing user actions or predefined events. This approach will provide a more comprehensive understanding of user engagement and behavior within the application, facilitating better decision-making and user experience optimization.

Another limitation of our app relates to the dataset used for training and testing our sentiment analysis model. It is important to note that this dataset does not directly reflect the conversational style of elderly individuals, who may use distinct communication patterns. Therefore, the predictive capabilities of the model may not demonstrate the desired robustness when applied to sentiment analysis of interactions among the elderly.

Looking ahead, to address this limitation, we plan to explore methods to augment the dataset with more representative elderly conversation data. By incorporating such data, we aim to enhance the model's accuracy and applicability in analyzing sentiment within conversational contexts involving elderly users. Nonetheless, despite this inherent limitation, the current dataset effectively serves as a reference for training and evaluating our model, providing a benchmark for its effectiveness.

References

1. Chin, S.F.: S'pore's Population Aging Rapidly: Nearly 1 in 5 Citizens is 65 Years and Older. The Straits Times (2022). https://str.sg/wrey
2. Teo, J.: Robotic Nursing Assistants Can Take Patients' Vital Signs, Freeing Up Nurses for Other Tasks. The Straits Times (2022). https://str.sg/w9Cq
3. Winn, Z.: Robots Help Patients Manage Chronic Illness at Home. MIT News (2019). https://news.mit.edu/2019/catalia-health-mabu-robot-1011
4. LOVOT. https://lovot.life/en/
5. Chaturvedi, R., Verma, S., Das, R., Dwivedi, Y.K.: Social companionship with artificial intelligence: recent trends and future avenues. Technol. Forecast. Soc. Chang. **193**, 122634 (2023). https://doi.org/10.1016/j.techfore.2023.122634
6. Wood, L.J., Zaraki, A., Robins, B., Dautenhahn, K.: Developing Kaspar: a humanoid robot for children with Autism. Int. J. Soc. Robot. **13**(3), 491–508 (2021). https://doi.org/10.1007/s12369-019-00563-6
7. Zhou, L., Gao, J., Li, D., Shum, H.-Y.: The design and implementation of XiaoIce, an empathetic social chatbot. Comput. Linguist. **46**(1), 53–93 (2020). https://doi.org/10.1162/coli_a_00368
8. Introducing ChatGPT. https://openai.com/index/chatgpt/
9. Berridge, C., Zhou, Y., Robillard, J.M., Kaye, J.: Companion robots to mitigate loneliness among older adults: perceptions of benefit and possible deception. Front. Psychol. **14**, 1106633 (2023). https://doi.org/10.3389/fpsyg.2023.1106633

Investigating the Human Longevity Properties of Human Umbilical Cord Tissue in the Prevention and Cure of Age-Related Diseases

Damien Cheng Yi Kiun[1](✉), Ong Crystal[1], Oh Ren Kai Jayden[1], Ong Chee Tian[2], Neela Muthu[2], Jeyakumar Masilamani[2], Mohamed Shirhan Mohamed Atan[2,4], and Phan Toan Thang[2,3]

[1] Raffles Institution, 1 Raffles Institution Ln, Singapore 575954, Singapore
damien.cykiun@gmail.com
[2] Grace Global Raffles Building, CellResearch Corporation, 137 Market Street, #08-02, Singapore 048943, Singapore
[3] Yong Loo Lin School of Medicine, 10 Medical Dr, Singapore 117597, Singapore
[4] Temasek Polytechnic, 21 Tampines Ave 1, Singapore 529757, Singapore

Abstract. Ageing is a prevalent issue in today's world. Age-related diseases currently lack effective treatment methods. Mesenchymal stem cells (MSCs) are known to have human longevity properties and could serve to treat age-related diseases. However, usage of MSCs from conventional sources faces various limitations as they lack abundance, have low proliferation rate, and require invasive harvesting procedures. Alternatively, human umbilical cord lining MSCs (CLMSCs) are found to also possess anti-ageing properties and could be better suited for medical use due to their abundance, low immunogenicity, and non-invasive harvest. Nonetheless, the anti-ageing molecular mechanism of CLMSCs remains underexplored and is thus not readily used for the treatment of age-related diseases.

Therefore, our study aims to:

I. Evaluate the anti-ageing effects of CLMSCs on animals
II. Detect the presence of GDF-11, IL-24 and CD47 in human umbilical cord tissues

so that we can understand the anti-ageing molecular mechanism of CLMSCs, and thus be able to safely and effectively utilise CLMSCs in regenerative medicine.

In our research, rats were injected with CLMSCs to test for anti-ageing effects and immunohistochemistry was conducted on human umbilical cord tissue to identify biomarkers that could be responsible for the human longevity effects of CLMSCs. Our results demonstrated the anti-ageing effects of CLMSCs and identified the presence of GDF-11 and IL-24 biomarkers in human umbilical cord lining and vessels, which are rich in CLMSCs, and likely confer CLMSCs its human longevity properties. We conclude that CLMSCs from human umbilical cord tissue have great potential for use to prevent and cure age-related diseases.

Keywords: Anti-Ageing · Stem Cells · Umbilical Cord

H. Guo et al. (Eds.): IRC-SET 2024, PTT, pp. 56–65, 2025.
https://doi.org/10.1007/978-981-96-3770-6_7

1 Introduction

The world faces a rapidly ageing population, a trend projected to worsen over the next three decades. Ageing is the predominant risk factor for most diseases and conditions that limit healthspan [6]. These diseases are largely the result of genomic instability, telomere shortening, and cellular senescence [14]. However, effective methods to treat age-related diseases are lacking, with current treatment methods mainly focusing on palliative relief, symptomatic relief, or preventing their onset and progression.

It is known that mesenchymal stem cells (MSCs) have anti-ageing and pro-regeneration potencies [13]. Conventional sources of MSCs include peripheral blood, bone marrow, and adipose tissues. Nevertheless, the use of MSCs from these sources is limited by several disadvantages: (i) a very low number of MSCs in adult tissues, (ii) low proliferation rate, and (iii) invasive procedures needed to obtain MSCs [10]. To overcome these challenges, we are studying a nascent source of MSCs derived from human umbilical cord, known as human umbilical cord lining MSCs (CLMSCs), which has been similarly found to possess human longevity properties [3]. CLMSCs are characterised by faster self-renewal and proliferation, stable proliferation and long-term storage, abundance in source, low incidence of graft-versus-host disease, low risk of viral transmission, and lack of tumorigenicity [3]. Moreover, harvesting of human umbilical cord does not face ethical concerns as its procedure is non-invasive, and thus safe for both infant and mother [19]. These unique advantages of CLMSCs make them a valuable asset for novel applications in regenerative medicine, thus being a promising solution in treating age-related diseases.

However, the anti-ageing molecular mechanism of CLMSCs remains underexplored in stem cell therapy despite its potential.

We identified 3 biomarkers: Growth Differentiation Factor 11 (GDF-11), Interleukin 24 (IL-24), and Cluster of Differentiation 47 (CD47), which we believe are present in CLMSCs and play a key role in the mechanisms underpinning its anti-ageing properties. To date, no research has definitively confirmed the presence of these markers within CLMSCs. GDF-11 is a protein found to reverse age-related dysfunction in muscle, nervous and cardiovascular systems [17]. Human studies conducted also highlighted decreasing levels of circulating GDF11 with age [1]. IL-24 acts in an autocrine and paracrine fashion to coordinate re-epithelialisation, re-vascularisation, dermal fibroblast proliferation, and collagen deposition to restore the damaged tissue to homeostasis, hence playing a key role in tissue repair and regeneration [12]. CD47 is a cell surface ligand that plays an integral role in various immune responses as well as autoimmunity, by sending a potent "don't eat me" signal to prevent phagocytosis, granting the cell immunoprivilege [5].

Thus, our study aims to:

I. Evaluate the anti-ageing effects of CLMSCs on animals
II. Detect the presence of GDF-11, IL-24 and CD47 in human umbilical cord tissues

So that we can understand the anti-ageing molecular mechanism of CLMSCs, and thus be able to safely and effectively utilise CLMSCs in regenerative medicine.

2 Hypothesis

I. Rats injected with CLMSCs will observe lesser ageing effects than control rats.
II. GDF-11, IL-24, and CD47 biomarkers are present in regions of the human umbilical cord tissue that are rich in CLMSCs.

3 Methods

3.1 Injection of CLMSCs in Rats

30 Male Wistar rats of age 45 weeks are used in the experiment.

The control group of 10 rats are subjected to monthly subcutaneous injection of media for 22 months while the treatment group of 20 rats are subjected to monthly subcutaneous injection of CLMSCs from the umbilical cord lining with a dosage of 10 million CLMSCs/kg suspended in 1 ml media for 22 months.

Blood samples are collected via the tail vein or jugular vein before the experiment and monthly to the end. Physiological parameters such as body weight and feed and water consumption are recorded weekly. Body conditioning scores (BCS) and Ageing Scores are recorded fortnightly. After the experiment, the organs brain, lung, heart, thymus, kidneys, liver, muscle, skin, and reproductive glands will be harvested. Tissues will be formalin-fixed, dehydrated and embedded in paraffin, snap-frozen in liquid nitrogen, and stored at −70 °C. Results analysis will be done by detection of the fate of CLMSCs via immunohistochemical and histological analysis and genomic analysis for ageing and anti-ageing.

3.2 Immunohistochemistry (IHC) for Biomarkers

Paraffin sections of human umbilical cord tissue are obtained from CellResearch Corporation Tissue Bank. The paraffin sections of human umbilical cord tissue are deparaffinised and rehydrated by soaking in various solutions (Fig. 1).

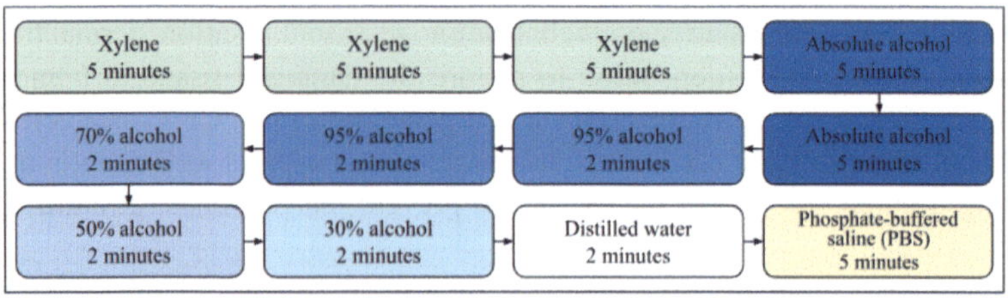

Fig. 1. Solutions and durations used for rehydration of the sections

Antigen retrieval is then performed on the sections. 1 ml of hydrochloric acid is mixed with 100 ml of 70% ethanol to make the 1% Acid Alcohol Solution. Sodium citrate buffer is preheated to a boil and the pH of the buffer is adjusted to pH 6.2 by the dropwise addition of 1% Acid Alcohol Solution. Soak the slides in the sodium citrate buffer (pH

6.2) and microwave for 15 min. The slides should never dry so check every minute and add sodium citrate buffer when necessary. Let the slides cool to room temperature.

Then, wash the slides in PBS for 5 min and shake and wipe off excess PBS. Circle all sections with a Hydrophobic Barrier Pap Pen. Next, block endogenous peroxides by incubating the sections in BLOXALL Blocking Solution for 10 min. Wash the sections in Tris Buffered Saline-TWEEN (TBS-Tween) for 5 min. Incubate the sections in 2.5% Normal Horse Serum for 20 min then tip off excess serum from the sections.

Dilute each rabbit polyclonal GDF-11, IL-24, and CD47 primary antibody individually in specified amounts of their respective antibody diluents (Table 1).

Table 1. Antibody dilutents for CD47, IL-24, and GDF-11 primary antibody respectively

Primary Antibody	Antibody concentration	Volume of antibody /ul	Volume of 2.5% HSS /ul	Volume of TBS-Tween /ul	Total volume /ul
GDF-11	1:300	1	180	119	300
IL-24	1:200	1.5	180	118.5	300
CD47	1:200	1.5	180	118.5	300

Apply 100 ul of the diluted antibody to each section. Only add one type of antibody to each section. Incubate the sections overnight in a humid chamber.

Then, wash the sections in TBS-Tween for 5 min and incubate the sections in R.T.U VECTASTAIN Elite ABC reagent for 30 min. Wash the sections in TBS-Tween for 5 min again. Prepare 3,3′-diaminobenzidine HRP substrate (DAB) by mixing 5 ml of distilled water, 2 drops of buffer stock solution, 4 drops of DAB, and 2 drops of hydrogen peroxide, and immediately apply to the slides and wait for 5 min for the colour change to occur. Drain the slides and rinse under distilled water for 2 to 3 min. Thereafter, counterstain with 100 ul of Gill's No.3 Hematoxylin per section and rinse with distilled water.

Next, perform dehydration on the sections by soaking them in various alcohol solutions (Fig. 2).

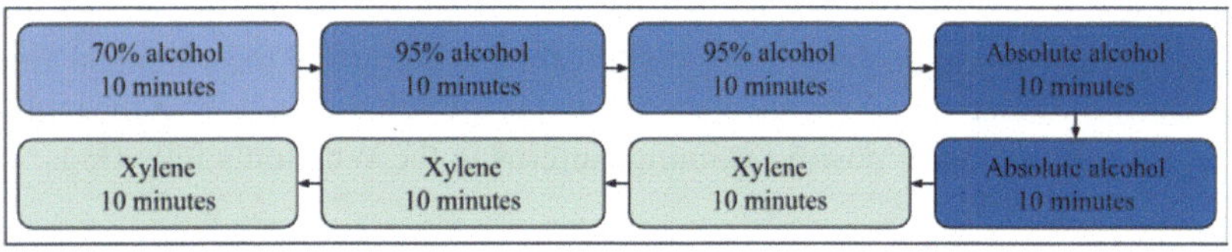

Fig. 2. Solutions and durations used for dehydration of the sections

Mount the sections using xylene-based mounting medium DEPEX. Lastly, view the mounted slides under an Olympus IX 73 microscope under 10× magnification to observe the staining.

4 Results

4.1 Injection of CLMSCs in Rats

The treatment group rats receiving CLMSCs were observed to have better fur coats and lower body mass despite consuming equal amounts of feed and water. The Ageing Score of the rats in the treatment group were lower and better than the control group (Table 2).

Table 2. Ageing Scores for the Rats in the Control and Treatment Groups

Time into experiment /months	Group	Ageing Score
10.5	Control	6.11
	Treatment	6.063
16.5	Control	6.11
	Treatment	6.07

A total of 8 rats died during the course of the experiment, 1 from the control group and 7 from the treatment group. Autopsy was performed to determine the cause of death. For the control group, the rat developed a lump which resulted in an infection that antibiotics failed to reduce. In the treatment group, 5 rats developed overgrown lower teeth, causing drastic weight loss due to difficulty in eating. These rats were put down according to IACUC protocol. 1 rat developed a large lump on the back of neck underneath the skin which was filled with clotted and unclotted blood. 1 rat had a stomach distended with digested food with obstruction in the large intestine with hardened faecal matter. Blood sample analysis and post-experiment tissue analysis have yet to be done as the experiment is still ongoing.

4.2 IHC for Biomarkers

GDF-11 and IL-24 Positive Staining
The slides incubated with GDF-11 and IL-24 antibodies are observed to have positive dark brown staining mainly on the umbilical cord lining and the umbilical vein and arteries (Figs. 3, 4 and 5). This shows that GDF-11 and IL-24 are present in human umbilical cord tissue, especially in the umbilical cord lining and the umbilical vessels. Lower levels of both biomarkers are also detected in the Wharton's Jelly of the umbilical cord, as seen in the sparse positive staining patterns in the Wharton's Jelly (Figs. 4 and 5).

CD47 Negative Staining
However, the slides incubated with CD47 antibodies were observed to be stained light brown similarly throughout and positive dark brown staining was not present at all, (Fig. 6), indicating the absence of CD47 in the human umbilical cord tissue.

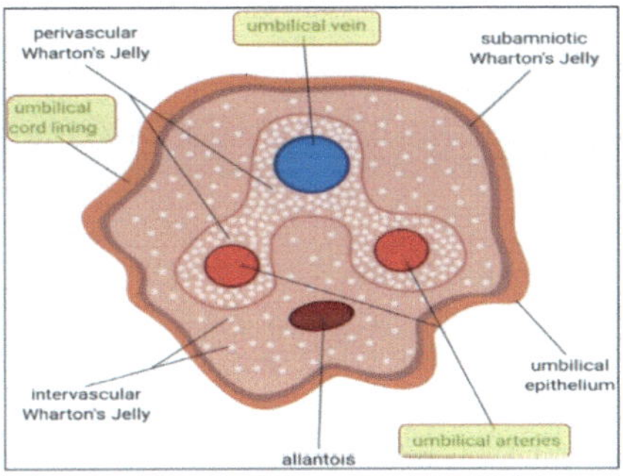

Fig. 3. Labelled cross-section diagram of the human umbilical cord (highlighted regions show positive staining for GDF-11 and IL-24) (Stefańska et al., 2020).

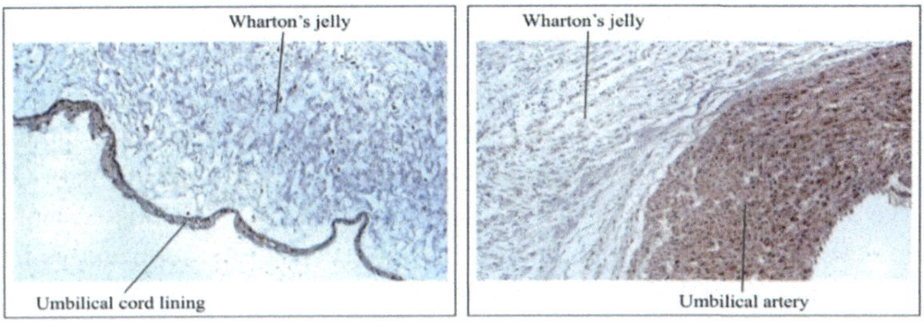

Fig. 4. Positive dark brown staining for GDF-11 found on umbilical cord lining (left) and umbilical artery (right) under 10× microscope

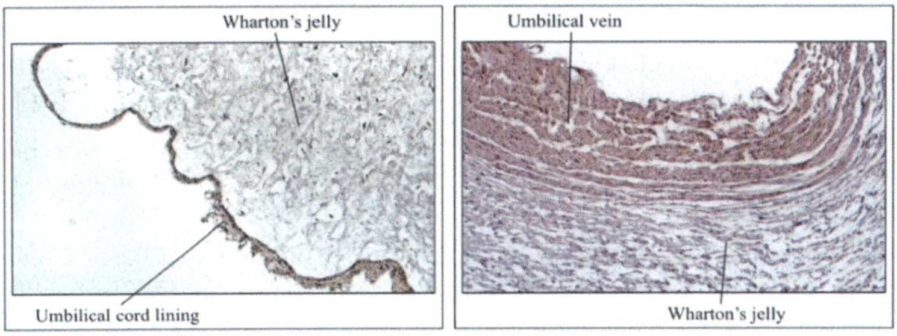

Fig. 5. Positive dark brown staining for IL-24 found on umbilical cord lining (left) and umbilical vein (right) under 10× microscope

5 Discussion of Results and Implications

5.1 Injection of CLMSCs in Rats

Our results show that the rats in the treatment group fared better in the physiological parameters and Ageing Scores, showing lesser ageing effects. Lower body mass in treatment group rats could be due to higher metabolism and the overgrowth of lower teeth resulting in death could be a sign of active production of growth hormones and

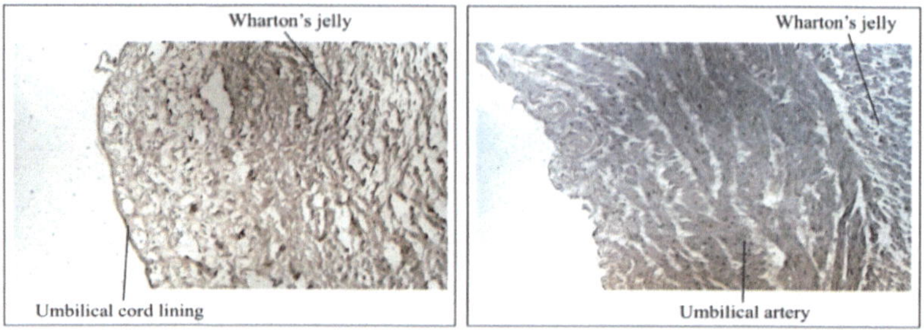

Fig. 6. Negative staining for CD47 on human umbilical cord tissue under 10× microscope

generation of new cells, which are characteristic of younger rats. The development of the lump and the large intestine obstruction resulting in death could be due to natural factors. In the control group rat, the failure of the infection to reduce resulting in death could be due to the weakened immune system of the rat brought about by ageing.

Treatment group rats displayed lesser ageing effects than control group rats. We attributed these observed effects to the CLMSCs which were introduced via the injections. Overall, our results agree with our hypothesis that rats injected with CLMSCs will observe lesser ageing effects, hence suggesting that CLMSCs possess anti-ageing effects. However, animal experiments often do not translate into replications in human trials, so future human trials is needed to confirm our findings.

5.2 IHC for Biomarkers

Our results show that GDF-11 and IL-24 are present in human umbilical cord tissue, with the highest concentrations in the umbilical cord lining and umbilical cord vessels. However, CD47 is not detected in these tissues. Since umbilical cord lining and vessels mainly consist of CLMSCs [16], our findings support our hypothesis that GDF-11 and IL-24 are present in regions of the human umbilical cord tissue rich in CLMSCs, suggesting CLMSCs likely contain these biomarkers. However, our results refute our hypothesis that CD47 are present.

Extensive research demonstrates the anti-aging properties of GDF-11 and IL-24. GDF-11 and IL-24 are proven to reduce the onset of age-related brain health issues. In mice, GDF-11 improves cognition, increase VEGF expression around brain blood vessels, increase cerebral blood flow, and reduce expression of inflammatory proteins (GFAP and Iba-1) in the brain. In a rodent model of ischemic stroke, GDF-11 treatment 7 days after stroke increased the proliferation of stem cells and endothelial cells, the expression of BDNF and VEGF, and the number of new neurons [15]. Likewise, IL-24 can limit inflammatory responses while promoting the expression of immunosuppressive and neuroprotective mediators, regulating or resolving central nervous system inflammation following bacterial infection to limit neuronal damage [2]. Furthermore, GDF-11 and IL-24 are proven to reduce the onset of age-related muscular diseases, in particular, cardiovascular diseases like pathological cardiac hypertrophy (PCH). GDF-11 is an anti-PCH factor and reduces cardiac hypertrophy, dilation, tacrolimus (TAC)-induced interstitial fibrosis, and depressed cardiac function induced by pressure overload [8]. Similarly,

IL-24 protects against H_2O_2-mediated endothelial cell damage and provides a novel therapeutic strategy for treatment of cardiovascular disease. IL-24 decreases cellular oxidative damage, reducing apoptosis, and decreasing the expression of cardiovascular disease-related factors [20].

Unexpectedly, our results show that CD47 are absent in the human umbilical cord tissue. We expected CD47 to be present as umbilical cord blood MSCs, closely related to umbilical cord tissue MSCs, express CD47 [9]. A possible explanation is that IHC may not be suitable for detecting CD47. Flow cytometry, a more sensitive immunostaining method, could better detect the presence of CD47 as it analyses cells individually for protein expression. In contrast, IHC examines the tissue as a whole. Other possible explanations for this negative result include: (i) the antibody used was not complementary to CD47, (ii) the antibody concentration (1:200) was too low for visible staining, (iii) CD47 levels are very low in young umbilical cord tissue as neonatal cells have lower basal surface CD47 levels [11].

5.3 Overall Discussion and Future Implications

Overall, our results conclude that CLMSCs in human umbilical cord tissue have anti-ageing properties and likely contain GDF-11 and IL-24 which confer CLMSCs its human longevity properties and enable it to prevent and cure age-related diseases.

This supports the use of the novel CLMSCs as a better candidate in regenerative medicine. CLMSCs have many advantages over bone marrow-derived MSCs (BMSCs), which are the most widely used source of MSCs. The harvesting procedure is non-invasive, and umbilical cords are in abundant supply. Furthermore, CLMSCs are capable of differentiating into mesenchymal lineages, and the number of fibroblast colony-forming units is significantly higher in CLMSCs than in BMSCs. CLMSCs also have anti-apoptotic potential, and are well-tolerated by the immune system [4].

Our research has limitations. Although many studies support the anti-aging effects of GDF-11, some recent studies have refuted these benefits [7]. Also, further immunohistochemistry staining for the proteins on CLMSCs should be conducted on the observed positive staining regions to confirm that CLMSCs express both biomarkers. Overall, understanding this mechanism offers a novel approach to utilise the human umbilical cord to more safely and effectively prevent and cure age-related diseases.

6 Conclusion

In conclusion, our study showed that CLMSCs have anti-ageing effects and likely contain GDF-11 and IL-24, which likely confer CLMSCs its human longevity properties. CLMSCs from human umbilical cord tissue have great potential to be used to prevent and cure age-related diseases. More research has to be done to study the anti-ageing action of IL-24 and GDF-11 in CLMSCs so that we can be certain of the pathways responsible for the human longevity properties of CLMSCs. The relatively high death rate of the rats in the CLMSCs treatment group is also a concern we have to keep in mind before future applications of CLMSCs as a therapeutic option. Since CD47 was unexpectedly absent, other methods such as flow cytometry can also be used to test for CD47 to detect its absence in human umbilical cord tissue.

References

1. Añón-Hidalgo, J., et al.: Circulating GDF11 levels are decreased with age but are unchanged with obesity and type 2 diabetes. Aging **11**(6), 1733–1744 (2019). https://doi.org/10.18632/aging.101865
2. Burmeister, A.R., Johnson, M.B., Yaemmongkol, J.J., Marriott, I.: Murine astrocytes produce IL-24 and are susceptible to the immunosuppressive effects of this cytokine. J. Neuroinflammation (2019). https://doi.org/10.1186/s12974-019-1444-1
3. Chang, L., Fan, W., Pan, X., Zhu, X.: Stem cells to reverse aging. Chin. Med. J. **135**(8), 901–910 (2022). https://doi.org/10.1097/CM9.0000000000001984
4. Ding, D.-C., Chang, Y.-H., Wu, K.-C., Liu, H.-W., Chu, T.-Y.: Human umbilical cord-derived mesenchymal stem cells reduce monosodium iodoacetate-induced apoptosis in cartilage. Tzu Chi Med. J. **30**(2), 71 (2018). https://doi.org/10.4103/tcmj.tcmj_23_18
5. Eladl, E., et al.: Role of CD47 in hematological malignancies. J. Hematol. Oncol. (2020). https://doi.org/10.1186/s13045-020-00930-1
6. Franceschi, C., et al.: The continuum of aging and age-related diseases: common mechanisms but different Rates. Front. Med. (2018). https://doi.org/10.3389/fmed.2018.0006
7. Harper, S.C., et al.: Is growth differentiation factor 11 a realistic therapeutic for aging-dependent muscle defects? Circ. Res. **118**(7), 1143–1150 (2016). https://doi.org/10.1161/CIRCRESAHA.116.307962
8. Harper, S.C., et al.: GDF11 decreases pressure overload-induced hypertrophy, but can cause severe cachexia and premature death. Circ. Res. **123**(11), 1220–1231 (2018). https://doi.org/10.1161/circresaha.118.312955
9. Kim, G.-H., et al.: Positively correlated CD47 activation and autophagy in umbilical cord blood-derived mesenchymal stem cells during senescence. Stem Cells Int. **2021**, 1–13 (2021). https://doi.org/10.1155/2021/5582792
10. Kita, K., Gauglitz, G.G., Phan, T.T., Herndon, D.N., Jeschke, M.G.: Isolation and characterization of mesenchymal stem cells from the sub-amniotic human umbilical cord lining membrane. Stem Cells Dev. **19**(4), 491–502 (2010). https://doi.org/10.1089/scd.2009.0192
11. Lawrence, D.W., King, S.B., Frazier, W.A., Koenig, J.M.: Decreased CD47 expression during spontaneous apoptosis targets neutrophils for phagocytosis by monocyte-derived macrophages. Early Human Dev. **85**(10), 659–663 (2009). https://doi.org/10.1016/j.earlhumdev.2009.09.005
12. Liu, S., et al.: A tissue injury sensing and repair pathway distinct from host pathogen defense. Cell (2023). https://doi.org/10.1016/j.cell.2023.03.031
13. Li, T., et al.: Human umbilical cord-derived mesenchymal stem cells ameliorate skin aging of nude mice through autophagy-mediated anti-senescent mechanism. Stem Cell Rev. Rep. **18**(6), 2088–2103 (2022). https://doi.org/10.1007/s12015-022-10418-9
14. Li, Z., et al.: Aging and age-related diseases: from mechanisms to therapeutic strategies. Biogerontology **22**(2), 165–187 (2021). https://doi.org/10.1007/s10522-021-09910-5
15. Lu, L., et al.: Growth differentiation factor 11 promotes neurovascular recovery after stroke in mice. Front. Cell. Neurosci. (2018). https://doi.org/10.3389/fncel.2018.00205
16. Nagamura-Inoue, T.: Umbilical cord-derived mesenchymal stem cells: their advantages and potential clinical utility. World J. Stem Cells **6**(2), 195 (2014). https://doi.org/10.4252/wjsc.v6.i2.195
17. Rochette, L., Mazini, L., Meloux, A., Zeller, M., Cottin, Y., Vergely, C., Malka, G.: Anti-aging effects of GDF11 on skin. Int. J. Molecular Sci. **21**(7), 2598 (2020). https://doi.org/10.3390/ijms21072598
18. Stefańska, K., et al.: Human Wharton's Jelly—cellular specificity, stemness potency, animal models, and current application in human clinical trials. J. Clin. Med. **9**(4), 1102 (2020). https://doi.org/10.3390/jcm9041102

19. Stubbendorff, M., et al.: Immunological properties of extraembryonic human mesenchymal stromal cells derived from gestational tissue. Stem Cells Dev. **22**(19), 2619–2629 (2013). https://doi.org/10.1089/scd.2013.0043
20. Wang, Z., Wang, Y., Chen, Y., Lv, J.: The IL-24 gene protects human umbilical vein endothelial cells against H2O2-induced injury and may be useful as a treatment for cardiovascular disease. Int. J. Mol. Med. **37**(3), 581–592 (2016). https://doi.org/10.3892/ijmm.2016.2466

Simultaneous Bioremediation of Wastewater and Generation of Electricity Using Laccase Enzyme from *Trametes Versicolor*

Tessa Yap(✉), Lim Si En Alvena, and Yong Qing Ya Ranice

Raffles Institution, Singapore, Singapore

Abstract. Industrial effluents are one of the main sources of water pollution in Singapore, and dye pollution is prevalent in neighbouring countries Malaysia and Indonesia. As textile industries heavily dispose of their dye in wastewater with the increase in textile production demand, Singapore's waters stand a high risk of even more dye pollution. Therefore, it is essential for textile industries and even wastewater treatment facilities to treat dye-polluted water more efficiently before disposing of them. As Singapore is small, it lacks natural energy resources, and it is thus important to find renewable and sustainable sources of electricity. This study explored a possible prototype in which laccase enzyme from *Trametes versicolor* was identified for simultaneous bioremediation of wastewater and electricity generation. The effect of the concentration of fabric dyes and different pH levels of the solution was investigated by determining enzymatic activity. Different concentrations of dye were used in the first experiment in which the average voltage and absorbance level of each setup was measured. Different pH levels of the solution were used in the second experiment in which the average voltage and absorbance level of each setup was also measured. Our findings showed that the laccase enzyme can bioremediate wastewater at dye concentrations of up to 20% and at pH 5.0, while simultaneously generating electricity. These results can potentially be utilised by textile industries to bioremediate wastewater and generate electricity in optimum environments.

Keywords: Water pollution · Laccase · Bioremediation · Electricity generation

1 Introduction

Industrial effluents are one of the main sources of water pollution in Singapore, and the textile industry accounts for about 20% of global water pollution from dyeing and finishing products [1, 2]. Many factories in the textile industry use a significant amount of water, up to 200 tonnes per tonne of textiles, and a large portion of this water is discharged back into the environment, containing a harmful mixture of dyes and hazardous chemicals [3]. Dye pollution has been seen before in Singapore, where reddish water filled with dye was spotted at Jurong West in 2016 [4]. It is especially common in neighbouring country Indonesia, where more than 1000 textile factories in West Java's Citarum River area release significant amounts of toxic waste into the river, resulting

H. Guo et al. (Eds.): IRC-SET 2024, PTT, pp. 66–75, 2025.
https://doi.org/10.1007/978-981-96-3770-6_8

in its classification as one of the most polluted rivers globally [5]. Singapore's waters can be easily polluted with dye both domestically and internationally as it is an island surrounded by water. The current main method of treating dye-polluted wastewater is through a nanofiltration membrane, which relies on physical separation to remove contaminants from water [6]. This, however, requires large amounts of energy and numerous replacements [6]. As Singapore is small, she lacks natural energy resources, and it is thus important to find renewable and sustainable sources of electricity. Therefore, it is essential for textile factories and wastewater treatment facilities to treat dye-polluted water more efficiently before disposing of them.

Our proposed solution is through the bioremediation of the phenolic pollutants present in the wastewater [7]. Specifically, laccase from *Trametes versicolor*, a type of white rot fungi, aids in decolourizing the fabric dye through its bioremediation process [8, 9]. Laccase, an oxidoreductase enzyme, catalyses the one-electron oxidation of the targeted phenolic compound by degrading down or modifying its ring cleavage and catalyses the reduction of oxygen to water [10, 11]. Simultaneously, the above redox reaction involves a transfer of electrons which act as mobile charge carriers, thus electricity is generated [12]. The immobilisation of the laccase enzyme enhances its stability compared to its free form [13, 14]. This is especially under extreme conditions such as the pH of wastewater, which ranges from 3 to 11 due to agents like respiration, photosynthesis and algae [15].

This project is driven by the underlying need to find sustainable sources of electricity as well as increase our water security by keeping our waters free from pollution. To maximise the potential of our limited resources, laccase from *Trametes versicolor*, which is commercially available in various food industry processes, was selected to perform bioremediation and electricity generation [16]. To find out when the inhibitory effect kicks in for laccase enzyme activity, varying concentrations of fabric dye were used to represent the different levels of pollution in a typical wastewater setting from the textile industry, which will be beneficial for textile factories to maximise the efficiency of bioremediation [17]. Varying pH levels of solutions were also used to represent the different conditions of wastewater ranging from 3 to 11 as wastewater pH levels constantly fluctuate due to agents like respiration, photosynthesis and algae [15]. Our study investigates three research questions: designing and prototyping our setup; determining the optimum concentration of fabric dyes for laccase enzymatic activity; and determining the optimum pH level to maximise this bioremediation process. Laccase enzymatic activity through bioremediation is measured through the rate of decolourization of the fabric dye, along with the average voltage generated by each setup. As the current nanofiltration method of bioremediating wastewater requires a lot of energy, laccase could be used to supplement the nanofiltration technique [6].

2 Hypothesis

Research Goal 1 (Design of Enzyme Setup). To design a setup that allows the laccase enzyme to successfully bioremediate wastewater by breaking down phenolic compounds from fabric dye and generating electricity simultaneously.

Research Goal 2 (Effect of Fabric Dye Concentration on Enzyme Activity). To determine the optimum fabric dye concentration for laccase enzyme activity before the inhibitory effect sets in. This will be measured through the rate of decolourisation of fabric dye using absorbance as well as the average voltage generated.

Research Goal 3 (Effect of pH on Enzyme Activity). To determine the optimum pH for the bioremediation of fabric dye and electricity generation for laccase enzyme activity. This will also be measured through the rate of decolourisation of fabric dye using absorbance as well as the average voltage generated.

3 Methodology

Research Goal 1 (Design of Enzyme Setup). Initially, we considered utilising the Nanobiokit from the Institute of Bioengineering and Nanotechnology, as shown in Fig. 1. It is a DIY setup with an enzyme-immobilised carbon electrode, requiring the substrate to be stored in a limited space. This method only allowed for electricity generation to be measured, and it was not feasible to add the fabric dye solution to serve our purposes of bioremediation as well.

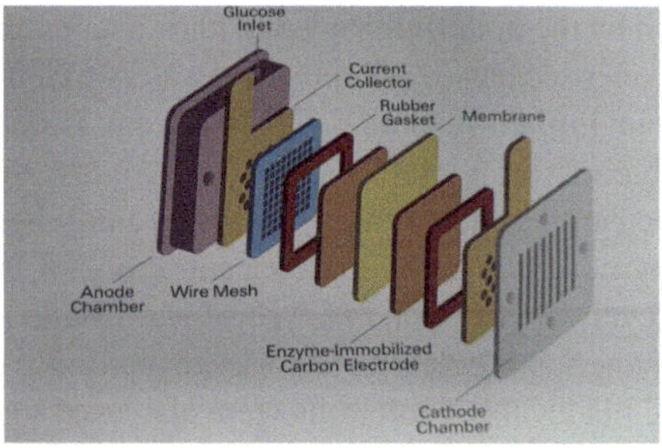

Fig. 1. Design of Nanobiokit setup where users assemble the parts themselves, with the membrane-substrate in a limited square formed in the middle and using an unknown immobilised enzyme carbon electrode to generate electricity.

Therefore, we decided to design our own prototype which consisted of immobilised laccase-alginate beads suspended in a solution of fabric dye and water, as shown in Fig. 2. As the immobilisation of the laccase enzyme enhances its stability compared to its free form, it would be more efficient [13, 14]. We then tested the prototype with trypsin enzyme and albumin substrate, which subsequently succeeded in producing electricity.

Research Goal 2 (Effect of Fabric Dye Concentration on Enzyme Activity). To create 5 setups in each procedure, laccase enzyme was immobilised using 1.5% sodium alginate and 2% calcium chloride. 2.5% laccase enzyme suspension consisted of 0.625 g of laccase enzyme and 25 ml of distilled water. The laccase enzyme-alginate mixture was formed by 5 ml of sodium alginate and 5 ml of laccase enzyme suspension. The mixture was then added dropwise to 50 ml of calcium chloride solution to form immobilised

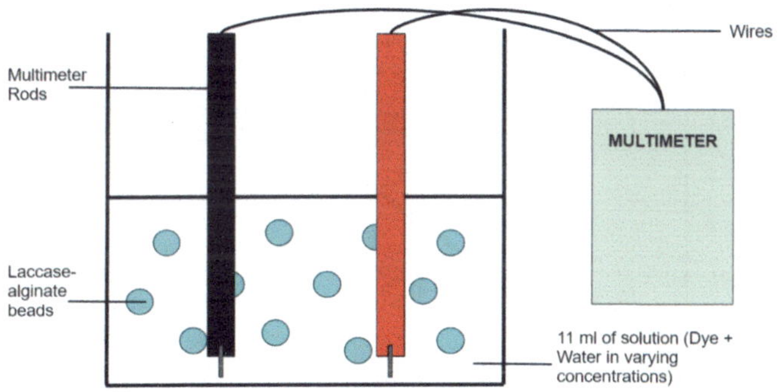

Fig. 2. Our prototype setup we designed and tested. Dye solutions of different concentrations are mixed with immobilised laccase-alginate beads in a beaker. Rods connected to a multimeter are then inserted into the solution to measure the voltage generated.

laccase-alginate balls. Dye solutions of concentrations 0%, 5%, 10%, 20%, and 25% were prepared by mixing Arto's Batik dye, which we used as our fabric dye for the experiment, from Krafers' Paradise with distilled water respectively. The immobilised laccase-alginate balls were then added to the dye solutions to form 5 setups. Setups were measured for masses on days 0 and 7 and voltage using a multimeter on days 0, 1, 2, 5, 6, and 7. The initial absorbance of each setup was measured on day 0 using UV-1280 Shimadzu Spectrophotometer at a wavelength of 520 nm. [18] On day 7, each setup was replaced with the amount of mass lost using distilled water and then diluted with distilled water by a factor of 10 before measuring the final absorbance. The average voltage generated and change in absorbance of setups were tabulated in a line graph and bar graph respectively. All experiments were replicated three times for reliability.

Research Goal 3 (Effect of pH on Enzyme Activity). For the 5 setups in each procedure, all methods remained the same, apart from varying the pH of the dye solutions, with pH solutions of levels 3, 5, 7, 9 and 11, rather than varying the dye concentrations. The different pH levels were controlled using the respective pH buffer solutions.

4 Results and Discussion

Research Goal 2 (Effect of Fabric Dye Concentration on Enzymatic Activity). Figures 3 and 4 show the effects of various concentrations of fabric dye on the average electricity generated and change in absorbance for each set up, respectively.

In Fig. 3, the average electricity generated increased from 7.09 mV at 0% to 20.1 mV at 10%, before decreasing to 7.01 mV at 20%. The electricity generated by the 10% setup was the highest, peaking at 20.1 mV. The graph shows that all dye concentrations are able to result in electricity generation, hence the laccase enzyme works to break down the phenolic compounds in the fabric dye at all concentrations up to 20%, which suggests that the inhibitory effect does not set in at fabric dye concentrations up to 20%.

Although the error bars overlap in the graph above, to further confirm the statistical significance of our results, we carried out the Kruskal-Wallis Test, which produces a H-value and a p-value. For the average electricity generated per day of each concentration,

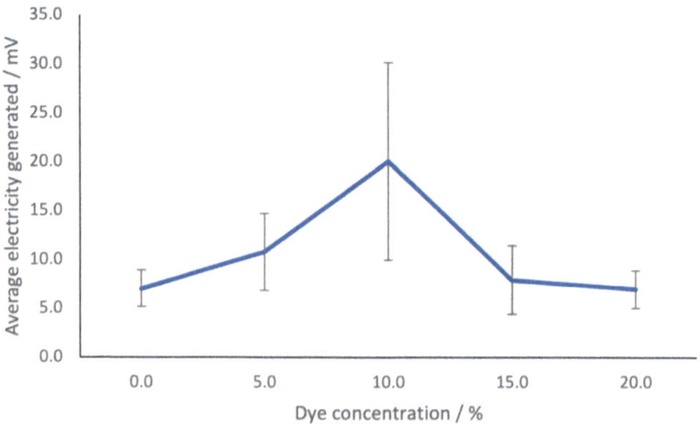

Fig. 3. Average electricity generated per day in mV (measured for 5 consecutive minutes daily over 7 days) by laccase breakdown of phenolic compounds in 0.0–20% concentrations of fabric dye. Error bars show the standard deviation of the mean of 3 replicates

the H-value was calculated as 13.8533, while the p-value was calculated as 0.00778, where the results are statistically significant at $p < 0.05$. Hence, it may be determined that at dye concentration of 10.0%, the highest rate of breakdown of phenols by laccase occurred and is thus the optimum pH which produces the most electricity out of all the set-ups.

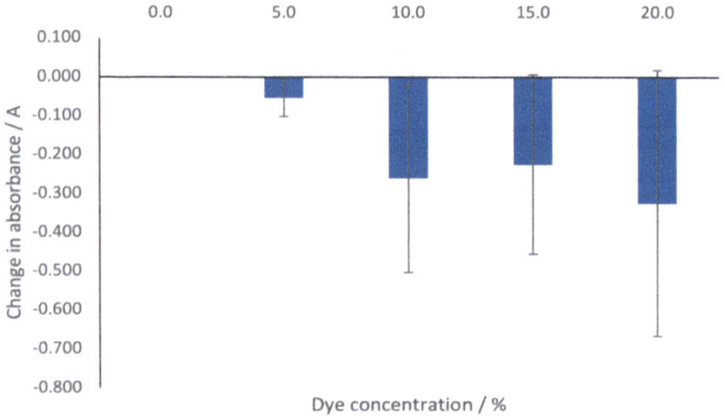

Fig. 4. Average change in absorbance in absorbance units (AU) (measured 3 times after 7 days) by laccase breakdown of phenolic compounds in 0.0–20% concentrations of fabric dye. Error bars show the standard deviation of the mean of 3 replicates.

In Fig. 4, the change in absorbance was negative for all setups, which tallies with the expected results, since the laccase enzyme breaks down the phenolic compounds in the fabric dye, decolourising it. A general downward trend in the change of absorbance can be seen, from 0.000 A at 0% to −0.324 A at 20%. Although the error bars are large and overlapping, the H-value produced after carrying out the Kruskal-Wallis Test was calculated as 25.5913, while the p-value was calculated as 0.00004, where the results are statistically significant at $p < 0.05$. Hence, it may be determined that at dye concentrations of 10.0% and 20.0%, the highest rates of breakdown of phenols by laccase occurred, where the rates of decolourisation of dye solution occurred the fastest. This also suggests that the inhibitory effect does not set in at fabric dye concentrations up to 20%.

Research Goal 3 (Effect of pH on Enzymatic Activity). Figures 5 and 6 show the effects of various pH of fabric dye on the average electricity generated and change in absorbance for each set up, respectively.

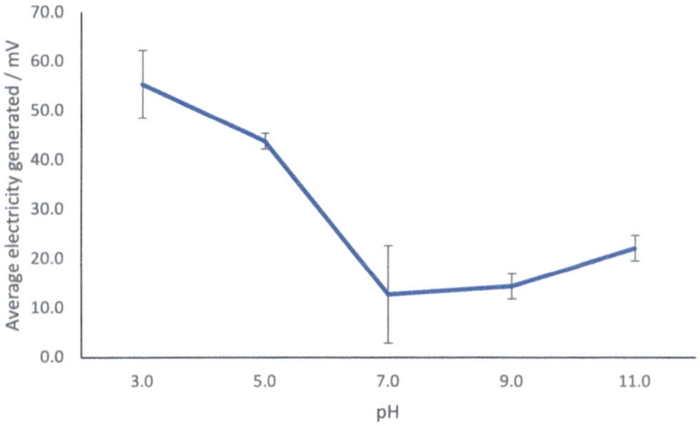

Fig. 5. Average electricity generated per day in mV (measured for 5 consecutive minutes daily over 7 days) by laccase breakdown of phenolic compounds in pH 3.0–11 fabric dye solutions. Error bars show the standard deviation of the mean of 3 replicates.

Figure 5 shows the electricity generated by each of the set-ups, which is an indication of the rate of reductive-oxidative reaction between laccase and phenols as it shows the electron transfer within the system.

In Fig. 5, the average electricity generated decreased from 55.41 mV at pH 3.0 to 12.84 mV at pH 7.0, before stagnating and slightly fluctuating at pH 7.0 to pH 11. This suggests that laccase enzymatic activity is highest at lower pHs. Although the error bars overlap for pH 7.0, 9.0 and 11.0, to further confirm the statistical significance of our results, we carried out the Kruskal-Wallis Test, which produced a H-value of 33.7496 and a p-value of less than 0.00001. Hence, it may be determined that at pH 3.0, the highest rate of breakdown of phenols by laccase occurred and is thus the optimum pH which produces the most electricity out of all the set-ups.

In Fig. 6, the highest decrease in absorbance after 7 days was observed in the pH 3.0 solution, at −0.392 A, while pH 5.0 and 11.0 both had similar results, at −0.283 A and −0.266 A respectively. Previous studies have found that the optimum pH for most fungal laccases is pH 3.0 to 5.5 [19]. This can be observed in our results for electricity generation, but not in change in absorbance. This may be due to other methods in which laccase breaks down phenolic compounds, such as modifying the structure of azo bonds which does not necessarily result in a high rate of electron transfer, hence may not produce a large voltage [10, 11]. However, more research is needed to confirm this

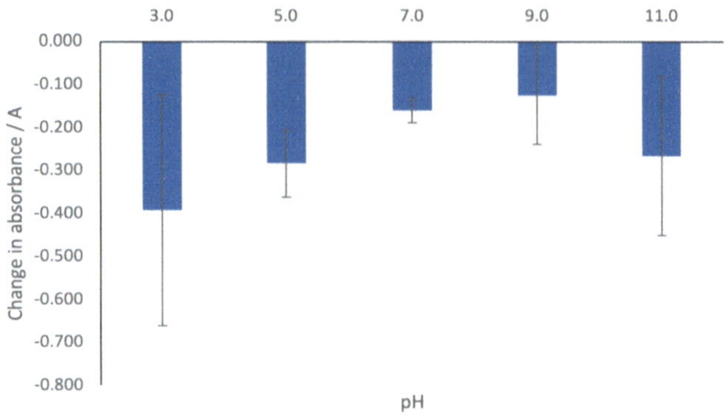

Fig. 6. Average change in absorbance in absorbance units (AU) (measured 3 times after 7 days) by laccase breakdown of phenolic compounds in pH 3.0–11 fabric dye solutions. Error bars show the standard deviation of the mean of 3 replicates.

phenomenon. It may also be speculated that this observation is due to the inhibitory effect of laccase enzyme activity.

The Kruskal-Wallis Test produced a H-value of 11.1484 and a p-value of 0.02495, where the results are statistically significant at $p < 0.05$. It may therefore be concluded that pH 3.0 and 11.0 had the highest rates of reductive-oxidative reactions where the dye solutions decolourised at the highest rates.

5 Limitations

The cost of the laccase enzyme, SGD$128 per gram, limited the number of setups and replicates of the experiments. This could have resulted in the large error bars that were recorded in Figs. 3, 4, 5 and 6. We were also unable to collect more data points due to time constraints.

6 Conclusion

The first set of results shows that up to the 20.0% concentration, the laccase enzyme is able to break down the phenolic compounds in the fabric dye to generate electricity. More specifically, it may be concluded that 10.0% dye concentration may be the optimum concentration of dye at which breakdown of phenols by laccase enzyme takes place at the highest rate. However, since the highest concentration we investigated was 20%, it would be prudent to carry out further investigation on the laccase enzymatic activity on higher substrate concentrations, to confirm the optimum concentration and saturation point, as well as at which concentration the inhibitory effect of laccase enzyme activity sets in, as that might lie above the range covered in our experiment.

The second set of results reflects that the laccase enzyme can break down phenolic compounds more effectively at acidic pHs, with the optimum pH being pH 3.0, followed by pH 5.0. However, as existing regulations surrounding the acidity of wastewater limit the pH of the wastewater to between 5.0 and 8.5 [20], pH 5.0 would be the most practical pH to use for the simultaneous bioremediation of wastewater and electricity generation using the laccase enzyme.

Across our 30 setups, the average electricity generated for each data point ranged from 7.01 mV to 55.41 mV. This amount of electricity generated at such a small scale would not be practical for industrial usage. At a larger scale, such as in wastewater treatment facilities, more electricity would be generated, which could be useful. When equipped with a low-voltage booster multiplier, our setup would also be able to generate larger amounts of electricity and store it for commercial use [21].

Our findings are beneficial to the real world as with further investigation, companies in the textile industry, as well as wastewater treatment facilities, can utilise the results on the optimum concentration and pH range to simultaneously bioremediate wastewater and generate electricity using the readily available laccase enzyme.

Overall, practically, our results suggest treating the fabric dyes at up to 20% concentration at pH 5.0 with a 2.5% laccase enzyme concentration at room temperature for optimum bioremediation and electricity generation to meet wastewater treatment regulations. To better tackle the issue of water pollution by the textile industry and simultaneously generate electricity for the sustainability of our energy resources, our findings which lead up to future investigations, along with the laccase enzyme, can be utilised.

Appendix

See Tables 1, 2, 3 and 4.

Table 1. Average electricity generated by solutions against dye concentrations

Dye concentration/%	Average electricity generated/mV	Standard deviation
0.0	7.09	±1.90
5.0	10.84	±3.92
10	20.11	±10.1
15	7.96	±3.53
20	7.01	±1.94

Table 2. Change in absorbance of solution in setup against dye concentrations

Dye concentration/%	Change in absorbance/A	Standard deviation
0.0	0.000	±0.000
5.0	−0.054	±0.0480
10	−0.260	±0.243
15	−0.225	±0.231
20	−0.324	±0.342

Table 3. Average electricity generated by solutions against pH

pH	Average electricity generated/mV	Standard deviation
3.0	55.4	±6.91
5.0	43.8	±1.61
7.0	12.8	±9.93
9.0	14.4	±2.57
11	22.2	±2.56

Table 4. Change in absorbance of solution in setup against pH

pH	Change in absorbance/A	Standard deviation
3.0	−0.392	±0.269
5.0	−0.283	±0.0785
7.0	−0.159	±0.0285
9.0	−0.124	±0.114
11	−0.266	±0.185

References

1. National Environment Agency. https://va.ecitizen.gov.sg/CFP/CustomerPages/NEA_google/displayresult.aspx?MesId=1069761&Source=Google&url=va.ecitizen.gov.sg-:~:text=Answer%3A,bothsuspendedanddissolvedsolids. Accessed 05 June 2023
2. European Parliament. https://www.europarl.europa.eu/topics/en/article/20201208STO93327/the-impact-of-textile-production-and-waste-on-the-environment-infographics. Accessed 05 June 2023
3. Good On You. https://goodonyou.eco/textile-dyes-pollution/. Accessed 05 June 2023
4. The Straits Times. https://www.straitstimes.com/singapore/environment/red-water-in-jurong-west-canal-is-organic-solvent-and-dye. Accessed 05 June 2023

5. Fabric of the World. https://www.fabricoftheworld.com/post/citarum-river-is-polluted-by-factories-that-make-your-clothes#:~:text=Over%201000%20textile%20factories%20dump,water%20to%20visibly%20change%20colour. Accessed 05 June 2023
6. Singapore Institute of Technology. https://www.singaporetech.edu.sg/news/novel-membranes-better-wastewater-treatment. Accessed 05 June 2023
7. Bala, S., Garg, D., Thirumalesh, B.V., et al.: Recent strategies for bioremediation of emerging pollutants: a review for a green and sustainable environment. Toxics **10**, 484 (2022). https://doi.org/10.3390/toxics10080484
8. Jebapriya, R., Gnanadoss, J.J.: Bioremediation of textile dye using white rot fungi: a review. Int. J. Curr. Res. Rev. **5**, 1 (2013)
9. Çifçi, D.İ, Atav, R., Güneş, Y., Güneş, E.: Determination of the color removal efficiency of laccase enzyme depending on dye class and chromophore. Water Sci. Technol. **80**, 134–143 (2019). https://doi.org/10.2166/wst.2019.255
10. Legerská, B., Chmelová, D., Ondrejovič, M.: Degradation of synthetic dyes by laccases – a mini-review. Nova Biotechnologica et Chimica **15**, 90–106 (2016). https://doi.org/10.1515/nbec-2016-0010
11. Jeon, J., Baldrian, P., Murugesan, K., Chang, Y.: Laccase-catalysed oxidations of naturally occurring phenols: from in vivo biosynthetic pathways to green synthetic applications. Microb. Biotechnol. **5**, 318–332 (2011). https://doi.org/10.1111/j.1751-7915.2011.00273.x
12. Chaijak, P., et al.: Enhancing Electricity generation using a laccase-based microbial fuel cell with yeast Galactomyces reessii on the cathode. J. Microbiol. Biotechnol. **28**, 1360–1366 (2018). https://doi.org/10.4014/jmb.1803.03015
13. Varga, B., Meiczinger, M., Somogyi, V.: Immobilization of laccase in alginate beads. Hung. J. Ind. Chem. (2019). https://doi.org/10.3327/hjic-2019-16
14. Mogharabi, M., et al.: Immobilization of laccase in alginate-gelatin mixed gel and decolorization of synthetic dyes. Bioinorg. Chem. Appl. **2012**, 1–6 (2012). https://doi.org/10.1155/2012/823830
15. Triplepoint Environmental. https://lagoons.com/blog/wastewater-lagoon-ph/#:~:text=High%20effluent%20pH%20is%20almost%20always%20due%20to%20algae.,during%20the%20hottest%20summer%20months. Accessed 05 June 2023
16. Mayolo-Deloisa, K., González-González, M., Rito-Palomares, M.: Laccases in food industry: bioprocessing, potential industrial and biotechnological applications. Front. Bioeng. Biotechnol. (2020). https://doi.org/10.3389/fbioe.2020.00222
17. Casella, L., et al.: Enzymatic and spectroscopic studies on the activation or inhibition effects by substituted phenolic compounds in the oxidation of aryldiamines and catechols catalyzed by Rhus vernicifera laccase. J. Inorg. Biochem. **100**, 2127–2139 (2006). https://doi.org/10.1016/j.jinorgbio.2006.07.005
18. Chemistry LibreTexts. https://chem.libretexts.org/Bookshelves/Analytical_Chemistry/Supplemental_Modules_(Analytical_Chemistry)/Analytical_Sciences_Digital_Library/Contextual_Modules/End_Creek%3A_Spotted_Frogs_and_Aquatic_Snails_in_Wetlands__A_Water_Quality_Investigation/09_Instructors_Guides/06_Analysis_of_Anions_by_Spectrophotometry. Accessed 05 June 2023
19. Yin, Q., et al.: The first fungal laccase with an alkaline pH optimum obtained by directed evolution and its application in indigo dye decolorization. AMB Express (2019). https://doi.org/10.1186/s13568-019-0878-2
20. Bari, M.L., Kawamoto, S.: PROCESS HYGIENE | Types of Sterilant, pp. 216–225. Elsevier eBooks (2014). https://doi.org/10.1016/b978-0-12-384730-0.00273-1
21. Koffi, N.J., Okabe, S.: High voltage generation from wastewater by microbial fuel cells equipped with a newly designed low voltage booster multiplier (LVBM). Sci. Rep. (2020). https://doi.org/10.1038/s41598-020-75916-7

Synthesis of Zeolite for CO2 Gas Adsorption with Singaporean and Korean Crustacean

Zhu Han(✉) and Ng Yan Ting Jolin

National Junior College, Singapore, Singapore
zhu_han@students.edu.sg

Abstract. Over 10 million tons of mollusc shells are produced each year, which have devastating effects on marine ecosystems. Our study aims to target the prominent clamshell waste issue by using clamshells to create enhanced zeolites which act as carbon sinks. We utilised clamshells from both Singapore and South Korea by conducting ion exchange with the Zeolite X material and the calcium chloride solution we obtained from dissolving the clamshells. We hypothesised that the zeolites that had gone through ion exchange would have superior gas absorption abilities compared to untreated zeolites. Using X-Ray Diffraction (XRD), Scanning Electron Microscope (SEM) and Energy-Dispersive X-ray spectroscope (EDX), we were able to observe the structural and elemental composition of our zeolites, which supported our hypothesis where ion-exchanged zeolites surpasses untreated counterparts in gas absorption capacity. An Ion chromatography machine was also used to get an accurate representation of the calcium concentration in our samples. In conclusion, our study presents another purpose for clamshell waste, which can help alleviate environmental impacts, while also offering a sustainable solution for creating pollutant-absorbing materials, reducing the amount of clamshell waste in the process.

Keywords: Clamshell · Zeolites · Carbon sinks · Ion exchange · Calcium · Adsorption

1 Introduction

Clamshell waste has become increasingly prominent in recent years. Over 10 million tons of mollusc shells are produced each year with 70% being represented by oyster, clam, scallop and mussel shells. In Singapore, clamshells are disposed of through incineration which is the most conventional way of discarding clamshell waste. Clamshells cannot be discarded into the ocean due to the potential water pollution they cause, possibly modifying soils, waters and marine ecosystems if their disposal is uncontrolled [10]. In Korea, about 0.4 million tons of shellfish are produced, and about 40% are discarded annually [11]. Thus the problem of disposing of this waste is considered serious in both countries, and hence we chose to conduct research with these two countries.

Zeolites' microporous and three-dimensional structure (Fig. 8 refer to appendix) allows for the absorption of pollutants [4, 8]. It is believed that the presence of heavy

H. Guo et al. (Eds.): IRC-SET 2024, PTT, pp. 76–84, 2025.
https://doi.org/10.1007/978-981-96-3770-6_9

metals also assists in the absorption of pollutants acting as polar organic solvents [6]. It was also found that the calcium exchanged zeolite had the ability to absorb carbon dioxide very well [7]. Also, Majority of clamshell waste consists of high concentration of calcium carbonate (CaCO3) [1, 7]

Zeolites may be ion-exchanged to remove specific ions from the zeolite depending on its intended purpose, and this is favourable as ion-exchanged zeolites have shown to produce better CaX zeolite, the calcium ion-exchanged zeolite, functions as a selective adsorbent to separate ethane, ethylene, and methane from one another at low temperatures and pressures [5].

Thus, we will obtain a calcium chloride solution (with other metal ions) through the clamshell and use it to exchange ions in the zeolite. As mentioned earlier, calcium-exchanged zeolite has several advantages over ordinary zeolite, and other metals contained in the shell can to absorb more carbon dioxide as compared to the original zeolite X.

We hypothesise that the Zeolite X that has gone through ion exchange is able to absorb more CO2 gas as compared to the Zeolite that has not gone through ion exchange.

2 Methodology

The Singaporean clamshells were purchased from a supermarket, and the clams bought are endemic to the waters in and around Singapore. For the Korean clams, we collected them from the Gwangalli beach in Busan. A larger variety of clamshells is optimal as it allows for our study to be better represented on a larger scale. Zeolite 13X was purchased from Sigma Aldrich, and Hydrochloric acid (Assay 35%) and Sodium Hydroxide were obtained from Duksan Pure Chemicals. Methanesulfonic acid was purchased from Fluka, and Dionex six cation-II standard was obtained from Thermo Scientific.

The shells were then washed with deionised water and scrubbed to remove dust or impurities such as sand or flesh. 5.00 g of the Singaporean and Korean shells were each crushed into powder using a mortar and pestle. A strong acid, hydrochloric acid, was used to dissolve the clam powder to obtain a calcium chloride solution [2]. 20 mL of 4.93 M hydrochloric acid (HC*l*) was used to dissolve the shells and was left for 24 hours. The equation shows the reaction that occurred when the shells were dissolved (Fig. 1)

$$CaCO_3 + 2HCl n\ CaCl_2 + CO_2 + H_2O$$

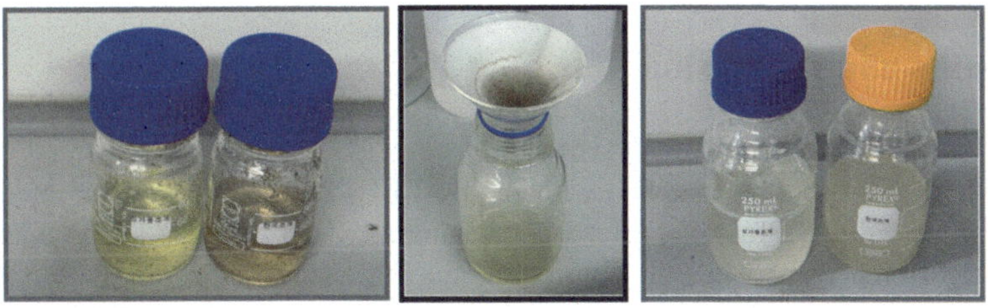

Fig. 1. Pictures of 1) shell-dissolved solutions, 2) filtration, and 3) after neutralisation

2 conical tubes were then filled with the clamshell solution, one containing the dissolved Singaporean shells while the other contained the dissolved Korean shells. There was also a control zeolite made using a pure $CaCl2$ solution. The zeolite 13X beads were then put into the solutions and the tubes were left for a week with the solution being changed every 12 hours. As seen in Fig. 2.

Fig. 2. Picture of ion-exchanging

The Quanta 2000 Scanning Electron Microscope (SEM) was used to observe structure of the zeolite samples. The Bruker Quantax200 Energy Dispersive X-Ray (EDX) was used to observe the elemental composition of the zeolites that were created, allowing us to better understand the structure of the zeolite that was produced and observe the changes that occurred. The Micromeritics ASAP 2020 Surface Area and Porosity Analyzer was used for the characterization of zeolites. After the sample went through degassing which took 1 day with the use of liquid nitrogen, it was able to be put into the analysis part of the ASAP. The Rigaku MiniFlex 600 X-Ray diffractometer (XRD) was used to analyse the structure of the material using the scattering patterns produced. The zeolite was crushed into powder using the mortar and pestle which was then put onto a slide and into the machine. Dionex ICS-1000, an Ion chromatography machine, was then used to calculate the concentration of each clamshell dissolved solution and to measure metal ion concentration. The clamshell-dissolved solutions were diluted 1000 times with DI water. Standard solutions were prepared with 10,20,30,40 ppm to draw the calibration curve by serial dilution. 0.2 M methanesulfonic acid solution was used as an eluent.

3 Results and Discussion

Figure 3 shows us the XRD analysis result. The peaks of the ion-exchanged zeolites (Korean and Singaporean) matched with zeolite 13X (control), even if the intensity of each peak was slightly lowered after ion-exchanging. Thus, we deduce no significant structural differences between the control and the ion-exchanged zeolites.

Table 1 shows the SEM pictures taken showing the structure of each zeolite. It can be observed that all 3 zeolites have different concentrations of different metallic elements.

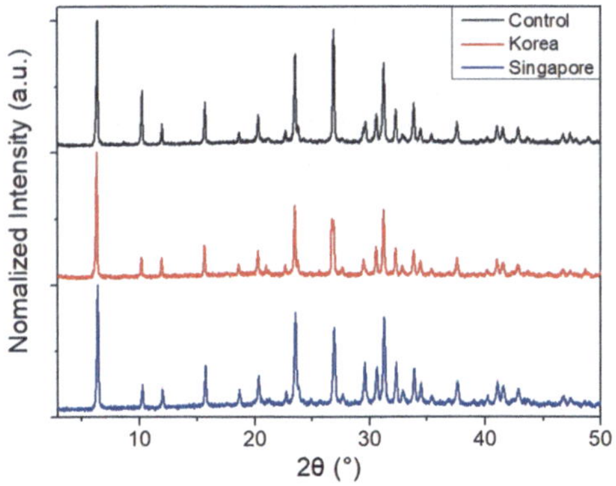

Fig. 3. XRD analysis

Table 2 shows the results from the EDX, it shows us the elemental composition of each zeolite. The portion of calcium in the control was 0.47 ± 0.16 %. The portion of calcium in the zeolite increased to 4.12 ± 0.20 % (Korean zeolite), and 2.22 ± 0.29 % (Singaporean zeolite) after ion-exchanging, indicating the success in ion exchange in the solutions.

Table 1. SEM, zeolite structure

Control zeolite	Singapore zeolite	Korean zeolite
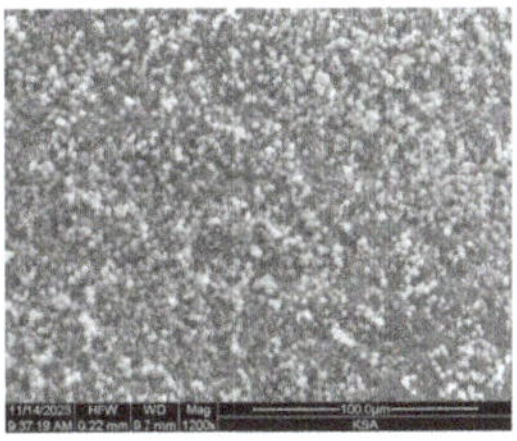	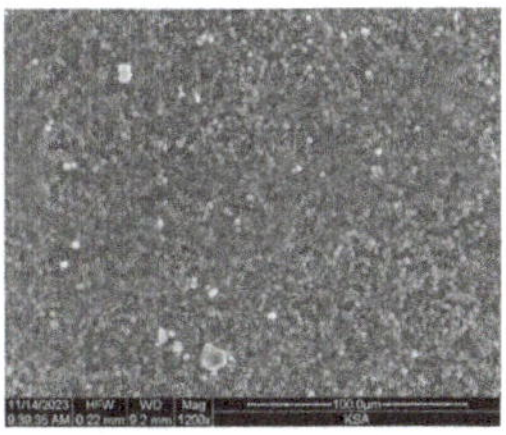	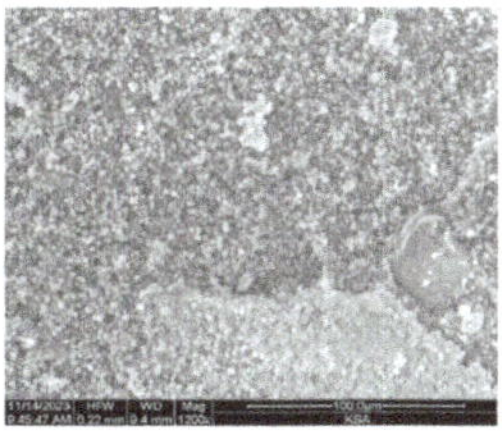

To find out the physical properties such as surface area and porosity, Brunauer–Emmett–Teller (BET) surface area analysis was conducted for all the samples, and surface areas were calculated to be 389.8077, 321.8059, and 311.9951 m^2/g for the control, Korean, and Singaporean samples respectively Fig. 4. In the pore volume distribution graphs, the pore volumes are increased by following the decrement of pore diameter until 20 Å, therefore, we assume that the pore size of zeolites is less than 20 Å, respectively. The known pore size of zeolite X is 13 Å, which aligns with our results. The ion-exchanged zeolites with Singaporean clamshell-dissolved solution show peaks at about 38 Å and 181 Å. These peaks imply large pores exist in the ion-exchanged zeolites with Singaporean clamshell-dissolved solution and might be created by structural collapses during the ion-exchanging process. This assumption can also explain why the

Table 2. EDX, elemental composition of each zeolite

Control zeolite

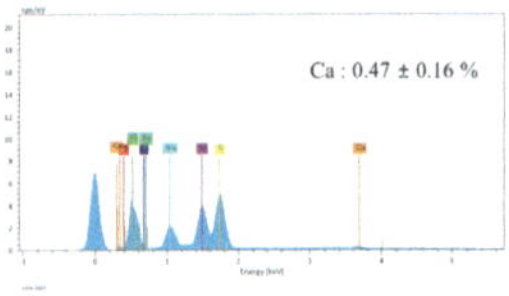

Element	At. No.	Mass [%]	Mass Norm. [%]	Atom [%]	abs. error [%] (3 sigma)
Nitrogen	7	7.34	9.72	13.23	3.50
Oxygen	8	34.59	45.84	54.62	12.36
Fluorine	9	1.62	2.15	2.16	1.06
Sodium	11	6.83	9.05	7.51	1.36
Aluminium	13	9.91	13.14	9.28	1.47
Silicon	14	13.87	18.38	12.48	1.82
Calcium	20	0.75	0.99	0.47	0.16
Iron	26	0.55	0.72	0.25	0.17
		75.46	100.00	100.00	

Singapore zeolite

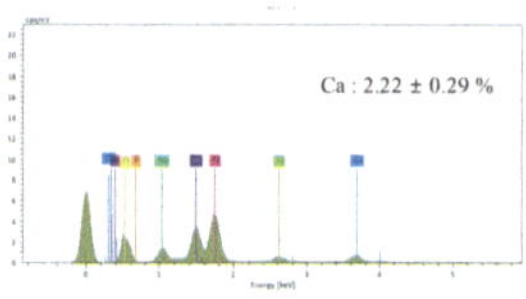

Element	At. No.	Mass [%]	Mass Norm. [%]	Atom [%]	abs. error [%] (3 sigma)
Nitrogen	7	5.44	10.86	14.82	2.79
Oxygen	8	23.76	47.40	56.64	8.78
Fluorine	9	0.58	1.16	1.17	0.54
Sodium	11	3.52	7.03	5.85	0.77
Aluminium	13	5.80	11.57	8.20	0.92
Silicon	14	7.09	14.14	9.62	0.99
Chlorine	17	1.01	2.02	1.09	0.19
Calcium	20	2.33	4.65	2.22	0.29
Iron	26	0.59	1.17	0.40	0.16
		50.13	100.00	100.00	

Korean zeolite

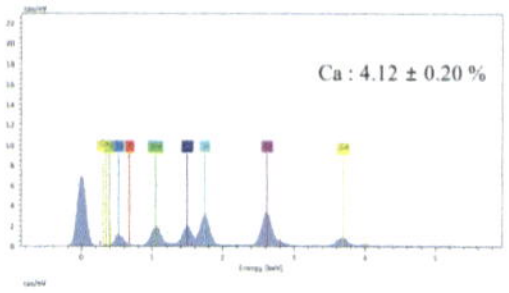

Element	At. No.	Mass [%]	Mass Norm. [%]	Atom [%]	abs. error [%] (3 sigma)
Nitrogen	7	1.16	6.82	10.42	0.89
Oxygen	8	6.32	37.07	49.59	2.65
Sodium	11	2.03	11.89	11.07	0.47
Aluminium	13	1.15	6.76	5.36	0.25
Silicon	14	1.90	11.13	8.48	0.32
Chlorine	17	2.95	17.32	10.46	0.38
Calcium	20	1.32	7.71	4.12	0.20
Iron	26	0.22	1.30	0.50	0.11
		17.05	100.00	100.00	

ion-exchanged zeolites with Singaporean clamshell-dissolved solution have a smaller BET surface area than Koreans (Fig. 5).

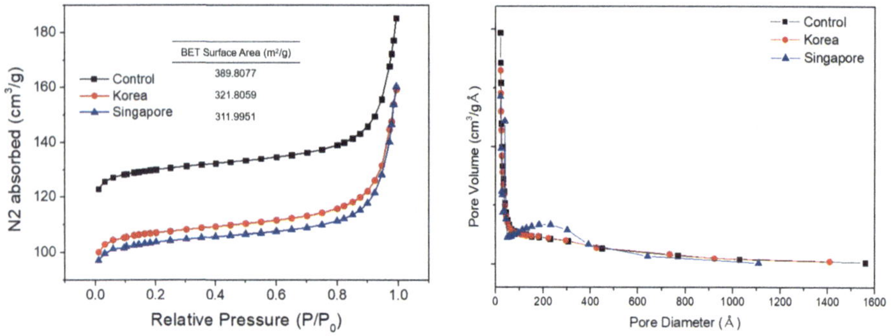

Fig. 4. Nitrogen Adsorption Isotherm Plots and BET Surface Areas of Zeolites, and Pore Volume Distribution from Adsorption Isotherms of Zeolites

In the ion chromatogram, six peaks represent six cations in the standard solution. The identity of about 5.93 min). Calibration curves were drawn by using chromatograms of standard solutions (10, 20, 30 and 40 ppm).

We assume that the concentration of the ion is directly proportional to the area of the peaks, as you increase the concentration of an ion, the height and area of the peak corresponding to that ion will increase [12]. We get the relationship between peak area

Control	Singapore	Korean
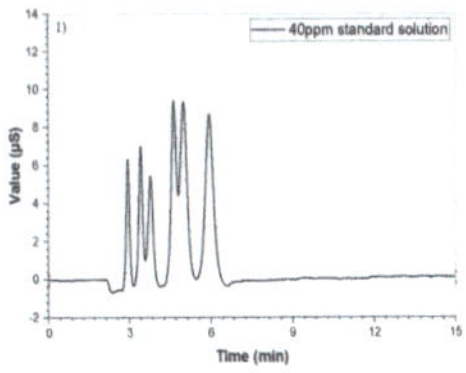	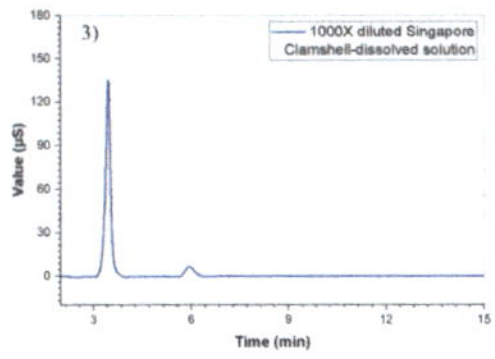	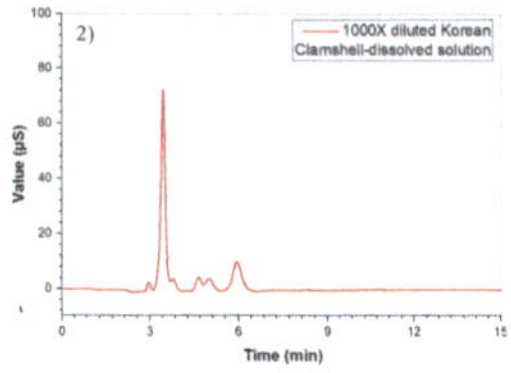

Fig. 5. Ion chromatogram

and Ca^{2+} concentration, solution to be 0.041523 M (Korean sample), and 0.032587 M (Singaporean sample).

NaOH powder being used in the neutralisation process . In the research on the chemical compositions of seashells, there is no Na compound in the shells. Thus, we deduce that the Na^{+} ion detected in the ion chromatography is not from the clamshells at all(Fig. 6).

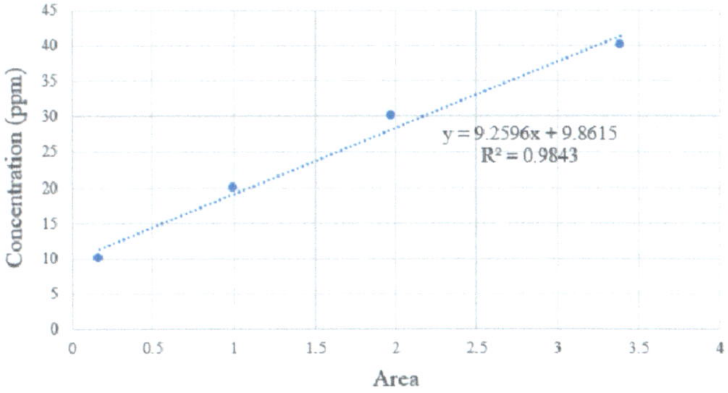

Fig. 6. Calibration Curve of Ca^{2+} cation

Figure 7 shows an abnormal isotherm plot where the amount of CO2 adsorption decreases as absolute pressure increases. The CO2 adsorption experiments were repeated more than 5 times with different samples, however, every isotherm plot exhibited the same trend as Fig. 7. The samples and degassing conditions were the same as the N2 adsorption experiments, thus, we concluded that there might be instrumental flaws that can lead to the abnormal isotherm plots. Hence, despite the fact that our other results strongly point to CO2 adsorption properties of ion-exchanged zeolites being greatly enhanced, we could not directly compare the CO2 adsorption properties of our zeolites,

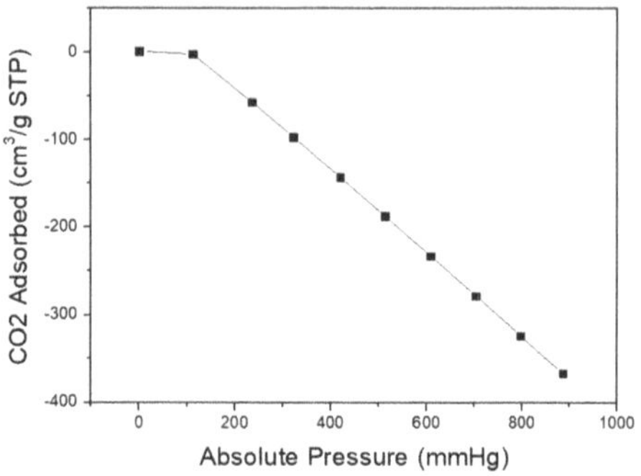

Fig. 7. A carbon Dioxide Adsorption Isotherm plot of Korean ion-exchanged zeolite

4 Conclusion

To conclude, we determined that the concentrations of calcium cations in the clamshell-dissolved solutions are 0.041523 M (Korean sample), and 0.032587 M (Singaporean sample) by ion chromatography. Using these solutions, the Zeolite 13X underwent ion-exchange and was characterised with EDX, XRD, and BET surface area analysis. The portion of calcium in zeolite increased to 4.12 $\pm$ 0.20 %(Korean zeolite) and 2.22 $\pm$ 0.29 %(Singaporean zeolite) after ion-exchanging, which implies the ion-exchanging of zeolites was successful. In the XRD analysis, we deduce there are no significant structural differences between the control and the ion-exchanged zeolites. In Brunauer–Emmett–Teller (BET) surface area analysis, surface areas were calculated to be 389.8077, 321.8059, and 311.9951 m^2/g for the control, Korean, and Singaporean samples respectively. We anticipate that there are two major reasons for differences in calcium concentrations of clamshell-dissolved solution and zeolite properties between Korean and Singaporean. Firstly, there are species differences due to us collecting available clamshells from the beach rather than from a particular species, so our samples consisted of various types of clamshells that represent the ecosystem. Due to this difference in species, the portions of the metal compound in each clamshell was different. This inequality can lead to differences in Ca^{2+}. Secondly, in the sea, organic compounds in the clamshells can be eliminated, which allows the clamshell to dissolve better in strong acids. Korean clamshells were already dead, collected from the beach, so it is likely that the Korean clamshells dissolved better than the Singaporean clamshells which were alive until they were processed.

Appendix

Figure 9 and Table 3.

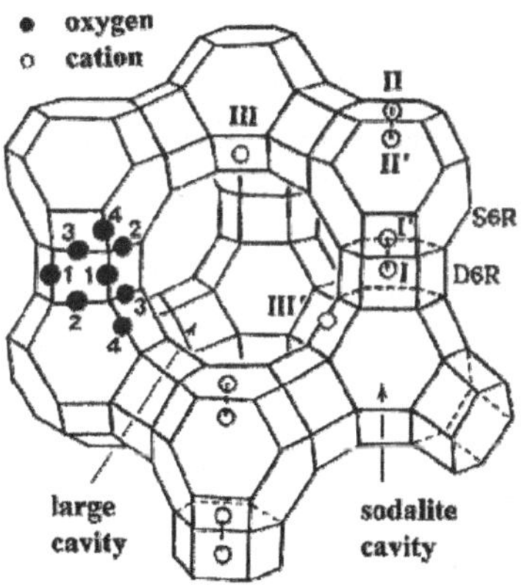

Fig. 8. The structure of zeolite [4]

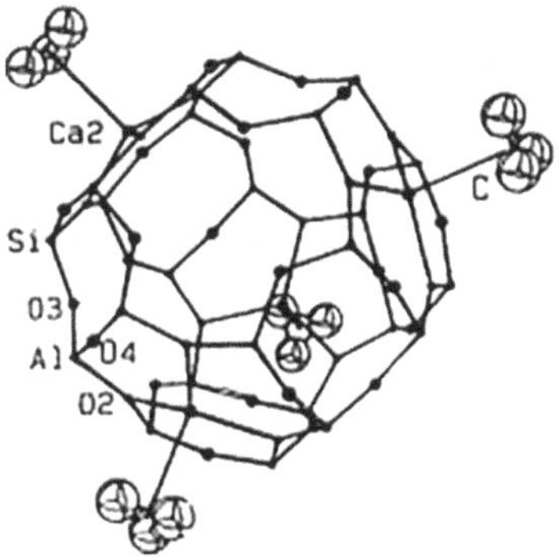

Fig. 9. A stereoview of a sodalite cavity of calcium-exchanged zeolite [4]

Table 3. The chemical composition compositions of seashells [13].

Chemical composition (%)	Type of seashells/Portland cement							
	Portland Cement[a]	Clam[b]	Mussel[b]	Oyster[b]	Periwinkle[c]	Snail[c]	Cockle[d]	Clam[e]
SiO_2	21.20	0.84	0.73	1.01	26.26	10.20	1.6	0.39
Al_2O_3	5.47	0.14	0.13	0.14	8.79	4.81	0.92	0.28
Fe_2O_3	3.31	0.06	0.05	0.07	4.82	3.15	-	0.02
CaO	65.52	53.99	53.38	53.59	55.53	61.95	51.56	67.70
MgO	1.97	0.08	0.03	0.46	0.40	0.18	1.43	-
K_2O	1.71	0.03	0.02	0.02	0.20	0.05	0.06	-
Na_2O	0.46	0.39	0.44	0.23	0.25	0.04	0.08	-
SO_3	1.90	0.16	0.34	0.75	0.18	0.03	-	-
Cl	-	0.02	0.02	0.01	-	-	-	-
SO_4	-	0.06	0.11	0.43	-	-	-	-
$CaCO_3$	-	96.8	95.6	96.8	-	-	-	-

Source: [a][16], [b][8], [c][17], [d][18], [e][19]

References

1. Kao, C.Y., Huang, Y.C., Chiu, S.Y., Kuo, K.L., Hwang, P.A.: Bacteriostatic effect of a calcined waste clamshell-activated plastic film for food packaging. Materials **11**(8), 1370 (2018)
2. Hussein, A.I., Ab-Ghani, Z., Che Mat, A.N., Ab Ghani, N.A., Husein, A., Ab. Rahman, I.: Synthesis and characterization of spherical calcium carbonate nanoparticles derived from cockle shells. Appl. Sci. **10**(20), 7170 (2020)
3. Van Speybroeck, V., Hemelsoet, K., Joos, L., Waroquier, M., Bell, R.G., Catlow, C.R.A.: Advances in theory and their application within the field of zeolite chemistry. Chem. Soc. Rev. **44**(20), 7044–7111 (2015)
4. Choi, E.Y., Kim, Y., Seff, K.: Crystal structure of a mesitylene sorption complex of dehydrated fully Ca2+-exchanged zeolite X. sorbed mesitylene appears to be significantly nonplanar. The Journal of Physical Chemistry B **106**(23), 5827–5832 (2002)
5. Hosseinpour, S., Fatemi, S., Mortazavi, Y., Gholamhoseini, M., Ravanchi, M.T.: Performance of CaX zeolite for separation of C2H6, C2H4, and CH4 by adsorption process; capacity, selectivity, and dynamic adsorption measurements. Sep. Sci. Technol. **46**(2), 349–355 (2010)
6. Inglezakis, V.J., Loizidou, M.D.: Ion exchange of some heavy metal ions from polar organic solvents into zeolite. Desalination **211**(1–3), 238–248 (2007)
7. Mohammad, W.A.S.B.W., et al.: A review on seashells ash as partial cement replacement. In: IOP Conference Series: Materials Science and Engineering, Vol. 271, No. 1, p. 012059. IOP Publishing (2017)
8. Abdi, M.R., Shakur, H.R., Saraee, R.E., K., & Sadeghi, M.: Effective removal of uranium ions from drinking water using CuO/X zeolite based nanocomposites: effects of nano concentration and cation exchange. J. Radioanal. Nucl. Chem. **300**(3), 1217–1225 (2014)
9. Oddy, S., Poupore, J., Tezel, F.H.: Separation of Co2 and CH4 on CaX zeolite for use in Landfill gas separation. The Canadian Journal of Chemical Engineering **91**(6), 1031–1039 (2013)
10. Topić Popović, N., Lorencin, V., Strunjak-Perović, I., Čož-Rakovac, R.: Shell Waste Management and utilisation: Mitigating organic pollution and enhancing sustainability. MDPI (2023). https://www.mdpi.com/2076-3417/13/1/623#:~:text=Every%20year%2C%20close%20to%208,soils%2C%20waters%20and%20marine%20ecosystems
11. 2019 Fishery Production Survey, Statistics Korea, (21 february 2020), retrived 1 April 2023 from https://kostat.go.kr/board.es?mid=a20102080000&bid=11715&tag=&act=view&list_no=382496&ref_bid=
12. Ion Chromatography, n.a., (2023, 29 August), retrieved 23 October 2023, https://chem.libretexts.org/Bookshelves/Analytical_Chemistry/Supplemental_Modules_(Analytical_Chemistry)/Analytical_Sciences_Digital_Library/Contextual_Modules/Effects_of_Acid_Rain_on_Atlantic_Salmon_Populations/06_Instructors_Guide/05_Ion_Chromatography#:~:text=As%20you%20increase%20the%20concentration%20of%20an%20ion%2C%20the%20height,to% 20that%20ion%20will%20increase
13. chemical composition of seashells of portland cement, research gate, (november 2017), retrieved 10 april from 2023. https://www.researchgate.net/figure/Chemical-composition-of-seashells-and-Portland-cement_tbl1_321659334

Detecting Defects in High Hydration Samples: A Comparative Study of Shearography Imaging and Terahertz Broadband Spectral Imaging

Verity Qi Wan Yi Lim[1], Zi Wen Jovis Tham[2], Lei Zhang[2](✉), Zi Xiao Zhang[3], and Yong Bo Zhu[3]

[1] Raffles Institution, One Raffles Institution Lane, Singapore 575954, Republic of Singapore
[2] Institute of Materials Research and Engineering (IMRE), Agency for Science, Technology and Research (A*STAR), 2 Fusionopolis Way, Innovis #08-03, Singapore 138634, Republic of Singapore
zhangl@imre.a-star.edu.sg
[3] SingTera Pte. Ltd., 60 Paya Lebar Road #11-19 Paya Lebar Square, Singapore 409051, Republic of Singapore

Abstract. Detecting defects in high hydration level biological samples is challenging due to the high transparency in visible wavelength, high absorption of water in electromagnetic wave, strong scattering and refraction, heterogeneous composition as well as its inherent dynamic nature. Shearography has been effectively used for defect imaging in various industries but less reported in biological tissues imaging. In this study, we explore the application of shearography technique for detecting surface and deep-level cracks or scratches in contact lenses. Additionally, we investigate the use of pulsed terahertz (THz) broadband spectroscopy and imaging for the same contact lens samples for comparison. The results from both imaging techniques were compared, highlighting the strengths and limitations of each method in characterizing defects in high hydration level biological samples.

1 Introduction

High water content samples can obscure the presence of defects, making traditional inspection and imaging techniques less effective, particularly in biomedical fields [1–3]. This issue is significant in samples such as corneal tissue, brain tissue, cartilage, amniotic fluid, and skin. Identifying the exact location and nature of defects allows for precise treatment planning, whether it's surgical intervention, targeted therapy, or other medical procedures. Understanding the nature and distribution of defects in biological tissues contributes to scientific knowledge and the development of new medical technologies and treatments. The inherent difficulty in distinguishing between normal and defective regions in high hydration samples necessitates the development of more advanced and sensitive detection methods to improve accuracy and reliability [4, 5].

High-Resolution Imaging Optical Coherence Tomography (OCT) is widely used for biological sample imaging due to its ability to provide high-resolution cross-sectional

H. Guo et al. (Eds.): IRC-SET 2024, PTT, pp. 85–92, 2025.
https://doi.org/10.1007/978-981-96-3770-6_10

images of tissue. However, OCT has a limited field of view and is easily disturbed by motion and noise. It is difficult to quantify defect density, depth, and size, and it is impossible to understand the chemical composition of defects. New technologies are still under development to improve the accuracy of detecting defect density, depth, and chemical composition [6–8].

Shearography is a full-field optical imaging method suitable for fast defect inspection. It operates by subtracting two interferometric images obtained from an object under stress and without stress. Traditional shearography techniques apply stress through vacuum, thermal, and vibration excitations. In this work, air coupled ultrasonic waves for stress loading, offering rapid inspection speed and deep penetration, ideal for efficient defect characterization of materials and structures. This technique has not yet been applied to biological samples. We are using it for the first time on hydration samples of contact lenses, marking a first step toward biological applications [9–11].

THz imaging is a technique that uses terahertz radiation to create detailed images of materials. In hydration biology samples, THz imaging can measure the water content and distribution within the samples. This is because terahertz waves are sensitive to water molecules, making it possible to see differences in hydration levels. The technique is non-invasive and can provide important information about the structure and condition of biological samples, such as tissues or cells, based on their water content [12–15].

In this work, defects on contact lens have been created by a fine-tipped needle. A deep scratch has been formed on the one side of the contact lens. The microscopic images confirmed the locations of the scratches in the contact lens. Both shearography imaging and THz imaging techniques have been attempted to investigate the corneal scarring tissue density and depth profile. Our experiment demonstrated the potential applications of shearography and THz imaging for detecting surface and subsurface defects in contact lens samples. Future research should explore their applications further in biological imaging contexts.

2 Experimental Sections

2.1 Sample Information

Contact lenses used in this study were sourced from Bosch & Lomb, known for their high-quality lenses designed for optimal hydration and comfort. Each contact lens was inspected for any pre- existing defects or irregularities using a standard light microscope to ensure the lenses were initially defect-free.

2.2 Surface Scratch Introduction

Controlled scratches were introduced to the surface of the contact lenses using a fine-tipped needle. Scratches were made from surface to a level of depth to simulate the defect scenarios for imaging analysis. After scratching, the lenses were carefully rinsed with sterile saline solution to remove any debris from the scratching process. The lenses can maintain its hydration level relatively constant for the experimental period. This is used for mimicking the conditions of biological tissues.

2.3 Shearography Imaging Experimental Setup

Shearography operates by subtracting two interferometric images obtained from an object under stress and without stress. During testing, coherent laser lights illuminate the sample surface to create an interferometry pattern via shearography optics. Changes in this pattern, caused by ultrasonic stress loading, are captured by a camera in reference to the unstressed pattern, allowing for nearly real-time detection of both surface and subsurface defects. In this study, a SE2 sensor (isi-sys GmbH, Germany) was used, and systematic tests with piezoelectric ultrasonic transducers were conducted on composite specimens.

2.4 THz Imaging Experimental Setup

TERA K15 (Menlo Systems GmbH, Germany) and in house built XYZ rotating stage were used in this experiment. In the system, the femtosecond fibre laser with 250 MHz repetition rate and 90 femtosecond laser pulses with around 1.56 μm central wavelength was used to excite two photoconductive antennas, one was used as emitter and another was used as receiver, which were fibre-coupled. The THz pulse was generated with the coverage of bandwidth up to 3.5 THz. The THz beam was focused with TPX lenses. An XYZ rotating stage with sample holder was used for sample movement for C-scan imaging.

3 Results and Discussion

3.1 Microscopic Images of Contact Lens with Defects

Figure 1 presents microscopy images of a contact lens. The image on the left displays a pristine contact lens with a smooth, uniform surface, highlighting its clear, defect-free structure. The image on the right shows the same type of contact lens but with an artificially created defect. This defect is not easily visualized unless the contact lens is viewed from a specific angle under specialized lighting conditions. The red color highlights indicate the exact location of the defect created.

3.2 Shearography Imaging

Figure 2 depicts the Shearography Imaging setup in reflection mode, specifically focused on a contact lens. In this configuration, coherent laser light is directed onto the contact lens surface. The reflected light creates an interferometric pattern, which is captured by a camera through shearography optics. Air-coupled ultrasound creates stress on the contact lens surface. A video is recorded as the stress is applied and then removed, capturing the changes with the shearography camera.

Figure 3 (a) shows the initial setup where the contact lens is securely placed on the holder and ready for inspection. Figure 3 (b) shows the optical picture of defects created on the contact lens surface. Despite the presence of defects, the optical images fail to reveal any significant details about the defects. Optical methods are generally limited to surface observations and may not capture subtle variations caused by defects.

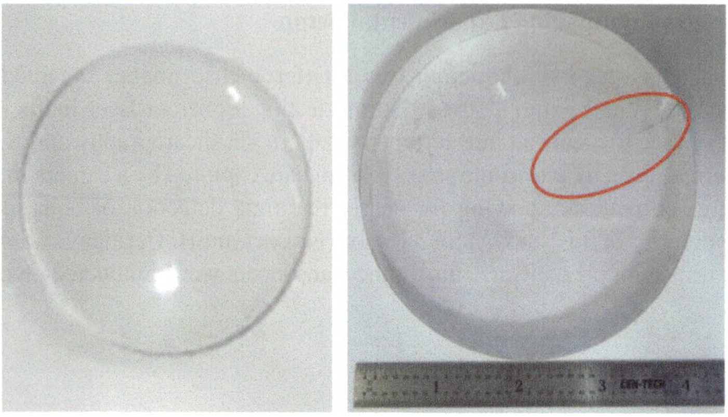

Fig. 1. Microscope results for the contact lens (left) and contact lens with defect (right).

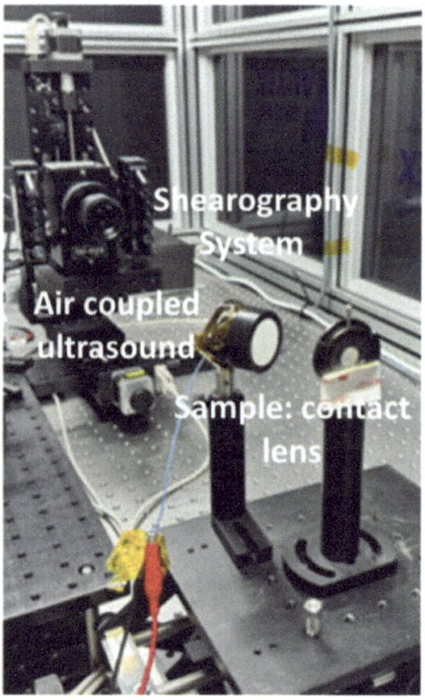

Fig. 2. Shearography technique for contact lens defect detection experimental setup.

The limitations of optical imaging in capturing subsurface or subtle surface anomalies are evident here. Figure 3 (c) shows the shearography recorded image which shows the defect on the contact lens surface. In contrast to the optical images, the shearography image successfully highlights the defects on the contact lens surface.

Shearography is a non-destructive testing method that uses laser speckle patterns to detect changes in surface strain. When stress is applied to a material, any defects present will cause localized variations in strain. In this experiment, we carefully tuned the ultrasound frequency and power in order to capture these variations. The results demonstrated that it is effective for identifying the surface defects in high hydration content samples.

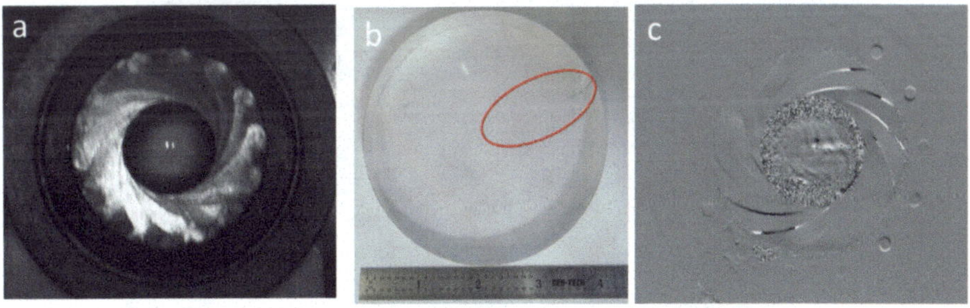

Fig. 3. (a) contact lens sample mounted on the sample holder. (b) optical pictures of defect created on the contact lens surface. (c) Shearography recorded image shows the defect on the contact lens surface.

3.3 Terahertz Time Domain Imaging

The samples were maintained in the same direction and position to establish a potential spatial correspondence between the THz images and slit lamp images. The THz system was calibrated by maximizing reflectivity off a strip of aluminium. Samples were placed on an aluminium stage attached to an XYZ translation stage, and the contact lens samples were further flattened on the sample stage. A thin-film Mylar window was applied to mitigate surface contours of the samples. A two-dimensional scan was performed over a 2 cm by 2 cm area for each corneal sample with a step size of 0.1 mm. The scan time for each 2D image was approximately 5 minutes, with a sampling rate of 75 samples per second.

Figure 4 illustrates the terahertz (THz) imaging setup for the contact lens. The THz time-domain spectroscopic (THz-TDS) imaging system operates within a wavelength range of 0.1 THz to 4 THz, featuring a scan resolution step of 100 micrometres and a beam size of approximately 1 millimetre. Imaging is performed using a reflection scheme, where the contact lens is pressed to a relatively flat surface to facilitate accurate imaging. This setup allows for detailed examination of the contact lens, capturing both surface and subsurface features with high precision.

The same lenses were then imaged using THz time-domain spectroscopy. The THz pulses penetrated the lenses, and the resulting data were used to reconstruct images. The time-domain curves consistently reveal broadened peaks, indicating deep THz wave penetration through the contact lens with delayed reflections back to the detector. We selected three typical delay points and reconstructed three B-scan images, shown at the bottom of Fig. 5. The figure shows that areas with defects are clearly visible in the first reconstructed image, corresponding to the lens surface. The second and third images are reconstructed from longer delay points, representing deeper levels of the lens. The images, despite appearing blurrier, successfully outline the defect locations and shapes created on the contact lens surface. This clarity in defect identification is crucial for

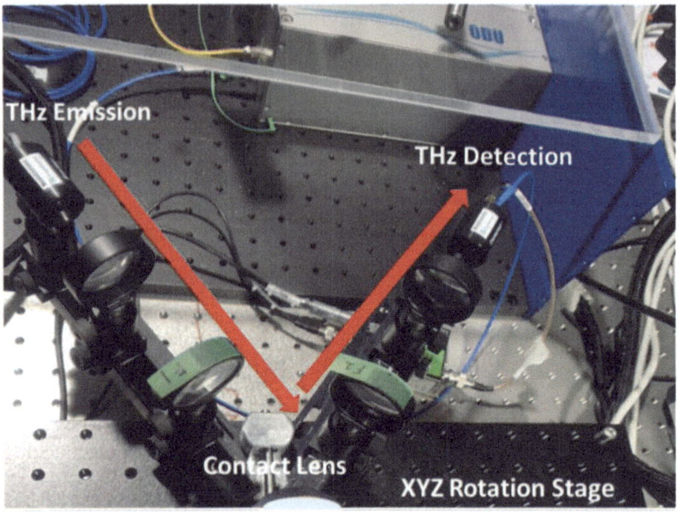

Fig. 4. Terahertz (THz) imaging setup for the contact lens defect imaging.

assessing the quality and integrity of the contact lenses. Furthermore, spectroscopic analysis provided absorption coefficient spectra and refractive index changes, offering additional insights into the material properties at the defect sites. This detailed analysis will be subject to another publication.

4 Discussion

The acoustic shearography technique excels in detecting subsurface defects due to its ability to penetrate deeply into various materials. In this study, for applications like contact lenses, the acoustic shearography technique requires further fine-tuning and parameter adjustments to achieve higher resolutions. Future work should focus on fine-tuning the external stress executed on biological samples to better visualize the distribution of the defects. However, for highly hydrated biological real applications, acoustic shearography's deeper penetration is a significant advantage, making it promising for future real biological tissue applications.

THz imaging achieved high resolution, enabling visualization of surface and subsurface features in high hydration biological samples. However, its raster scan approach due to THz detector limitations results in longer imaging times. THz waves have limited penetration depth in water- rich or dense materials, restricting their utility for deep tissue imaging. THz spectroscopy can further provide detailed information on absorption coefficients and refractive index changes, which offer insights into material properties and enhance differentiation between tissue types for medical applications, although not reported in this study.

Our experiment demonstrated the potential applications of shearography and THz imaging for detecting surface and subsurface defects in contact lens samples. Future research should explore their applications further in biological imaging contexts.

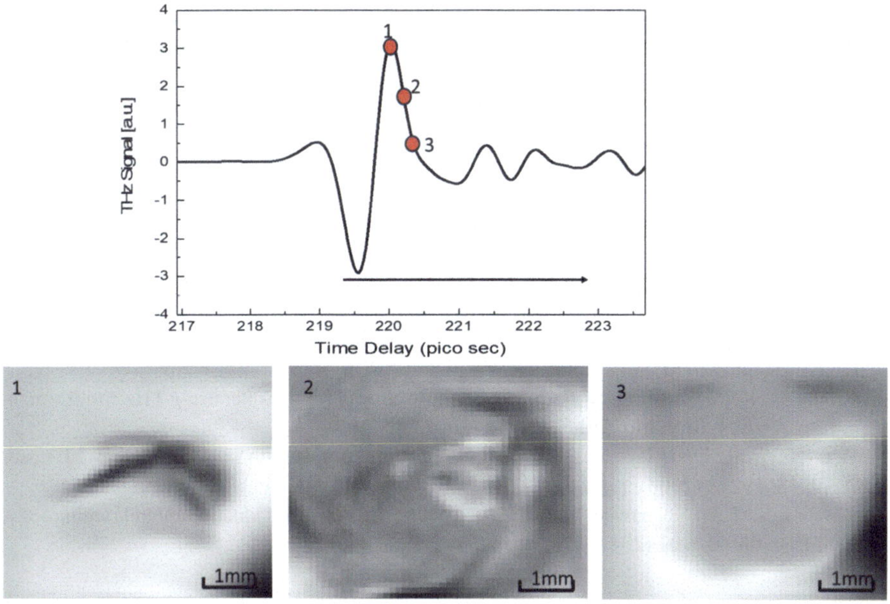

Fig. 5. Top: Typical THz time-domain signal collected from the contact lens. 5b: B-scan images reconstructed at the peak intensity delay times for each corneal sample, illustrating the distribution and intensity of defects.

5 Conclusion

In summary, both shearography and THz spectroscopy exhibited sufficient contrast in assessing defect densities and locations within contact lenses, indicating potential applications in visualizing the spatial distribution of defects. The advantages and disadvantages of each technique have been evaluated. However, significant further research is needed for these techniques to be effectively applied in biological tissue imaging. For more complex systems, combining both techniques could enhance the characterization of defect density and distribution profiles from the collected spectra.

References

1. Pierre, B., Laurent, C.: On some current challenges in high-resolution optical bioimaging. ACS Photonics **9**(8), 2538–2546 (2022)
2. Burns, D.H., Callis, J.B., Christian, G.D., Davidson, E.R.: Strategies for attaining superresolution using spectroscopic data as constraints. Appl. Opt. **24**(2), 154–161 (1985)
3. Nguyen, P.K., Riegler, J., Wu, J.C.: Stem cell imaging: from bench to bedside. Cell Stem Cell **14**, 431–444 (2014)
4. Owens, E.A., Henary, M., Fakhri, G.E., Choi, H.S.: Tissue-specific near-infrared fluorescence imaging. Acc. Chem. Res. **49**, 1731–1740 (2016)
5. Owens, E.A., Lee, S., Choi, J., Henary, M., Choi, H.S.: NIR fluorescent small molecules for intraoperative imaging. Wiley Interdiscip. Rev. Nanomed. Nanobiotech **7**, 828–838 (2015)

6. Braunmühl, T.: Hautarzt **66**(7), 499–503 (2015). https://doi.org/10.1007/s00105-015-3607-z
7. Fujimoto, J.G.: Optical coherence tomography for ultrahigh resolution in vivo imaging. Nat. Biotechnol. **21**, 1361–1367 (2003)
8. Drexler, W., Morgner, U., Kartner, F.X., Pitris, C., Boppart, S.A., Li, X.D., et al.: In vivo ultrahigh- resolution optical coherence tomography. Opt. Lett. **24**, 1221–1223 (1999)
9. Zhang, L., et al.: Defect imaging in carbon fiber composites by acoustic shearography. Compos. Sci. Technol. **223**, 109417 (2022)
10. Sampath, S.K., et al.: Non- contact measurements of residual stress distribution and grain size in titanium alloys with laser ultrasonic system. Int. J. Mech. Sci. **264**, 108809 (2024)
11. Zhang, L., et al.: In-situ real- time imaging of subsurface damage evolution in carbon fiber composites with shearography. Composites Communications **32**, 101170 (2022)
12. Cong, M., et al.: Biomedical application of terahertz imaging technology: a narrative review. Quant. Imaging Med. Surg. **13**(12), 8768–8786 (2023)
13. Rønne, C., Keiding, S.R.: Low frequency spectroscopy of liquid water using THz-time domain spectroscopy. J. Mol. Liq. **101**, 199–218 (2002)
14. Berry, E., et al.: Do in vivo terahertz imaging systems comply with safety guidelines? J. Laser Appl. **15**, 192–8 (2003)
15. Burford, N.M., El-Shenawee, M.O.: Review of terahertz photoconductive antenna technology. Opt. Eng. **56**, 010901 (2017)

Enhanced Protection Against Oxidative Stress in a Retinal Epithelium Model Through Pre-Absorption of Lutein

Siska Anjeliqa Binte Imran(✉)

National Junior College, 37 Hillcrest Rd, Singapore 288913, Singapore
siskaanjeliqa1480@gmail.com

Abstract. The Retinal Pigment Epithelium (RPE) supports retinal health and enables the maintenance of basic retinal functions. However, it is susceptible to oxidative damage induced by Reactive Oxidative Species (ROS). This increases its liability to degenerative retinal diseases as damaged RPE cells cannot be replaced spontaneously. Dietary carotenoids, like lutein, can protect the RPE from oxidative damage. However, the specific effectiveness of lutein against H_2O_2, a primary ROS in the retina, have not been explored. Hence this experiment determines if lutein pre-absorption enhances the protection of RPE cells against H_2O_2 by measuring cell viability. ARPE-19 cells were thawed, cultured, and split into wells before being treated with differing concentrations of lutein (2 µM, 1 µM, 0.2 µM, 0.1 µM, 0 µM) and exposed to H_2O_2 before undergoing the CCK-8 cell viability assay. The assay shows that there is no statistically significant difference in the average cell viability between 0 µM and 0.1 µM, and 0 µM and 0.5 µM lutein respectively. This suggests that lutein at small concentrations between 0.1 µM and 0.5 µM respectively are inadequate to act as a shield against H_2O_2. A statistically significant difference in the average cell viability between lutein concentrations higher than 1 µM and 0 µM, showing they can efficiently guard RPE cells against H_2O_2. However, increments in lutein concentrations do not always entail an increased protective effect. This could be attributed to errors while preparing for the experimentation.

Keywords: Oxidative Damage · Lutein · Retinal Pigment Epithelium

1 The Importance of Protecting the Retinal Epithelium (RPE) from Oxidative Damage

1.1 Anatomy and Functions of the Human Retina

The human retina has four essential parts - the fovea, macula, neurosensory retina, and retinal pigment epithelium (RPE). The macula lies in the centre, surrounded by major vascular vessels. It contains the smaller circular fovea where photoreceptor rods are tightly packed to sharpen central vision [1]. The neurosensory retina has three layers – the photoreceptor rods responsible for light sensitivity, the photoreceptor cones that

H. Guo et al. (Eds.): IRC-SET 2024, PTT, pp. 93–100, 2025.
https://doi.org/10.1007/978-981-96-3770-6_11

allow for colour detection and the ganglion cells that transfer visual signals to the optic nerve. Underlying the neurosensory retina is the closely associated RPE layer. This layer consists of polygonal RPE cells [2]. It assists with photoreceptor support and nourishment by transporting nutrients, ions, and oxygen from the choroidal blood vessel. The RPE also carries out the phagocytosis of shed photoreceptor outer segments to prevent retinal debris that impedes the clarity of the visual pathway. The RPE also directly contributes to the immune privilege of the eye [3]. Overall, RPE cells are crucial in regulating the regular visual functions and health of the retina [4] (Fig. 1).

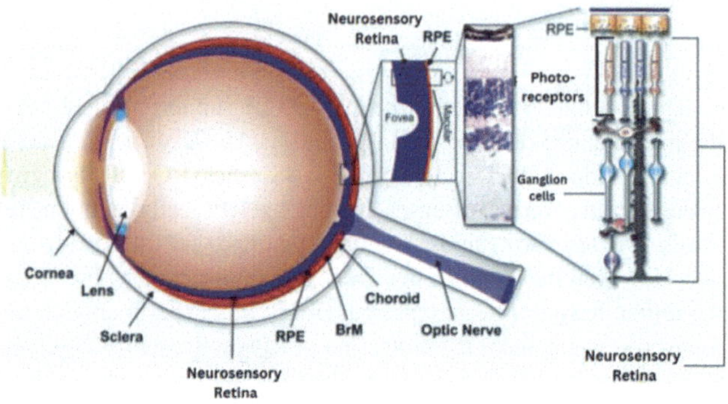

Fig. 1. Anatomy and cellular organisation of the human retina (Adapted from Hunt et al., 2018).

1.2 Oxidative Damage of the RPE and Its Impacts

The onset of oxidative stress is triggered by an accumulation of Reactive Oxidative Substances (ROS) which are formed in low amounts under normal enzymatic and oxidative processes. However, the retina has an especially ideal environment for the accumulation of ROS, thus a higher probability of incurring oxidative damage upon the RPE. Firstly, the retina is constantly subjected to high levels of oxygen consumption [5]. Secondly, the photoreceptors that are close to the RPE are rich in polyunsaturated fatty acids that can easily trigger lipid peroxidation [6, 7]. Thirdly, the regular phagocytosis of photoreceptors results in the release of an oxidative substance, H_2O_2 [8]. H_2O_2 in particular is particularly damaging to the mitochondria in human RPE cells [9]. While the RPE contains antioxidants that protect against oxidative stress, its susceptibility to oxidative stress multiplies with age [3, 10, 11].

Oxidative damage is thought to be the precursor to a variety of retinal degenerative diseases, including early-stage age-related macular degeneration (Early AMD), which is the leading cause of blindness of persons aged 50 and over globally [12, 13]. The dysfunction of RPE cells caused by such diseases results in the disruption of the regular visual functions and health of the retina [4]. For example, RPE degradation and oxidative damage disrupt the anti-inflammatory nature of the retinal microenvironment [3, 5]. The resulting inflammation triggers the dysregulation of several protective mechanisms of the retina, damaging the sensitive blood barrier retina which causes vision loss [14]. As damaged RPE cells cannot be replaced spontaneously, antioxidants are vital in protecting the viability of the RPE and preventing vision loss. [15].

1.3 Lutein as a Protective Substance for RPE Cells

Carotenoids are natural pigments with well-established antioxidant characteristics [16]. Among them is lutein, a major carotenoid found in the macula and prevalently available in eggs and dark green leafy vegetables [17]. Lutein can quench both singlet oxygen and lipid peroxy radicals that are usually present within the retina, assisting in maintaining the health of RPE cells and preventing oxidative stress and damage [18]. The functions of lutein as a supplementary dietary antioxidant for the retina to maintain and improve overall visual acuity have been explored in clinical trials. In one such trial, lutein intake resulted in an improvement in macular pigment optical density and visual contrast sensitivity in 90 AMD patients [19].

1.4 Experimental Aim

Prior works have scoped their research on the antioxidant properties of Lutein to specific cellular and molecular targets or enactors of oxidative stress, such as the one conducted by Yu et al. [20] Likewise, this research explores the protective effects of lutein specifically against H_2O_2, given its major role in damaging RPE cells.

In this experiment, ARPE-19 cells will be treated with differing concentrations of lutein before exposure to H_2O_2 to test if lutein pre-absorption can enhance the protection of RPE cells against H_2O_2 and its efficacy at differing concentrations. By using average cell viabilities as a proxy, the proportion of functional cells that have not undergone oxidative damage due to H_2O_2 can be estimated. It is hypothesised that when ARPE-19 cells are treated with a higher concentration of lutein, the average cell viability will be higher upon exposure to H_2O_2.

2 Materials and Methods

2.1 ARPE-19 Cell Thawing, Culturing and Passaging

ARPE-19 cells (CRL-2302, ATCC) were frozen in liquid nitrogen vapour. Two weeks before experimentation, they were thawed and cultured on a 75 ml T flask using 10 ml Fetal Bovine Serum (FBS) (ThermoScientific) as the culture media. The flask was stored in an incubator at 37 °C and 5% $CO2$. After 5 days, the cells were confirmed to be at least 80% confluent via microscope observation. The media was disposed of, and the flask was rinsed with 10 ml Phosphate Buffered Saline (PBS) (ThermoFisher Scientific). Once PBS was aspirated, 2 ml of Trypsin (ThermoFisher Scientific) was added and incubated at 37 °C for 5 minutes to detach the cells from the flask. Afterwards, FBS was added to make a cell suspension that was then transferred to a centrifuge tube that was spun at 125 xg for 5 minutes. The supernatant was aspirated before the cells were resuspended. The cell suspension was then seeded into two 96-well plates (~15 000 cells per well) with FBS before incubation again. After 6 days, the cells in the wells became confluent and deemed appropriate for bioactive compound testing.

2.2 Treatment of RPE Cells with Lutein and Exposure to Hydrogen Peroxide

2000 μM of lutein (Merck) solution was diluted using FBS into 2 ml solutions of 2 μM, 1 μM, 0.2 μM and 0.1 μM concentrations. Each cell plate was split into 6 groups with different concentrations of lutein (2 μM, 1 μM, 0.2 μM, 0.1 μM, 0 μM, 0 μM) that were added in 100 μl volumes. Any wells without cells were marked as "blank" and were not utilised for the experiment.

The plates were incubated for 2 days to allow for the absorption of lutein by the cells. Afterwards, 3% hydrogen peroxide (Merck) was diluted to 800 μM using FBS. 100 μl of the resultant solution was added to each group except for one 0 μM group (100 μl of FBS was added instead to make it a control group). The cells were exposed to H_2O_2 for 2 hours.

2.3 Cytotoxicity Assay

After the media used for the exposure of hydrogen peroxide was aspirated, 90 μl of FBS and 10 μl of WST-8 from a Cell Counting Kit-8 (CCK-8) (Dojindo) were added to each well. Following that, the plates were left to incubate for 2 hours to allow for the colourimetric reaction to take place. The absorbance values of the wells were obtained by a plate reader.

2.4 Statistical Analysis

The absorbance values were analysed in Excel to obtain the average cell viability (%) for each group. To determine if there was a statistically significant difference between the cell viabilities of each H_2O_2 treated group, a One Way ANNOVA and Tukey's multiple comparisons test were done in Prism GraphPad (Fig. 3.2) Additionally, to check if there is a statistically significant difference between the cell viabilities of the control and each H_2O_2, treated group, a Dunnet's multiple comparisons test was done (Fig. 2).

2.5 Risk Assessment & Safety Procedures

The Human ARPE-19 cell line is categorized as BSL1 (CRL-2302, ATCC Product sheet) and was only handled in a sterile Biological Containment Level 2 biohood in a BSL2 Lab. 3% Hydrogen Peroxide is deemed as corrosive and an irritant to the eyes and skin (88597, Sigma-Aldrich SDS). There are no clear toxicological effects of the CCK-8 Reagent stated in its SDS (CK04, Dojindo SDS). Any liquid, biological or chemical waste was aspirated and bleached before disposal in a sink. Disposable equipment was disposed of in their respective biological waste or sharps bins. PPE was always worn during experimentation and there was frequent disinfection of equipment before and after use.

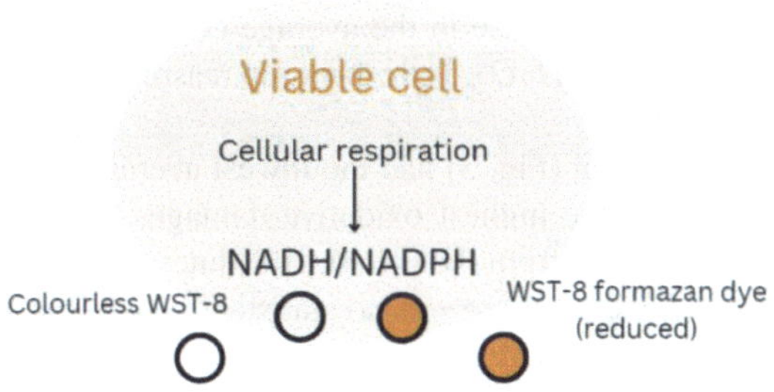

Fig. 2. The mechanism of the CCK-8 cell viability assay

3 Results and Discussion

The CCK-8 cell viability assay involves the reduction of colourless WST-8, which produces an orange WST-8 Formazan dye. Unlike other cell viability assays (i.e. MTT assay), the CCK-8 assay involves most of the dehydrogenase (NADH/NADPH produced from cellular respiration) in a cell which gives it enhanced sensitivity. It also has one of the shortest handling times across assay types [21]. The absorbance value obtained is dependent on the intensity of the WST-8 Formazan dye – a higher absorbance value indicates a higher cell viability.

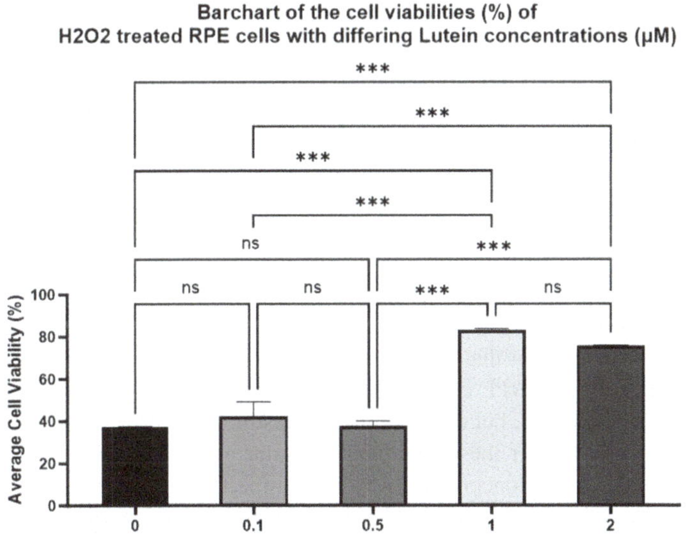

Fig. 3. **(Bottom)**: Bar graph showing the cell viabilities of H_2O_2, treated RPE cells in comparison with differing Lutein concentrations. (***p $<$ 0.001, no statistically significant difference (ns))

The average cell viability values in the figure above are normalised to the control (i.e. control's average cell viability is taken as 100%). The results of Dunnet's multiple comparisons test indicate that between each group of H_2O_2, treated RPE cells and the control, there is a significant decrease in the average cell viability. This confirms that the exposure of the RPE cells to H_2O_2, is the primary reason for cell cytotoxicity in the experiment, not other factors.

The group with 0 μM lutein (Fig. 3) had the lowest average cell viability, meaning that its cells had undergone the highest oxidative damage. While there was a visual increase in average cell viability from 0 μM to 0.1 μM lutein, there was a decrease from 0.1 μM to 2 μM lutein. However, there was no statistically significant difference in the average cell viability between 0 μM and 0.1 μM, 0 μM and 0.5 μM and, 0.1 μM and 0.5 μM lutein respectively. Given the already low average cell viability of the 0 μM lutein group, the results suggest that lutein at the concentrations between 0.1 μM and 0.5 μM respectively are too small to protect the RPE cells from oxidative damage. When comparing the average cell viabilities of 0 μM, 0.1 μM and 0.5 μM lutein respectively to 1 μM lutein. The latter had a significantly higher average cell viability. This suggests that the concentration of 1 μM is high enough to allow lutein to effectively protect the cells from oxidative damage and maintain their cell viability. Interestingly, when the concentration of lutein was increased from 1 μM to 2 μM, there was a visually notable, but not statistically significant, drop in the average cell viability. However, when comparing 2 μM lutein to the other lower concentrations, it still holds that the former has a significantly higher average cell viability. This means that 2 μM lutein can maintain but not increase the efficiency of the protection of RPE cells against H_2O_2, when compared to 1 μM lutein.

The drop in average cell viability from 1 μM to 2 μM lutein, where an increase was expected instead based on the comparison of 0.5 μM and 1 μM lutein, could be attributed to interference or errors during absorbance reading. The formation of bubbles in wells can easily affect the actual light absorbed by the plate reader to obtain the absorbance values. Additionally, before the CCK-8 assay, dark threadlike structures were observed in a few wells involved in the 2 μM group. These threads have presumably risen from accidental contamination during the process of changing the cell media during the cell splitting phase of the preparation and had interfered with the absorbance reading.

4 Conclusion, Future Works and Limitations

Overall, it has been proven that at high enough concentrations (> 1 μM), lutein can enhance the protection of RPE cells against H_2O_2,. Additionally, the increment in the concentration of lutein can, but does not always result, in an increase in the protective effects of lutein. To further these findings, a similar experiment can be run across a wider spectrum of lutein concentrations to determine how to maximise its protective effects and if there is a concentration where it induces cytotoxicity instead, although it is unlikely [21]. Studying the effects of combining lutein with other dietary antioxidants to evaluate if they can complement each other's functions can also help to maximise the protective benefits of lutein on RPE cells.

However, there are some limitations to this experiment. Firstly, the *in vitro* exposure time of RPE cells to H_2O_2, is short. This may not reflect the real-life *in vivo* damaging effects of H_2O_2, where RPE cells are exposed to it over long periods of time. Additionally, the oxidative damage can be exacerbated by other ROS *in vivo*. Secondly, as explored briefly before, there could have been small errors in the absorbance readings due to random errors and accidental contamination that could have negated statistically significant differences between the different concentration groups.

Additionally, it is acknowledged that solely relying on the CCK8 assay to quantify the overall cell viability of the RPE cells may not be sufficient. While the cells are to able respirate, there is a chance that the H_2O_2 had induced morphological changes that remain undetectable by the assay. This in turn affects their arrangement in the RPE monolayer required for their full functionality, especially *in vivo conditions*. As such, future works can be supplemented by immunofluorescence assays to observe any morphological changes and nuclear-staining assays to confirm cell apoptosis.

References

1. Hunt, N.C., Hallam, D., Chichagova, V., Steel, D.H., Lako, M.: The application of biomaterials to tissue engineering neural retina and retinal pigment epithelium. Advanced Healthcare Materials [Internet]. [cited 2023 Dec 7] **7**, 1800226 (2018). https://doi.org/10.1002/adhm.201800226
2. von der Emde, L., et al.: Histologic cell shape descriptors for the retinal pigment epithelium in age-related macular degeneration: a comparison to unaffected eyes. Translational Vision Science & Technology [Internet]. [cited 2023 Dec 8] **11**, 19–19 (2022). https://doi.org/10.1167/tvst.11.8.19
3. Taylor, A., Hsu, S., Ng, T.F.: The role of retinal pigment epithelial cells in regulation of macrophages/microglial cells in retinal immunobiology. Frontiers in Immunology 12 (2021)
4. Naylor, A., Hopkins, A., Hudson, N., Campbell, M.: Tight junctions of the outer blood retina barrier. Int. J. Molecular Sci. 21 (2020)
5. Masuda, T., Shimazawa, M., Hara, H.: Retinal diseases associated with oxidative stress and the effects of a free radical scavenger (Edaravone). In: Hrelia, S. (ed.) Oxidative Medicine and Cellular Longevity [Internet], pp. 9208489 (2017). https://doi.org/10.1155/2017/9208489
6. Nita, M., Grzybowski, A.: The role of the reactive oxygen species and oxidative stress in the pathomechanism of the age-related ocular diseases and other pathologies of the anterior and posterior eye segments in adults. In: Venditti, P. (ed.) Oxidative Medicine and Cellular Longevity [Internet] 3164734 (2016). https://doi.org/10.1155/2016/3164734
7. Jarrett, S.G., Boulton, M.E.: Consequences of oxidative stress in age-related macular degeneration. Molecular Aspects of Medicine [Internet] **33**, 399–417 (2012). https://www.sciencedirect.com/science/article/pii/S0098299712000386
8. Miceli, M.V., Liles, M.R., Newsome, D.A.: Evaluation of oxidative processes in human pigment epithelial cells associated with retinal outer segment phagocytosis. Experimental Cell Research [Internet] **214**, 242–9 (1994). https://www.sciencedirect.com/science/article/pii/S0014482784712547
9. Ballinger, S.W., Van Houten, B., Conklin, C.A., Jin, G.-F., Godley, B.F.: Hydrogen Peroxide Causes Significant Mitochondrial DNA Damage in Human RPE Cells. Experimental Eye Research [Internet] **68**, 765–72 (1999). https://www.sciencedirect.com/science/article/pii/S0014483598906613
10. Rohrer, B., Bandyopadhyay, M., Beeson, C.: Reduced Metabolic Capacity in Aged Primary Retinal Pigment Epithelium (RPE) is Correlated with Increased Susceptibility to Oxidative Stress. In: Bowes Rickman, C., LaVail, M.M., Anderson, R.E., Grimm, C., Hollyfield, J., Ash, J. (eds.) Retinal Degenerative Diseases, pp. 793–8. Springer International Publishing, Cham (2016)

11. Yang, S., Zhou, J., Li, D.: Functions and diseases of the retinal pigment epithelium. Frontiers in Pharmacology 12 (2021)
12. Cai, J., Nelson, K.C., Wu, M., Sternberg, P., Jones, D.P.: Oxidative damage and protection of the RPE. Progress in Retinal and Eye Research [Internet] **19**, 205–21 (2000). https://www.sciencedirect.com/science/article/pii/S1350946299000099
13. Steinmetz, J.D., et al.: Causes of blindness and vision impairment in 2020 and trends over 30 years, and prevalence of avoidable blindness in relation to VISION 2020: the Right to Sight: an analysis for the Global Burden of Disease Study. The Lancet Global Health [Internet]. [cited 2023 Dec 8] **9**, e144–60 (2021). https://doi.org/10.1016/S2214-109X(20)30489-7
14. Mohan, K.V., et al.: Immunological consequences of compromised ocular immune privilege accelerate retinal degeneration in retinitis pigmentosa. Orphanet Journal of Rare Diseases [Internet] **17**, 378 (2022). https://doi.org/10.1186/s13023-022-02528-x
15. Jin, Z.-B., et al.: Stemming retinal regeneration with pluripotent stem cells. Progress in Retinal and Eye Research [Internet] **69**, 38–56 (2019). https://www.sciencedirect.com/science/article/pii/S1350946217301234
16. Young, A.J., Lowe, G.L.: Carotenoids—Antioxidant Properties. Antioxidants 7 (2018)
17. Olaf Sommerburg, J.E.E.K., Bird, A.C., van Kuijk, F.J.G.M.: Fruits and vegetables that are sources for lutein and zeaxanthin: the macular pigment in human eyes. Br J Ophthalmol [Internet] **82**, 907 (1998). http://bjo.bmj.com/content/82/8/907.abstract
18. Böhm, F., Edge, R., Truscott, G.: Interactions of dietary carotenoids with activated (singlet) oxygen and free radicals: Potential effects for human health. Molecular Nutrition & Food Research [Internet]. [cited 2023 Dec 13] **56**, 205–16 (2012). https://doi.org/10.1002/mnfr.201100222
19. Richer, S., et al.: Doublemasked, placebo-controlled, randomized trial of lutein and antioxidant supplementation in the intervention of atrophic age-related macular degeneration: the Veterans LAST study (Lutein Antioxidant Supplementation Trial). Optomet. J. American Optomet. Ass. [Internet] **75**, 216–29 (2004). https://www.sciencedirect.com/science/article/pii/S1529183904700494
20. Yu, C.-C., Nandrot, E.F., Dun, Y., Finnemann, S.C.: Dietary antioxidants prevent agerelated retinal pigment epithelium actin damage and blindness in mice lacking $\alpha v\beta 5$ integrin. Free Radical Biology and Medicine [Internet] **52**, 660–70 (2012). https://www.sciencedirect.com/science/article/pii/S0891584911012081
21. Cai, L., Qin, X., Xu, Z., Yiyan, S., Jiang, H., Wu, Y., et al.: Comparison of cytotoxicity evaluation of anticancer drugs between real-time cell analysis and CCK-8 method. ACS Omega **4**, 12036–42 (2019)

Qubit Dynamics Simulation and Optimal Control

Linxi Liu[1(✉)] and Wenzu Zhang[2]

[1] National Junior College, 37 Hillcrest Road, Bukit Timah 288918, Singapore
linxiinsg@gmail.com

[2] Institute of High Performance Computing (IHPC), 1 Fusionopolis Way #16-16 Connexis (North Tower), Singapore 138632, Singapore

Abstract. Quantum simulation coupled with quantum optimal control enables one to design a control protocol, which manipulates a quantum system to achieve a targeted quantum operation. This is essential for the realization of quantum computers and related technologies such as quantum sensing. However, susceptibility to noise critically limits the performance and capabilities of noisy intermediate-scale quantum (NISQ) devices, as well as any future applications of quantum technologies. Effectively manipulating quantum hardware in the presence of imperfect devices and systems is thus a central challenge in realizing useful quantum computers (Ball, Harrison 2020). In addition, although much work is done experimentally to eliminate the unwanted quantum noise and its resultant interactions with the quantum system, the coupling between the system of interest and the external environment remains. By utilizing open-source framework: QuTiP (Quantum Toolbox in Python) (J. R. Johansson, P. D. Nation, and F. Nori 2013), Krotov python package (M. H. Goerz et al. 2019), we studied an open and closed quantum system and implemented optimised control field. An optimal control protocol to implement a quantum operation is developed in the study.

Keywords: Quantum Dynamics · Optimal Control

1 Introduction

1.1 Quantum Technology

In quantum technology, we utilize the quantum properties of a qubit, namely superposition and entanglement (Schleich, Wolfgang P 2016). A quantum bit (qubit) is a basic unit of quantum information, physically realized with a well-defined two-state quantum system, such as the two distinct spin states of an electron or atomic nuclei, charge states of quantum dots, atomic energy levels, vibrational states of groups of atoms, polarization states of photons and so on. A quantum state of a qubit is a superposition of two basis states, Quantum computers are typically built on qubits and take advantage of quantum features like superposition and entanglement to allow algorithms such as prime factorization to run much faster than a classical computing (Shor, Peter 1998). These technologies rely on the ability to accurately prepare an arbitrary quantum state (Koch, Christiane P

H. Guo et al. (Eds.): IRC-SET 2024, PTT, pp. 101–114, 2025.
https://doi.org/10.1007/978-981-96-3770-6_12

2022), such as all the states on the Bloch sphere shown in Fig. 1 for two-level systems. In addition, these technologies also rely on the act of performing quantum operations on the qubit, which allows for the usage of qubit in areas of quantum technology (Koch, Christiane P 2022). The motivation behind the development quantum technology is also due to quantum operations that can be applied to achieve new algorithms or enable one to conduct quantum programming and find ways to implement the properties of superposition. This also proves to be an advantage for optimisation problems in machine learning and areas in cyber security too. (Keplinger, Keegan 2018).

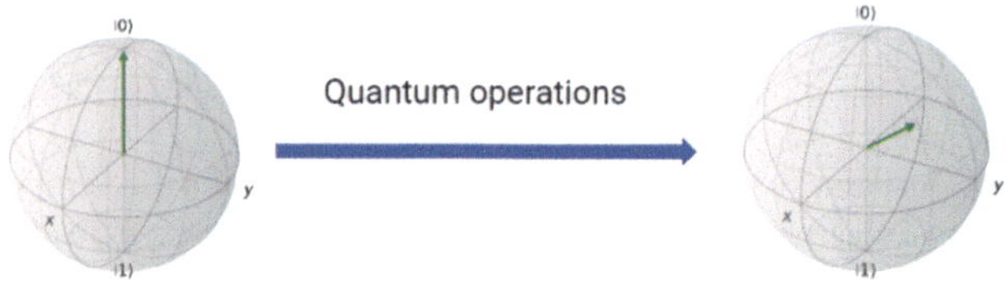

Fig. 1. Bloch sphere visualization of state-to-state transfer

1.2 Optimal Control of Quantum System

In quantum optimal control, we look to prepare some specific state and transfer from state-to-state through some unitary transformation (or gate) on a quantum system. Quantum optimal control theory (QOCT) refers to a set of methods to devise and implement external control fields that manipulate quantum dynamical processes at the atomic or molecular scale in the best way possible. (Koch, Christiane P 2022). For instance, in Nuclear Magnetic Resonance (NMR) for quantum information processing, it has been an established method to control the spin dynamics of a nuclei in a strong constant magnetic field with a weak oscillating magnetic field. This is usually done by varying the frequencies of the magnetic component of an interacting microwave, until the oscillation frequency of the magnetic field matches the intrinsic frequency of the nuclei, leading to nuclear magnetic resonance (Suter, Dieter, and T. S. Mahesh 2008). In search of optimal characteristics of our control fields, current numerical methods such as the gradient based optimization methods like the Krotov method (M. H. Goerz et al., Krotov 2019) and Gradient Ascent Pulse Engineering (GRAPE) can prepare individual target states by optimizing the parameters of our control field (Petruhanov, Vadim N 2023).

1.3 Noise in Quantum System

All quantum devices in the open quantum system will be subjected to noise coming from their environment. The effect of an environment on the closed system of interest induces stochastic transitions between energy levels and introduces uncertainty of quantum states of the system (J. R. Johansson, P. D. Nation, and F. Nori 2013). As examples, the interactions between thermal bath and quantum system or a magnetic field required to trap the system, all of which we shall classify as quantum noise. Interactions with the noise will cause the decoherence of the quantum system and the quantum state becomes mixed states where there is a classical probability associated with each quantum state.

1.4 Control Pulse Optimization

Control pulse optimization methods such as Krotov method use first order updates to guarantee a local maximum during optimization, these optimal control pulses which can generate accurate and precise quantum operations. While second order gradient methods like GRAPE could potentially generate higher fidelity quantum operations for state transitions, they need to calculate the second order hessian matrix which requires huge computational power and much needed time.

1.5 Objectives

In this research we attempt to numerically describe and simulate the closed and open quantum system using QuTiP while also implementing a python implementation of Krotov method to conduct a state-to-state transition in the closed quantum system as well.

1.6 Hypothesis

The hypothesis of this research will be that the quantum system simulated will fit in the real-world context of qubit dynamics and the Krotov method introduced to conduct optimized state to state transfer will yield a highly optimized result.

2 Methodology and Results

The main goal is for the dynamical system under study to operate optimally. To design a satisfactory optimal control field, first we need to simulate the quantum system as close as possible to the real physically realized quantum system. In general, there are two main groups of dynamical models: unitary and non-unitary. In unitary evolution, the state of the system remains in the pure state. In non-unitary, or dissipative, systems, it does not, instead in a mixed state. This leads to two kinds of quantum systems: open systems that is interacting with a larger environment and closed systems that do not (J. R. Johansson, P. D. Nation, and F. Nori 2013). Next, we attempt to add the time varying control fields in the open and closed systems to simulate the qubit dynamics. We then attempt to mathematically describe the optimisation process of Krotov method and conduct a state-to-state transition with Krotov Python package.

2.1 Closed Quantum System

The dynamics of the closed quantum system is governed by the Schrödinger equation, The time evolution of a closed quantum system's dynamics is governed by the partial differential Eq. (1). The Schrödinger equation governs the time evolution of a pure quantum state, and it does so by the Hamiltonian. Thus, once the Hamiltonian is known, the dynamics of the closed quantum system are in principle known. The Hamiltonian, H of a two-level quantum system is subjected to a static magnetic field B_0 and an external controlled magnetic field B_1 perpendicular to B_0, which rotates about B_0 with an angular velocity of ω. The static magnetic field is usually several magnitudes greater in strength and B_1 is timing varying. The Hamiltonian of each magnetic field can be described by H = μ · B, where μ is the spin, and $\text{H} = B_0\sigma_z + B_1\left((\sigma_x \cos \omega t + \theta_0) + \sigma_y \sin \omega t + \theta_0)\right)$ thus the final lab frame Hamiltonian $H_{total}(t)$ (2),

$$\frac{\partial}{\partial t}|\Psi(\text{t})\rangle = -\frac{\text{i}}{\hbar}H_{\text{total}}(t)|\Psi(\text{t})\rangle \tag{1}$$

$$H_{\text{total}}(t) = \frac{\omega_0}{2}\hbar\sigma_z + A\hbar\cos(\omega t + \theta_0)\sigma_x + A\hbar\sin(\omega t + \theta_0)\sigma_y \tag{2}$$

where ω_0 is the transitional angular frequency between the ground state and the excited state, and *A*/GHz the describing the strength of the magnetic field. Lastly σ_x, σ_y and σ_z are spin operators Pauli matrices around the x, y and z axis respectively.

In simulations, we set the following parameters as $\omega_0 = 2\pi f$, $where f = 1$ GHz, $\hbar = 1$, ω is the angular velocity and θ_0 is the phase difference of the field. Reasons for why plank constant is equals to 1 is due to it being the common factor in the Schrodinger equation (1).

2.1.1 Lab Frame to Rotating Frame

All that remains is to plug the lab form Hamiltonian Eq. (2) into the Schrödinger equation and solve for the system state as a function of time. However, before one does so, it must be noted that the Schrödinger equation must be transformed into the rotating frame from the laboratory frame. Such a form can allow us to easily evaluate the effects of resonance. This brings us from the Schrödinger picture into the interaction picture. Before we change the frame of reference, we look into a simplified version of the Schrödinger equation using the unitary evolution. This unitary evolution of a quantum state is defined by the unitary matrix, $U_{control}$ where $U_{control} = e^{iH_{total}(t)t}$. The unitary function can also be used to transition from one state to another (3).

$$U_{control}|\psi(0)\rangle = |\psi(t)\rangle \tag{3}$$

Before our simulation, we attempted to change our frame of reference from lab frame to rotating frame. The final Schrodinger equation governing the dynamics of the system is defined by Eq. (4.3).

$$U_{rotating\ frame\ transformation}|\psi(t)\rangle = |\varphi(t)\rangle \tag{4.1}$$

$$\overline{\overline{H}} = \frac{\omega_0 - \omega}{2}\sigma_z + A_1\sigma_X + A_2\sigma_y \tag{4.2}$$

$$U_{\text{control}}{}'|\varphi(0)\rangle = |\varphi(t)\rangle \tag{4.3}$$

$$\overline{\overline{H}} = A_1\sigma_X + A_2\sigma_y \tag{4.4}$$

Control fields are added into the system as $\epsilon(t)H_i$, $\epsilon_i(t)$ describes a set of controllable parameters such as the amplitude (strength), A/GHz of the magnetic field. While H_0

describes the system Hamiltonian such as the two-level quantum system of an atom. H_i typically represents the Pauli matrices spin operators to define the direction of our control fields.

$$\overline{\overline{H}} = H_0 + \sum_{i=1}^{\infty} \epsilon_i(t) H_i \tag{5}$$

$$\epsilon_1(t)H_1 = \mathrm{A}_1\sigma_x \tag{5.1}$$

$$\epsilon_2(t)H_2 = \mathrm{A}_2\sigma_y \tag{5.2}$$

$$\mathrm{A}_1(t) = Ag\cos(\theta_0) \tag{5.3}$$

$$\mathrm{A}_2(t) = Ag\sin(\theta_0) \tag{5.4}$$

Physically, the control fields might be the amplitudes of a RF pulse for the control of radio-frequency fields for nuclear magnetic resonance. *A/GHz* is the constant amplitude of the control field, g-factor is defined to be 2 and θ_0 to be $\frac{\pi}{4}$.

2.1.2 Time Evolution of the Closed Quantum System

The simulation of the closed quantum system is important for us to understand the dynamics of the quantum system due to corresponding changes in the angular velocity, ω of the magnetic field. Additionally, simulation allows us to find a better initial approximation for the control field before applying optimization methods, potentially cutting down on the iterations required to achieve an optimal control pulse. We too perform a projection operator to obtain the population of the initial state and target state and watch its behavior evolve through time. This will be important in understanding the long run pattern of a state-to-state transition.

Now that we have obtained the effective rotating frame Hamiltonian, we start by simulating the time evolution of the closed quantum system under influence of a control pulse. We define the initial state to be $|\Psi(0, 0)\rangle$ and the target state to be $\left|\Psi\left(\frac{\pi}{2}, \pi\right)\right\rangle$ and the time range of the simulation to be 0–20 ns with increment of 0.1 ns every instance.

The Hamiltonian used in simulation is defined by Eq. (4.4), where the amplitude of the control field, $\mathrm{A} = 0.2$, $g = 2$, $\theta_0 = \frac{\pi}{4}$ *and* $\omega_0 = 2\pi f$ *where* $f = 1$ GHz. Next, we proceed to input 3 different values of angular velocity, ω to account for the change in the dynamics of the evolution of a quantum state.

To evaluate the effectiveness of our control pulse, we need to generate the fidelity between the target state and final state generated by the dynamic equation. Fidelity quantifies the closeness between the target and final state with values from 0 to 1, The fidelity 1 represents the closest between the final state and target state.

First, we measure the fidelity without the presence of a control field and obtained 0.70711 as shown in Fig. 2(a). In the next two figures, we adjust the angular velocity of our control field, and measure the fidelity values above 0.70711 and take its mean. In Fig. 2(b), $\omega = 1$, $\omega = \omega_0$ where the special case of resonance would occur. In Fig. 2(c),

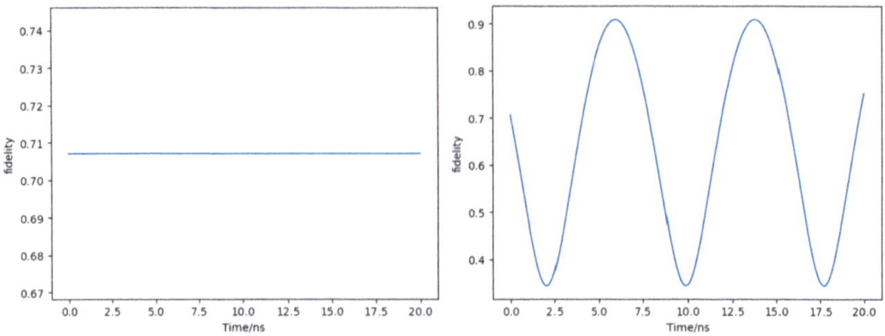

(a). Fideility at $A = 0$ on the left and (b). Fideility at $\omega = 1$ on the right

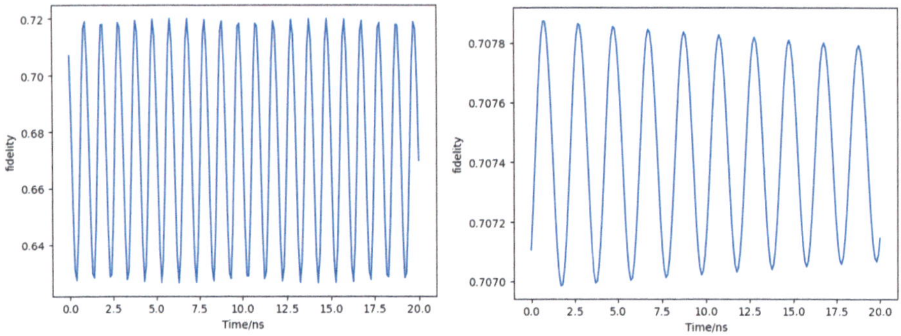

(c). Fideility at $\omega = 0.001$ on the left and (d). Fideility at $\omega = 100$ on the right

Fig. 2. (a) Fideility at $A = 0$ on the left (b) Fideility at $\omega = 1$ on the right (c) Fideility at $\omega = 0.001$ on the left and (d) Fideility at $\omega = 100$ on the right

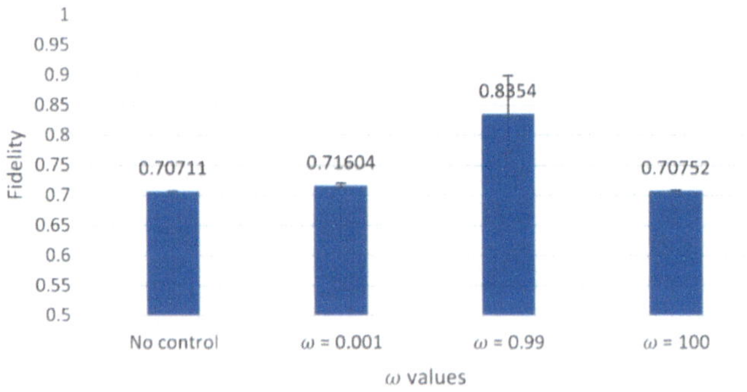

Fig. 3. Evaluation of fidelity associated with different ω values

we adjust $\omega = 0.001$, $\omega \ll \omega_0$ where the control field would be considered as a small and negligible perturbation. In Fig. 2(d), we attempt to adjust $\omega = 100$, $\omega \gg \omega_0$ (Fig. 3).

From the results presented in the above figures, the mean fidelity is 0.70711 when no control field is applied to the system. Figure 2(b) yields a mean fidelity of 0.8354, and a standard deviation of 0.06321, whilst Fig. 2(c) yields us a mean fidelity of 0.71604 and a standard deviation of 0.0037. Similarly, Fig. 2(d) yields us a mean fidelity of 0.70752 and a standard deviation of 0.00024, which does not prove useful as compared to Fig. 2(b) where the fidelity is highest. We can hypothesize that for ω far from ω_0, the effects on the quantum state are negligible. This allows us to make the conclusion that our selected angular velocity of the rotating magnetic field around the z-axis would be approximately close to the value of ω_0, where resonance would most likely occur.

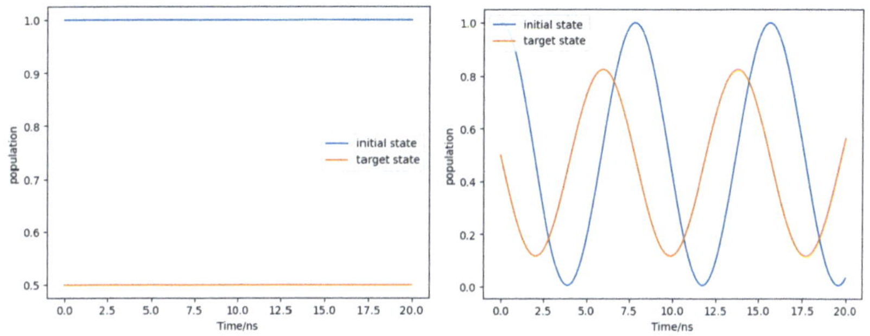

(a). Population of state at $A = 0$ on the left and (b). Population of state at $\omega = 1$ on the right

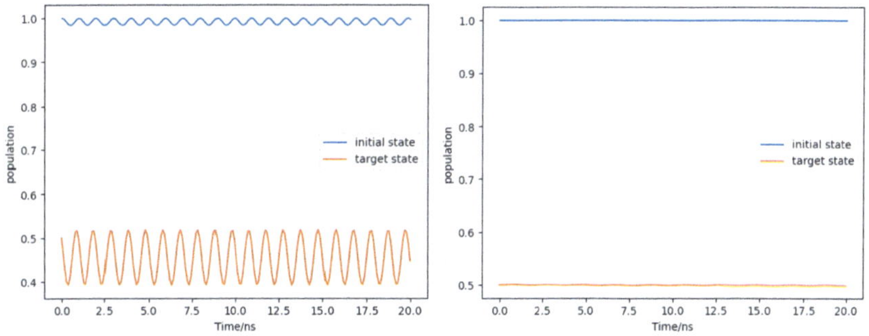

(c). Population of state at $\omega = 0.001$ on the left and (d). Population of state at $\omega = 100$ on the right

Fig. 4. (a) Population of state at $A = 0$ on the left (b) Population of state at $\omega = 1$ on the right (c) Population of state at $\omega = 0.001$ on the left and (d) Population of state at $\omega = 100$ on the right

Though the fidelity of a given quantum state can give us insights into the performance of the guess control field, we attempt to simulate the change in the population of target and desired state due to the guess control field to obtain a more comprehensive view of the effects of the control field. Figure 4(a), (b), (c), and (d) correspond to the ω values used to generate Figs. 2(a), (b), (c) and (d) respectively. We apply the projection operator of the target state and desired state on the quantum state generated by the dynamics equation to obtain the data. We notice that at certain time, t where the population of the final state reaches the population state of the initial state at time = 0, the fidelity is the highest at 1. The converse is true where we can observe that for population state of final state in Fig. 4(a), (c) and (d) furthest from the initial state's population at time = 0, the fidelity is lowest at 0.70711.

2.2 Open Quantum System

When we are modelling an open system, or an ensemble of systems, the use of the density matrix is mandatory. First, we define our Hamiltonian of the open system to be $H(t) = H_{total}(t) + H_{noise}(t)$, where $H_{noise}(t)$ is defined in the master equation captures the effect of the noise Hamiltonian through the Lindblad master Eq. (6). We adopt the framework of Lindblad Master Equation which governs the time-evolution in the open quantum system where the mixed states, consisting of states to go through the non-unitary evolution. We will be looking into using Lindblad Master Eq. (6). The Eq. (6) has four key assumptions:

I. Separability. *At* $\rho(0)$, where the system is not coupled with the environment quantum state.
II. Born approximation. The state of the quantum system does not vary significantly due to interactions with the environment. The Hamiltonian system and environment remain separable throughout evolution.
III. Markov approximation. The time scale of decay is much smaller than the time scale of evolution.
IV. Secular approximation. Fast rotating terms in the interaction picture are ignored.

$$\dot{\rho}(t) = -\frac{i}{\hbar}[H_{\text{total}}(t), \rho(t)] + \sum_{n} \frac{1}{2}\left[2C_n\rho(t)C_n^{\dagger} - \rho(t)C_n^{\dagger}C_n - C_n^{\dagger}C_n\rho(t)\right] \quad (6)$$

The above equation has two parts in its rights hand side. The first term represents the Liouville-von Neuman equation with $H'_{\text{total}}(t)$ being the previously defined "rotating frame" Hamiltonian in the close system and the latter being mainly the Lindblad super operator which is defined by a list of collapse operators or Lindblad operator C_n and its adjoints. The terms involving Lindblad operators in (6) represent the dissipative effects due to the interaction with the environment, and the non-unitary evolution of the system. The Collapse operators are thus used to model the collapse of the state vector that can occur due to noise. The dissipative environment is also defined by $C_n = \sqrt{\gamma}A_n$, where $A_n = \sigma_x, \sigma_y$ *and* σ_z is the spin operator which couples the system with the environment and γ is the dissipative rate. In robust control, we attempt to increase the fidelity between the non-unitary matrix and the identity operator as seen in Eq. (7). This is such that the resultant density matrix generated from the master Eq. (6) is not influenced by noise to a large extent.

2.2.1 Time Evolution of the Open Quantum System

In the open quantum system, dissipation occurs and could lead to the decoherence of a quantum state in the long run, we visualize and understand its effects through the simulation in QuTiP. We define the initial state to be $|\Psi(0, 0)\rangle$ and the target state to be $\left|\Psi\left(\frac{\pi}{2}, \frac{\pi}{2}\right)\right\rangle$ and the time range of the simulation to be 0–20 ns with increment of 0.1 ns every instance. The Hamiltonian used in simulation is defined by Eq. (4.4), where the amplitude of the control field, A = 0.2, $g = 2$, $\theta_0 = \frac{\pi}{4}$ *and* $\omega_0 = 2\pi f$, *where* $f =$ 1 GHz. Next, we proceed to input 3 different values of angular velocity, ω to account for the change in the dynamics of the evolution of a quantum state in the open system. Additionally, we define the dissipative rate to be γ = 1/24, and the Hamiltonian to be the effective "rotating frame" Hamiltonian in Eq. (4.4). Figure 5(a), (b), (c), and (d) correspond to the ω values used to generate Figs. 2(a), (b), (c) and (d) respectively under quantum noise.

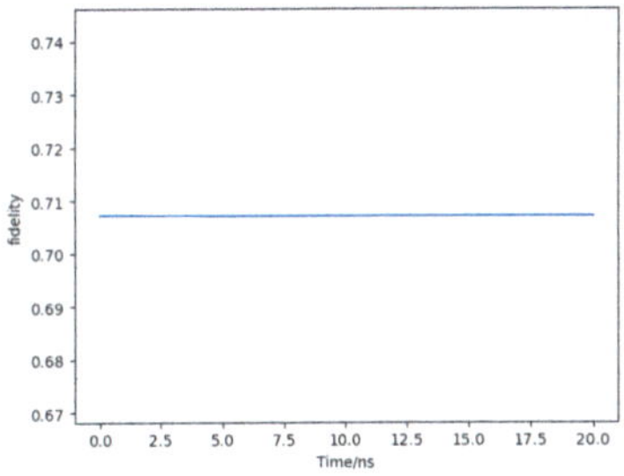

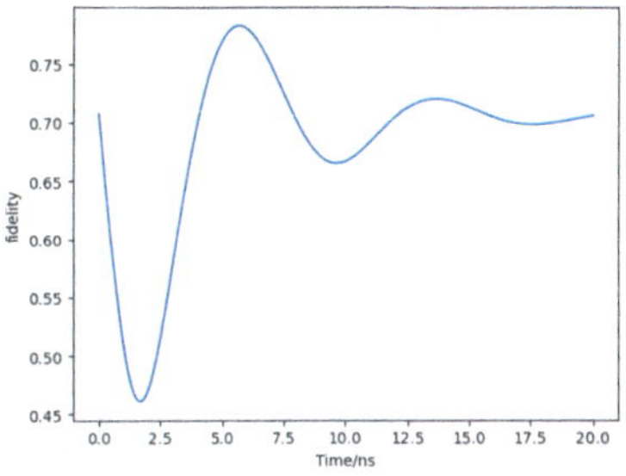

(a). Fidelity in open system, when $A = 0$ on the left and right

(b). Fidelity in open system, when $\omega = 1$ on the

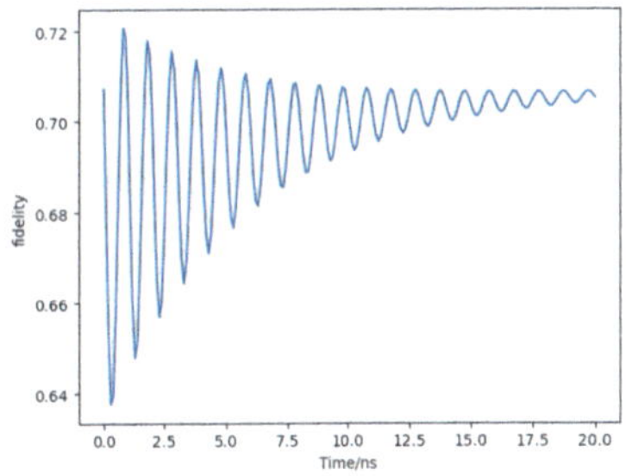

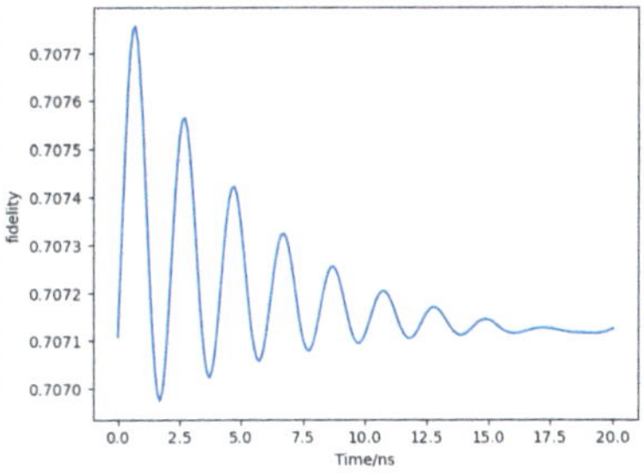

(c). Fidelity in open system, when $\omega = 0.001$ on the left and (d). Fidelity in open system, when $\omega = 100$ on the right

Fig. 5. (a) Fidelity in open system, when $A = 0$ on the left (b) Fidelity in open system, when ω = 1 on the right (c) Fidelity in open system, when ω = 0.001 on the left and (d) Fidelity in open system, when ω = 100 on the right

In the figures above, we can see the effects of quantum noise due to the Lindblad master equation. The effect is vividly displayed by the convergence of the fidelity towards 0.70711, as the quantum state collapses down towards the origin of the Bloch sphere.

Though the fidelity of a given quantum state can give us insights into the performance of the control field, we attempt to simulate the change in the population of target and desired state due to the control field to obtain a more comprehensive view of the effects of the control field. Figure 6(a), (b), and (c) correspond to the ω values used to generate

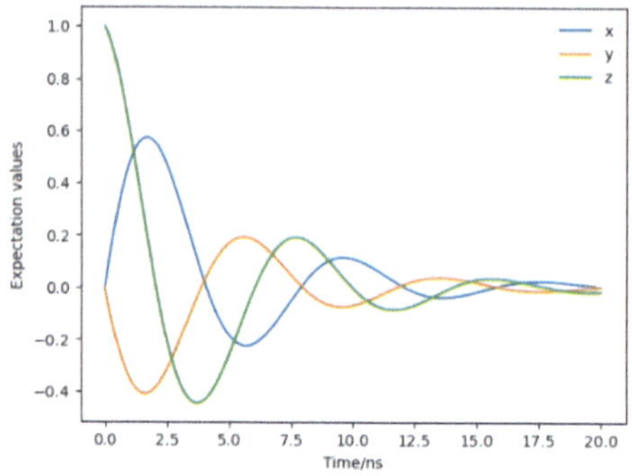

(a). Expectation values in open system, when $\omega = 1$

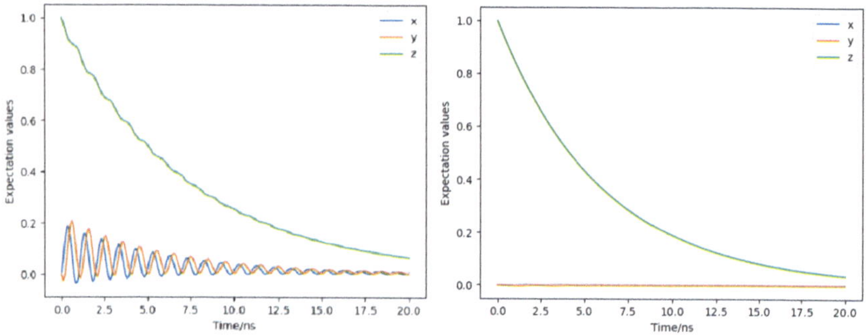

(b) on the left and (c) on the right

Fig. 6. (a) Expectation values in open system, when $\omega = 1$ (b) on the left and (c) on the right

Fig. 5(b), (c) and (d) respectively. We apply the projection operator of the target state and desired state on the quantum state generated by the dynamics to obtain the data. From the given data in Fig. 6(a), (b) and (c), we observe that the quantum noise has led the quantum state to converge towards the origin of the Bloch sphere, thereby effectively simulating the effects of quantum noise too.

2.3 Krotov Method

In general, Krotov method is a first order iterative update method to determine the local minimum of the optimization problem. However, it can only be effective to a certain extent in the context of closed quantum systems. Methods that introduce Quasi Newton Methods to do a second order iterative update has been proved useful in cases requiring higher fidelity (Eitan, Reuven & Mundt 2011).

2.3.1 Optimization Functional

An optimization functional is a key concept in optimization problems. In the context of optimization, a functional is a mathematical expression that depends on a function or a set of functions. The optimization functional is the quantity that either maximize or minimize during the optimization process. Mathematically, optimization ***functional*** J is the function or set of functions under consideration. The goal is to find the function or functions that either maximize or minimize in optimization problems, this functional represents a measure of the performance, or cost associated with the chosen function(s). The choice of whether to maximize or minimize the functional depends on the specific problem and the desired outcome.

2.3.2 Defining Functionals

Given that our closed quantum system evolves in discrete time steps. Fundamentally, we can rely on the minimization of a functional, J (7), which includes the final time functional J_T and running cost on the control fields, $g_a(\epsilon_i(t))$. The condition for minimizing is then for the first order derivative of functional J to reach as close to zero (Koch, Christiane P 2022).

$$J = J\_T + g_a(\epsilon_i(t)) \tag{7}$$

The final time functional J_T is defined by the infidelity between our target quantum state and final quantum state generated by the set on control pulses. Infidelity is a measure of closeness between two independent states in the Bloch sphere and is important in defining how effective our control pulse is in steering the quantum state to its desired state as seen in Fig. 7. The infidelity is minimized to reach the maximum effectiveness of our control pulses. Infidelity is defined in (8).

$$J\_T = \mathbf{1} - \left|\left\langle \psi_{\text{actual}} \middle| \psi_{\text{target}} \right\rangle\right|^2 \tag{8}$$

The running cost functional $g(\epsilon_i(t))$ is important as it serves to provide iterative control update for every set of control parameters, which we have arrived at (M. H. Goerz et al., Krotov 2019). The (inverse) Krotov step width λ can be used to determine the overall magnitude of $\Delta\epsilon(t)$. A larger λ would entail a smaller change in $\epsilon(t)$ in each iterative update and vice versa. Thus, a larger λ would lead to a slower convergence of the running cost functional however a smaller λ would lead to a faster convergence and may result in numerical instability if λ is too small. (t) is the update shape which helps make (9) well defined when t = 0 and t = t_{final} (M. H. Goerz et al., Krotov 2019).

$$g_a(\epsilon_i(t)) = \frac{\lambda}{S(t)}(\Delta\epsilon_i(t))^2 \tag{9}$$

$$\Delta\epsilon_i(t) \equiv \epsilon_i(t) - \epsilon_{i-1}(t) \tag{10}$$

Now that the optimizational functional J_T is defined, we utilize Krotov's iterative step algorithm in giving us iterative updates in control pulse. In the forward propagation step, the Schrödinger equation is solved to evolve the quantum state forward in time using the current control fields. In the backward propagation step, the adjoint equation (backward Schrödinger equation) (11) is solved to calculate the gradients of the objective function with respect to the control fields. (M. H. Goerz et al., Krotov 2019)

$$|\varphi(0)\rangle = U'^{\dagger}_{control}|\varphi(t)\rangle \tag{11}$$

Additionally, Krotov method makes sure that the condition in Eq. (12) is satisfied before iterating another time to achieve a lower infidelity (M. H. Goerz et al., Krotov 2019). This ensures that the optimization functional approaches the lowest fidelity and does not diverge from its path to convergence.

$$J\_T\{\epsilon_i(t), |\Psi_i\rangle\} < J\_T\{\epsilon_{i-1}(t), |\Psi_{i-1}\rangle\} \tag{12}$$

2.3.3 Implementation

Parameters:
See Table 1.

Table 1. Simulation parameters

Hamiltonian, H	$\overline{\overline{H}} = \frac{\omega_0 - \omega}{2}\sigma_z + Ag\cos(\theta_0)\sigma_X + Ag\sin(\theta_0)\sigma_y$, where $\hbar = 1, \omega_0 = 1$, $g = 2, \omega = 0.99$, A $= 0.2, \theta_0 = \frac{\pi}{4}$
Initial state	$\lvert\psi\rangle = \cos\frac{\theta}{2}\begin{pmatrix}1\\0\end{pmatrix} + e^{i\phi}\sin\frac{\theta}{2}\begin{pmatrix}0\\1\end{pmatrix}, \theta = 0, \phi = 0$
Target state	$\lvert\Psi\rangle = \cos\frac{\theta}{2}\begin{pmatrix}1\\0\end{pmatrix} + e^{i\phi}\sin\frac{\theta}{2}\begin{pmatrix}0\\1\end{pmatrix}, \theta = \frac{\pi}{2}, \phi = \pi$
Update shape S(t) for magnetic field in the x direction	**S(t) = 0.4 *cos*(0.25π)**
Update shape W(t) for magnetic field in the y direction	**S(t) = |0.4 *sin*(0.25π)|**
Lambda λ	**10**
Time steps, t	**Time 0–20, with 200steps in total, each step being 0.1 time step**

In Fig. 7, we can see that with each iteration of changing the control pulse amplitude, we achieve lower and lower final infidelity from the optimization protocol. This provides us with a control field that can bring us closer and closer to the final desired target state. This infidelity decreases at a decreasing rate as the value of the infidelity converges towards a value near 0. In Fig. 8(a) and (b), we can observe that there is a significant change in the amplitude of the control field. Obtaining 400 values of A_1 *and* A_2 for one iteration. The values of A_1 *and* A_2 is changing periodically, suggesting that the amplitude is time dependent and varying. In Fig. 8(c), throughout the 11 iterations of the optimized control pulse, we plot the population change of the target state and final state and find that the optimized control field can bring our population of the desired state near the population of the target state.

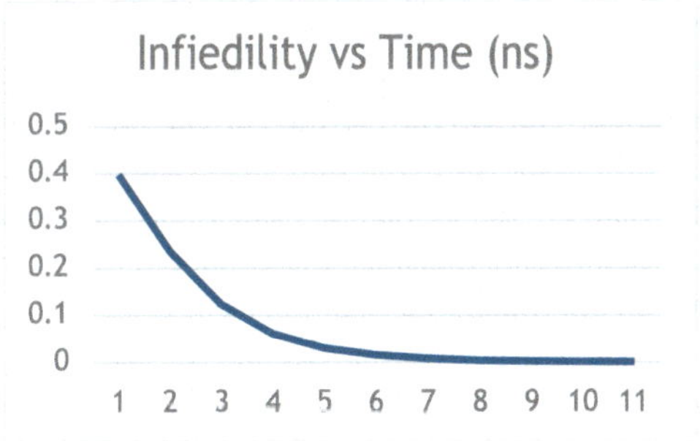

Fig. 7. Infidelity between target state and initial state

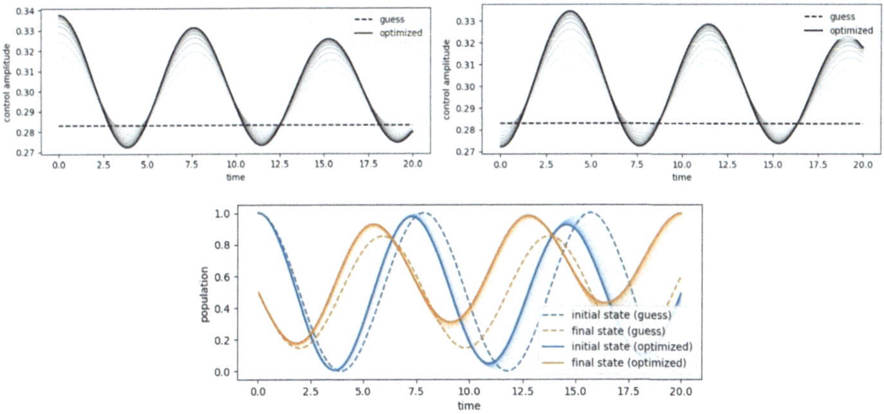

Fig. 8. (a) Optimized control pulse around the x-axis. (b) Optimized control pulse in the y-axis. (c) Population transfer of the initial state (Orange line) and the final state (Blue line). Results of (c) are generated by projecting the initial state and final state to state generated by the dynamic close quantum system evolution.

3 Discussion

The simulation of the closed and open two-level quantum system has been successful with the mathematical description of the system using quantum mechanics. The oscillatory motion displayed by the qubit dynamics is most likely since this arises from the wave-like nature of quantum particles. The fundamental concept behind this oscillatory behavior may be due to the solutions to the Schrödinger equation often involve oscillatory terms, such as complex exponentials where is the angular frequency. Why the oscillatory terms are used may be due to the wave-particle duality, a principle of quantum mechanics that suggests particles, such as electrons and photons, can exhibit both wave-like and particle-like properties. In an attempt to utilize the Krotov's method to reach optimized control pulse, what was unexpected is that there were only slight changes in amplitude as seen in Fig. 8(a) and (b), however upon changing the amplitude of the control pulse to a constant value of 0.4, we see that Krotov method immediately changes the amplitude to one that is oscillating. To enable slower convergence and better results for infidelity, we can decrease the step width by increasing the lambda value from 10 to higher values. If one wishes to enable a faster convergence, there is a risk involved such as numerical instability and may cause the values of the optimizational functional to diverge and cause a slower convergence in the long run. One may also realize that there is potential for reinforcement learning to be implemented due to fact that optimization in the quantum world introduces many factors and decisions to be made for convergence, In particular, reinforcement learning, an approach that maximizes a (long-term) reward by means of the actions taken by an agent in a given environment, can allow one for optimizing parameters and potentially reach a global minimal in the optimization problem.

4 Conclusion

In this research paper, we have successfully hypothesized and simulated the open and closed quantum system which will prove useful in context of optimal quantum control or for future experiments. In addition, we have managed to conduct control field optimization for state-to-state transfer on the Bloch sphere.

References

Ball, H., et al.: Software Tools for Quantum Control: Improving Quantum Computer Performance Through Noise and Error Suppression. arXiv.org, 9 July 2020. https://doi.org/10.48550/arXiv.2001.04060

Eitan, R., Mundt, M., Tannor, D.: Optimal control with accelerated convergence: combining the Krotov and quasi-Newton methods. Phys. Rev. A. **83** (2011). https://doi.org/10.1103/PhysRevA.83.053426

JR Johansson PD Nation F Nori 2013 QuTiP 2: a Python framework for the dynamics of open quantum systems Comput. Phys. Commun.. Phys. Commun. 184 1234 https://doi.org/10.1016/j.cpc.2012.11.019

Keplinger, K.: Is quantum computing becoming relevant to cyber-security?. Netw. Secur. (2018). https://doi.org/10.1016/S1353-4858(18)30090-4

Koch, C.P., et al.: Quantum optimal control in quantum technologies. Strategic report on current status, visions and goals for research in Europe. EPJ Quantum Technol., 20 July 2022. https://doi.org/10.1140/epjqt/s40507-022-00138-x

Goerz, M.H., et al.: Krotov: a Python implementation of Krotov's method for quantum optimal control. SciPost Phys. **7**, 080 (2019). https://doi.org/10.48550/arXiv.1902.11284

VN Petruhanov AN Pechen 2023 GRAPE optimization for open quantum systems with time-dependent decoherence rates driven by coherent and incoherent controls J. Phys. a: Math. Theor. 56 30 305303

Schleich, W.P., et al.: Quantum technology: from research to application. Appl. Phys. B, 27 April 2016. https://doi.org/10.1007/s00340-016-6353-8

Shor, P.W.: Quantum computing. Doc. Math. **1**(1000), 1 (1998). https://www2.math.upenn.edu/~ted/210S14/References/Shor.MAN.pdf

Suter, D., Mahesh, T.S.: Spins as qubits: quantum information processing by nuclear magnetic resonance, 5 February 2008. https://pubs.aip.org/aip/jcp/article-abstract/128/5/052206/775529/Spins-as-qubits-Quantum-information-processing-by

A Novel Algorithm to Improve Bounding Box Regression for Corner Casting Detection Using Zero-Shot Detection

Elven Kee(✉), Jun Jie Chong, Zi Jie Choong, and Michael Lau

Faculty of Science, Agriculture and Engineering, SIT Building at Nanyang Polytechnic Singapore, Newcastle University in Singapore, Singapore 567739, Singapore
E.Kee@newcastle.ac.uk

Abstract. Accurate detection of corner castings is pivotal for enhancing the performance of automated crane systems for containers, to streamline the loading and unloading operations of ships. Bounding box regression, a fundamental component of object detection, plays a critical role in precisely identifying and localizing these corner castings. In this study, we have developed an algorithmic framework aimed at integrating annotation outputs from both single-shot detection (SSD) and zero-shot detection (ZSD) methodologies. We introduce a novel technique for computing deviations in the spatial centrality of bounding boxes relative to their respective detected objects. Our approach enhances bounding box regression, resulting in significantly improved Best-Fit Bounding Box (BFBB). Our findings highlight substantial reductions in bounding box size ranging from 17.72% to 36.47%, and centreness accuracy improved, ranging from 7.49 pixels to 25.17 pixels. This demonstrates enhanced BFBB performance due to better generalization of the model with ZSD annotation, underscoring our contribution to advancing automated container port operations.

Keywords: zero-shot detection · single-shot detection · centreness · bounding box · SSD MobileNet

1 Introduction

The integration of intelligent automated crane systems in container port automation has become essential. Accurate detection of corner castings of containers is crucial for enhancing the performance of these systems, which in turn streamlines the process of loading and unloading ships. This highlights the increasing significance of precise and real-time corner casting detection. Despite advancements in generic object detection, there is still a noticeable gap in performance when it comes to detecting small objects compared to larger ones. In automated crane systems, accurately identifying corner casting holes, particularly from a top-down view as a spreader approaches, is a challenging task.

In conventional object detection scenarios [1], the common approach involves the localization of objects using horizontal Bounding Boxes (BB). The best-fit bounding

H. Guo et al. (Eds.): IRC-SET 2024, PTT, pp. 115–125, 2025.
https://doi.org/10.1007/978-981-96-3770-6_13

box (BFBB) is the smallest bounding box among all the possible orientations for the same shape [2]. Large-scale object detection datasets (e.g., MS-COCO) aim to define the ground truth bounding boxes as clearly as possible. However, ambiguities are still introduced when labeling the bounding boxes, such as inaccurate labeling, occlusion, and unclear object boundaries caused by occlusion. There are issues of localization misalignment, particularly when the classification score is high but the object is not entirely labeled.

We aim to develop an advanced algorithm that seamlessly integrates annotations from both single-shot detection (SSD) and zero-shot detection (ZSD) methods, enhancing the accuracy and completeness of object annotations in AI systems. Additionally, we propose methods to calculate deviations in bounding box size and centreness. These methods provide quantitative metrics to evaluate and improve the precision of object detection, ensuring more reliable spatial localization and size estimation in AI-driven vision applications.

This study builds upon our previous works and focuses on advanced corner casting detection, as illustrated in Fig. 1. A custom dataset is utilized to train an SSD model, which is then deployed on a Raspberry Pi version 5. The Raspberry Pi runs two key deep learning models: SSD MobileNet V2 FPN320-Lite and GroundingDINO. The SSD MobileNet V2 FPN320-Lite model handles object detection, specifically targeting corner castings. GroundingDINO refines the bounding boxes and the ZSD annotations produced by this process are fed back into the SSD model to derive the Best-Fit Bounding Boxes (BFBB). This integration of deep learning techniques, SSD and GroundingDINO, plays a crucial role in accurate object detection as the precise hole center coordinates ('centreness') is provided to the control system, enabling the activation of the spreader lifting mechanism.

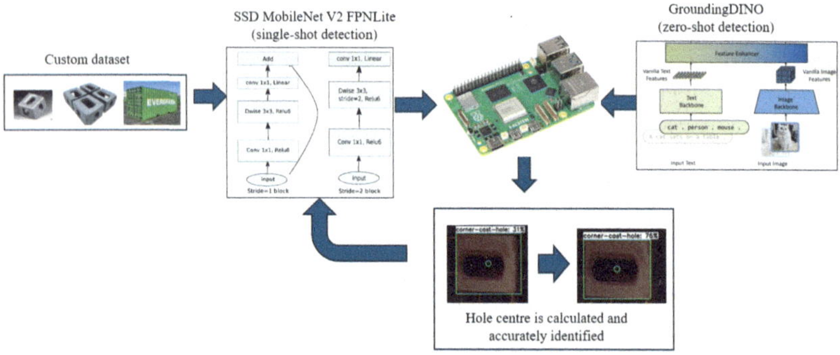

Fig. 1. Integration of ZSD and SSD model to produce best-fit bounding box

In the architecture of SSD MobileNetV2 FPN-Lite, MobileNetV2 is used as the base network, SSD as the detection network, and FPN-Lite as the feature extractor. GroundingDINO is a zero-shot detection model that extracts fundamental image and text features using dedicated backbones, then channels them into a feature enhancer network for cross-modality fusion, producing final bounding boxes based on text prompts. Our methodology aligns closely with that of Du et al. [3], who advanced low-light object detection through zero-shot adaptation. We address these challenges with minimal adjustments to their deep learning architectures, mirroring the approach adopted by Xiao et al. [4].

2 Related Works

There is continuous interest in addressing the performance gap in detecting objects in challenging settings, such as small sizes or low illumination. Sun et al. [5] proposed quantizing labels into a grid and formulated two-hot ground truth labels to improve the detection of small objects. Du et al. [3] introduced a zero-shot day-night method for illumination-invariant object detection, while Di Feng et al. [6] incorporated label uncertainty to enhance probabilistic object detection. Expanding on this concept, Wang et al. [7] developed a generative model to estimate label uncertainties and proposed the Jaccard IoU (JIoU) metric for evaluating probabilistic object detectors. He et al. [8] further contributed by proposing a bounding box regression loss to improve bounding box transformation and localization variance.

Building on the research of these authors, our contributions are as follows:

- Develop an algorithmic framework aimed at integrating the annotation outputs derived from both single-shot detection (SSD) and zero-shot detection (ZSD) methodologies.
- Propose a methodology to quantitatively assess the variation or deviation in the dimensions of bounding boxes generated during object detection task
- Introduce a novel technique for computing the deviation in the spatial centrality of bounding boxes relative to their respective detected objects.

3 Detailed Methodology: Zero-Shot Annotation for Corner Casting System

3.1 Control and Test Dataset

The project employs a custom dataset consisting of images sourced from the internet and online repositories. This dataset encompasses two distinct categories of workpieces: corner castings and containers, as depicted in Fig. 2. The corner castings and containers exhibit varying colors and sizes, thereby introducing complexity and variability into the dataset. Additionally, the dataset varies in terms of the quantity of corner castings and shipping containers depicted in each image, ranging from singular instances to multiple objects.

Fig. 2. Dataset of corner casting and container

Figure 3 illustrates the dataset and training methodology. The dataset consists of two identical subsets of 80 images each, augmented to a total of 560 images. The key difference lies in the labeling approach: the test subset includes 10 images labeled via zero-shot learning (ZSD), while the control subset contains manually labeled annotations for all images. We recognize that the ZSD labeling is automatic and not within user control, whereas the manual labeling for SSD is user-driven, which may lead to some deviations. To address this concern, we have ensured that the manual labeling process is conducted with strict adherence to guidelines and has undergone multiple rounds of review to minimize variation and ensure consistency.

The training process compared the performance of two labelling methods, Manual and ZSD over different epochs: 150, 225, and 375. The performance metrics observed included classification loss, localization loss, regularization loss, and total loss.

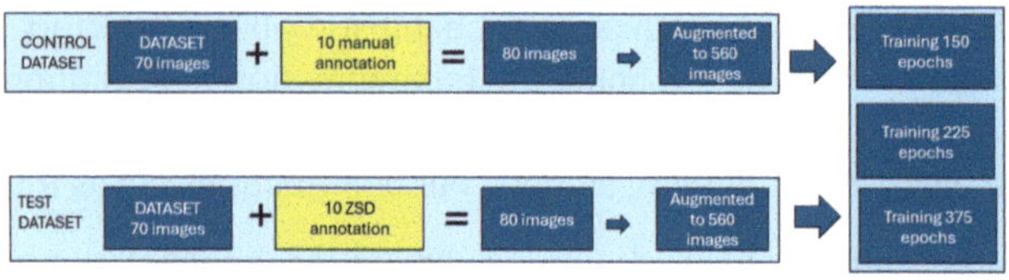

Fig. 3. Dataset used and training method

3.2 Preprocessing and Augmentation Techniques

The preprocessing techniques applied to the custom dataset include auto-orienting images based on metadata to ensure correct orientation and resizing them to fit within a 320 × 320 pixel resolution for the SSD model while maintaining the aspect ratio. For EfficientDet-Lite, images are resized to a fixed resolution of 512 × 512 pixels.

By augmenting both datasets to the same number of images and training them under identical conditions, any observed differences in model performance can be more confidently attributed to the annotation methods rather than dataset size or training regimen. Augmentation techniques involve flipping images horizontally and vertically, rotating them between -30° and $+ 30^\circ$, and applying horizontal and vertical shearing up to $\pm 30^\circ$. Additionally, the images undergo saturation adjustments between -30% and $+ 30\%$, brightness modifications within the same range, and the introduction of random noise affecting up to 3.03% of the pixels.

3.3 Training Process

Similar to our previous work on single-shot detection, we employ the following deep learning training process. We use a batch size of 32 for faster training, maintain the default learning rate of 0.08, utilize the ReLU activation function, and set the image size to 320 × 320 pixels. After training the model on Google Colab, it is converted to the ".tflite" format and deployed on a Raspberry Pi 5 for control and inference. For image

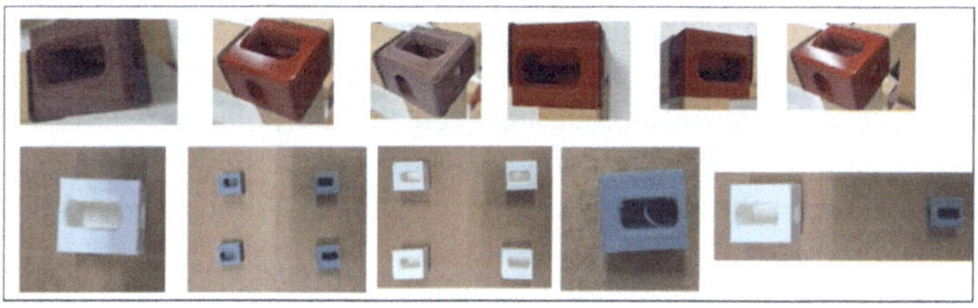

Fig. 4. Corner castings as ground truths

inference, we utilize 19 custom 3D-printed corner castings, as depicted in Fig. 4. The smallest corner casting in our setup is scaled to 1:40 of the actual size.

Figure 5 shows an example of the annotation in XML format, where xmin, xmax, ymin, and ymax represent the coordinates of the bounding box. The manual annotation is performed using Roboflow, while the ZSD annotation is generated by GroundingDINO.

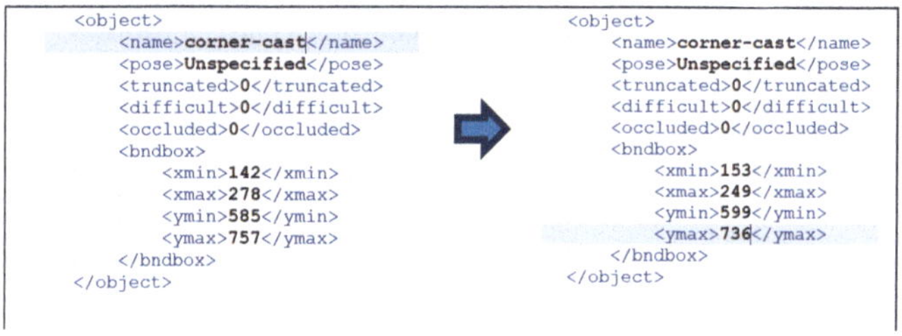

(a) (b)

Fig. 5. Dataset used and training method: (a) manual annotation; (b) ZSD annotation

3.4 Algorithm to Obtain Bounding Box Size and Centreness

As opposed to the conventional approach of using the corners of bounding box, we use the "centreness" approach [9], where centreness denotes the center of a given bounding box. The algorithm is structured as follows:

1) Obtain the ZSD annotation file using a text prompt.
2) Obtain the SSD annotation file using manual annotation.
3) Substitute the ZSD coordinate in SSD ".xml" file.
4) Train the model for 150, 225 and 375 epochs.
5) Select the number of epochs to provide the optimised mAP results with training loss factor.
6) Calculate the reduction of BB size (see Formula 1).
7) Calculate the deviation of the centreness (see Formula 2–4)

Here is the definition of the terms used in the formulas:

- Xsize_manual = the size of bounding box in x-axis with manual annotation
- Ysize_manual = the size of bounding box in y-axis with manual annotation

- Xsize_zsd = the size of bounding box in x-axis with ZSD annotation added
- Ysize_zsd = the size of bounding box in y-axis with ZSD annotation added
- xmid_dev_manual = the difference between middle point in x-axis with manual annotation and the actual middle point in x-axis
- ymid_dev_manual = the difference between middle point in y-axis with manual annotation and the actual middle point in y-axis
- xmid_dev_zsd = the difference between middle point in x-axis with ZSD annotation added and the actual middle point in x-axis
- ymid_dev_zsd = the difference between middle point in y-axis with ZSD annotation added and the actual middle point in y axis

To calculate the reduction in bounding box size due to ZSD labeling compared to manual labeling at each epoch, we compare the areas of the bounding boxes for each epoch.

$$\text{Reduction in BB size} = \frac{(\text{Xsize\_manual} * \text{Ysize\_manual}) - (\text{Xsize\_zsd} * \text{Ysize\_zsd})}{\text{Xsize\_manual} * \text{Ysize\_manual}} \times 100\% \tag{1}$$

The formulas to calculate the deviation of centreness are as follows:

$$\text{Deviation\_manual} = \sqrt{\text{xmid}_{\text{dev}_{\text{manual}}} * \text{ymid}_{\text{dev}_{\text{manual}}}} \tag{2}$$

$$\text{Deviation\_zsd} = \sqrt{\text{xmid}_{\text{dev}_{\text{zsd}}} * \text{ymid}_{\text{dev}_{\text{zsd}}}} \tag{3}$$

$$\text{Deviation\_centreness} = \text{Deviation\_manual} - \text{Deviation\_zsd} \tag{4}$$

4 Experimental Results

4.1 Training Loss

Table 1 compares the performance of models trained using control and test dataset across different epochs. For control dataset, the total loss decreases from 21.19% to 20.21% as the number of epochs increases from 150 to 375, with classification, localization, and regularization losses showing minor variations. With test dataset, shown in Table 2, the total loss significantly decreases from 25.74% to 18.78%, resulting in a 6.96% reduction compared to a 0.98% reduction for manual annotation. This shows that by incorporating ZSD annotation, the model learns to generalize more effectively and perform more accurate object detection.

4.2 Mean Average Precision (mAP)

For mAP for the control and test datasets, we use the widely used Intersection Over Union (IOU) 0.5:0.95 metric. As shown in Table 3, ZSD annotation at 375 epochs (32.96%) has the highest overall mAP compared to manual annotation and other ZSD configurations. The mAP of corner-cast-hole improved using ZSD annotation, achieving the highest mAP (33.33%) at 375 epochs. Figure 6 shows that while manual annotation performs well initially, the ZSD method's performance improves significantly with more training.

Table 1. Training loss of manual annotation (control dataset)

Training Loss	150 epochs (%)	225 epochs (%)	375 epochs (%)
Classification loss	4.99	5.45	5.18
Localisation loss	2.35	3.49	3.59
Regularization loss	13.83	13.02	11.43
Total loss	21.19	21.96	20.21

Table 2. Training loss with ZSD annotation added (test dataset)

Training Loss	150 epochs (%)	225 epochs (%)	375 epochs (%)
Classification loss	6.16	4.13	4.83
Localisation loss	5.66	2.94	2.49
Regularization loss	13.91	13.04	11.46
Total loss	25.74	20.11	18.78

Table 3. Mean Average Precision with IOU 0.5:0.95

Mean Average Precision (%)	Manual 150 epochs	Manual 225 epochs	Manual 375 epochs	ZSD 150 epochs	ZSD 225 epochs	ZSD 375 epochs
Container	31.21	27.91	29.95	27.45	26.48	32.59
Corner-cast-hole	25.71	34.88	35	31.24	30.89	33.33
Overall mAP	28.46	31.39	32.47	28.24	28.69	32.96

4.3 Overall Improvement of Bounding Box and Centreness

Figure 7 illustrates real-time detection results using SSD MobileNet FPNLite. We observe that the bounding box accuracy improves significantly after incorporating ZSD annotations, with a particularly tighter bounding box evident at 150 epochs. The centreness accuracy also shows marked improvement with ZSD at 375 epochs, where the green circle previously offset to the right, is more accurately centered. This consistent enhancement is evident throughout the training process, at 150, 225, and 375 epochs.

4.4 Deviation of Bounding Box Size

Table 4 presents a comparison of the ground-truth bounding box (230 $\times$ 200 pixels) with the results obtained from the SSD model trained for 150, 225, and 375 epochs. All units are in pixels. The ground-truth bounding box has an area of 46,000 pixels2. The results show a significant improvement after adding ZSD annotation.

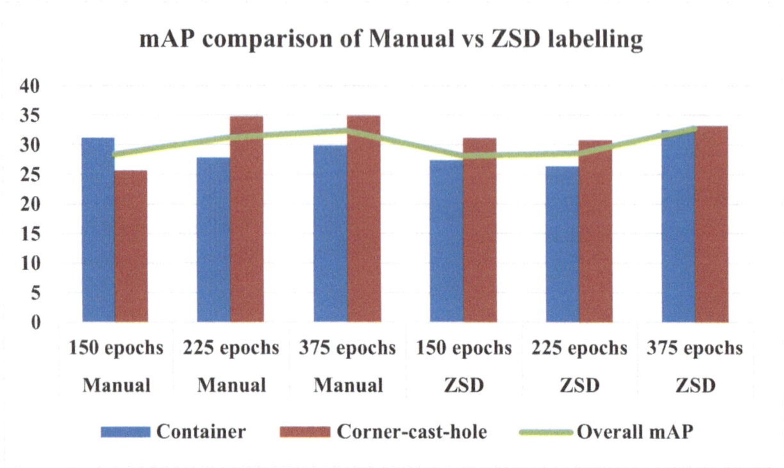

Fig. 6. Comparison of Manual and ZSD annotation using IOU 0.5:0.95

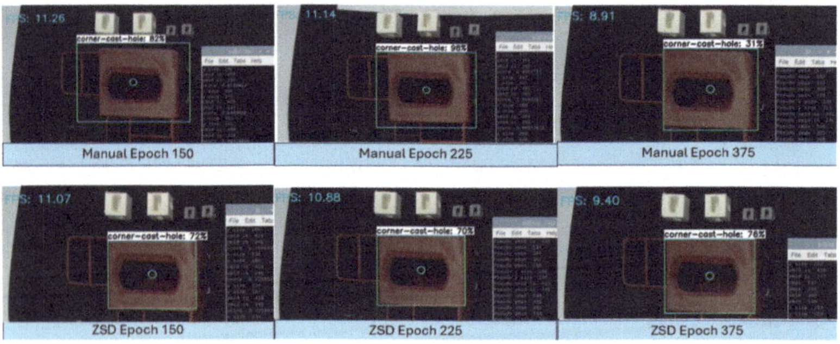

Fig. 7. Real-time detections results

Table 4. Size of Bounding Box

Specification	Manual label 150 epochs	Manual label 225 epochs	Manual label 325 epochs	ZSD label 150 epochs	ZSD label 225 epochs	ZSD label 325 epochs
X size	335	288	249	234	236	233
Y size	243	228	232	185	197	204
Area	81405	65664	57768	51714	46492	47532

Table 5 shows that ZSD labeling significantly reduces the bounding box size compared to manual labeling at all epochs. The percentage reduction is substantial, ranging from 17.72% to 36.47%. This demonstrates that ZSD labeling contributes to better model generalization and efficiency by reducing the bounding box size.

Table 5. Reduction of Bounding Box to ZSD labelling

Specification	150 epochs	225 epochs	325 epochs
Reduction of BB Area (pixel^2)	29691	19172	47532
Reduction (%)	36.47%	29.19%	17.72%

4.5 Deviation of Bounding Box Size

Table 6 presents the X and Y coordinates of the midpoints for corner casting, as predicted by the SSD model at different epochs (150, 225, 325) under two different labeling methods: manual labeling and ZSD labeling. Table 7 shows that the deviations in both X and Y midpoints and the overall centreness deviation are significantly lower for the ZSD label compared to the manual label across all epochs.

Table 6. Midpoint of corner casting

Specification	Manual label 150 epochs	Manual label 225 epochs	Manual label 325 epochs	ZSD label 150 epochs	ZSD label 225 epochs	ZSD label 325 epochs
X midpoint	407	461	432	419	419	416
Y midpoint	237	268	240	246	246	247

Table 7. Deviation of Centreness

Specification	Manual label 150 epochs	Manual label 225 epochs	Manual label 325 epochs	ZSD label 150 epochs	ZSD label 225 epochs	ZSD label 325 epochs
X midpoint	13	41	12	1	1	4
Y midpoint	13	18	10	4	4	3
Centreness	13	27.17	10.95	2	2	3.46

Table 8 highlights the deviation of centreness in pixels for different epochs, comparing ZSD labelling with manual labeling. After adding the ZSD annotation, the deviation of centreness has substantial reduction, ranging from 7.49 pixels to 25.17 pixels. This reduction indicates that by incorporating ZSD annotation, the SSD model shows improved precision in locating the centre of the objects.

5 Validation with Other SSD Models

To validate the effectiveness of our approach, we conduct a comparison of mean Average Precision (mAP) and centreness using SSD MobileNet V2 FPNLite against other SSD architectures, namely MobileNet-V2 and EfficientDet-Lite0, based on results from the

Table 8. Reduction of Centreness deviation

Specification	150 epochs	225 epochs	325 epochs
Centreness deviation	11	25.17	7.49

test dataset at 150 epochs. As shown in Table 9, SSD MobileNet V2 FPNLite outperforms EfficientDet-Lite0 by 8.85% and MobileNet-V2 by 0.26% in terms of mAP. Additionally, our mAP is 6.04% higher than the COCO2017 validation dataset's 22.2%.

Table 9. Comparison of Mean Average Precision (150 epochs)

mAP	SSD MobileNet FPNLite (%)	MobileNet (%)	EfficientDet-Lite0 (%)
Container	27.45	27.87	27.83
Corner-cast-hole	31.24	28.10	10.95
Overall mAP	28.24	27.98	19.39
Comparison	-	+ 0.26	+ 8.85

As shown in Table 10, the SSD MobileNet V2 FPNLite model outperforms both MobileNet and EfficientDet-Lite0 in centreness, exhibiting the lowest deviation in centreness scores and achieving more accurate predictions of object centering within bounding boxes.

Table 10. Deviation of centreness (150 epochs)

Specification	Ground truth (pixel)	SSD MobileNet V2 FPNLite (pixel)	MobileNet (pixel)	EfficientDet-Lite0 (pixel)
X midpoint	420	419	444	431
Y midpoint	250	246	243	234
Deviation of centreness	-	2	12.96	13.27

6 Conclusion

In this paper, we have developed an advanced algorithm that seamlessly integrates annotations from both single-shot detection (SSD) and zero-shot detection (ZSD) methods. We used the "centreness" approach and a detailed, comprehensive structured algorithm to calculate the best-fit bounding box and midpoint deviation.

We reduced bounding box size by 17.72% to 36.47% and improved centreness accuracy by up to 25.17 pixels, demonstrating better best-fit bounding box performance. Incorporating ZSD annotation allowed the SSD Model to generalize more effectively across classes, leading to a much lower training loss. A model trained with ZSD annotation saw a 6.96% total loss reduction over 375 epochs, compared to just 0.98% for the control model with manual annotation.

We validated our SSD MobileNet V2 FPNLite model against other SSD models and found it significantly outperforms them, surpassing the baseline efficientDet-lite0 by 8.85% and MobileNet V2 by 0.26%. This research advances container port automation by improving spatial localization and size estimation in AI-driven vision applications.

References

1. Zhao, Z.-Q., Zheng, P., Shou-tao, X., Xindong, W.: Object detection with deep learning: a review. IEEE Trans. Neural Networks Learn. Syst. **30**(11), 3212–3232 (2019)
2. Yuan, B., et al.: Finding the Best-Fit Bounding-Boxes. In: Bunke, H., Spitz, A.L. (eds.) Document Analysis Systems VII. Lecture Notes in Computer Science, vol 3872. Springer, Berlin, Heidelberg (2006). https://doi.org/10.1007/11669487_24
3. Du, Z., Shi, M., Deng, J.: Boosting object detection with zero-shot day-night domain adaptation. Proceedings of the IEEE/CVF Conference on Computer Vision and Pattern Recognition (2024)
4. Xiao, Z., et al.: Theoretically achieving continuous representation of oriented bounding boxes. Proceedings of the IEEE/CVF Conference on Computer Vision and Pattern Recognition (2024)
5. Sun, H., et al.: Confidence-driven Bounding Box Localization for Small Object Detection. arXiv preprint arXiv:2303.01803 (2023)
6. Feng, D., et al.: Labels are not perfect: improving probabilistic object detection via label uncertainty. arXiv preprint arXiv:2008.04168 (2020)
7. Wang, Z., et al.: Inferring spatial uncertainty in object detection. 2020 IEEE/RSJ International Conference on Intelligent Robots and Systems (IROS). IEEE (2020)
8. He, Y., et al.: Bounding box regression with uncertainty for accurate object detection. Proceedings of the IEEE/CVF conference on computer vision and pattern recognition (2019)
9. Filipiak, D., et al.: Polite teacher: semi-supervised instance segmentation with mutual learning and pseudo-label thresholding. IEEE Access (2024)

Investigating the Use of Mushrooms to Make Antimicrobial Hydrogels for Plasters

Calix Zheng Xuan Chua(✉), Charis Qi En Lim, and Fellicia Mei Hui Tan

National Junior College, Singapore, Singapore
2007cczx@gmail.com

Abstract. It has been found that mushrooms contain phenolic compounds, aldehydes and ketones that render them antimicrobial properties. Its antimicrobial activity can be exploited to make antimicrobial plasters, which can kill bacteria on wound surfaces and block bacteria from entering wounds and causing infections. Mushroom extracts were made from *Pleurotus Ostreatus*, *Lentinula Edodes*, *Agaricus Bisporus* and brown and white *Hypsizygus Tessulatus* in ethanol with the ratio of 18 g: 50 ml and was evaporated and stored in 33% ethanol. Their antimicrobial activities were tested using the disk diffusion method, and we narrowed down the most effective mushroom. Next, we tested the different parts and dryness of the selected mushroom before incorporating the best extract into hydrogels. We further tested the hydrogels' absorption, breathability and ability to kill and block bacteria. The ethyl acetate extract of fresh whole Brown Hypsizygus *Tessulatus* displayed the best antimicrobial properties, while Brown *Hypsizygus Tessulatus* showed the best results overall in both blocking and killing bacteria–fresh whole Brown *Hypsizygus Tessulatus* extract was able to block 100% of bacteria while dried cap Brown *Hypsizygus Tessulatus* extract killed 96.1% of bacteria. The hydrogel made with extracts was also able to absorb water and allowed oxygen to pass through.

Keywords: Mushrooms · Antimicrobial · Hydrogel

1 Introduction

1.1 The Skin and Wounds

The skin is the largest organ in the human body. There are millions of bacteria on the human skin, the most common ones being *Staphylococcus Epidermidis* and *Staphylococcus Aureus* [1]. When a person has a wound, bacteria can enter the body through the wound and cause an infection. There have been developments in recent years in which antimicrobial plasters have been produced, and these plasters can kill microbes on the skin that can cause an infection and prolong the wound healing process. It is also found that at the beginning of a chronic wound formation, Gram-positive bacteria, especially *Staphylococcus aureus*, are dominant. Gram-negative species, such as *Escherichia coli* (E. coli) and *Pseudomonas*, are detected and tend to invade deeper layers of skin causing significant tissue damage [2]. If these infections do not get treated, they can cause life threatening issues [3].

H. Guo et al. (Eds.): IRC-SET 2024, PTT, pp. 126–138, 2025.
https://doi.org/10.1007/978-981-96-3770-6_14

1.2 Antimicrobial Plasters

To prevent wound infections, antimicrobial plasters have been used to kill the bacteria that is around wounds and block bacteria from entering the wound. They are commonly made of benzalkonium chloride, studies have also been conducted to find out if silver nanoparticles can be used as an alternative [4]. Used in other antibacterial products, benzalkonium chloride is a common disinfectant. However, it has also been found that benzalkonium chloride is an irritant, and can even cause allergic contact dermatitis, which is not ideal [5, 6]. Benzalkonium chloride has also been found to be possibly lethal to aquatic animals, and may even invoke antimicrobial resistance [7], further reinforcing the fact that it is not ideal to use benzalkonium chloride in antimicrobial plasters. Furthermore, silver nanoparticles are also very expensive to procure and produce, which is not ideal for making plasters which would be frequently replaced.

1.3 Mushrooms

Mushrooms have been playing a big role in the medical field due to their rich amounts of chemical compounds such as polysaccharides, terpenes, proteins and steroids that help to boost the human immunity system and offer many health benefits [8]. In recent studies, mushrooms have been found to possess antimicrobial properties [9–12]. Mushrooms contain phenolic compounds such as flavonoids, which are effective against many bacteria such as Staphylococcus and E. coli [13]. Aldehydes and ketones that were found in mushrooms are also known to prevent the growth of microbial organisms by disrupting electron transport chains, interfering with biological processes that involve them, and reacting with proteins or nucleic acids due to their electronegative nature [7]. Due to its antimicrobial nature, it is possible that mushroom extracts can be used to make antimicrobial plasters to prevent wound infections. This experiment is hence conducted to find out if mushrooms can be used as an alternative to the commercially available antimicrobial plasters, as mushrooms are relatively low cost and can even be obtained from such that are rejected from supermarkets and farms, therefore also reducing wastage. Mushrooms have very few allergens, and most of the allergens are in the spores [14]. Therefore, if handled correctly, plasters made of mushroom extracts will not cause a lot of allergies, which makes it a potential replacement for benzalkonium chloride. The environmental impact of using mushrooms can then be measured in the future, to further evaluate if mushrooms can be a viable option to produce antimicrobial plasters.

1.4 Hydrogels

An ideal plaster has enough pores for oxygen to pass through to aid the healing process, one that can absorb any wound drainage to prevent any sogginess, and one that has antimicrobial properties to kill any bacteria on the wound whilst also blocking bacteria from entering the wound. With all this, price and environmental friendliness is also another factor to take note of. We would want our plaster to be low cost, easily made, and environmentally friendly.

Scientists have performed extensive research focusing on the design and methods to synthesize wound dressing materials. In recent years, hydrogels have been identified as a potential wound dressing material [15]. Hydrogel wound dressings are three-dimensional polymer network structures with the ability to absorb large amounts of water in their networks while retaining their structure [16]. Due to their properties such as oxygen permeability, ability to absorb and prevent loss of bodily fluids, similarity to the natural extracellular matrix, and ability to prevent wound infections, hydrogels have proven to be a promising wound dressing material [17, 18]. Wound dressings should prevent infections by blocking bacterial infections or delivering antimicrobial agents to protect wounds from infection. Hydrogels can prevent bacterial infections as the superposition of porous layers blocks bacteria from reaching the wound surface through the hydrogel and the porous structure provides a more comfortable environment for the wound [19]. Hydrogels can load and release the antimicrobial agents in a controlled manner. Sustaining the release of these agents can help to prevent the colonization of bacteria at the wound site, protecting the wound from infections [20].

Studies show that it is possible for antimicrobial hydrogels to be made with certain molecules like silver nanoparticles or gold nanoparticles [21–23], but there has not been evidence of mushroom extracts displaying antimicrobial properties when incorporated into hydrogels. Therefore, this experiment is aimed at finding if the antimicrobial properties will remain in hydrogels, and to find the mushroom with the highest antimicrobial activity.

2 Materials and Methods

2.1 Preparation of Mushroom Extracts

Fresh mushrooms were procured, and dried mushrooms were prepared by drying the fresh mushrooms at 50 °C in an oven for 48 h, and then ground into a powder. Mushroom extracts were prepared by adding 18 g of whole fresh and dried *Pleurotus Ostreatus*, *Agaricus Bisporus*, *Grifola Frondosa*, *Hypsizygus Tessulatus* into 50 ml of ethyl acetate and ethanol separately. The mushrooms were then extracted by sonication for 1 h at room temperature and then filtered. The mushroom extract was then dried in a rotary evaporator and dissolved in 1 ml of ethanol and 2 ml of water. Ethanol and ethyl acetate were used as the solvents for extraction, as it has been found that the compounds contributing to the antimicrobial properties of mushrooms were phenols, aldehydes and ketones, thus polar solvents were used and no other non-polar organic solvents. Fresh and edible mushrooms (*Pleurotus Ostreatus, Agaricus Bisporus, Grifola Frondosa, Hypsizygus Tessulatus*) were procured, and dried mushrooms were prepared by drying the fresh mushrooms at 50 °C in an oven for 48 h, and then ground into a powder. Mushroom extracts were prepared by adding 18 g of whole fresh and dried mushrooms into 50 ml of ethyl acetate and ethanol separately. The mushrooms were then extracted by sonication for 1 h at room temperature and then filtered. The mushroom extract was then dried in a rotary evaporator and dissolved in 1 ml of ethanol and 2 ml of water. Ethanol and ethyl acetate were used as the solvents for extraction, as it has been found that the compounds contributing to the antimicrobial properties of mushrooms were phenols, aldehydes and ketones, thus polar solvents were used and no other non-polar organic solvents.

2.2 Testing of Extracts

Luria broth (LB) agar plates were prepared by mixing 12.5 g of LB and 5 g of agar powder in 500 ml water, and it was autoclaved at 121 °C for 1 h. It was then poured into plastic petri dishes and let to set at room temperature. 50 μl E. coli was then spread onto the agar. The bacteria were left to settle down for 30 min, and 6 mm paper disks were soaked in the extracts, ampicillin, and 33% ethanol and then placed onto the agar plates. The agar plates were then incubated for 24 h at 30 °C, and the zones of inhibition were measured. The cap and stems of the mushroom which had the largest zone of inhibition were then extracted again separately. The extracts were tested again to find the zone of inhibition and were also used when the hydrogels were made.

2.3 Preparation of Hydrogels

To make the cellulose for the structure of the hydrogel, 1 g of grass was digested in 75 ml of 1 mol dm^{-3} sodium hydroxide for 6 h at 121 °C. It was then filtered and then bleached in 2% sodium hypochlorite for 1 h at 96 °C. The hydrogels were prepared by adding 3.75 g of polyvinyl alcohol (PVA) into 40 ml of warm water, and they were heated at 135 °C for 1 h. 0.2 g of cellulose was then added into the PVA, which was let to cool. After the PVA mixture was at room temperature, 1 ml of 6 extracts and ampicillin was added into the PVA mixture, and it was dried at 50 °C in an oven for 48 h. One hydrogel was also made without any extracts or ampicillin as controls.

2.4 Testing of Hydrogels

The hydrogel's ability to absorb moisture was determined by evaluating the amount of water that can be absorbed by it; they were soaked in water and weighed every 30 min to measure their absorption over time. The hydrogel's permeability to oxygen was then measured using an oximeter by placing the probe on the hydrogel. The hydrogel's ability to block bacteria from wounds was tested by pipetting 25 l of E. coli that was diluted 100 times from the stock solution onto the top of the hydrogel, which was placed on an agar plate and incubated for 24 h. The number of colonies of E. coli was then counted and recorded. An agar plate was also plated with the diluted bacteria as a control, to help us find out if hydrogels that were not made with any extracts or ampicillin could block bacteria from going through the hydrogel. The hydrogels were then disinfected, and new bacteria was spread on the bottom of the hydrogel. They were then placed on new agar plates, which were incubated for another 24 h to find out how much bacteria were killed by the hydrogel. The number of colonies of bacteria was counted and recorded and a control test like the control done for the blocking of the bacteria was also performed.

3 Results and Analysis

3.1 Zone of Inhibition of Extracts

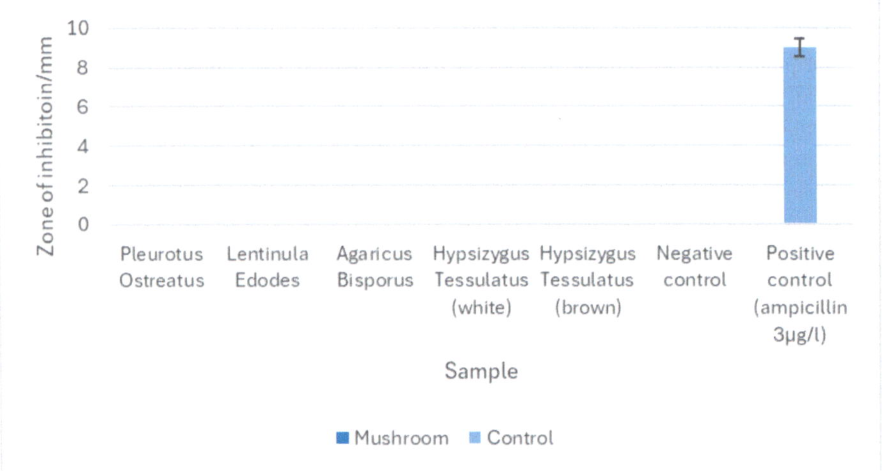

Fig. 1. Graph of zone of inhibition/mm of the ethanolic extracts of the different mushrooms

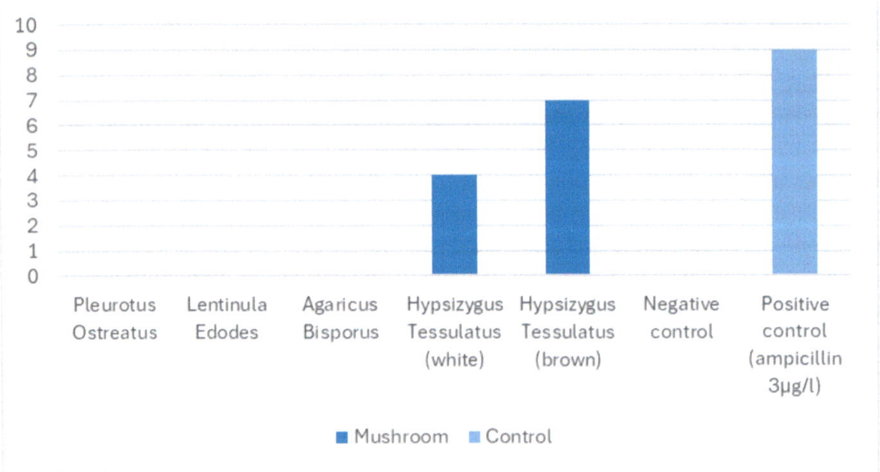

Fig. 2. Graph of zone of inhibition/mm of the ethyl acetate extracts of the different mushrooms

According to Fig. 1, the ethanol extracts from all the different types of mushrooms did not display any antimicrobial properties. According to Fig. 2, the ethyl acetate extracts of *Pleurotus Ostreatus, Lentinula Edodes and Agaricus Bisporus* did not display any antimicrobial properties. This does not correspond with other reports [24, 25]. This could be because the method of extraction was not ideal to extract the compounds that contribute to the antimicrobial properties of these mushrooms, and studies can be conducted in the future to investigate the optimal extraction method and conditions. Ethyl acetate extracts for both white and brown *Hypsizygus Tessulatus* had a zone of inhibition of 4 mm and 7 mm respectively, showing that it has antimicrobial properties. This was observed in other studies [26, 27]. As the ethyl acetate extract of the Brown *Hypsizygus Tessulatus* showed the highest zone of inhibition with 7 mm, it was used in making the hydrogels.

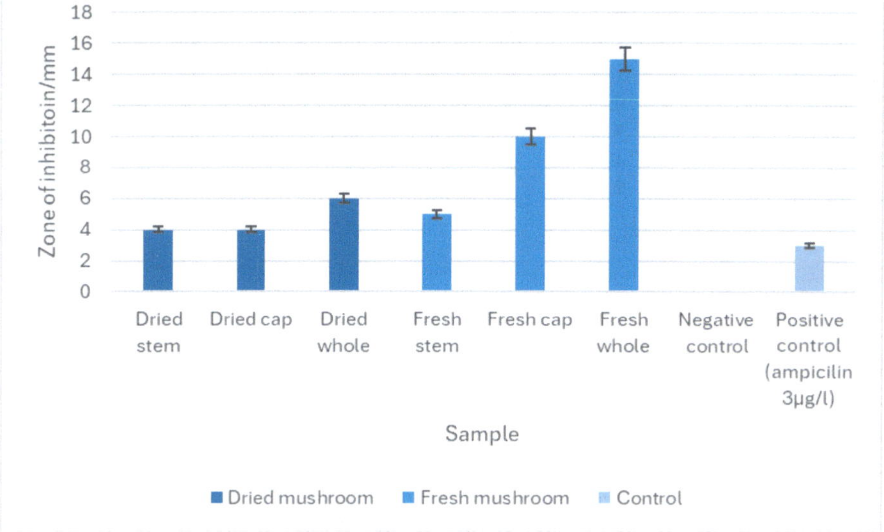

Fig. 3. Graph of zone of inhibition/mm of the ethyl acetate extracts of the different parts and dryness of Brown *Hypsizygus Tessulatus*

According to Fig. 3, all the extracts for the different parts and dryness of the brown *Hypsizygus Tessulatus* were able to display antimicrobial properties towards E.Coli with the extract of the whole fresh brown *Hypsizygus Tessulatus* having the highest zone of inhibition with 15 mm. This trend was also observed in a study that was conducted [26]. Caps of *Hypsizygus Tessulatus* were found to perform better than the stems, contradicting another report [28]. It is observed that positive control had a lower zone of inhibition than that in Figs. 1 and 2 above. This could be because some of the ampicillin degraded by the time the test was conducted. As all extracts of the different parts and dryness of *Hypsizygus Tessulatus* were able to display antimicrobial properties, its ability to display its antimicrobial properties in hydrogel plasters was investigated.

3.2 Absorption of Hydrogel

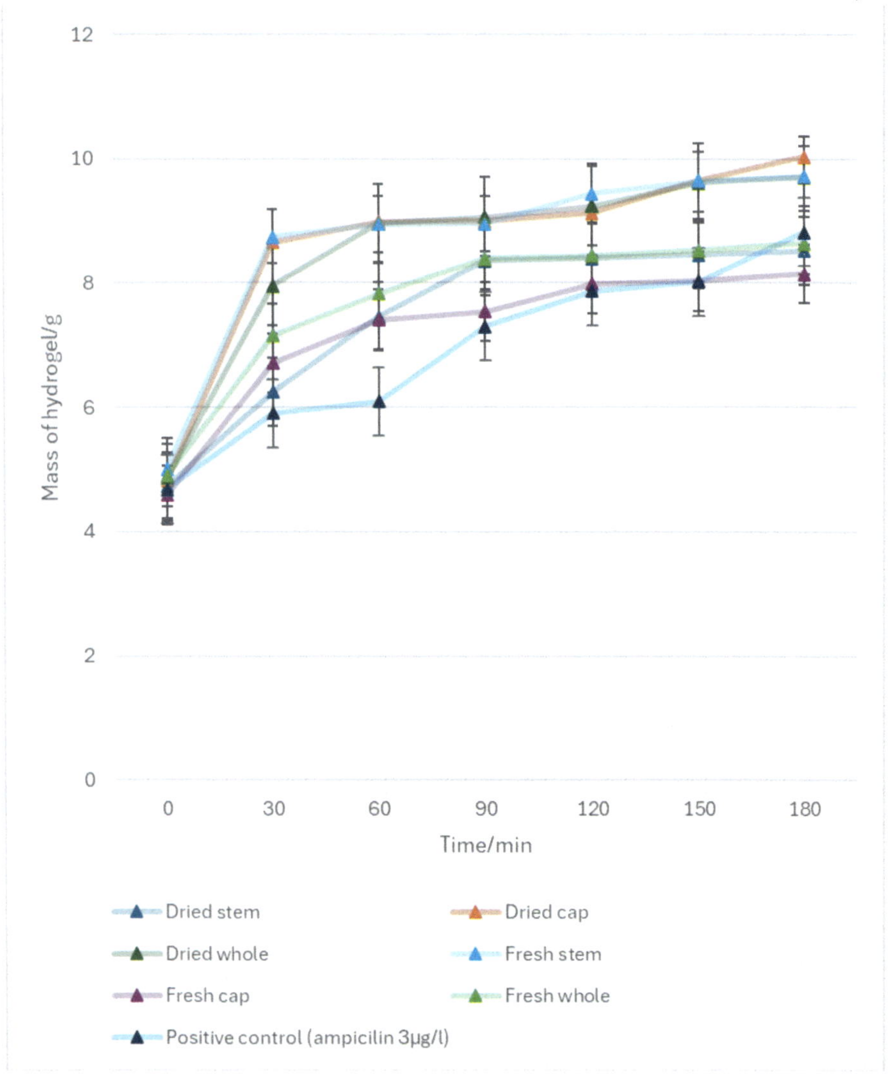

Fig. 4. Graph of mass/g of the hydrogels made with different extracts with respect to time being fully soaked in water

According to Fig. 4, all hydrogels show a common trend of increase in mass over time as they were soaked in water. The hydrogels were able to absorb around 90% of their mass, and the total absorption of the hydrogels made with or without extracts were comparable, showing that the hydrogels can generally absorb water, regardless of whether they are made with extracts or not. This shows that the hydrogels made with

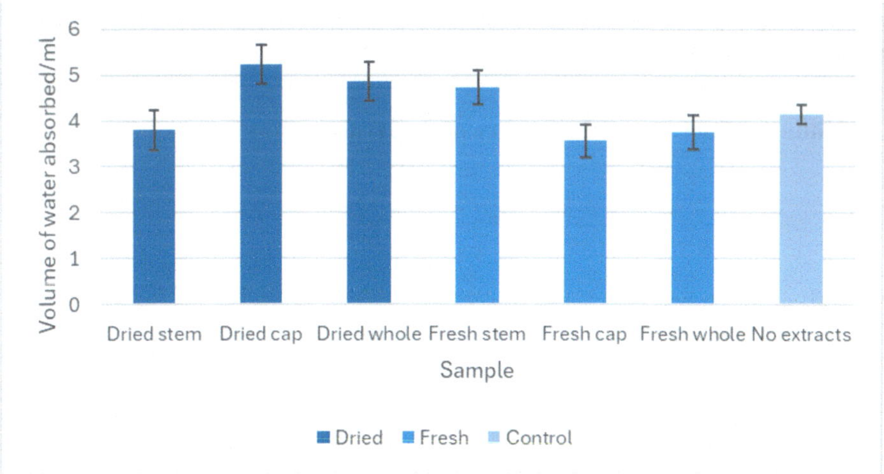

Fig. 5. Graph of the volume of water absorbed/ml of the different hydrogels made with the different extracts

extracts are generally able to absorb any wound drainage that might be present on the wounds, making it a potential material for making antimicrobial plasters (Fig. 5).

3.3 Oxygen Permeability

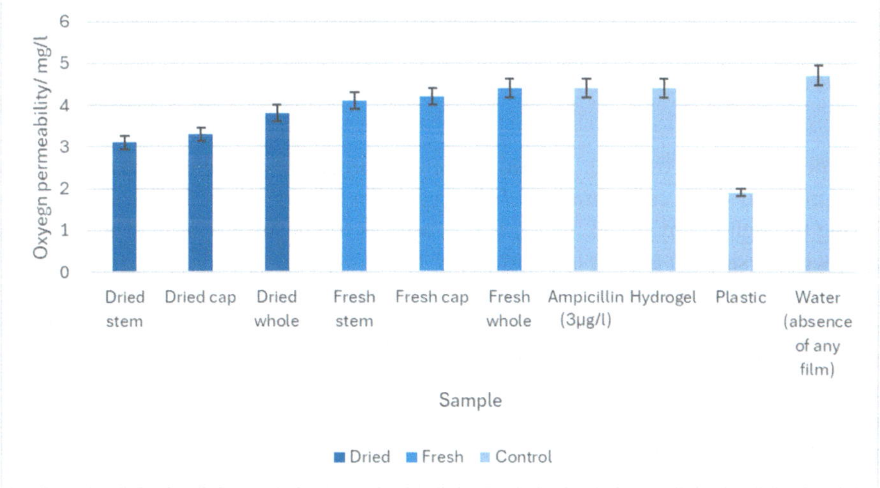

Fig. 6. Graph of the amount of oxygen (mg/l) passing through the different hydrogels and plastic and water

According to Fig. 6, The hydrogels made with fresh whole extract of *Hypsizygus Tessulatus* allowed the same amount of oxygen to pass through as the hydrogels with no extract. Compared to the positive control of water (the absence of any film), it was 6.4% less permeable to oxygen.

3.4 Killing of Bacteria on the Surface of the Hydrogel

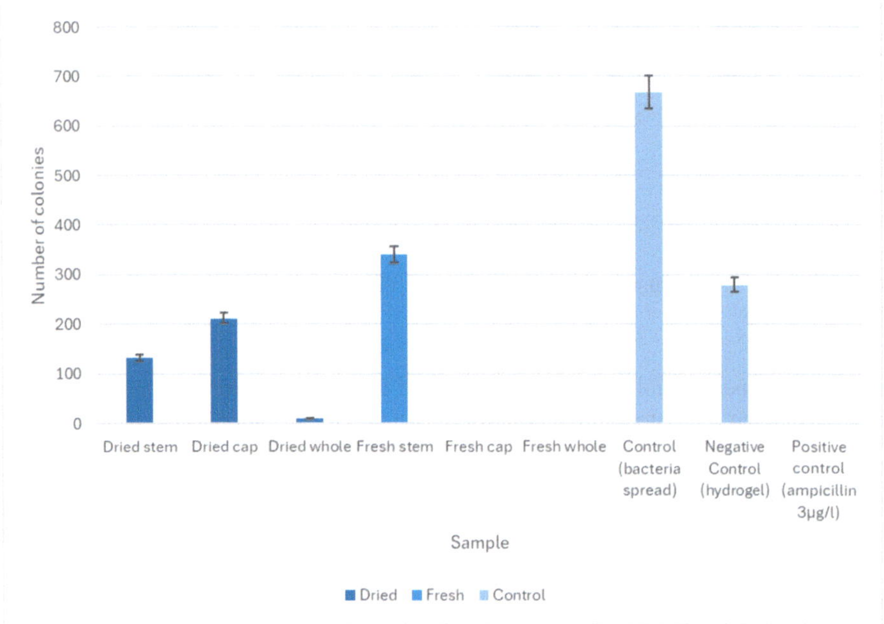

Fig. 7. Graph of the number of colonies of bacteria that grew with the different hydrogels

According to data from Fig. 7, hydrogels made with fresh whole extract of *Hypsizygus Tessulatus* and fresh cap extract of *Hypsizygus Tessulatus* blocked all bacteria from passing through the hydrogel, which is the same as the positive control. The hydrogel made with fresh stem extract of *Hypsizygus Tessulatus* had 61 more colonies than the hydrogel made with no extracts or ampicillin, which might be a sign of there being a source of energy for the bacteria to grow. It is also found that hydrogels that are made with no extracts or ampicillin can block 55.8% of bacteria. As the hydrogels made with extracts were generally able to block bacteria, they are suitable for making antimicrobial hydrogels. Hydrogels made with fresh cap and fresh whole extracts are the best at blocking bacteria.

3.5 Killing of Bacteria on the Surface of the Hydrogel

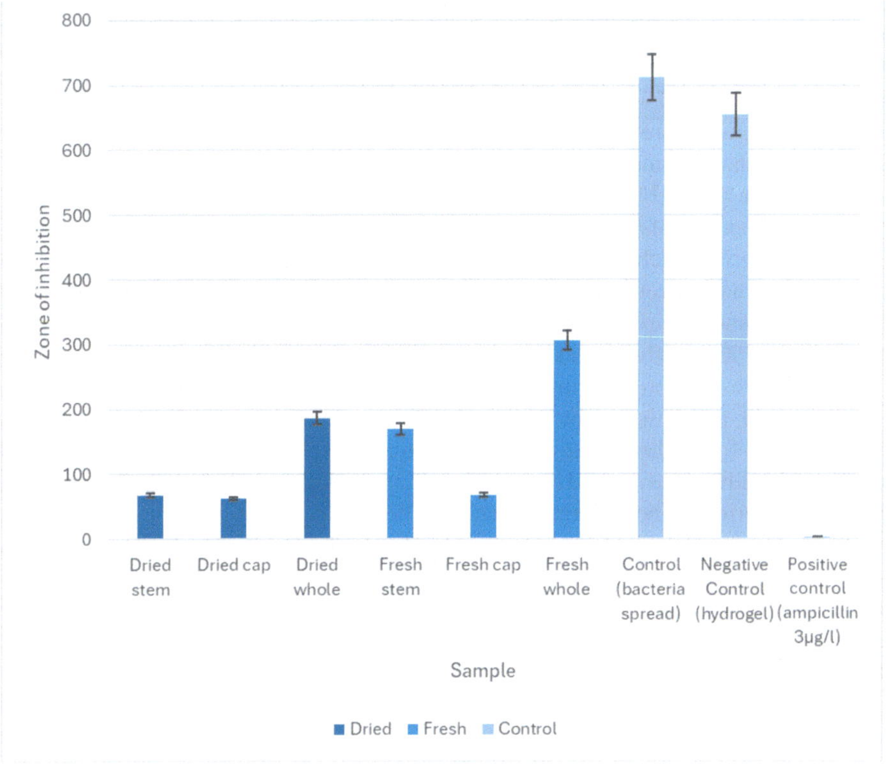

Fig. 8. Graph of the number of colonies of bacteria that grew with the different hydrogels

According to the data from Fig. 8, hydrogels made with the dried cap extract of *Hypsizygus Tessulatus* were able to kill the most bacteria, killing 96.1% of bacteria. Hydrogels made with ampicillin, the positive control, were found to be the best at killing bacteria, killing 3.7% more bacteria than the hydrogels made with the dried cap extract of *Hypsizygus Tessulatus*. However, ampicillin is not ideal to be used in making antimicrobial plasters, especially for chronic wounds, as it can result in the bacteria developing antibiotic resistance [29, 30]. Hence the dried cap extract of *Hypsizygus Tessulatus* is a good alternative to be used in a low-cost plaster. It is also observed that the hydrogels made with fresh whole and dried whole *Hypsizygus Tessulatus* did not kill bacteria as well as its ability to block bacteria, and the trend for the blocking of bacteria was not the same as the killing of bacteria.

4 Conclusion and Future Works

Our results show that among ethanol and ethyl acetate extracts of *Pleurotus Ostreatus, Lentinula Edodes, Agaricus Bisporus* and white and brown *Hypsizygus Tessulatus*, the ethyl acetate extract of brown *Hypsizygus Tessulatus* has the best antimicrobial activity. When extracts of different parts and dryness of brown *Hypsizygus Tessulatus* was impregnated into the hydrogel, it was found that hydrogels with extracts did better at blocking bacteria from going through the hydrogel and killing bacteria than normal hydrogels. The hydrogel that was made with the extract of fresh whole and cap of *Hypsizygus Tessulatus* was the best at blocking bacteria, blocking all the bacteria. The hydrogel made with dried cap extract of *Hypsizygus Tessulatus* was able to kill the most bacteria, killing 96.1% of bacteria, which is 3.7% less than hydrogels made with ampicillin and 37.5% more than normal hydrogels. It was also found that whole mushroom extract did better than the stem and cap extract at killing and blocking bacteria. Hydrogels were able to absorb around 90% of their mass and are permeable to oxygen.

In the future, the optimal extraction methods and conditions for the mushrooms can be investigated, and possible interactions between cap and stem extract can be investigated as it was found that whole mushroom extracts displayed a stronger antimicrobial property. The difference in trend between blocking and killing bacteria can also be investigated. Furthermore, the experiment can also be conducted on gram positive bacteria and the thickness of the hydrogel can be varied to find out if the plaster is effective on gram positive bacteria, and if the blocking of the bacteria could be due to the hydrogel being too thick. The active component of the mushroom that is contributing to the antimicrobial properties of the mushrooms can be isolated using high performance liquid chromatography and elucidated using nuclear magnetic resonance so that it can be used in making the hydrogel.

References

1. Davis, C.P.: Normal flora. Medical Microbiology, 4th edition (1996)
2. Liang, Y., Liang, Y., Zhang, H., Guo, B.: Antibacterial biomaterials for skin wound dressing. Asian Journal of Pharmaceutical Sciences (2022)
3. Leaper, D., Assadian, O., Edmiston, C.E.: Approach to chronic wound infections. Br. J. Dermatol. **173**(2), 351–358 (2015)
4. Maneewattanapinyo, P., Monton, C., Pichayakorn, W., Dangmanee, N., Wunnakup, T., Suksaeree, J.: Plaster Gel Loaded with Silver Nanoparticle-Mediated Ganoderma applanatum: from Fabrication to Evaluation. AAPS PharmSciTech **24**(5), 105 (2023)
5. Basketter, D.A., Marriott, M., Gilmour, N.J., White, I.R.: Strong irritants masquerading as skin allergens: the case of benzalkonium chloride. Contact Dermatitis **50**(4), 213–217 (2004)
6. Wentworth, A.B., Yiannias, J.A., Davis, M.D., Killian, J.M.: Benzalkonium chloride: a known irritant and novel allergen. Dermatitis **27**(1), 14–20 (2016)
7. Barber, O.W., Hartmann, E.M.: Benzalkonium chloride: A systematic review of its environmental entry through wastewater treatment, potential impact, and mitigation strategies. Crit. Rev. Environ. Sci. Technol. **52**(15), 2691–2719 (2022)
8. Liu, P., Li, Y.: Uses of Mushrooms in Traditional Medicine in China. In Wild Mushrooms and Health, pp. 155–179. CRC Press (2023)

9. Alves, M.J., et al.: A review on antimicrobial activity of mushroom (Basidiomycetes) extracts and isolated compounds. Planta medica, 1707–1718 (2012)
10. Kosanic, M., Rankovic, B., Dasic, M.: Antioxidant and antimicrobial properties of mushrooms. Bulgarian J. Agr. Sci. **19**(5), 1040–1046 (2013)
11. Mondal, T.: Studies on antioxidant and antimicrobial properties of some common mushrooms. J. Today's Biolog. Sci. Res. Rev. **2**, 60–67 (2013)
12. Tinrat, S.: Antimicrobial activities and synergistic effects of the combination of some edible mushroom extracts with antibiotics against pathogenic strains. Int. J. Pharmaceut. Sci. Rev. Res. **35**(2), 253–262 (2015)
13. Barros, L., Calhelha, R.C., Vaz, J.A., Ferreira, I.C., Baptista, P., Estevinho, L.M.: Antimicrobial activity and bioactive compounds of Portuguese wild edible mushrooms methanolic extracts. Eur. Food Res. Technol. **225**, 151–156 (2007)
14. Horner, W.E., Helbling, A., Salvaggio, J.E., Lehrer, S.B.: Fungal allergens. Clin. Microbiol. Rev. **8**(2), 161–179 (1995)
15. Zhang, X., et al.: The fabrication of antibacterial hydrogels for wound healing. Eur. Polymer J. **146**, 110268 (2021)
16. Peppas, N.A., Slaughter, B.V., Kanzelberger, M.A.: Hydrogels. Polymer Science: A Comprehensive Reference, 385–395 (2012)
17. Higa, O.Z., Rogero, S.O., Machado, L.D.B., Mathor, M.B., Lugão, A.B.: Biocompatibility study for PVP wound dressing obtained in different conditions. Radiat. Phys. Chem. **55**(5–6), 705–707 (1999)
18. Sudhakar, C.K., et al.: Hydrogels—Promising Candidates for Tissue Engineering. Nanotechnology Applications for Tissue Engineering, 77–94 (2015)
19. Pan, H., Duan, Z., Zhu, C., Fu, R., Li, X.: Non-stick hemostasis hydrogels as dressings with bacterial barrier activity for cutaneous wound healing. Mater. Sci. Eng., C **105**, 110118 (2019)
20. Aliakbar Ahovan, Z., et al.: Antibacterial smart hydrogels: New hope for infectious wound management. Materials Today Bio **17**, 100499 (2022)
21. Yang, K., et al.: Antimicrobial hydrogels: promising materials for medical application. Int. J. Nanomed. 2217–2263 (2018)
22. Salomé Veiga, A., Schneider, J.P.: Antimicrobial hydrogels for the treatment of infection. Pept. Sci. **100**(6), 637–644 (2013)
23. Ng, V.W., et al.: Antimicrobial hydrogels: A new weapon in the arsenal against multidrug-resistant infections. Adv. Drug Deliv. Rev. **78**, 46–62 (2014)
24. Asri, R.M., Yahya, H., Rehan, M.M., Yahya, H.N.: Antibacterial properties of ethanolic extract of mushrooms sold in malaysian local market. East African Scholars Journal of Agriculture and Life Sciences (2019)
25. Roy, D.N., Azad, A.K., Sultana, F., Anisuzzaman, A.S.M.: In-vitro antimicrobial activity of ethyl acetate extract of two common edible mushrooms. The Journal of Phytopharmacology **5**(2), 79–82 (2016)
26. Shah, S.R., Ukaegbu, C.I., Hamid, H.A., Alara, O.R.: Evaluation of antioxidant and antibacterial activities of the stems of Flammulina velutipes and Hypsizygus tessellatus (white and brown var.) extracted with different solvents. Journal of Food Measurement and Characterization **12**, 1947–1961 (2018)
27. Chowdhury, M.M.H., Kubra, K., Ahmed, S.R.: Screening of antimicrobial, antioxidant properties and bioactive compounds of some edible mushrooms cultivated in Bangladesh. Ann. Clin. Microbiol. Antimicrob. **14**, 1–6 (2015)

28. Ishmael, U.C., Rashid, S.S.: Biological characterization of water extracts from the caps and stalks of white Hypsizygus Tessellatus (Bunapi Shimeji) and Flammulina Velutipes (Enoki) Mushrooms. Innovat International Journal Of Medical & Pharmaceutical Sciences **2**(3) (2017)
29. Punjataewakupt, A., Napavichayanun, S., Aramwit, P.: The downside of antimicrobial agents for wound healing. Eur. J. Clin. Microbiol. Infect. Dis. **38**, 39–54 (2019)
30. Pirvanescu, H., Balasoiu, M., Ciurea, M.E., Balasoiu, A.T., Manescu, R.: Wound infections with multi-drug resistant bacteria. Chirurgia **109**(1), 73–9 (2014)

A Fast Way to Compute Matrix Multiplication

Ananya Kharbanda[1(✉)], Chua Zhong Ding[2], and Shu Jian Jun[3]

[1] Raffles Institution, 1 Raffles Institution Ln, Singapore 575954, Singapore
24YANAN481Z@student.ri.edu.sg

[2] River Valley High School, 6 Boon Lay Ave, Singapore 649961, Singapore
chua zhong_ding@students.edu.sg

[3] School of Mechanical and Aerospace Engineering, Nanyang Technological University, 50 Nanyang Avenue, Singapore 639798, Singapore
mjjshu@ntu.edu.sg

Abstract. Matrix multiplication plays a pivotal role in various domains, including image processing and artificial intelligence, making its optimization a crucial endeavor. This research paper delves into enhancing the efficiency of matrix multiplication by specifically focusing on reducing the number of additions required in the process. Traditional methods typically necessitate between 15 to 18 additions; this study aims to lower this count, thereby streamlining computational requirements.

The methodology encompasses two primary stages:

1. Initially, we assess the practicality of this optimization approach by comparing it with existing matrix multiplication techniques. This comparative analysis not only benchmarks our approach but also highlights potential areas for improvement.
2. Subsequently, we employ a brute force algorithm to derive a novel formula that minimizes the number of additions. This stage is critical in evolving a theoretical concept into a practical application.

Through this multi-faceted approach, the paper aims to contribute to the field, potentially leading to more efficient algorithms in image processing and AI applications.

Keywords: Matrix · Computing · Math

1 Introduction

1.1 Reasons for Undertaking This Study

Matrix multiplication is a critical concept in computer science, essential for a wide range of applications. Its role is especially significant in deep learning, where it is the key to calculating outputs in neural networks. In image processing, matrix multiplication is extensively used for manipulating and transforming images. Despite its essential role, matrix multiplication is computationally demanding (Fig. 1).

H. Guo et al. (Eds.): IRC-SET 2024, PTT, pp. 139–145, 2025.
https://doi.org/10.1007/978-981-96-3770-6_15

$$Where\ A\ is\ a\ m \times n\ matrix\ and\ B\ is\ a\ m \times p\ matrix$$

$$(AB_{ij}) = \sum_{k=1}^{n} A_{ik} B_{kj}$$

Fig. 1. Naive Matrix Multiplication

The standard process of multiplying two matrices of size N typically requires $2(N/2)^3$ operations, involving eight multiplications. This process can be quite time-consuming, especially in complex calculations. As such, there is a need to improve the efficiency of matrix multiplication, such that research in these fields can be sped up.

1.2 Strassen Algorithm

To overcome this challenge, various methods have been developed to speed up matrix multiplication, significantly improving computational efficiency in these important areas (Fig. 2).

$$A = \begin{bmatrix} A_{11} & A_{12} \\ A_{21} & A_{22} \end{bmatrix}, \quad B = \begin{bmatrix} B_{11} & B_{12} \\ B_{21} & B_{22} \end{bmatrix}, \quad C = \begin{bmatrix} C_{11} & C_{12} \\ C_{21} & C_{22} \end{bmatrix},$$

$$With\ A \times B = C$$

$$M_1 = (A_{11} + A_{22}) \times (B_{11} + B_{22})$$

$$M_2 = (A_{21} + A_{22}) \times B_{11}$$

$$M_3 = A_{11} \times (B_{12} - B_{22})$$

$$M_4 = A_{22} \times (B_{21} - B_{11})$$

$$M_5 = (A_{11} + A_{12}) \times B_{22}$$

$$M_6 = (A_{21} - A_{22}) \times (B_{11} + B_{12})$$

$$M_7 = (A_{12} - A_{22}) \times (B_{21} + B_{22})$$

$$\begin{bmatrix} C_{11} & C_{12} \\ C_{21} & C_{22} \end{bmatrix} = \begin{bmatrix} M_1 + M_4 - M_5 + M_7 & M_3 + M_5 \\ M_2 + M_4 & M_1 - M_2 + M_3 + M_6 \end{bmatrix}.$$

Fig. 2. Strassen Algorithm

Strassen Algorithm [1] is a key development in optimizing matrix multiplication. It effectively reduces the number of necessary multiplications from 8 to 7, markedly decreasing the computational complexity. This is particularly advantageous for complex and large-scale calculations. However, this efficiency does not come without a trade-off. While Strassen Algorithm cuts down on multiplications, it increases the number of additions from 4 to 18. Although additions are less computationally intensive, requiring only $(N/2)^2$ operations each, this increase still contributes to a higher overall computation time.

1.3 Winograd Form

$$t = A_{11} \times B_{11}$$

$$u = (A_{21} - A_{11}) \times (B_{21} - B_{22})$$

$$v = (A_{21} + A_{22}) \times (B_{21} - B_{11})$$

$$w = t + (A_{21} + A_{22}) \times (B_{21} - B_{11})$$

$$\begin{bmatrix} a & b \\ c & d \end{bmatrix} \begin{bmatrix} A & C \\ B & D \end{bmatrix} = \begin{bmatrix} t + b \times B & w + v + (a + b - c - d) \times D \\ w + u + d \times (B + C - A - D) & w + u + v \end{bmatrix}$$

Fig. 3. Winograd Form

The Winograd Form [2] of Strassen's Algorithm reduces the number of additions in matrix multiplication from 18 to just 15. The number of multiplications remains at 7 (Fig. 3).

1.4 Limitations

Despite our best efforts in further reducing the number of additions in matrix multiplication, the computational complexity of finding such an algorithm was too difficult for our computers to handle in the brute force approach.

1.5 Preliminaries

This section will define all significant terms used in this report.

Matrix Multiplication Is a binary operation that takes a pair of matrices, and produces another matrix. Each element of the resulting matrix is computed as the dot product of the corresponding row of the first matrix and the column of the second matrix.

Time Complexity Refers to a computational analysis that estimates how the runtime of an algorithm grows as the size of the input increases.

Square Matrices Are matrices that have the same number of rows and columns.

Recursion Is a method where the solution to a problem depends on solutions to smaller instances of the same problem.

NxN MATRIX Is a matrix consisting of n rows and n columns (Fig. 4).

$$\begin{bmatrix} A_{11} & A_{12} & A_{13} \\ A_{21} & A_{22} & A_{23} \\ A_{31} & A_{32} & A_{33} \end{bmatrix}$$

Fig. 4. An N x N Matrix where N = 3

2 Aims/Objectives

This paper sets out to explore the potential for further optimization in matrix multiplication, specifically targeting a reduction in the number of additions without adding extra computational burdens. Our research is grounded in the hypothesis that an optimal algorithm exists which can achieve this balance. Our study involves, firstly, examining the practicality of minimizing additions in the matrix multiplication process, and secondly, developing an algorithm that can effectively identify and apply a more efficient formula for matrix multiplication. This exploration is driven by the premise that such an algorithm can be realized without necessitating additional computations, thereby enhancing the overall efficiency of the process, a critical factor in numerous applications within the realms of computer science.

3 Methodology

3.1 Comparing Effectiveness

In order to determine how effective reducing the number of additions in matrix multiplication will be, we will be using a programme to analyse the time taken for matrix multiplication for different sizes of matrices, for square matrices sizes N x N. We will compare the time taken for 3 algorithms, naive multiplication, Strassen algorithm, and the winograd form of Strassen algorithm. We will also analyse how much time an algorithm with only 14 and 13 additions will take. The code for this is shown below.

3.2 Naive Brute Force Approach

To identify an algorithm capable of performing matrix multiplications with 14 or fewer additions, we developed a naive algorithm with two key components.

The Combination Function is the first component of this algorithm. It is a function which takes in 2 arrays of size N, designed to determine a combination of k additions and m multiplications that yields a specified value. This function memorizes the results of different combinations it has previously calculated. By storing these results, the function avoids redundant computations, thus conserving the counts of additions and multiplications used. By iterating through each combination it has already come up with, this algorithm takes $O((k + m)!\ N^2)$ time.

The Matrix Subarray Testing Function, as the second key component of our algorithm, methodically evaluates every viable combination of additions and multiplications within the four subarrays of the matrix. Specifically, it explores combinations that result in exactly K additions and M multiplications. This function efficiently utilizes the results previously computed and stored by the combination function, effectively reusing them when possible to avoid redundant calculations. This strategic reuse of computations not only enhances efficiency but also streamlines the exploration process. As there are K! possible combinations for additions and M! possible combinations for multiplication, the total time complexity for this algorithm would be $O(K! + M!)$.

When our Matrix Subarray Testing Function targets a combination of 14 additions and 7 multiplications, it initially faces 364 * 20 = 7,280 potential operations. To optimize this, we've introduced minimum and maximum constraints for the additions and multiplications in each subarray. Specifically, the range for additions is set between 1 and 5, while for multiplications, it's limited to 1 to 3 per subarray. This adjustment significantly narrows down the possibilities to 224 combinations for additions and 16 for multiplications, cumulatively reducing the loop count in our function to 3,584.

Moreover, this refinement not only lessens the loop iterations but also decreases the frequency of running our combination function. Considering the worst-case scenario where the number of additions and multiplications are at their maximum (5 and 3, respectively), the combination function would need to run at most (5 + 3)! * 22 * 2 = 322,560 times. Consequently, in the most demanding scenario, our algorithm would need to execute 322,560 * 3,584 = 1,156,055,040 times in total.

This strategic approach significantly streamlines the algorithm, ensuring more efficient computation even in the worst-case scenario.

4 Results

4.1 Effectiveness

Table 1 demonstrates the average time taken to execute the Naive, Strassen, and Winograd Algorithm. These algorithms are already well established algorithms.

Table 1. Time taken to execute each pre existing algorithm

N	Algorithms Time Taken/s		
	Naive	*Strassen*	*Winograd*
2^4	8.958e^{-02}	5.244e^{-01}	4.495e^{-01}
2^5	2.709e^{-02}	1.257e^{-01}	1.237e^{-01}
2^6	2.087e^{-01}	9.198e^{-01}	8.914e^{-01}
2^7	1.454	5.286	5.196
2^8	1.149e^{01}	3.769e^{01}	4.020e^{01}

Table 2 demonstrates the average time taken to execute an algorithm with 14 and 13 additions respectively. These algorithms are ones we are aiming to search for.

Table 2. Time taken to execute each hypothetical algorithm

N	Algorithms Time Taken/s	
	14 Additions	*13 Additions*
2^4	4.405e^{-01}	4.309e^{-01}
2^5	1.193e^{-01}	1.182e^{-01}
2^6	9.252e^{-01}	8.651e^{-01}
2^7	5.173	5.168
2^8	3.679e^{01}	3.666e^{01}

Figure 5 presents a graphical comparison of each algorithm's performance.

The collected data reveals that, in general, Winograd's algorithm outperforms Strassen's algorithm in matrix multiplication. Notably, configurations using 14 additions tend to be more efficient than Winograd's version with 15 additions, and similarly, those with 13 additions usually surpass the 14-addition setups in performance. Interestingly, the naive algorithm demonstrates the best performance across these comparisons. This superior efficiency is likely attributable to the overhead associated with the recursive nature of the Strassen algorithm. Furthermore, as the size of the matrices increases, the performance gap between the 13-addition method and the naive algorithm narrows, suggesting that the impact of recursion diminishes with larger data sets.

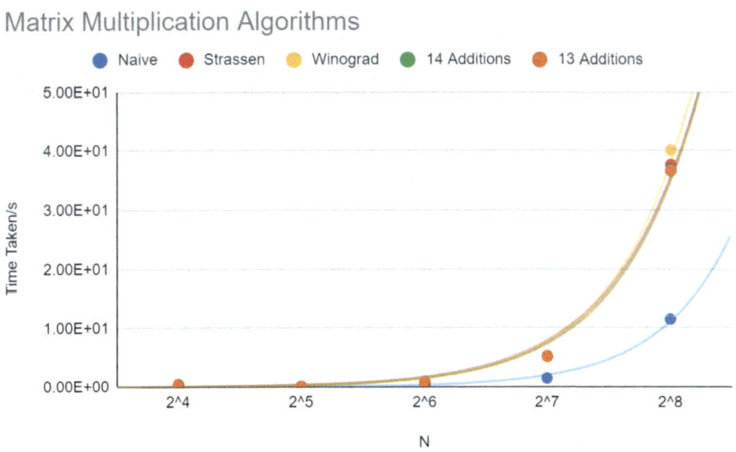

Fig. 5. Graph of each algorithm and it's time taken

5 Conclusion

In conclusion, achieving a reduction in the number of additions required for matrix multiplication stands as a significant milestone in the ongoing quest to optimize this crucial computational process. While this study has not yet pinpointed the specific algorithm capable of minimizing additions, our proposed brute force approach lays the groundwork for future discovery. This method holds the promise of eventually uncovering the optimal algorithm, though it is contingent upon the availability of additional time and resources. The pursuit of such an algorithm is not just an academic endeavour; it has the potential to substantially impact various fields reliant on efficient computational processes.

Acknowledgements. We would like to thank our project supervisor Shu Jian Jun for guiding us in this project. We would also like to thank our NRP teachers in charge from River Valley High School and Raffles Institution.

References

1. Strassen, V.: Gaussian elimination is not optimal. Numer. Math. **13**(4), 354–356 (1969). https://doi.org/10.1007/bf02165411
2. Coppersmith, D., Winograd, S.: Matrix multiplication via arithmetic progressions. J. Symb. Comput. **9**(3), 251–280 (1990). https://doi.org/10.1016/s0747-7171(08)80013-2

Tackling Pharmaceutical Pollution: Jackfruit Derived Biochar-Alginate Composite Hydrogel Beads as a Novel Approach for Ofloxacin Removal

Le Qi Ho, Yan Tong Chen, Esther Suraya Puk, Grace Lim(✉), and Abigayle Ng

Raffles Institution, 1 Raffles Institution Lane, Singapore, Singapore
grace.lim@ri.edu.sg

Abstract. Ofloxacin (OFX), a popular fluoroquinolone antibiotic, is hard to metabolise and can cause harm to ecosystems and humans when released into the environment. Although Biochar proves effective in OFX removal, its unsustainability arises from challenges in removing it from the aquatic environment and the high costs associated with an energy-intensive pyrolysis process and expensive feedstocks. In this study, jackfruit derived biochar-alginate composite hydrogel beads (BAB) were synthesised to investigate the adsorption of OFX from aqueous solution. Three different samples were prepared by encapsulating biochar in alginate hydrogel, with the biochars encapsulated being unpyrolysed (B0) and pyrolysed at temperatures of 200 °C (B2) and 500 °C (B5). BAB exhibited satisfactory performance in OFX adsorption, with B2 reaching a maximum adsorption capacity of 11.4 mg g^{-1} indicating the better performance of BAB with biochar of a lower pyrolysis temperature. A good fit of the pseudo-second-order model ($R^2 \geq 0.979$) indicated chemisorption. The adsorption isotherms were described by the Freundlich model ($R^2 \geq 0.930$) with n values of more than 1, indicating a multilayer adsorption process and adsorption of OFX onto BAB being favourable. For swelling behaviour, B0 exhibited the highest degree of swelling (118%). Furthermore, swelling increased after 2 cycles of swelling in water, indicating potential for reusability in different applications.

Keywords: Ofloxacin · Biochar · Pharmaceutical Pollution · Adsorption Kinetics · Adsorption Isotherms · Alginate Hydrogel

1 Introduction

Pharmaceutical active contaminants are contaminants of emerging interest in the field of water treatment due to their potential impacts on ecosystems and human health. Notably, Ofloxacin (OFX) has been gaining a lot of attention as humans and animals alike greatly consume it as a fluoroquinolone antibiotic, but the human body absorbs less than 10% and excretes the rest into wastewater [1]. When released into the environment, OFX can inhibit plant growth and encourage the growth of non-target microorganisms, possibly resulting in micro-ecosystem imbalance and inevitably harming human health through

H. Guo et al. (Eds.): IRC-SET 2024, PTT, pp. 146–155, 2025.
https://doi.org/10.1007/978-981-96-3770-6_16

food chains [2]. OFX does not degrade easily and remains relatively unchanged when it enters most water systems [3], where conventional treatment methods are insufficient for its removal.

Biochar as an adsorbent has generated great interest due to its abundance of surface functional groups, porous structure, and specific surface area. Currently, studies show that biochar can effectively remove OFX [4, 5]. However, biochar cannot be readily separated from the aquatic environment and its high cost associated with an energy-intensive pyrolysis process and expensive feedstocks makes large-scale adaptation of it unviable, making it difficult for real-life application [6].

To improve viability, sodium alginate can be used. Sodium alginate is a biodegradable biopolymer that forms water-insoluble gels in the presence of divalent cations such as Ca^{2+} [7]. However, its poor mechanical strength limits their practical applicability. Enhancing alginate hydrogels with organic fillers like biochar improves both adsorption properties and mechanical strength [8]. Hence, alginate beads encapsulating biochar have gained traction for water remediation in recent years [9]. However, there is no study on the effectiveness of a jackfruit derived biochar-alginate hydrogel composite on the removal of OFX from aqueous solution. Jackfruit peel, which is bitter and usually discarded, has excellent mechanical strength [10], showing great potential as a cost-effective precursor to biochar to remove OFX when encapsulated in hydrogels.

Hence, in this study, we aim to create cost-effective and environmentally viable jackfruit derived biochar-alginate composite hydrogel beads and investigate its effectiveness in removing the pharmaceutical pollutant Ofloxacin.

2 Materials and Methods

2.1 Preparation of Adsorbent and Adsorbate

Preparation of Jackfruit-Derived Biochar. Discarded jackfruit peels were acquired from a local fruit stall, washed, cut and oven dried at 60 °C overnight. They were then ground and sieved to 63–125 μm. Pyrolysis was performed on two out of three portions of jackfruit peel powder at 200 °C and 500 °C respectively for 2 h.

Preparation of Jackfruit Derived Biochar - Alginate Composite Hydrogel Beads. Alginate solution was prepared by dissolving 1 g of sodium alginate (GCE Laboratory Chemicals) in 100 mL of deionised (DI) water and shaken at 160 rpm for 24 h at room temperature. 2.0 wt/v% biochar-alginate solution was prepared and added dropwise into a 0.2 M aqueous $CaCl_2$ solution, using biochar pyrolysed at 200 °C (B2), 500 °C (B5), and unpyrolysed biochar (B0). After 3 h of curing, the beads were washed with DI water to remove residual $CaCl_2$ solution and dried overnight at 50 °C.

Preparation of Ofloxacin (OFX). 250 mg of OFX powder (Sigma-Aldrich) was dissolved in DI water to obtain 1000 mg L^{-1} of OFX. Hydrochloric acid was added to promote dissolution. The resultant solution was stored under dark conditions at 4 °C in the refrigerator for later use. Other concentrations were obtained by diluting the 1000 mg L^{-1} OFX solution with DI water.

2.2 Characterisation of Jackfruit-Derived Biochar

The surface morphology of the biochar particles were analysed using a JEOL JSM-6100LA scanning electron microscope (SEM) in the Science Center Singapore lab. BET (Brunauer–Emmett–Teller) analysis of resolution 0–50 nm was done using Quantachrome Autosorb 1 analyzer in a certified lab (Latech Scientific Supply, Singapore).

2.3 Swelling Behaviour Experiments

To investigate the swelling properties of each BAB sample, 0.15 g of dry BAB were placed in 20 mL of DI water at room temperature. At stipulated time intervals (10, 15, 30, 45, 60, 90, 120, 150, 165, 180 and 1440 min), the beads were filtered, patted dry and weighed with an electronic mass balance. This is considered the first swelling cycle of our re-swelling studies.

After the first swelling cycle, the same BAB samples were dried overnight at 50 ºC and were left again in DI for 24 h, weighed, and dried at 50 °C (Fig. 1(a)). 3 swelling cycles were conducted in total.

2.4 Batch Adsorption Experiments

Adsorption capacity for all experiments was analysed using a UV-VIS spectrophotometer (Shimadzu UV-1280) at 294 nm (Fig. 1(b)). Standard OFX solutions (2, 4, 6, 8, 10 mg L^{-1}) were then prepared to create a OFX calibration curve.

For kinetic studies, 0.15 g of each BAB sample (B0, B2, B5) were added into 3 separate 200 mL samples of 30 mg L^{-1} OFX standard solution. Samples were then shaken at 150 rpm at room temperature at fixed timings of 15, 30, 60, 90, 120, 180, 240, 300, 360, 1440 and 2880 min.

For adsorption isotherm studies, 0.03 g of each BAB sample (B0, B2, B5) was added into separate bottles containing 40 mL of OFX solution with the initial concentrations of 10, 20, 30, 40 and 50 mg L^{-1} at room temperature for 24 h at 150 rpm. This procedure was performed in triplicate, and the reported value is the average of the three replicates.

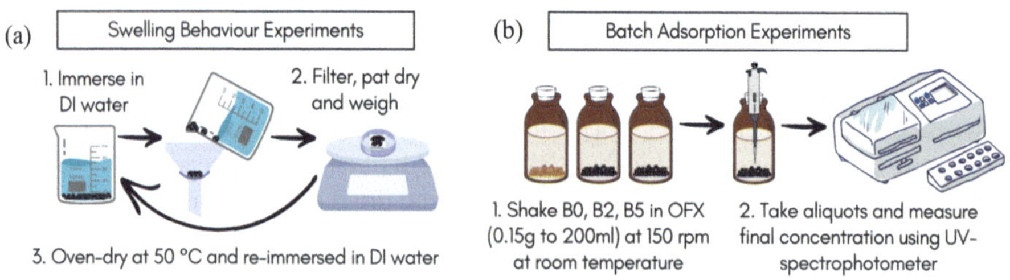

Fig. 1. Visual representation of (a) Swelling Behaviour and (b) Batch Adsorption experiments conducted (Image source: Canva)

2.5 Data Analysis

Swelling Behaviour. Swelling behaviour was analysed using its swelling ratio (SR):

Swelling ratio:

$$SR(\%) = \left[\frac{W_s - W_d}{W_d}\right] \times 100\%$$

where W_s and W_d is the weight of the swollen hydrogels after each swelling cycle and the dried hydrogels before each swelling cycle respectively. Higher SR shows better swelling properties. Consistent SR across swelling cycles indicates reusability.

Adsorption Kinetics. The kinetic studies were analysed using the Pseudo-First and Second-Order kinetic models:

Pseudo-First-Order model(PFO):	*Pseudo-Second-Order model(PSO):*
$qt = q_e\left(1-e^{-(k_1t)}\right)$	$q_t = \frac{k_2q_e^2t}{1+k_2q_et}$

Where q_t (mg g^{-1}) is the adsorption capacity of the BAB sample at a given time t, q_e is the equilibrium adsorption capacity (mg g^{-1}) of the BAB sample, and k_1 and k_2 is the rate constants of PFO and PSO respectively. For PFO, A graph of $\ln(q_e - q_t)$ against t was plotted, where the gradient of the slope is $-k_1$ and the vertical intercept is $\ln(q_e)$. For PSO, a graph of $\frac{t}{q_t}$ against t was plotted, where the gradient of the slope is $\frac{1}{q_e}$ and the vertical intercept is $\frac{1}{k_2q_e^2}$. Higher correlation coefficient (R^2) values for PFO and PSO indicate physisorption and chemisorption respectively.

Adsorption Isotherm Studies. The adsorption isotherm studies were analysed using the Langmuir and Freundlich isotherm models:

Langmuir adsorption isotherm model:	*Freundlich adsorption isotherm model:*
$\frac{C_e}{q_e} = \frac{1}{q_e}(C_e) + \frac{1}{q_mk_L}$	$ln\, q_e = ln\, C_e + \frac{1}{n} ln\, kf$

Where C_e (mg L^{-1}) is the concentration of OFX at equilibrium, q_e is the equilibrium adsorption capacity (mg g^{-1}) of the BAB sample, q_m is the maximum adsorption capacity (mg g^{-1}) of the BAB sample, k_L is the Langmuir adsorption constant, k_f is the Freundlich constant and relates to adsorption capacity, and n is the Freundlich adsorption intensity. For the Langmuir model, A graph of $\frac{C_e}{q_e}$ against C_e was plotted, where the gradient of the slope is $\frac{1}{q_e}$ and vertical intercept is $\frac{1}{q_mk_L}$. For the Freundlich model, a graph of $\ln q_e$ against $\ln C_e$ was plotted. The gradient of the slope is $\frac{1}{n}$ (typically 0–1), and the vertical intercept is given by $\ln k_f$. Higher R^2 values for Langmuir and Freundlich describe monolayer and multilayer adsorption respectively.

3 Results and Discussion

3.1 Characterisation of Jackfruit-Derived Biochar

SEM images show the biochar samples to have rough irregular surfaces (Fig. 2).

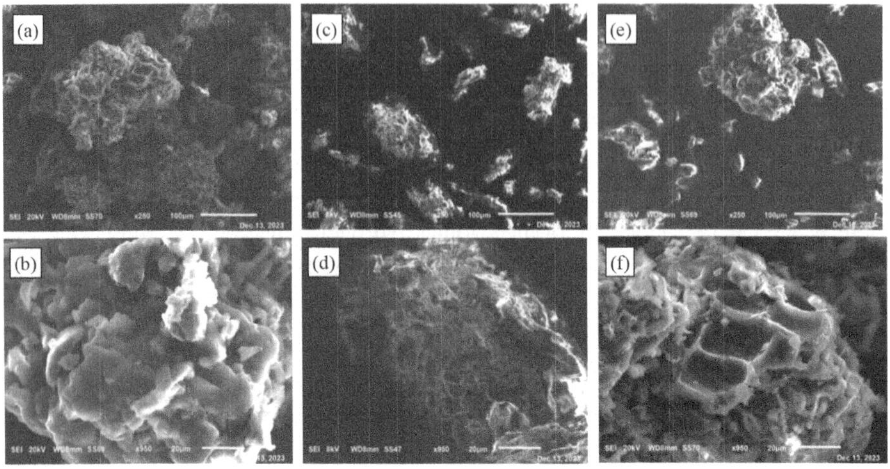

Fig. 2. Surface morphology of biochar incorporated in: (a) B0 at 100 μm, (b) B0 at 20 μm, (c) B2 at 100 μm, (d) B2 at 20 μm, (e) B5 at 100 μm, (f) B5 at 20 μm

Porous structure can be seen on the B2 and B5's biochar, but not on the unpyrolysed B0 biochar, with pore sizes increasing with pyrolysis temperature.

There was an increase in BET surface area from B0 to B2 (Table 1). This may be due to thermal decomposition of hemicellulose, which begins breaking down at around 200 °C, while the decrease in surface area from B2 to B5 biochar may be due to further breakdown of cellulose, resulting in larger pores and lower surface area [11]. A lower surface area could mean fewer adsorption sites available for OFX adsorption. Pore sizes of the biochar vary from 7.87–10.7 nm, indicating a structure of mainly mesopores (2–50 nm).

Table 1. BET pore volume, pore width and surface area of biochar incorporated in B0, B2 and B5

Sample	Surface area/m^2 g^{-1}	Total pore volume/cm^3 g^{-1}	Average pore diameter/nm
B0	0.4477 ± 0.0070	0.0010	7.9120
B2	0.5087 ± 0.0081	0.0009	7.8665
B5	0.2563 ± 0.0400	0.0013	10.6611

3.2 Swelling Behaviour of BAB

All samples showed a dramatic swelling in initial times and then plateauing at around 150 min, indicating equilibrium has been reached (Fig. 3(a)). The SR of the BAB samples was consistently in the order B0 > B5 > B2, with equilibrium SR (Q_S) of 118%, 74% and 77% respectively in the first swelling cycle. The higher swelling ratio of B0 compared to B5 and B2 may be due to the removal of hydrophilic groups due to high temperatures during pyrolysis [12]. Additionally, B5's higher SR compared to B2 may be due to the more porous biochar structure and bigger pore diameters formed from higher pyrolysis temperatures [13].

Q_s showed an increase of 19%, 5% and 7% for B0, B2 and B5 respectively from the first to the second swelling cycle before showing a decrease of 21%, 3% and 1% for B0, B2, B5 respectively in the third swelling cycle (Fig. 3(b)). This increase in Q_s from the first to second cycle is consistent with other hydrogels [14]. The decrease in the third cycle may be due to degradation or structural collapse after repeated use [15]. The increase in Q_s in the second cycle shows stability in mechanical performance and has potential of reusability in aqueous media.

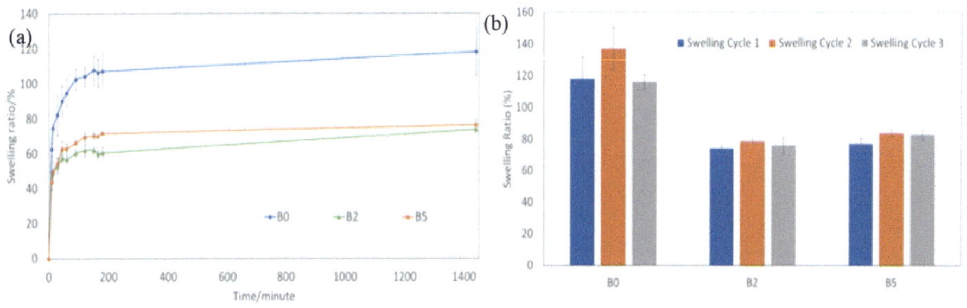

Fig. 3. (a) Swelling kinetics graph of the three kinds of BAB, (b) Maximum swelling ratio (%) reached by the three kinds of BAB in the first, second and third cycle of swelling

3.3 Adsorption Kinetics

The kinetic studies (Fig. 4(a)) indicate a general increase in adsorption of OFX with increased contact time. However, there is a drastic overshooting behaviour observed at 180 min for all samples, before plateauing off at the 1440 min mark, indicating equilibrium has been reached. Performance of the BAB samples followed the order B0 > B5 > B2 before the 1200 min mark and B2 > B5 > B0 after the 1200 min mark with maximum adsorption values of 11.4 mg g^{-1}, 9.16 mg g^{-1}, 8.37 mg g^{-1} respectively. The better performance of B2 suggests that biochar pyrolysed at lower temperatures is more effective in removing OFX. This is consistent with other studies and may be due to the higher surface area shown in BET analysis and presence of more functional groups on the surface of biochars pyrolysed at lower temperatures [5].

The overshooting behaviour may be caused by the presence of Ca^{2+} ions in the solution. Grishkewich et al. (2020) noted the leaching of free Ca^{2+} ions from the alginate hydrogel [16]. The Ca^{2+} ions can form strong complexes with OFX, reducing their adsorption affinity to the BAB samples, causing desorption of OFX molecules, and a subsequent increase in OFX concentration [17]. Additionally, there could be possible screening of electrostatic interaction by the Ca^{2+} ions, causing a further increase in OFX concentration [18]. OFX concentration then decreases to equilibrium as some OFX continues to be adsorbed onto the BAB samples.

The R^2 values for PSO (Table 2) were all ≥ 0.979, indicating that OFX adsorption in BAB fits the PSO model better than the PFO model. This is consistent with existing literature on the adsorption of OFX by other kinds of biochar hydrogel composites and carbon-based adsorbents [17, 19, 20]. The high R^2 values for PSO indicates that OFX undergoes chemisorption related to the exchange or sharing of electrons between OFX and the binding sites on the BAB samples.

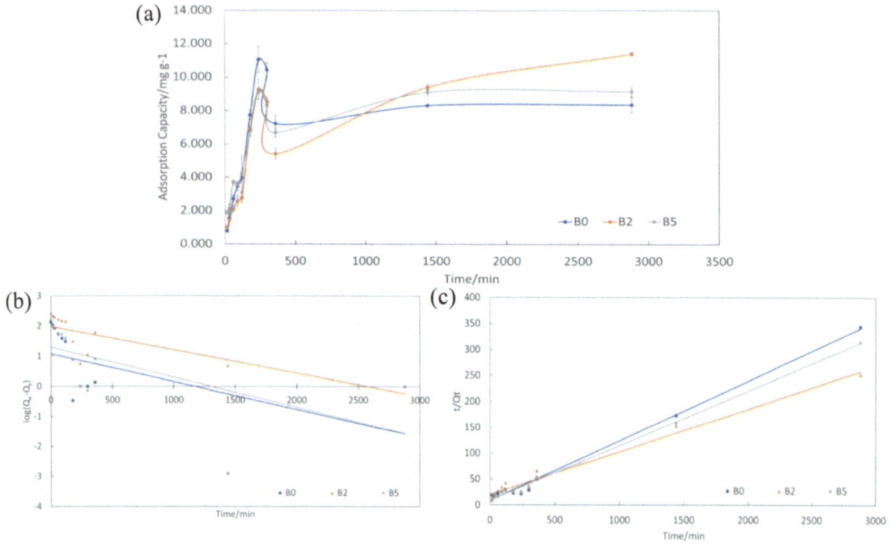

Fig. 4. (a) Raw OFX adsorption kinetics, kinetics of OFX adsorption using (b) pseudo-first-order, (c) pseudo-second-order

Table 2. Pseudo-Order kinetic parameters for OFX adsorption

Sample	Experimental q_e/mg g^{-1}	Pseudo-First-Order			Pseudo-Second-Order		
		q_e/mg g^{-1}	k_1/min^{-1}	R^2	q_e/mg g^{-1}	k_2/g mg^{-1} min^{-1}	R^2
B0	8. 2	2.926	0.0009	0.283	8.688	0.00147	0.993
B2	11.429	7.306	0.0008	0.659	12.121	0.000328	0.979
B5	9.161	3.662	0.0001	0.343	9.506	0.00115	0.997

3.4 Adsorption Isotherms

The adsorption of OFX onto the BAB samples increases in a non-linear manner as the concentration of adsorbate increases (Fig. 5(a)). At lower concentrations, the equilibrium adsorption capacity follows the order B2 > B0 > B5. At higher concentrations of above

38 mg L^{-1}, the equilibrium adsorption increases with pyrolysis temperature, in the order B5 > B2 > B0.

The Freundlich model is a better fit for all samples with higher R^2 values (0.930–0.963) compared to Langmuir (0.253–0.810) (Table 3). This shows that the surface of the BAB samples were heterogeneous and the adsorption of OFX on the BAB samples were multilayer [21]. This was consistent with other studies where Ofloxacin adsorption mechanisms were generally Langmuir or Freundlich [22]. Furthermore, the n values of all BAB samples were more than 1, implying support of physical process in the adsorption so that OFX was easily adsorbed onto the BAB samples [4].

Additionally, SEM analysis revealed a rough inhomogeneous porous surface, where the distribution of functional groups are random such that the adsorption sites possess variable adsorption energies that lead to multilayer adsorption [22], corroborating the fit with Freundlich.

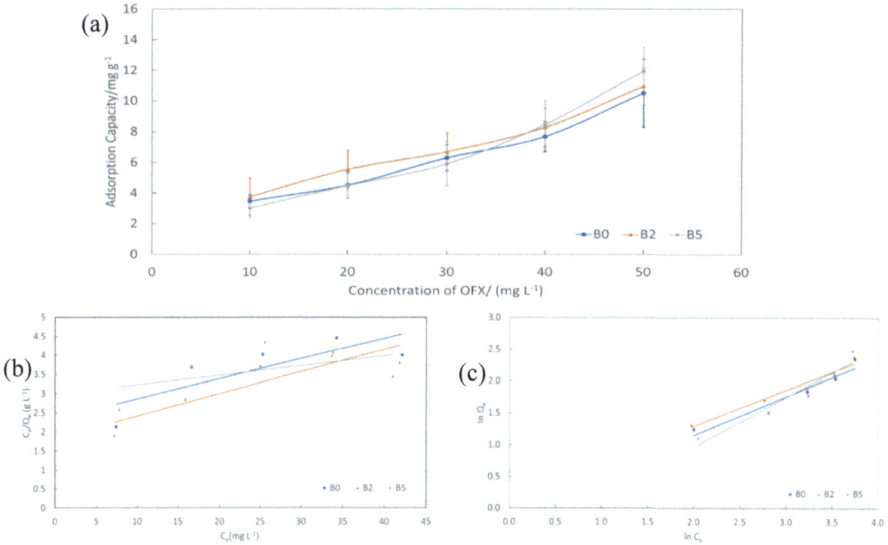

Fig. 5. (a) Graph of OFX batch adsorption using concentrations of 10–50 mg L^{-1}, modelling of OFX batch adsorption using (b) the Langmuir model, (c) the Freundlich model

Table 3. Langmuir and Freundlich parameters for OFX adsorption

Sample	Langmuir Model			Freundlich Model		
	K_L/mg L^{-1}	Q_m/mg g^{-1}	R^2	n	k_f/mg g^{-1} (L $mg^{-1})^{1/n}$	R^2
B0	0.0226	18.939	0.658	1.640	0.931	0.930
B2	0.0317	17.182	0.810	1.762	1.178	0.963
B5	0.0085	39.370	0.253	1.277	0.548	0.933

4 Conclusion

Jackfruit derived biochar-alginate composite hydrogel beads were successfully synthesised to remove OFX from aqueous solution. The adsorption of OFX was found to fit the Freundlich and Pseudo-Second Order kinetic model. B2 was most effective in removing OFX with a maximum adsorption capacity of 11.4 mg g^{-1}, showing that a lower pyrolysis temperature of biochar incorporated produces a more effective adsorbent. This makes BAB less energy intensive, aligning with our aims of producing a cost-efficient adsorbent. Furthermore, discarded jackfruit peels could potentially be reutilised as biochar for treating wastewater, reducing waste. Overall, jackfruit biochar, when encapsulated in alginate hydrogel beads, shows great potential in removing Ofloxacin and can also be easily removed from the aquatic environment as beads, making it environmentally viable. Further research can be done to investigate its potential in removing other pollutants like textile dyes, other pharmaceutical pollutants, and heavy metal pollutants.

References

1. Georgin, J., Franco, D.S., Ramos, C.G., Piccilli, D.G.A., Lima, E.C., Sher, F.: A review of the antibiotic ofloxacin: Current status of Ecotoxicology and scientific advances in its removal from aqueous systems by adsorption technology. Chem. Eng. Res. Des. **193**, 99–120 (2023). https://doi.org/10.1016/j.cherd.2023.03.025
2. Liu, D., Zhang, Y., Yang, Q., Li, Y., Li, J., Liao, X.: Fate of ofloxacin in rural wastewater treatment facility: Removal Performance Pathways and Microbial Characteristics. Bioresource Technology **371**, 128611 (2023). https://doi.org/10.1016/j.biortech.2023.128611
3. Patel, M., Kumar, R., Kishor, K., Mlsna, T., Pittman, C.U., Mohan, D.: Pharmaceuticals of emerging concern in aquatic systems: chemistry, occurrence, effects, and removal methods. Chem. Rev. **119**(6), 3510–3673 (2019). https://doi.org/10.1021/acs.chemrev.8b00299
4. Wang, Y., Lu, J., Wu, J., Liu, Q., Zhang, H., Jin, S.: Adsorptive removal of fluoroquinolone antibiotics using bamboo biochar. Sustainability **7**(9), 12947–12957 (2015). https://doi.org/10.3390/su70912947
5. Akhtar, L., Ahmad, M., Iqbal, S., Abdelhafez, A.A., Mehran, M.T.: Biochars' adsorption performance towards Moxifloxacin and ofloxacin in aqueous solution: Role of pyrolysis temperature and biomass type. Environ. Technol. Innov. **24**, 101912 (2021). https://doi.org/10.1016/j.eti.2021.101912
6. Zilberman, D., Laird, D., Rainey, C., Song, J., Kahn, G.: Biochar supply-chain and challenges to commercialization. GCB Bioenergy **15**(1), 7–23 (2022). https://doi.org/10.1111/gcbb.12952

7. Turunen, S., et al.: Future solutions for osteoarthritis using 3D bioprinting of articular cartilage. Handbook of Surgical Planning and 3D Printing, 335–369 (2023). https://doi.org/10.1016/b978-0-323-90850-4.00004-1
8. Cui, X., Dai, X., Khan, K.Y., Li, T., Yang, X., He, Z.: Removal of phosphate from aqueous solution using magnesium-alginate/chitosan modified biochar microspheres derived from Thalia Dealbata. Biores. Technol. **218**, 1123–1132 (2016). https://doi.org/10.1016/j.biortech.2016.07.072
9. Roh, H., Yu, M.-R., Yakkala, K., Koduru, J.R., Yang, J.-K., Chang, Y.-Y.: Removal studies of CD(II) and explosive compounds using buffalo weed biochar-alginate beads. J. Ind. Eng. Chem. **26**, 226–233 (2015). https://doi.org/10.1016/j.jiec.2014.11.034
10. Yusop, M.F., Abdullah, A.Z., Ahmad, M.A.: Adsorption of remazol brilliant blue R dye onto jackfruit peel based activated carbon: Optimization and simulation for mass transfer and surface area prediction. Inorg. Chem. Commun. **158**, 111721 (2023). https://doi.org/10.1016/j.inoche.2023.111721
11. Lu, G.Q., Lua, A.C., Liu, C.Y., Low, J.C.F.: Surface area development of sewage sludge during pyrolysis. Fuel and Energy Abstracts **36**(3), 221 (1995). https://doi.org/10.1016/0016-2361(95)93465-p
12. Adhikari, S., Mahmud, M.A.P., Nguyen, M.D., Timms, W.: Evaluating fundamental biochar properties in relation to water holding capacity. Chemosphere **328**, 138620 (2023). https://doi.org/10.1016/j.chemosphere.2023.138620
13. Gray, M., Johnson, M.G., Dragila, M.I., Kleber, M.: Water uptake in biochars: The roles of porosity and Hydrophobicity. Biomass Bioenerg. **61**, 196–205 (2014). https://doi.org/10.1016/j.biombioe.2013.12.010
14. Uysal, Y., Doğaroğlu, Z.G., Çaylali, Z., Makas, M.N.: Evaluation of swelling properties of different biochar-doped hydrogels. J. Soils Sediments **23**(11), 3787–3805 (2023). https://doi.org/10.1007/s11368-023-03594-9
15. Oyekunle, I.P., et al.: Recent advances in the adsorption of ofloxacin from Aqueous Media. J. Ind. Eng. Chem. **128**, 1–16 (2023). https://doi.org/10.1016/j.jiec.2023.07.049
16. Grishkewich, N., Mohammed, N., Tam, K.C.: Impact of cross-linking agents on the adsorptive performance of cellulose nanocrystal–alginate hydrogel beads. Cellulose **30**(11), 6933–6943 (2023). https://doi.org/10.1007/s10570-023-05300-x
17. Ma, Y., et al.: Iron/zinc and phosphoric acid modified sludge biochar as an efficient adsorbent for fluoroquinolones antibiotics removal. Ecotoxicol. Environ. Saf. **196**, 110550 (2020). https://doi.org/10.1016/j.ecoenv.2020.110550
18. Yadav, S., et al.: Methionine-functionalized graphene oxide/sodium alginate bio-polymer nanocomposite hydrogel beads: Synthesis, isotherm and kinetic studies for an adsorptive removal of fluoroquinolone antibiotics. Nanomaterials **11**(3), 568 (2021). https://doi.org/10.3390/nano11030568
19. Teo, C.Y., Jong, J.S., Chan, Y.Q.: Carbon-based materials as effective adsorbents for the removal of pharmaceutical compounds from aqueous solution. Adsorpt. Sci. Technol. **2022**, 1–17 (2022). https://doi.org/10.1155/2022/3079663
20. Nguyen, H.T., et al.: Low-cost hydrogel derived from agro-waste for veterinary antibiotic removal: Optimization, kinetics, and toxicity evaluation. Environmental Technology & Innovation **20**, 101098 (2020). https://doi.org/10.1016/j.eti.2020.101098
21. Freundlich, H.: Über die adsorption in lösungen. Z. Phys. Chem. **57U**(1), 385–470 (1907). https://doi.org/10.1515/zpch-1907-5723
22. Ureña-Amate, M.D., del Socias-Viciana, M., del Urbano-Juan, M., del García-Alcaraz, M.: Effects of ph and crosslinking agent in the evaluation of hydrogels as potential nitrate-controlled release systems. Polymers **15**(5), 1246 (2023). https://doi.org/10.3390/polym15051246

Predicting Stress Levels During a Mental Arithmetic Task from EEG Data

Yiming Kang[1], Sarah Tan Si Yu[1(✉)], Woon Qi Rui Ethan[1], Brian Premchand[2], and Tan Guoxian[1]

[1] Raffles Science Institute, Raffles Institution, 1 Raffles Institution Lane, Singapore 575954, Singapore
kangyiming2006@gmail.com, 24ysara954e@student.ri.edu.sg
[2] Institute for Infocomm Research (I2R), Agency for Science, Technology and Research (A*STAR), 1 Fusionopolis Way, #21-01 Connexis (South Tower), Singapore 138632, Republic of Singapore

Abstract. Stress is known to have a range of negative impacts – for example, occupational stress faced by uniformed personnel may harm their performance and mental health. The detection of stress from electroencephalography (EEG) data is of interest in stress management. In this project, we investigated different factors that affect the performance of models in predicting stress from EEG data. We analyzed a publicly available online repository of EEG recordings of 36 individuals in stressed and unstressed states, accessed through PhysioNet, extracting two features: Pearson's correlation coefficient and mutual entropy between channels. Nine classifiers from scikit.learn were trained and evaluated under different validation conditions (k-fold without pooling, k-fold after pooling, and leave-one-out). A trend was seen in performance, which decreased in the order of validation conditions listed above. Different classifiers emerged as the most accurate under different combinations of feature and validation methods.

Keywords: electroencephalography · stress · validation

1 Background and Purpose

Electroencephalography (EEG) is a non-invasive technique where the electrical activity of the brain is recorded using electrodes arranged on the subject's scalp [1]. Numerous authors have sought to measure mental stress by analyzing EEG signals [2]. Different features can be extracted from EEG data, such as power spectral density, measures of correlation between channels, Pearson's correlation, and more [3]. To predict stress levels from EEG states, various studies have used a range of classifiers, such as Support Vector Machine and k-nearest neighbours; different methods of validation have been used, such as k-fold cross-validation and leave-one- out validation [4]. Existing literature includes a study that induced stress in subjects through mental arithmetic. That study chose power band features, with a Support Vector Machine as the classifier, achieving 96% accuracy in two-level stress classification [5].

H. Guo et al. (Eds.): IRC-SET 2024, PTT, pp. 156–167, 2025.
https://doi.org/10.1007/978-981-96-3770-6_17

Occupational stress is "the harmful physical and emotional responses that occur when job demands exceed the capabilities of the worker" [6]. Uniformed personnel have an exceptionally stressful occupation, arising from the potentially psychologically and physically traumatic nature of their tasks [7–9]. Working under such conditions contributes to emotional and physical exhaustion [10], increasing risks of mental health and stress-related disorders [11–13]. This strains their health and well-being [14], on top of their quality of life [15] and performance levels [16]. Thus, the accurate detection of stress, especially among uniformed personnel, is necessary. EEG is preferable to stress measurement surveys due to its objectivity.

While the equipment is bulky, it can be more easily incorporated into the headgear worn by uniformed professionals, possibly increasing its appropriateness for use in stress monitoring. Hence, the development of models to detect mental stress in uniformed personnel from EEG data is relevant and timely. This project aims to compare the performance of models using different features and classifiers, to determine the most appropriate ones in stress monitoring. The project will also build on the literature by applying multiple validation methods, as the models' performances have different implications on their real-world implementation.

2 Materials and Methods

An online database of EEG recordings [17] was used, accessed through PhysioNet [18]. 36 participants were included in the recordings, with 3 min of rest state and 1 min of stress-inducing mental arithmetic recorded per participant. The recordings contained 21 channels. The sample rate was 500 Hz. From the literature, Pearson's correlation and mutual information between channels were identified as potential features for analysis. The Python library eeglib [19] was used in processing the data.

Testing was first done using Pearson's correlation. For each subject, each of the two EEG recordings was segmented into non-overlapping windows [20] of length 1000 samples, i.e. 2 s. For each of these 2-s segments, a 21-by-21 matrix was created, where each matrix contains the Pearson's correlation coefficient between the ith and j^{th} channels of the EEG recording. The Pearson's correlation coefficients were calculated. This matrix was then flattened into an array. A dataset was obtained, containing 36 tuples, one per individual, each containing a list of arrays and a list of Booleans indicating whether the array for that individual at the corresponding index was stressed. For all except one of the 36 subjects, the ratio of unstressed to stressed data was approximately 3:1. Hence, for every subject, two-thirds of only the unstressed data were discarded to prevent bias in the training data. The range of classifiers imported can be found in the Appendix.

2.1 Training and Testing on One Subject's Data at a Time with k-fold Cross Validation

Each classifier was trained on data from one subject at a time with k-fold cross- validation (k = 5), i.e. for each of the 36 individuals, different fifths of the data were used five times to test the model trained on the remaining four-fifths.

2.2 Training and Testing on All Subjects' Data Simultaneously with k-fold Cross-Validation

36 subject tuples were merged into a single tuple of two lists containing all the arrays and all the Boolean variables respectively. Each classifier was trained in turn on this merged dataset, again using k-fold cross-validation (k = 5).

2.3 Leave-One-Out Validation

The classifiers were trained on 35 subjects at a time and tested using the remaining one. A master dataset with two lists (one for EEG data in arrays and the other for Booleans) was created. Each classifier was trained on elements (0, n) and (1, n) of the master dataset and tested on the two lists in the nth tuple of the original dataset in part 2.1 above. Leave-one-out validation is conducted for only the best-performing classifier. The fraction of data identified by the model as stressed, from the stressed(= sensitivity) and unstressed(= 1 – specificity) recordings respectively, were recorded.

The process above, consisting of data processing, followed by steps 2.1, 2.2 and 2.3, was repeated using mutual information between channels.

Confusion Matrix	
True negatives (TN)	False positives (FP)
False negatives (FN)	True positives (TP)

Accuracy = (TN + TP)/(TN + TP + FN + FP)
Sensitivity = TP/(TP + FN) Specificity = TN/(TN + FP)
Positive predictive value (PPV) = TP/(TP + FP)
Negative predictive value (NPV) = TN/(TN + FN)

The accuracy, sensitivity, sensitivity, specificity, PPV and NPV of the models were calculated from a confusion matrix for each step, using the formulae above. Their average values and standard deviations were found, and relevant diagrams were created.

3 Results and Discussion

All tables mentioned can be found in the Appendix. For all figures titles, the accuracy, sensitivity, specificity, PPV and NPV values will be referred to as "5 metrics", Pearson's correlation as "PC", and mutual information as "MI".

Figure 1 suggests that most classifiers perform slightly better with Pearson's correlation than mutual information, possibly due to inaccuracies in the latter's calculation using a 10-by-10 histogram with low resolution. Despite this, the performance difference between the two features is minor, and especially for top-performing MLPClassifier, it falls well within the uncertainty range in respective metrics (Tables A1 and A2). Deleting two-thirds of unstressed data successfully prevents data bias, as sensitivity and specificity values across classifiers were similar. This safeguards against higher accuracy scores driven by consistent prediction of the majority class, or unstressed, avoiding

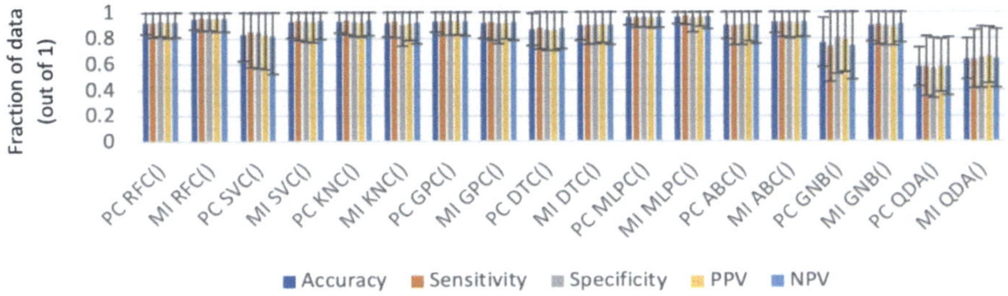

Fig. 1. 5 metrics from step 2.1. Error bars represent standard deviations.

skewed outcomes that compromise experiment validity. In both cases, the MLPClassifier performed best, indicating its suitability for EEG-based stress detection systems, possibly attributed to the MLP algorithm being said to be strong at "adaptive learning" [21] and to "[work] well with large input data" [22].

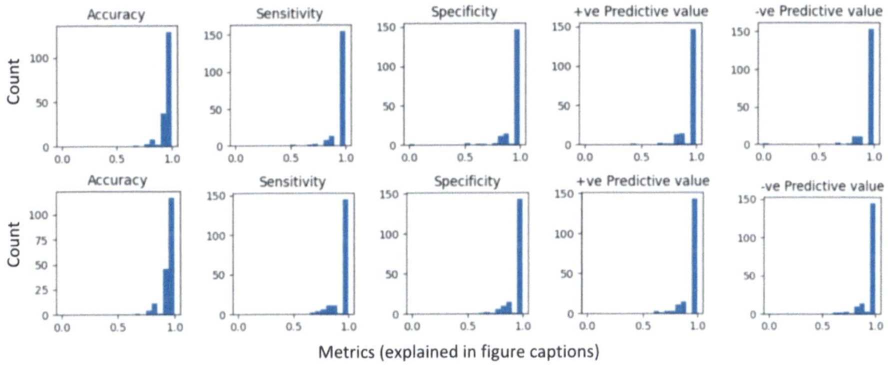

Fig. 2. Histograms showing distribution of the 5 metrics across 180 folds for MLPClassifier in part 2.1, using PC (top row) and MI (bottom row) as features.

Figure 2 shows the performance of the MLPClassifier was relatively consistent as almost all folds returned high values for the five metrics.

Comparing Fig. 3 to Fig. 1, using all subjects' data for training and testing in step 2.2 led to a general performance decline, potentially attributed to subject-specific idiosyncrasies. Training classifiers separately for each subject's stressed and unstressed states might allow them to ignore these idiosyncrasies, but pooling data from all subjects could confound classifiers, resulting in poorer performance. When pooling all subjects' data together, the best classifier is KNeighborsClassifier (Tables A3 and A4). Its better performance may be due to the nature of the algorithm, which classifies a new query point based on the known classification of its k nearest neighbours in the training data [24]. Given each subject's idiosyncrasies, the few closest neighbours to each query point are likely the training data from the same subject under the same condition. Switching from training and testing on one subject to doing so on all subjects may make little difference

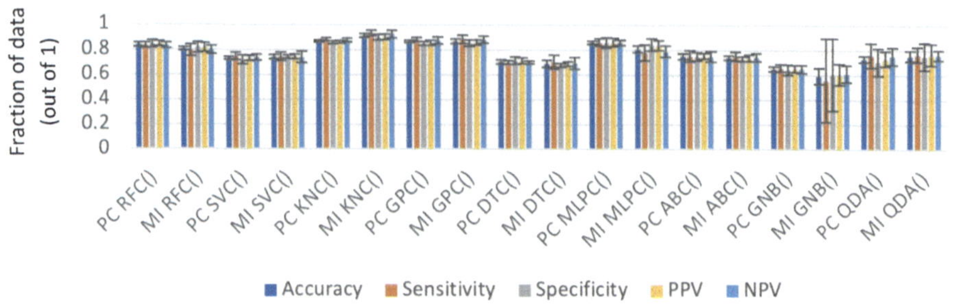

Fig. 3. 5 metrics from step 2.2. Error bars represent standard deviations.

to its performance. Another reason could be that this classifier is more suited for non-linear boundaries, since it is a form of unsupervised learning, where only the similarity to its neighbours are considered.

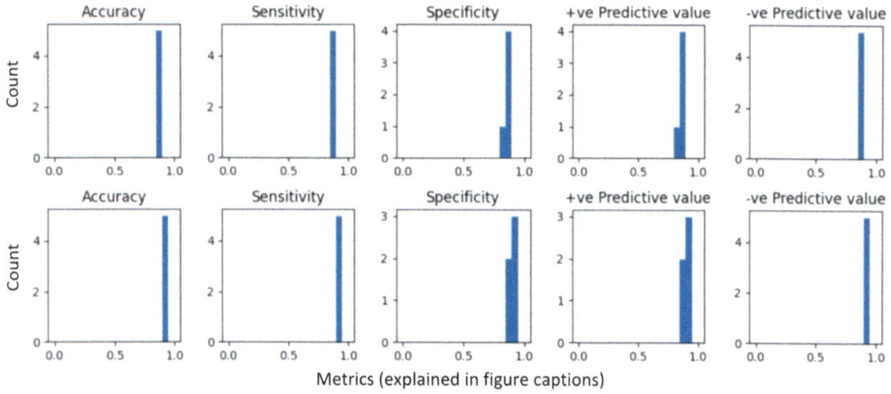

Fig. 4. Histograms showing the distribution of the 5 metrics across 5 folds for KNeighborsClassifier in part 2.2, with PC (top row) and MI (bottom row) as features.

Figure 4 shows that the performance of KNeighborsClassifier is consistent and balanced when used on the combined dataset described in part 2.2.

The poorer performance observed in step 2.3 than 2.2 may stem from the model's unfamiliarity with the idiosyncratic features of the excluded subject, as it lacks training on that subject's data. The less consistent performance, evident in Fig. 6, and higher standard deviations (Tables A5 and A6), may also be due to this. The notable drop in performance of the KNeighborsClassifiers (comparing Figs. 3 and 5), is thus explained. Nevertheless, most classifiers still fared better than chance which would have produced a value of 0.5 for all five metrics. For Pearson's correlation coefficient, the best-performing classifiers were RandomForestClassifier and AdaBoostClassifier (Table A5). Although the AdaBoostClassifier displayed a slightly lower average accuracy, it had more balanced predictions, with about equal average sensitivity and specificity. Its better performance could be due to the AdaBoostClassifier being better able to identify features of samples, due to its design in which "multiple classifiers [are combined] to increase the accuracy

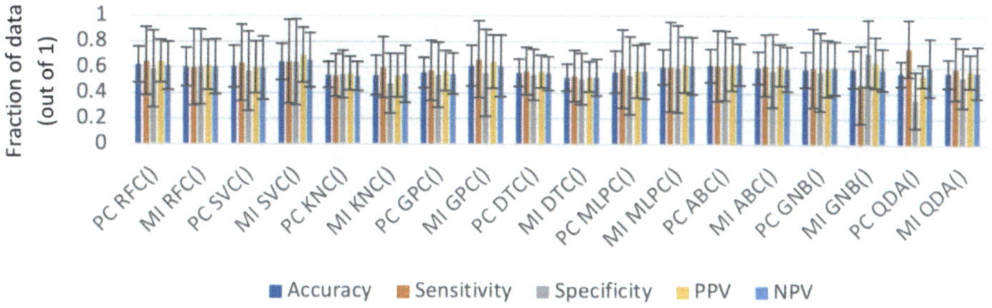

Fig. 5. 5 metrics from step 2.3. Error bars represent standard deviations.

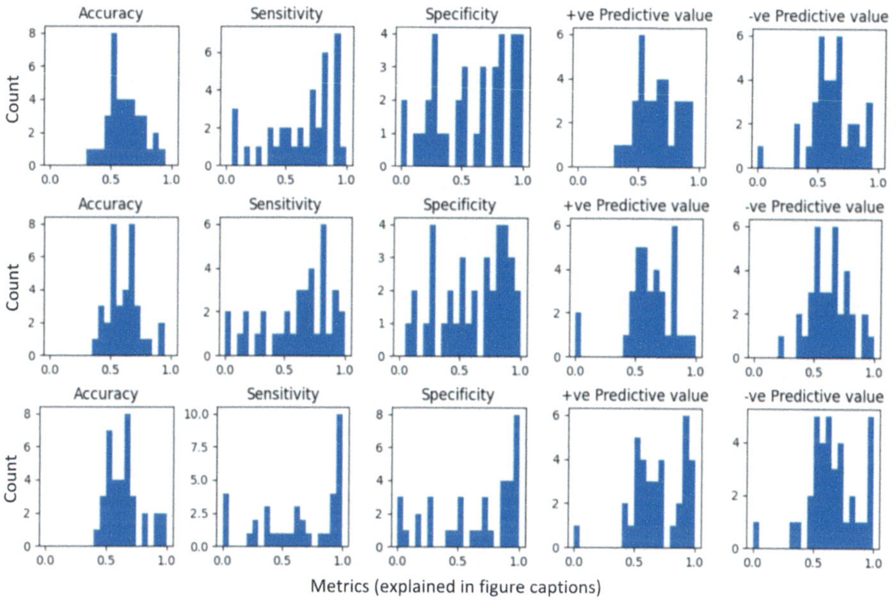

Fig. 6. Histograms showing the distribution of the 5 metrics across 36 tests in part 2.3. Top row: RandomForestClassifier on PC; middle row: AdaBoostClassifier on PC; bottom row: SVC(C = 0.025, kernel = 'linear') on MI.

of classifiers" [25]. For mutual information, SVC performed the best (Table A6). The specific reasons for the classifiers' performance variations remain unclear. However, the support vector machine algorithm has been observed in a study to classify EEG data more accurately than k-nearest neighbours or Naïve Bayes, albeit with the drawback of longer running time [5].

Despite poorer results from leave-one-out testing in part 2.3, Fig. 7 shows the most successful classifiers were still able to differentiate between stressed and unstressed data to some extent. On both diagrams, the blue box lies appreciably above the orange box and they lie on opposite sides of the horizontal through 0.5, which shows that the classifiers successfully labelled the majority of the stressed test EEG fragments as stressed, and only a minority of the unstressed test EEG fragments as stressed. The great range of the

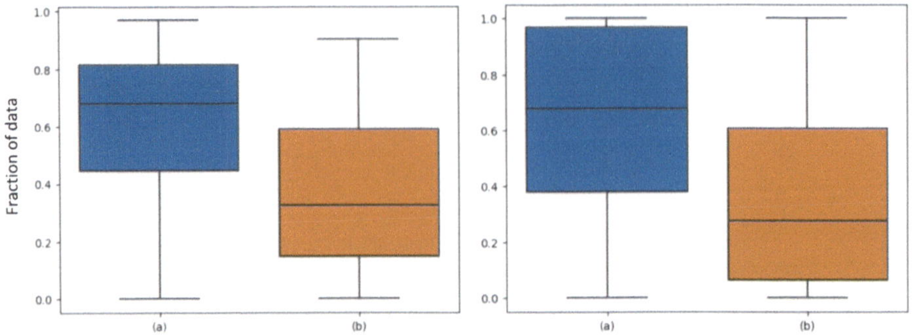

Fig. 7. Box-and-whisker plots (n = 36 each) for AdaBoostClassifier, PC (left) and SVC, MI (right). (a): fraction of data from the stressed recordings identified by the model as stressed (= sensitivity); (b): fraction of data from the unstressed recordings identified by the model as stressed (= 1 – specificity). Box: 25th, 50th i.e. median, 75th percentiles; whiskers: minimum and maximum.

whiskers in Fig. 7 and wide spread of the histograms in Fig. 6 imply that the performance of the models is inconsistent across subjects.

4 Conclusion

Varying performances across the three validation conditions have implications on the implementation of real-life EEG-based mental stress detection systems. Leave-one-out analysis in part 2.3 replicates how a classification model trained on a dataset of multiple subjects would respond to recordings from a new subject. The poorer performance implies that a real stress detection system's training dataset for the classifier should instead include recordings from each person using it. As performance in part 2.1 was the best, separate models trained for each user exclusively on that user's data may yield better performance.

Several areas for potential future research have been identified. Firstly, more features could be evaluated. As a rather rudimentary method was used to quantify mutual information here, future studies may consider other means to obtain a more accurate estimate, such as by using bi-dimensional histograms with greater numbers of bins on each axis. In addition, systematic investigations could be conducted concerning the combinations of EEG channels relevant in stress detection. Data from certain electrodes less affected by stress state could be omitted or assigned a lower weight. Moreover, more classifier algorithms could be tested and arguments passed to the classifiers used adjusted. The arguments chosen could have significant effects on classifier performance, especially the number of neighbours to consider in KNeighborsClassifier and the base estimator for AdaBoostClassifier. Furthermore, both the box-and-whisker plots in Fig. 7 show an area of overlap between the boxes (a) and (b), from 0.4 to 0.6. It may be meaningful to investigate whether this overlap arose coincidentally, due to the poor performance of the classifiers, or if it represents a third category of moderate stress. A multi-tiered system may provide better insight into triggers of particularly high stress levels for uniformed personnel, whose work results in a certain level of background stress, than a binary classification system that would constantly detect stress. Finally, different means of inducing stress other than mental arithmetic can be used in future studies, as the effects of mental arithmetic-induced stress on EEG recording features may not be identical to those due to other sources of stress, such as those faced in the occupational context.

Appendix

The classifiers used are as follows: RandomForestClassifier(), SVC(C = 0.025, kernel = 'linear'), KNeighborsClassifier(3), GaussianProcessClassifier(), DecisionTreeClassifier(), MLPClassifier(alpha = 1, max_iter = 1000), AdaBoostClassifier(), GaussianNB(), QuadraticDiscriminantAnalysis().

In the following tables, they will be referred to as "RFC()", "SVC()", "KNC()", "GPC()", "DTC()", "MLPC()", "ABC()", "GNB()" and "QDA()" respectively.

Table A1. Results from step 2.1 for Pearson's correlation coefficient (n = 180)

	Accuracy	Acc. Std. Dev	Sensitivity	Sens. Std. Dev	Specificity	Spec. Std. Dev	PPV	PPV Std. Dev	NPV	NPV Std. Dev
RFC()	0.95487	0.08294	0.95857	0.10206	0.95812	0.09554	0.95628	0.10489	0.95726	0.10575
SVC()	0.93223	0.13042	0.93825	0.15245	0.93444	0.16034	0.93213	0.16577	0.93854	0.14808
KNC()	0.92064	0.11166	0.93159	0.12945	0.90297	0.16399	0.91307	0.13270	0.92049	0.16389
GPC()	0.91822	0.12093	0.92666	0.14556	0.91281	0.15602	0.92201	0.13031	0.92518	0.14483
DTC()	0.90033	0.12068	0.90027	0.15113	0.90352	0.15035	0.90527	0.14163	0.90233	0.15250
MLPC()	0.96907	0.05721	0.97461	0.07067	0.95936	0.11213	0.96681	0.07975	0.96923	0.09802
ABC()	0.92582	0.08898	0.92885	0.12449	0.92317	0.12711	0.92599	0.12330	0.92936	0.11872
GNB()	0.89891	0.12584	0.90747	0.15729	0.89576	0.15326	0.89751	0.15631	0.90896	0.14227
QDA()	0.63656	0.15717	0.63843	0.22690	0.65183	0.23749	0.66843	0.21841	0.64414	0.22629

Table A2. Results from step 2.1 for mutual information (n = 180)

	Accuracy	Acc. Std. Dev	Sensitivity	Sens. Std. Dev	Specificity	Spec. Std. Dev	PPV	PPV Std. Dev	NPV	NPV Std. Dev
RFC()	0.92493	0.09074	0.92615	0.11694	0.93053	0.11216	0.92637	0.12446	0.92652	0.11705
SVC()	0.83445	0.20843	0.85357	0.27626	0.84903	0.28049	0.83209	0.26941	0.82000	0.29337
KNC()	0.93027	0.08661	0.93890	0.11229	0.92767	0.11399	0.92612	0.11482	0.93875	0.11713
GPC()	0.93082	0.08649	0.93324	0.11215	0.93415	0.11404	0.93410	0.10920	0.93173	0.11411

(continued)

Table A2. *(continued)*

	Accuracy	Acc. Std. Dev	Sensitivity	Sens. Std. Dev	Specificity	Spec. Std. Dev	PPV	PPV Std. Dev	NPV	NPV Std. Dev
DTC()	0.86594	0.12386	0.87610	0.16196	0.86267	0.16097	0.86195	0.16175	0.87638	0.16022
MLPC()	0.96167	0.06172	0.96165	0.08303	0.96314	0.08006	0.96055	0.08711	0.96393	0.08203
ABC()	0.90153	0.10942	0.90073	0.15445	0.90185	0.15530	0.90997	0.13559	0.90405	0.14893
GNB()	0.76612	0.19013	0.74235	0.28137	0.80625	0.28419	0.79952	0.26153	0.74109	0.26057
QDA()	0.57900	0.14916	0.58289	0.23190	0.56677	0.23105	0.58648	0.20372	0.58125	0.22392

Table A3. Results from step 2.2 for Pearson's correlation coefficient (n = 5)

	Accuracy	Acc. Std. Dev	Sensitivity	Sens. Std. Dev	Specificity	Spec. Std. Dev	PPV	PPV Std. Dev	NPV	NPV Std. Dev
RFC()	0.83965	0.01663	0.83455	0.02242	0.84494	0.02962	0.84643	0.02104	0.83388	0.02301
SVC()	0.73216	0.01099	0.74684	0.03172	0.71783	0.03603	0.72999	0.02120	0.73661	0.02472
KNC()	0.86766	0.00778	0.87988	0.01323	0.85505	0.01074	0.86040	0.01189	0.87493	0.01512
GPC()	0.86405	0.00942	0.88053	0.01672	0.84836	0.01462	0.85413	0.02182	0.87356	0.02687
DTC()	0.70462	0.01685	0.69692	0.01541	0.71218	0.02999	0.71153	0.02393	0.69798	0.01311
MLPC()	0.85773	0.01387	0.85911	0.03071	0.85584	0.04267	0.86033	0.03299	0.85787	0.02241
ABC()	0.74797	0.02333	0.75101	0.04560	0.74593	0.03310	0.75079	0.02169	0.74730	0.04225
GNB()	0.64906	0.01996	0.65612	0.02941	0.64201	0.04056	0.65127	0.02833	0.64719	0.03113
QDA()	0.73081	0.02554	0.75765	0.09881	0.70634	0.10932	0.73255	0.05947	0.75237	0.06856

Table A4. Results from step 2.2 for mutual information (n = 5)

	Accuracy	Acc. Std. Dev	Sensitivity	Sens. Std. Dev	Specificity	Spec. Std. Dev	PPV	PPV Std. Dev	NPV	NPV Std. Dev
RFC()	0.80534	0.01493	0.79535	0.04664	0.81691	0.04181	0.81665	0.03589	0.79873	0.03261
SVC()	0.74118	0.01796	0.74364	0.03370	0.74065	0.01333	0.74418	0.01832	0.73887	0.04524
KNC()	0.91327	0.01564	0.93033	0.02247	0.89720	0.02142	0.90170	0.02297	0.92485	0.03033
GPC()	0.86811	0.01928	0.88390	0.03397	0.85139	0.02690	0.85854	0.02165	0.87997	0.02917
DTC()	0.68836	0.02958	0.70082	0.05613	0.67680	0.01518	0.68711	0.01873	0.69158	0.04787
MLPC()	0.80803	0.02697	0.77598	0.06456	0.84146	0.04910	0.83542	0.03921	0.78874	0.04641
ABC()	0.73758	0.01513	0.74840	0.03646	0.72730	0.01849	0.73549	0.02661	0.74062	0.03156
GNB()	0.59172	0.06374	0.55695	0.33635	0.60423	0.28937	0.60616	0.08423	0.61130	0.06737
QDA()	0.75475	0.04101	0.76524	0.05983	0.75031	0.11032	0.76424	0.08639	0.75989	0.04000

Table A5. Results from step 2.3 for Pearson's correlation coefficient (n = 36)

	Accuracy	Acc. Std. Dev	Sensitivity	Sens. Std. Dev	Specificity	Spec. Std. Dev	PPV	PPV Std. Dev	NPV	NPV Std. Dev
RFC()	0.61934	0.13920	0.64785	0.26474	0.58485	0.29924	0.64862	0.16568	0.61149	0.18622
SVC()	0.60428	0.16150	0.63441	0.29614	0.56968	0.30780	0.60047	0.20274	0.59388	0.24467
KNC()	0.54019	0.10193	0.53674	0.16591	0.54352	0.18514	0.55252	0.13030	0.52925	0.11197
GPC()	0.55820	0.11854	0.57437	0.23482	0.53964	0.25320	0.57246	0.15608	0.54908	0.16058
DTC()	0.55760	0.10661	0.56989	0.18531	0.54199	0.19890	0.56572	0.12790	0.55142	0.13191
MLPC()	0.56496	0.16227	0.58871	0.30417	0.53650	0.30153	0.57002	0.20552	0.57272	0.21536
ABC()	0.61578	0.12877	0.61380	0.27524	0.61410	0.27058	0.62699	0.20771	0.62578	0.16149
GNB()	0.58604	0.13802	0.59767	0.30720	0.56863	0.30396	0.59488	0.22058	0.59754	0.20706
QDA()	0.55703	0.09670	0.75269	0.22128	0.35160	0.21712	0.53970	0.08736	0.60244	0.22423

Table A6. Results from step 2.3 for mutual information (n = 36)

	Accuracy	Acc. Std. Dev	Sensitivity	Sens. Std. Dev	Specificity	Spec. Std. Dev	PPV	PPV Std. Dev	NPV	NPV Std. Dev
RFC()	0.60119	0.15097	0.59946	0.29870	0.60040	0.29138	0.61857	0.19521	0.60308	0.21342
SVC()	0.64312	0.14087	0.64427	0.32637	0.63899	0.33234	0.69479	0.21352	0.65500	0.21090
KNC()	0.53825	0.15384	0.59946	0.23612	0.47290	0.23208	0.53651	0.17053	0.54686	0.21967
GPC()	0.61269	0.15708	0.66129	0.29949	0.55594	0.33769	0.64342	0.20653	0.61096	0.24001
DTC()	0.52218	0.10432	0.53136	0.19994	0.50977	0.19931	0.52721	0.11852	0.52296	0.14242
MLPC()	0.60284	0.13669	0.60663	0.34866	0.58840	0.33899	0.62347	0.21711	0.61350	0.22364
ABC()	0.59924	0.12475	0.61738	0.23985	0.57665	0.28430	0.62073	0.16112	0.59016	0.17507
GNB()	0.59191	0.14503	0.46953	0.30169	0.71426	0.25956	0.64195	0.19772	0.58607	0.16177
QDA()	0.55928	0.11037	0.59498	0.24478	0.52599	0.23497	0.57584	0.13229	0.56541	0.20087

References

1. Biasiucci, A., Franceschiello, B., Murray, M.M.: Primer: Electroencephalography. Current Biology **29**(3), 80–85 (2019). https://doi.org/10.1016/j.cub.2018.11.052
2. Katmah, R., et al.: A Review on Mental Stress Assessment Methods Using EEG Signals. Sensors **21**(15), 5043 (2021). https://doi.org/10.3390/s21155043
3. Darzi, A., Azami, H., Khosrowabadi, R.: Brain functional connectivity changes in long- term mental stress. J. Neurodevelopmental Cognition **1**(1), 16–41 (2022). https://doi.org/10.52547/jncog.2022.103423
4. Attallah, O.: An effective mental stress state detection and evaluation system using minimum number of frontal brain electrodes. Diagnostics **10**(5), 292 (2020). https://doi.org/10.3390/diagnostics10050292
5. Guo, J., Smitha, K.G.: EEG based stress level identification. In: 2016 IEEE International Conference on Systems, Man, and Cybernetics (SMC), Budapest (2016). https://doi.org/10.1109/SMC.2016.7844738

6. What Is Job Stress?. National Institute for Occupational Safety and Health. https://www.cdc.gov/niosh/docs/99-101/default.html#What%20Is%20Job%20Stress?
7. Cross, A.L., Ashley, L.: Police trauma and addiction: coping with the dangers of the job. FBI Law Enforcement Bulletin **73**(10), 24–32 (2004). https://www.ojp.gov/ncjrs/virtual-library/abstracts/police-trauma-and-addiction-coping-dangers-job
8. Duran, A., Montalban, F.M., Stangeland, P.: Burnout in police work: Incidence profile and influence of socio-demographic factors. Revista de Psicología Social **21**(1), 95–107 (2006). https://doi.org/10.1174/021347406775322205
9. Backteman-Erlanson, S., Padyab, M., Brulin, C.: Prevalence of burnout and associations with psychosocial work environment, physical strain, and stress of conscience among Swedish female and male police personnel. Police Pract. Res. **6**, 1–15 (2012). https://doi.org/10.1080/15614263.2012.736719
10. Johnson, O., Jaeckle, T.: Mitigating the emotional impact of stress on law enforcement personnel. Bureau of Justice Assistance (2018). https://bja.ojp.gov/library/publications/mitigating-emotional-impact-stress-law-enforcement-personnel
11. Violanti, J.M., Gehrke, A.: Police trauma encounters: precursors of compassion fatigue. Int. J. Emer. Mental Health **6**(2), 75–80 (2004). https://pubmed.ncbi.nlm.nih.gov/15298078/
12. Stansfeld, S., Candy, B.: Psychosocial work environment and mental health—a meta-analytic review. Scandinavian Journal of Work, Environment & Health **32**(6), 443–462 (2006). https://www.jstor.org/stable/40967597
13. Nieuwenhuijsen, K., Bruinvels, D., Frings-Dresen, M.: Psychosocial work environment and stress-related disorders, a systematic review. Occupational medicine (Oxford, England) **60**(4), 277–286 (2010). https://doi.org/10.1093/occmed/kqq081
14. Igboanugo, S., Bigelow, P.L., Mielke, J.G.: Health outcomes of psychosocial stress within firefighters: a systematic review of the research landscape. J. Occupational Health **63**(1) (2021). https://doi.org/10.1002/1348-9585.12219
15. Parsaei, R., Roohafza, H., Feizi, A., Sadeghi, M., Sarrafzadegan, N.: How Different Stressors Affect Quality of Life: An Application of Multilevel Latent Class Analysis on a Large Sample of Industrial Employees. Risk Management and Healthcare Policy **13**, 1261–1270 (2020)
16. Yu, A., et al.: Psychological contract breach and job performance of new generation of employees: Considering the mediating effect of job burnout and the moderating effect of past breach experience. Frontiers in Psychology **13** (2022). https://doi.org/10.3389/fpsyg.2022.985604
17. Zyma, I., et al.: Electroencephalograms during Mental Arithmetic Task Performance. Data **4**(1), 14 (2019). https://doi.org/10.3390/data4010014
18. Goldberger, A., et al.: PhysioBank, PhysioToolkit, and PhysioNet: Components of a new research resource for complex physiologic signals. Circulation **101**(23), e215–e220 (2000)
19. Cabañero-Gomez, L., Hervas, R., Gonzalez, I., Rodriguez-Benitez, L.: eeglib: A Python module for EEG feature extraction. SoftwareX **15** (2021). https://doi.org/10.1016/j.softx.2021.100745
20. Cabañero-Gomez, L., Hervas, R., Gonzalez, I., Rodriguez-Benitez, L.: eeglib: A library with tools for EEG analysis. Github (2022). https://github.com/Xiul109/eeglib
21. L'Sousa, R., Gupta, M.: Why MultiLayer Perceptron/Neural Network? https://courses.media.mit.edu/2006fall/mas622j/Projects/manu-rita-MAS_Proj/MLP.pdf
22. Akkaya, B., Çolakoğlu, N.: Comparison of multi-class classification algorithms on early diagnosis of heart diseases. In: y-BIS 2019 Conference, Istanbul (2019). https://www.researchgate.net/publication/338950098_Comparison_of_Multi-class_Classification_Algorithms_on_Early_Diagnosis_of_Heart_Diseases
23. Saba-Sadiya, S., Chantland, E., Liu, T., Ghassemi, M.M.: Unsupervised EEG Artifact Detection and Correction. Frontiers in Digital Health **2**, 608920 (2021). https://doi.org/10.3389/fdgth.2020.608920

24. K-Nearest Neighbors Algorithm. IBM. https://www.ibm.com/topics/knn
25. AdaBoost Classifier Algorithms using Python Sklearn Tutorial. www.datacamp.com. https://www.datacamp.com/tutorial/adaboost-classifier-python#adaboost-classifier

What's Under the Hood? Root Cause and Patch Analyses of Elevation of Privilege Vulnerabilities in the Windows Operating System

Tze Bin Bryan Pak(✉)

Raffles Institution, 1 Raffles Institution Lane, Singapore 575954, Singapore
bryanpak2006@gmail.com

Abstract. The Windows operating system implements many access control techniques in order to ensure its integrity, confidentiality and availability. Despite these access control techniques, studies have shown that the Windows operating system is still vulnerable to improper privilege management. This cybersecurity research project delves into three elevation of privilege Common Vulnerabilities and Exposures (CVEs), namely CVE-2020-1170, CVE-2021-21551, and CVE-2023-21768. The objective is to scrutinise the effectiveness of the respective vendor's patching strategies, specifically examining whether patches comprehensively address the root cause of the vulnerabilities or merely mitigate specific methods of replication and exploitation. The methodology uses a combination of static analysis tools like Ghidra, IDA and BinDiff, as well as dynamic analysis tools like Process Monitor (ProcMon) and WinDbg. Each vulnerability was replicated on a virtual machine (VM) to conduct a detailed analysis of its impact on the operating system and its root cause. Next, we evaluated if their patches are sufficient at fully mitigating the root cause of the vulnerability. In our study of CVE-2020-1170 and CVE-2021-21551, we found that the patches do not completely mitigate the vulnerability as there are ways that attackers can bypass them. In contrast, we found that CVE-2023-21768 exhibits successful patching that prevents elevation of privileges. While vendors patch certain vulnerabilities effectively, our research emphasises the need for more comprehensive measures to ensure robust cybersecurity.

Keywords: Cybersecurity · Dynamic Analysis · Patch Analysis · Reverse Engineering · Root Cause Analysis · Static Analysis · Vulnerability · Windows Operating System

1 Introduction

1.1 Background Information

The Windows operating system, developed by microsoft, is the most widely used operating system in the world, with 75.34% of users using it as of the end of 2022 [1]. However, it has the most attacks and system vulnerabilities to date due to its high number of users, which makes it the best operating system to attack or target [2]. In fact,

H. Guo et al. (Eds.): IRC-SET 2024, PTT, pp. 168–180, 2025.
https://doi.org/10.1007/978-981-96-3770-6_18

the total number of vulnerabilities reported by Microsoft has been increasing over the years, with a total of 1292 in 2022, a 7% increase from 2021 [3]. As such, the security of the Windows operating system is of paramount importance to both Microsoft and the 1.4 billion Windows computers worldwide [4].

The Windows operating system implements many access control techniques to ensure its integrity and confidentiality. This is done by restricting unprivileged users from accessing privileged data to prevent unauthorised access to sensitive system resources. Together with other security measures, this ensures a robust defence system against potential threats.

Even though there are protective measures, a paper by Bhurtel and Rawat [5] on the landscape of operating system vulnerabilities finds that Windows 10 and 11 are highly vulnerable to CWE-269, which refers to improper privilege management, that is, the Windows operating system does not properly assign, modify, track or check privileges of an actor, which creates an unintended sphere of control for that actor [6]. They recommend that specific security measures should be implemented to mitigate these vulnerabilities [5]. Despite the regular updates released by Microsoft for the Windows operating system, Softić and Vejzović's paper on vulnerability assessment and exploitation in the Windows 10 operating system argues that they are not effective at patching vulnerabilities. The study found that installing the latest version of Windows 10 did not guarantee complete protection against attacks [7]. This suggests that even after vendors apply security patches, vulnerabilities can still exist. Therefore, this project aims to analyse the patches that various vendors are taking to mitigate elevation of privilege vulnerabilities in the Windows operating system.

In this report, "elevation of privilege" is defined as a malicious user gaining access to a lower-level account and exploiting a weakness in the system to gain root-level access to a resource or system [8]. Contrary to Microsoft's security servicing criteria [9], We argue that administrator-to-kernel is a security boundary, because an administrator, despite having elevated privileges, would not have the same level of privileges as the kernel.

1.2 Purpose of Project

We have two main aims of this project. The first is to replicate these vulnerabilities in virtual machines (VMs), in order to understand their root causes. The second is to carry out patch analysis to determine whether patches released by respective vendors address the root cause of the vulnerability or if it misses out other threat scenarios that attackers can make use of.

1.3 Hypothesis

When a vulnerability is identified in the Windows operating system, various vendors attempt to patch it. However, the patch may not be comprehensive. We hypothesise that patches are typically designed to address a specific method of replication instead of targeting the root cause of the vulnerability. This means that an attacker can use a slightly different method of replication based on a similar concept to bypass the patch, which would render the patch ineffective against this new method of replication.

1.4 Methodology

We examined three vulnerabilities in drivers and executables found in various parts of the Windows operating system: CVE-2020-1170 in Windows Defender, CVE-2021-21551 in the Dell dbutil_2_3.sys driver and CVE-2023-21768 in the Windows Ancillary Function Driver (afd.sys), that led to elevation of privilege.

We conducted both static and dynamic analyses. For static analysis, we used Ghidra and IDA, which are reverse-engineering tools, as well as BinDiff which identifies the differences between two versions of the binary. For dynamic analysis, we used Process Monitor (ProcMon) to display file system activity, as well as WinDbg to debug activity taking place in the kernel.

2 CVE-2020-1170 (Elevation of Privilege leads to Arbitrary File Deletion Vulnerability in Windows Defender)

2.1 Background Information

CVE-2020-1170 is an elevation of privilege vulnerability in Microsoft Windows Defender which leads to arbitrary file deletion on the Windows operating system [10]. Microsoft has given this vulnerability a Common Vulnerability Scoring System (CVSS) score of 7.8 [10].

Windows Defender uses two log files, one of which is MpCmdRun.log. It is stored in the folder C:\Windows\Temp, where only administrators and SYSTEM have full access control. MpCmdRun.log records events such as signature updates. Signature updates can be manually triggered by any user using the Update-MpSignature command in Windows Powershell. When this happens, information is written by SYSTEM to MpCmdRun.log. As signature updates occur frequently, MpCmdRun.log can grow large over time. To prevent it from exceeding its maximum log size of 16 megabytes, a log rotation mechanism compresses, archives and deletes old logs. A malicious unprivileged user can create a junction link between the target folder that they want to delete and MpCmdRun.log.bak. Then, they perform a signature update to fill up the log file. Once 16 megabytes of arbitrary data have been written to MpCmdRun.log, the unprivileged user then performs a final signature update. This causes the junction MpCmdRun.log.bak to be deleted, in turn causing the target folder linked by the junction to be deleted. We note that the deletion is recursive, which means that the files in the target folder will be deleted as well. An attacker can link the junction to any folder they want, including sensitive system resources such as drivers and dynamic-link libraries. This can lead to system damage.

2.2 Materials and Methods

In order to narrow down the scope of static analysis, we first replicated the vulnerability in a vulnerable VM and obtained the necessary call stacks from ProcMon. From these call stacks, we determined the executables of interest. We exported them out of the vulnerable VM and into Ghidra on the host machine for static analysis. We then figured out how the function was called from the frame in the call stack to the deletion of the file.

To perform patch analysis, we replicated the vulnerability on a patched VM and used the same process to obtain the call stacks from ProcMon and the executable of interest. We then reverse-engineered the changes that were made to the new executable using Ghidra.

2.3 Root Cause Analysis

There were two parts in the replication of this vulnerability. The first part involved proving that a log rotation mechanism existed within MpCmdRun.log and the second part involved replicating the arbitrary file deletion. Both parts were done in a VM (Fig. 1).

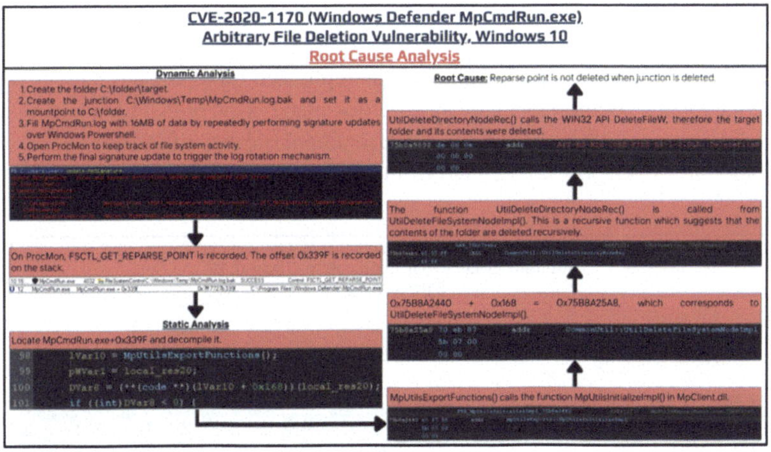

Fig. 1. A flow chart showing the steps taken during the root cause analysis of CVE-2020-1170

For the first part, we created the file MpCmdRun.log.bak in the path C:\Windows\Temp. Next, we filled MpCmdRun.log with arbitrary data such that its size was close to 16 megabytes. Then, we triggered a signature update. The file size of MpCmdRun.log became 3 kilobytes and the file size of MpCmdRun.log.bak became 16,384 kilobytes. This proved that a log rotation mechanism existed to prevent the size of MpCmdRun.log from exceeding 16 megabytes.

For the second part, we created a directory C:\ZZ_SANDBOX\target. Next, we created a junction called MpCmdRun.log.bak in C:\Windows\Temp and set it as a mountpoint to C:\ZZ_SANDBOX. Then, we filled MpCmdRun.log with arbitrary data such that its size was close to 16 megabytes. We opened ProcMon to keep track of file system activities and triggered a signature update. It was observed that the MpCmdRun.log.bak we created was deleted. The file size of MpCmdRun.log became 3 kilobytes and a new MpCmdRun.log.bak file was created with a file size of 16,384 kilobytes. By accessing C:\ZZ_SANDBOX, we found that the target folder did not exist. This proved that the signature update deleted the directory. From ProcMon, two call stacks that contained these operations were of interest: "Delete: True" and "FSCTL_GET_REPARSE_POINT".

From ProcMon, both processes were attributed to MpCmdRun.exe, so it was imported into Ghidra for static analysis. From the static analysis in Ghidra, it was ascertained that the function MpUtilsExportFunctions() called another dynamic link library MpClient.dll, so MpClient.dll was also imported into Ghidra for static analysis. It was found that MpClient.dll called the WIN32 API DeleteFileW, therefore the target folder and its contents were deleted.

Throughout the entire process, the information of the reparse point was not used even though it had been obtained earlier from FSCTL_GET_REPARSE_POINT. Since the junction was deleted, the reparse point of the junction should be deleted first to prevent the target folder and its contents from being deleted. Hence, the root cause of the vulnerability is that the reparse point is not deleted when the junction is deleted.

2.4 Patch Analysis

Next, the second part of the vulnerability was replicated on a patched VM. It was discovered that the log rotation mechanism still took place: MpCmdRun.log had a final file size of 1 kilobyte and MpCmdRun.log.bak had a final file size of 16,386 kilobytes. The target folder, however, was not deleted from C:\ZZ_SANDBOX (Fig. 2).

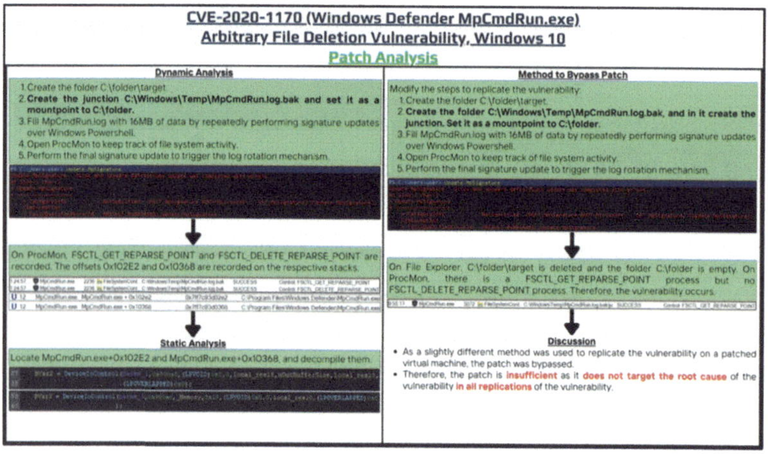

Fig. 2. A flow chart showing the steps taken during the patch analysis of CVE-2020-1170

From ProcMon, three call stacks were of interest: "Delete: True", "FSCTL_GET_REPARSE_POINT" and "FSCTL_DELETE_REPARSE_POINT". We analysed the similarities and differences in these three call stacks by loading MpCmdRun.exe from the patched VM into Ghidra. We found that if the attributes of the file was FILE_ATTRIBUTE_REPARSE_POINT, another function would run with two DeviceIoControl() statements: one for FSCTL_GET_REPARSE_POINT and one for FSCTL_DELETE_REPARSE_POINT. This suggests that the patched MpCmdRun.exe checks for a reparse point on MpCmdRun.log.bak. If there is a reparse point, MpCmdRun.exe deletes the reparse point. Next, if the file is not a directory, SetFileInformationByHandle() runs with the FileInformationClass of FILE_DISPOSITION_INFO to indicate that the file should be deleted. If the file is a directory, MpUtilsExportFunctions() runs which has the same outcome as in the vulnerable VM: the directory is deleted and a BAK file is created.

2.5 Method to Bypass Patch

The patched version of MpCmdRun.exe only checks if MpCmdRun.log.bak has a reparse point but does not check if the contents in MpCmdRun.log.bak have reparse points. A similar method of replicating the vulnerability on a patched VM would be to create the junction inside a folder MpCmdRun.log.bak. In order to bypass the patch, we first created a folder called C:\ZZ_SANDBOX\target. Next, we created a MpCmdRun.log.bak folder, and then created a junction inside it. This junction was set as a mountpoint to C:\ZZ_SANDBOX. Then, MpCmdRun.log was filled with arbitrary data such that its size was close to 16 megabytes. ProcMon was opened to keep track of file system activities, and a signature update was triggered. In File Explorer, the target folder was deleted. On ProcMon, there was no process with detail FSCTL_DELETE_REPARSE_POINT, so the reparse point of the junction was not deleted. This shows that it is possible to bypass the patch and delete the contents of the folder. Hence, only a specific method of replication, and not the root cause of the vulnerability, was addressed in the patch. The patch for CVE-2020-1170 is insufficient as it does not target the root cause of the vulnerability in all replications of the vulnerability.

3 CVE-2021-21551 (Elevation of Privilege leads to Arbitrary Write and Read Vulnerabilities in Dell Driver dbutil_2_3.sys)

3.1 Background Information

CVE-2021-21551 is a vulnerability that involves the dbutil_2_3.sys driver. Since 2009, Dell has released hundreds of millions of Windows devices worldwide which contain the vulnerable driver [11]. The driver contains an insufficient access control vulnerability, which leads to elevation of privileges [12]. Dell has given this vulnerability a CVSS score of 8.8 [12].

The vulnerability stems from the dbutil_2_3.sys driver's acceptance of IOCTL requests without the appropriate access control list (ACL) restrictions. This allows an unprivileged user to get a handle to the driver and trigger IOCTL operations. Notably, two of the operations will lead to memmove() where memory is being read or written. The unprivileged user can control the destination and input buffer, which means that they can read any privileged kernel memory and write to any kernel memory. This can be exploited using an arbitrary read or write as attackers can write to kernel memory to change their own privileges, or corrupt page table entries to spawn a privileged SYSTEM shell.

3.2 Materials and Methods

This CVE is split into two parts: arbitrary write and arbitrary read. For the arbitrary write vulnerability, we called the Application Programming Interface (API) DeviceIoControl()

to interact with dbutil_2_3.sys. We conducted static analysis on the driver to find the name of the string and the input/output control (IOCTL) code, which are used as parameters in DeviceIoControl(). We also reverse-engineered the code path to reach memmove() so that we can craft the input buffer to pass the checks that lead to it. Next, we wrote C code to call DeviceIoControl() with these parameters and used WinDbg to dynamically analyse the execution of the driver. This ensured that all conditional checks were passed, and information was written to the correct address. When the conditional checks failed, or the wrong address was written to, we modified the parameters to pass these checks.

For the arbitrary read vulnerability, we called DeviceIoControl() to interact with dbutil_2_3.sys. We conducted static analysis on the driver to find another code path to get to memmove(), as well as its IOCTL code and offset. We edited the C code used in our arbitrary write vulnerability with these new parameters, and printed the bytes read from kernel memory in the command prompt.

3.3 Root Cause Analysis of Arbitrary Write Vulnerability

For the analysis of the arbitrary write vulnerability, the first step was to perform static analysis in IDA. We wanted to obtain a handle to the driver using CreateFileA() and send data to the driver using DeviceIoControl(). By loading dbutil_2_3.sys in IDA, we located the name of the string used in IoCreateDevice(). It was \\Device\\DBUtil_2_3. This exact string allowed us to call CreateFileA() to create a handle to the driver so that we could interact with it. Next, based on previous research, we found that the vulnerable IOCTL code was 0x9B0C1EC8 [11], which we used as the IoControlCode parameter in DeviceIoControl(). We needed to locate where memmove() was within the function, as the movement of memory from source to destination could be exploited to perform the arbitrary write. Looking through the code logic, we found that there was a function call to sub_15294(). In sub_15294(), there was another function call to memmove(). However, there were many code branches from DriverEntry() to memmove(), and dynamic analysis was required to ascertain that the code followed the correct branches (Fig. 3).

The second step was dynamic analysis in WinDbg. This was split into four parts. The first part involved sending a QWORD to the kernel. We first wrote C code that called the CreateFileA() function to obtain a handle to the driver. We then created an unsigned long long variable called inBuf, which was 0x4141414141414141, or AAAAAAAA. This QWORD was 0x8 bytes. Last but not least, we called DeviceIoControl() with the driver handle, IOCTL code and QWORD as parameters. We then imported the POC code into the VM. We attached WinDbg to the VM and set a breakpoint at dbutil_2_3 + 0x5294, which was the start of the function sub_15294(). After hitting the breakpoint at dbutil_2_3 + 0x5294 using the POC code, we checked IDA for the values that each register will be taking. The RAX register would take the value of the IOCTL code, the RCX register would take the value of memmove() destination, the RDX register would take the value of memmove() source and the R8 register would take the value of buffer size. By displaying the bytes in the RCX register, we saw that the QWORD was written into the kernel. We then attempted to continue stepping into functions in order to get to memmove(). However, it failed a subsequent compare statement. This was because if the buffer size was less than 0x18, an error code was returned before we could get to memmove().

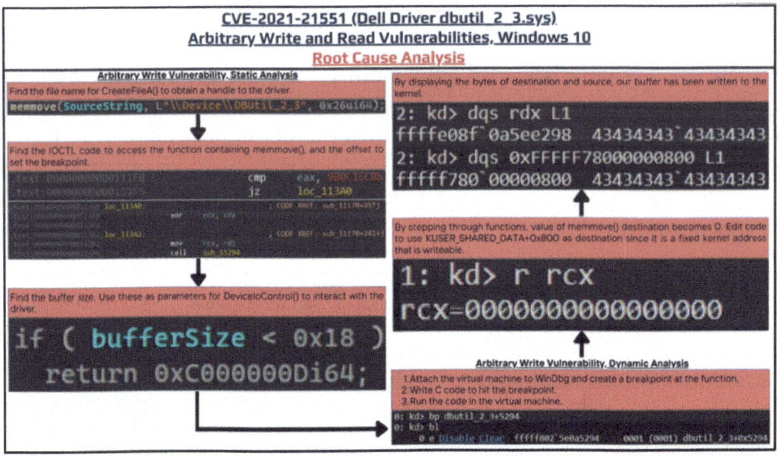

Fig. 3. A flow chart showing the steps taken during the root cause analysis of the arbitrary write vulnerability of CVE-2021-21551

The second part involved writing 0x18 bytes of data to the kernel to pass the compare statement. inBuf was changed to be a combination of three unsigned long longs: 0x4141414141414141, 0x4242424242424242 and 0x4343434343434343. By importing the new POC code into the VM, setting the breakpoint and hitting it, the compare statement was passed as expected. However, a few statements before hitting memmove(), the value of RCX changed from 0x18 to 0x0. This resulted in an invalid write address, so memmove() could not occur. Thus, we decided to specify a destination for memmove() to write to.

The third part involved writing 0x20 bytes of data to KUSER_SHARED_DATA + 0x800. We chose it because it is a kernel mode page that is mapped at a fixed virtual address in the kernel [13], and it has writable permissions [14]. We added KUSER_SHARED_DATA + 0x800 such that inBuf is now a combination of four unsigned long longs: 0x4141414141414141, 0xFFFFF78000000800, 0x4242424242424242 and 0x4343434343434343. By importing the new POC code into the VM, setting the breakpoint and hitting it, we found that the final address was 0xFFFFF78042424A42, which was the original address added with the lower 32-bits of the third unsigned long long. Hence, if the third unsigned long long was 0, then the final address would be 0xFFFFF78000000800.

The fourth part involved changing the third unsigned long long to 0x0000000000000000 to prevent it from changing the original address. By importing the new POC code into the VM, setting the breakpoint and hitting it, we found that the final address was 0xFFFFF78000000800. By displaying the values in the RCX and RDX registers, we saw that the unsigned long long 4343434343434343 was written into KUSER_SHARED_DATA. This completes the arbitrary write vulnerability.

3.4 Root Cause Analysis of Arbitrary Read Vulnerability

For the analysis of the arbitrary read vulnerability, the first step was also static analysis in IDA. We looked through the control flow graph to find other IOCTL codes that could reach memmove(). We also checked for its offset with respect to the code. By looking through the control flow graph, we found that the IOCTL code 0x9B0C1EC4 also led to memmove(). Its offset was at 0x111E5 (Fig. 4).

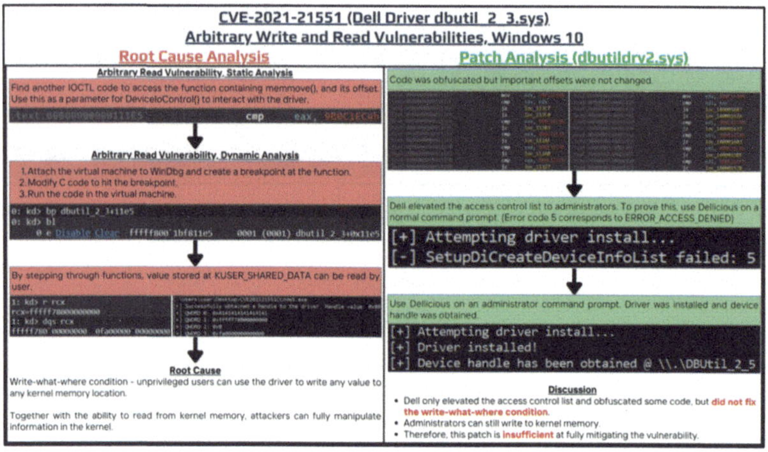

Fig. 4. A flow chart showing the steps taken during the root cause analysis of the arbitrary read vulnerability, and the patch analysis of CVE-2021-21551

The second step was dynamic analysis in WinDbg. After updating the C code to include the new IOCTL code, a second DeviceIoControl() and a for loop to print the QWORDS onto the command prompt, we hit the new breakpoint of 0x111E5. By repeatedly stepping through the function, we observed the changes in the RAX and RCX registers. We found that the first part of the buffer was written into the RAX register, followed by the second part, which was the KUSER_SHARED_DATA address. KUSER_SHARED_DATA was then moved into the RCX register, and its contents was 0x0FA0000000000000. This content could be read by the user on the command prompt. This completes the arbitrary read vulnerability.

Therefore, the root cause of the vulnerability is the write-what-where condition, as unprivileged users could use the driver to write any value to any kernel memory location. Together with the ability to read any kernel memory location, an attacker is able to fully manipulate the information in the kernel.

3.5 Patch Analysis

For the patch, Dell released a new driver dbutildrv2.sys that limited access to administrative users [15]. To prove this, we used Dellicious on both normal and administrator command prompts. On the normal command prompt, ERROR_ACCESS_DENIED was returned. On the administrator command prompt, the driver was installed and the device handle was obtained. This proved that the access control list was elevated.

Upon further inspection, it was found that Dell only elevated the access control list and obfuscated some code, but did not fix the write-what-where condition. A malicious administrator can still write to kernel memory, which is a security boundary. In fact, this led to CVE-2021-36276, where the same vulnerability took place in the dbutildrv2.sys driver [16]. Thus, only a specific method of replication, and not the root cause of the vulnerability, was addressed in the patch. The patch released for CVE-2021-21551 is insufficient at fully mitigating the vulnerability.

4 CVE-2023-21768 (Elevation of Privilege leads to Arbitrary Write Vulnerability in Windows Ancillary Function Driver)

4.1 Background Information

CVE-2023-21768 is a vulnerability in the Windows Ancillary Function Driver, which leads to elevation of privileges [17]. Microsoft has given this vulnerability a CVSS score of 7.8 [17].

The vulnerable function, AfdNotifyRemoveIoCompletion() has a line of code that writes 0x1 to a memory address supplied by a user buffer provided by a user. However, AfdNotifyRemoveIoCompletion() does not check if the address provided is a kernel land or a user land address. Hence, an unprivileged process is able to write to a kernel address and corrupt the system. Additionally, the decompiled code shows that AfdNotifyRemoveIoCompletion() checks for the user's privileges but still allows an unprivileged user to write a kernel address. This could be attributed to a code error made by the developers.

4.2 Materials and Methods

We first identified the differences between the vulnerable and patched afd.sys drivers by using BinDiff. In order to find out how these changes affected the afd.sys driver, we conducted static analysis on both the vulnerable and patched drivers. After finding the root cause of the vulnerability, we identified the vulnerable IOCTL code and wrote a program to trigger it. We also found additional prerequisites for the input buffer being sent, such as the size of the buffer and a configured IoCompletionObject. Finally, we called the NtDeviceIoControlFile() API to interact with the driver. We wrote C code with these parameters and used WinDbg to dynamically analyse the execution of afd.sys. When any checks failed, we modified the buffer being sent to pass these checks and improved our code accordingly before reaching a final working proof-of-concept (POC) code.

4.3 Root Cause Analysis

The first step was to load the two binaries in BinDiff to find the differences between the vulnerable and patched versions of afd.sys. BinDiff's results showed that only one function, AfdNotifyRemoveIoCompletion(), had been changed (Figs. 5 and 6).

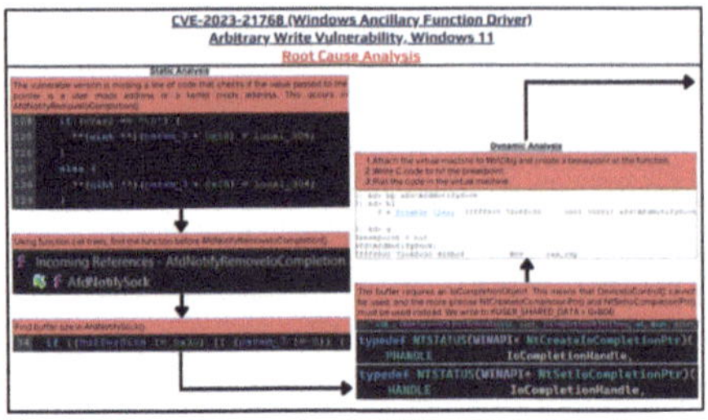

Fig. 5. A flow chart showing the steps taken during the first part of the root cause analysis of CVE-2023-21768

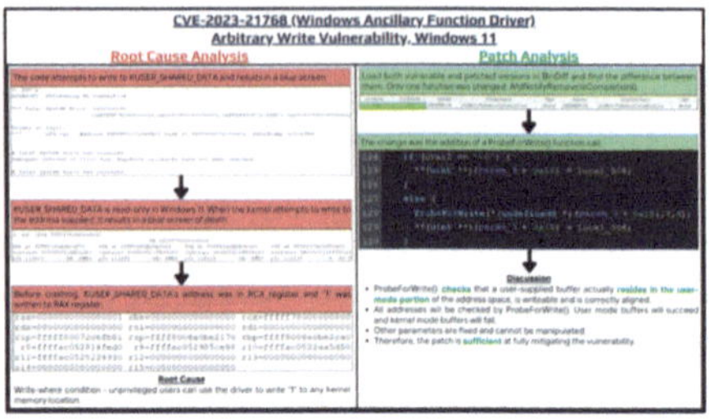

Fig. 6. A flow chart showing the steps taken during the second part of the root cause analysis, and the patch analysis of CVE-2023-21768

Next, both vulnerable and patched versions of the function were loaded into Ghidra. The decompiled code was examined to identify the changes made in the patch. We compared the lines of code in both versions of AfdNotifyRemoveIoCompletion() and found that the addition of ProbeForWrite() was the only change made. We found that ProbeForWrite() checks if the address in the buffer supplied to afd.sys resides in the user-mode portion of the address space [18]. The patched version of AfdNotifyRemoveIoCompletion() first checks for the user's privileges. If the caller comes from kernel-mode, it simply writes to the address provided. If the caller comes from user-mode, it uses ProbeForWrite() to validate the address provided. On the other hand, the vulnerable version did not check if the target write address provided by a user-mode caller resided in the user-mode address space, which resulted in the vulnerability. To verify this, we attempted to write to KUSER_SHARED_DATA at an offset of 0x800.

Through further static analysis in Ghidra, we identified that the size of the buffer being sent to KUSER_SHARED_DATA needed to be 0x30 bytes.

The next step was to find the IOCTL code needed to interact with afd.sys, which was obtained by finding the references of the vulnerable function and calculating the respective index number.

An initial POC code was written in C and compiled into an executable. WinDbg was attached to the VM, and a breakpoint was set up at afd!AfdNotifySock, the function that calls AfdNotifyRemoveIoCompletion(). By simultaneously using WinDbg and IDA to find the specific line the POC code had executed, we confirmed that the POC code had successfully passed through the buffer size check but failed at the IoCompletion check. This was because the POC code was missing an IoCompletion object.

In order to establish an IoCompletion object, the functions NtCreateIoCompletion() and NtSetIoCompletion() were called to create and set up an IoCompletion object. When the POC code was executed again, it passed through the check and headed to the target memory write line. When that line was reached, WinDbg showed that the VM encountered a fatal error, and a blue screen of death (BSoD) occurred.

By displaying the register values at the time of failure, it showed that the value in the RCX register was FFFFF78000000800, which meant that the memory address of KUSER_SHARED_DATA + 0x800 had been successfully loaded into the RCX register. Additionally, the value of the RAX register was 0000000000000001, signifying that the value to be written, "1", had also been successfully loaded into the RAX register. Hence, all preliminary steps before the actual moving of the written value into the memory address had been completed.

We used WinDbg to display the permissions of KUSER_SHARED_DATA and found out that it had been changed to a read-only address. On further research, we found that the page permissions for KUSER_SHARED_DATA had changed from writable to read-only in recent versions of Windows [19]. Since the kernel tried to write to a read-only address, a BSoD occurred and the VM crashed.

Therefore, the root cause of the vulnerability was the write-where condition, as unprivileged users could use the driver to write "1" to any kernel memory location.

4.4 Patch Analysis

The patch released by Microsoft was sufficient in preventing future replications of the vulnerability. The additional ProbeForWrite() function call is sufficient as it ensures that the memory address provided by the buffer resides in the user-mode portion of the address space, is writeable and is correctly aligned [18]. The address supplied by the user would be checked by ProbeForWrite(), with all user mode buffers succeeding and all kernel mode buffers failing. This prevents any user from providing a kernel address to write to, which addresses the write-where condition. Furthermore, the other parameters of this function are fixed as "4" and cannot be manipulated. Hence, the root cause of the vulnerability was addressed in the patch, and the patch for CVE-2023-21768 is sufficient at fully mitigating the vulnerability.

5 Conclusion

In conclusion, by conducting patch analysis, we found that some patches released by vendors, such as CVE-2023-21768, are comprehensive because they target the root cause of the vulnerability. At the same time, many patches, such as CVE-2020-1170 and CVE-2021-21551, are not comprehensive. They do not target the root cause of the vulnerability, but merely target a specific method of replication. Thus, there still exists similar methods of replication that can bypass the patch, rendering these patches ineffective.

We believe that this project has a major impact on many stakeholders. Primarily, software developers can leverage this information to enhance their patching practices, by addressing the root cause of the vulnerability instead of specific methods of replication. They could also consider incorporating a dedicated monitoring period to promptly detect and respond to any exploitation following the implementation of patches. Additionally, end users can indirectly benefit from this information. This knowledge can be used to create more robust patches that offer broader protection against a range of potential vulnerabilities. The improvements in software security and system administration practices that result from this newfound information can lead to safer and more secure operating systems for end users.

References

1. Desktop Operating System Market Share Worldwide. StatCounter Global Stats. https://gs.statcounter.com/os-market-share/desktop/worldwide
2. Vander-Pallen, M.A., Addai, P., Isteefanos, S., Mohd, T.K.: Survey on Types of Cyber Attacks on Operating System Vulnerabilities Since 2018 Onwards. 2022 IEEE World AI IoT Congress (AIIoT) (2022). https://doi.org/10.1109/aiiot54504.2022.9817246
3. Versluis, D.: The 2023 Microsoft Vulnerabilities Report – 10th Anniversary Edition. Service Dynamics. https://blog.servicedynamics.co.nz/2023-microsoft-vulnerabilities-report-10th-anniversary-edition
4. Bott, E.: Is Windows 10 Too Popular for its Own Good? ZDNET (2023). https://www.zdnet.com/article/iswindows-10-too-popular-for-its-own-good
5. Bhurtel, M., Rawat, D.B.: Unveiling the landscape of operating system vulnerabilities. Future Internet **15**(7), 248 (2023). https://doi.org/10.3390/fi15070248
6. CWE-269: Improper Privilege Management. Common Weakness Enumeration. https://cwe.mitre.org/data/definitions/269.html
7. Softić, J., Vejzović, Z.: Windows 10 Operating System: Vulnerability Assessment and Exploitation. 2022 21st International Symposium INFOTEH-JAHORINA (INFOTEH) (2022). https://doi.org/10.1109/infoteh53737.2022.9751274
8. Schrader, D.: What is Elevation of Privilege and Why is it Significant? Netwrix (2018). https://blog.netwrix.com/2018/09/05/what-is-privilege-escalation
9. Microsoft Security Servicing Criteria for Windows. Microsoft (2018). https://www.microsoft.com/enus/msrc/windows-security-servicing-criteria
10. Microsoft Windows Defender Elevation of Privilege Vulnerability CVE-2020-1170 Security Vulnerability. MSRC Security Updates (2020). https://msrc.microsoft.com/update-guide/en-us/advisory/CVE-2020-1170
11. Dekel, K.: CVE-2021-21551 - Hundreds Of Millions Of Dell Computers At Risk Due to Multiple BIOS Driver Privilege Escalation Flaws. SentinelOne (2021). https://www.sentinelone.com/labs/cve-2021-21551-hundreds-of-millions-of-dell-computers-at-risk-due-to-multiple-bios-driver-privilege-escalation-flaws
12. CVE-2021-21551 Detail. National Vulnerability Database (2021). https://nvd.nist.gov/vuln/detail/cve-2021-21551
13. Randomizing the KUSER_SHARED_DATA Structure on Windows. MSRC Blog (2022). https://msrc.microsoft.com/blog/2022/04/randomizing-the-kuser_shared_data-structure-on-windows
14. McGarr, C.: Exploit Development: Leveraging Page Table Entries for Windows Kernel Exploitation (2020). https://connormcgarr.github.io/pte-overwrites
15. Toulas, B.: Dell Driver Fix Still Allows Windows Kernel-level Attacks. BleepingComputer (2021). https://www.bleepingcomputer.com/news/security/dell-driver-fix-still-allows-windows-kernel-level-attacks
16. CVE-2021-36276 Detail. National Vulnerability Database (2021). https://nvd.nist.gov/vuln/detail/CVE-2021-36276
17. CVE-2023-21768 Detail. National Vulnerability Database (2023). https://nvd.nist.gov/vuln/detail/CVE-2023-21768
18. ProbeForWrite Function (wdm.h). Microsoft Learn (2023). https://learn.microsoft.com/en-us/windows-hardware/drivers/ddi/wdm/nf-wdm-probeforwrite
19. McGarr, C.: Exploit Development: ASLR - Coming To A KUSER_SHARED_DATA Structure Near You! (2022). https://connormcgarr.github.io/kuser-shared-data-changes-win-11

Production of Recombinant Biologics Using Engineered Protein Shells

Angad Singh[1(✉)], Siddhesh Sujit Vaidya[2], Hrucha Shielesh Damle[2], Girish Vallerinteavide Mavelli[2], and Chester Lee Drum[2]

[1] Raffles Institution, 1 Raffles Institution Ln, Singapore 575954, Singapore
24yanga813@student.ri.edu.sg

[2] Department of Medicine, National University of Singapore, 21 Lower Kent Ridge Rd, Singapore 119077, Singapore

Abstract. Recombinant proteins play a pivotal role across industrial, biopharmaceutical, and agricultural sectors. *Escherichia Coli* stands out as the preferred host system for heterologous protein expression due to its simplicity, versatility, and cost-effectiveness. However, the challenge of protein misfolding persists, resulting from cellular stress and the prokaryotic host's inability to perform crucial post-translational modifications. Due to the high substrate loads present in recombinant expression contexts, chaperones are also oversaturated and unable to regulate the folding process effectively. Current solutions, relying on co-expression of chaperone proteins, often fall short due to undesirable side effects of chaperone overproduction. This study utilizes a novel approach featuring thermostable Exoshells (tES), engineered protein-based nanoparticles derived from Archaeoglobus Fulgidus ferritin, that encapsulate Transforming Growth Factor Beta 1 (TGF-β1). Engineered with unique technical capabilities, tES boasts several technical advantages over typical chaperone proteins. These capabilities were used to encapsulate TGF-β1, an ubiquitous growth factor prone to aggregation with extremely high production costs. tES has been used in previous studies and has boosted functional yields of various proteins between 2-fold to > 100-fold, making it a promising low-cost and simple method of production for such difficult to fold proteins.

Keywords: Thermostable Exoshells · Nanoparticles · TGF-β1 · Refolding

1 Introduction

In the rapidly advancing landscape of biotechnology, recombinant proteins have found various applications in the industrial, biopharmaceutical and agricultural fields (Demain and Vaishnav, 2011). *Escherichia coli* (*E. coli*) has remained the host system of choice for the heterologous expression of various functional proteins due to its relative simplicity, versatility, low cost, and high yields (Overton, 2014; Cornelis, 2000)

However, protein misfolding is a serious issue in E. coli production systems. The overproduction of the target protein causes thermal stress, often leading to protein aggregation (Villaverde & Carrió, 2003). Furthermore, the lack of adequate cellular machinery

H. Guo et al. (Eds.): IRC-SET 2024, PTT, pp. 181–188, 2025.
https://doi.org/10.1007/978-981-96-3770-6_19

for proper post-translational protein modifications such as disulfide linkage formation, frequently results in aggregation with much lower yield (Bhatwa et al., 2021).

Current solutions to this problem often rely on chaperone proteins, which are significant factors in protein quality control (Saibil, 2013). Since recombinant protein production demands high protein substrate amounts, naturally occurring chaperones may be oversaturated and unable to control proper protein folding (Martínez-Alonso et al., 2010). To increase chaperone concentration, certain chaperones can be overproduced via coexpression with target proteins. Commonly selected chaperone families for co-expression in *E. coli* systems include the HSP60; GroEL and HSP70; DnaK families (Martínez-Alonso et al., 2010).

However, the overexpression of these chaperones can have undesirable side effects on the production of proteins such as increased proteolysis rates (Martínez-Alonso et al., 2010). Moreover, the co-expression does not improve synthesis rates, yield or the stability of target proteins (Thomas & Baneyx, 1996).

Thermostable Exoshells (tES) are 12 nm diameter engineered protein-based nanoparticles derived from the thermophilic bacteria *A. fulgidus* ferritin (AfFtn) with an 8 nm aqueous cavity, a volume that can theoretically accommodate any molecules up to a volume of ~ 306 nm^3 (Deshpande et al., 2017). The shells, an assembly of 24 subunits contain 4.5 nm surface pores, which allow for easy permeation of solutes (Deshpande et al., 2017). tES features extra internal space due to the truncation of unstructured c-terminus peptide from AfFtn subunits by a single point mutation at residue 164 (Glutamine 164) as shown in Fig. 1. These unstructured peptides would otherwise protrude into the assembly's inner cavity, interfering with internalized protein folding and function (Deshpande et al., 2017; Sadeghi et al., 2021).

Fig. 1. Wild-type AfFtn (left) and tES C-Terminus truncated (right) (Image credit: Deshpande et al. 2017)

tES can also be made to encapsulate and release proteins via mild pH titrations where tES shells disassemble at pH < 6.0 and reassemble at pH > 7.0. This feature was accomplished through a single point mutation at Residue 116 in the 3-fold symmetry axis of tES, where the amino acid Phenylalanine was substituted by Histidine (Deshpande et al. 2017) (Fig. 2). This substitution is expected to create mutual charge repulsion with Histidine (Residue 104) on a neighboring subunit in mild acidic pH, leading to cage disassembly.

Fig. 2. F116H mutation (Image credit: Deshpande et al., 2017)

tES is also capable of encapsulating and folding diverse protein substrates (Fig. 3). AfFtn has a naturally occurring negatively charged internal cavity, allowing it to complement proteins with positive surface charges. By largely altering glutamate residues, shells with positively charged, tES (+), and neutral, tES (±), interior cavities were also formed, allowing charge complementation between the tES cavity and proteins with various surface charges (Deshpande et al., 2017).

tES has numerous technical advantages including extremely high thermal stability (stable at temperatures above 90 °C) and the ability to protect encapsulated target proteins against extreme pH changes or degrading or denaturing agents including proteases or solvents. It can protect encapsulated proteins from aggregation and promotes in-vitro folding of target proteins, which are difficult to fold. tES also allows easy scalability to gram levels using *E. coli* expression systems.

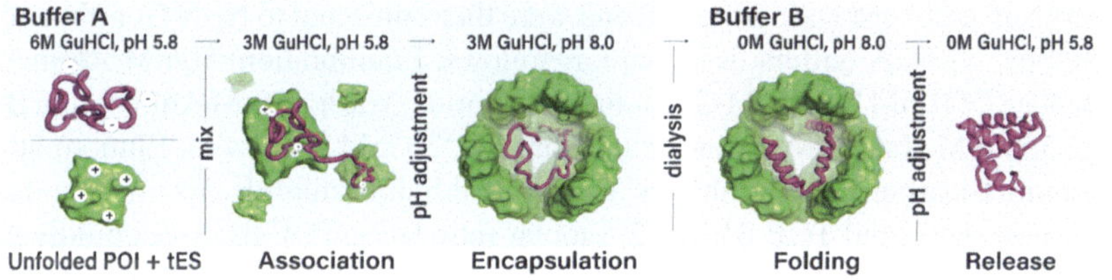

Fig. 3. Schematic representation of tES encapsulation and folding Protein of Interest (Image credit: Sadeghi et al. 2021)

Previous studies have shown that tES is able to encapsulate, fold and stabilize a wide variety of proteins (Deshpande et al. 2017). In this study, we examined the ability of tES to encapsulate and fold difficult to fold growth factors such as Transforming Growth Factor-β1 (TGF-β1). We hypothesize that tES will be able to successfully encapsulate and fold TGF-β1. Growth factors have various applications in serum free media, cell therapy and cultured meat production (Chase et al. 2010; Enriquez-Ochoa et al. 2020; Ahmad et al. 2023). TGF-β1 is a multifunctional dimeric cytokine that plays a key role in cellular processes such as cell growth, differentiation, apoptosis, and metabolism. The conventional recombinant expression of TGF-β1 is particularly costly (Soni et al. 2022).

Therefore, we have decided to focus on TGF-β1 as a protein that requires cheaper alternative production methods such as tES folding. TGF-β1 presents a significant challenge in folding due to its high cysteine density and unique structural fold, making it more prone to aggregation. Therefore, TGF-β1 folding will serve as a proof-of-concept study that tES can reliably fold and stabilize such proteins prone to aggregation.

2 Materials and Methods

pRSFDuet-1 plasmids with the TGF-β1 gene sequence were transformed into BL21(DE3) *E. Coli* cells and grown on Luria-Bertani (LB) agar plates with 50 μg/mL of kanamycin to isolate positive colonies. A starter culture from the positive colony was initiated through overnight culture at 37 ºC, 200 rpm on a shaker incubator. The overnight grown starter culture was used to initiate large flask cultures (1 L) under constant shaking until its absorbance at 600 nm reached approximately 0.4–0.5 as measured by a NanoDrop Spectrophotometer (DeNovix Inc., United States). IPTG at a concentration of 500 μM was then added to induce protein expression. After further incubation at 37 °C, the cells were centrifuged to obtain the cell pellet. The pellet was then suspended in a lysis buffer (50 mM Ammonium Bicarbonate, 10 mM Ethylenediaminetetraacetic acid, 2% Triton-X, 150 mM NaCl, 20 mM Dithiothreitol) and sonicated. Post centrifugation at 12,100 rpm, the pellet was collected and washed twice with a washing buffer (Wash 1: 50 mM Ammonium Bicarbonate, 10mM Ethylenediaminetetraacetic acid, 1% Triton-X, 1M NaCl, 20 mM Dithiothreitol, 2M Urea; Wash 2: 50 mM Ammonium Bicarbonate, 10 mM Ethylenediaminetetraacetic acid, 150 mM NaCl, 20 mM Dithiothreitol). The proteins solubilized overnight under shaking in a solubilization buffer (50 mM Ammonium Bicarbonate, 10 mM Ethylenediaminetetraacetic acid, 6M Guanidinium chloride, 150 mM NaCl, 20 mM Dithiothreitol) and were then subjected to Ni-NTA affinity chromatography. Ni-NTA buffers used were as follows: Equilibration: 50mM Ammonium Bicarbonate, 500 mM NaCl, 6M Guanidinium chloride; Wash: 50mM Ammonium Bicarbonate, 500 mM NaCl, 6M Guanidinium chloride, 20 mM Imidazole; Elution: 50 mM Ammonium Bicarbonate, 500 mM NaCl, 6M Guanidinium chloride, 500 mM Imidazole. Both tES(+) shells and TGF-β1 in a 2:5 molar ratio (tES: TGF-β1) were mildly acidified to pH 5.8. The mixture was placed in a dialysis cassette filled with a dialysis buffer. Size-exclusion chromatography was employed after dialysis to collect tES-encapsulated TGF-β1 and SDS-PAGE was used to validate encapsulation.

3 Results

TGF-β1 expression was investigated, and SDS-PAGE was performed to assess the expression pattern within bacterial inclusion bodies (Fig. 4). TGF-β1 is majorly expressed in the insoluble fraction of the lysate with a molecular weight between 10–15 kDa. Hence, the insoluble fraction was utilized for the subsequent purification with Ni-NTA chromatography to achieve almost 90% purity (Fig. 5).

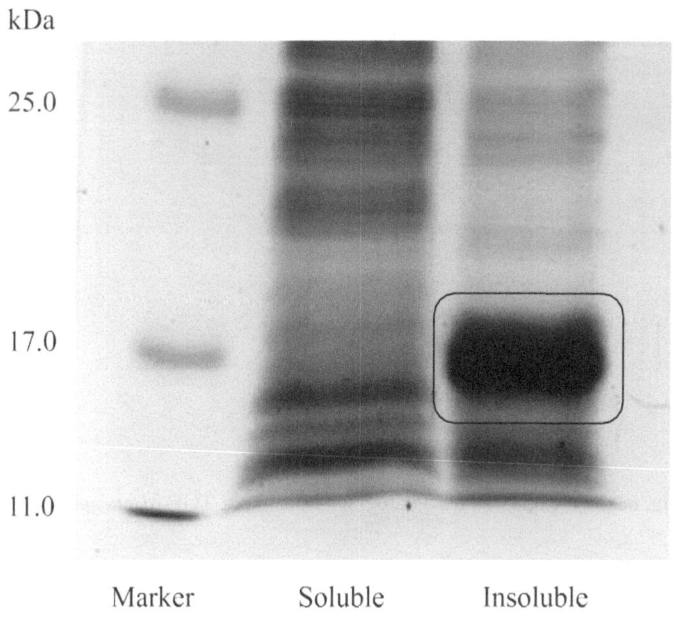

Fig. 4. SDS-PAGE analysis of TGFβ1 expression showing the expressed proteins in insoluble fraction.

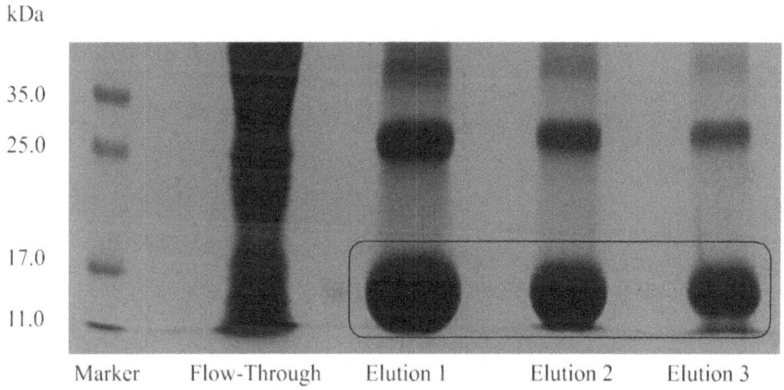

Fig. 5. SDS-PAGE analysis of TGF-β1 purification via Ni-NTA.

After being encapsulated by tES, the caged proteins were subjected to Size Exclusion Chromatography (SEC) (Fig. 6). Referring to Fig. 6, SEC gave 2 peaks. Figure 7 features a control band of TGF-β1 between 10–15 kDa and the fractions from Peak 1 and 2 display a band at 19 kDa. We suspected this 19 kDa band was tES-encapsulated TGF-β1. Western Blot analysis using a His-tagged antibody confirmed the presence of tES-encapsulated TGF-β1 in the fraction from Peak 2 (Fig. 8), and thus encapsulation was confirmed.

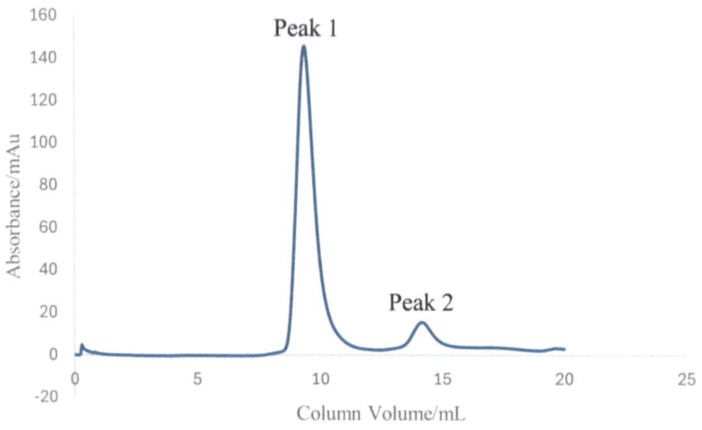

Fig. 6. Size-exclusion chromatography profile of TGF-β1 encapsulation

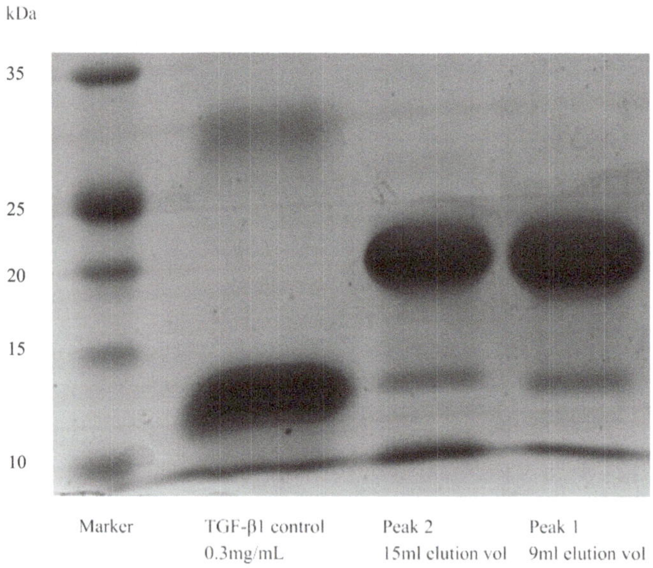

Fig. 7. SDS-PAGE analysis of TGF-β1 encapsulation

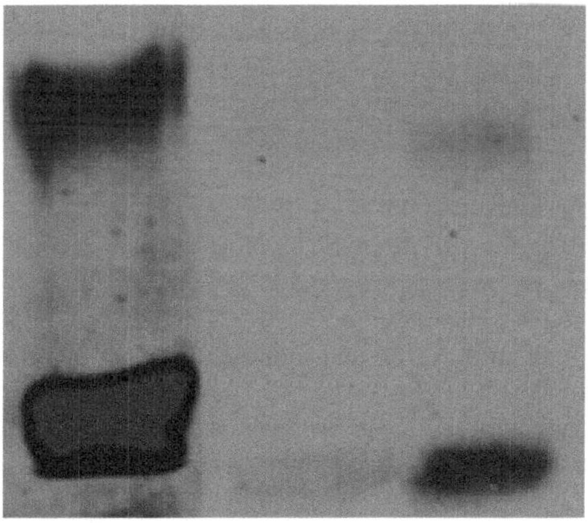

TGF-β1 0.5 mg/mL Control Peak 1 Peak 2

Fig. 8. Western Blot analysis of fractions collected at Peak 1 and 2

4 Discussion and Conclusion

TGF-β1 is a homodimeric growth factor with a cysteine knot structure that is linked together by intramolecular disulphide linkages and whose appropriate folding and dimerization is required for its biological activity (Hinck et al. 2016). Although it can be produced recombinantly in mammalian expression systems such as Chinese Hamster Ovary systems, the synthesis of active TGF-β1 is difficult and the yield obtained is relatively low (Soni et al. 2022). tES-based folding remains preferable to these conventional procedures because the former relies on in vitro folding from denatured proteins, which can be accomplished in high yields using bacterial expression systems, lowering overall production costs. Given that aggregation is a significant issue in protein folding, this study has demonstrated that effective encapsulation within exoshells can significantly minimize protein loss when compared to free TGF-β1, which precipitates when exoshells are absent.

In conclusion, the current study has created an improved methodology for encapsulating and stabilizing TGF-β1 from aggregation. TGF-β1 characterization will be studied further in the future using various structural and functional assays.

Acknowledgements. This project was supported using the NRF Central Gap Fund (NRF-CG2025-CG02-IG2-001002).

References

Demain, A.L., Vaishnav, P.: Production of recombinant proteins by microbes and higher organisms. Comp. Biotech. **3**, 333–345 (2011)

Overton, T.W.: Recombinant protein production in bacterial hosts. Drug Discov. Today **19**(5), 590–601 (2014)

Cornélis, P.: Expressing genes in different Escherichia coli compartments. Curr. Opin. Biotech. **1**(5), 450–4541 (2000)

Villaverde, A., Carrió, M.M.: Protein aggregation in recombinant bacteria: Biological role of inclusion bodies. Biotechnol. Lett. **25**(17), 1385–1395 (2003)

Bhatwa, A., et al.: Challenges associated with the formation of recombinant protein inclusion bodies in escherichia coli and strategies to address them for industrial applications. Front. Bioeng. Biotechnol. (2021). https://doi.org/10.3389/fbioe.2021.630551

Saibil, H.R.: Chaperone machines for protein folding, unfolding and disaggregation. Nat. Rev. Mol. Cell Biol. **14**(10), 630–642 (2013)

Martínez-Alonso, M., García-Fruitós, E., Ferrer-Miralles, N., Rinas, U., Villaverde, A.: Side effects of chaperone gene co-expression in recombinant protein production. Microb. Cell Fact. **9**(1), 64 (2010)

Thomas, J.G., Baneyx, F.: Protein misfolding and inclusion body formation in recombinant Escherichia coli cells overexpressing heat-shock proteins. J. Biol. Chem. **271**(19), 11141–11147 (1996)

Deshpande, S., et al.: Thermostable exoshells fold and stabilize recombinant proteins. Nat. Commun. (2017). https://doi.org/10.1038/s41467-017-01585-2

Sadeghi, S., et al.: A general approach to protein folding using thermostable exoshells. Nat. Commun. (2021)

Chase, L.G., Lakshmipathy, U., Solchaga, L.A., Rao, M.S., Vemuri, M.C.: A novel serum-free medium for the expansion of human mesenchymal stem cells. Stem Cell Res. Ther. (2010). https://doi.org/10.1186/scrt8

Enriquez-Ochoa, D., Robles-Ovalle, P., Mayolo-Deloisa, K., Brunck, M.E.G.: Immobilization of growth factors for cell therapy manufacturing. Front. Bioeng. Biotechnol. (2020). https://doi.org/10.3389/fbioe.2020.00620

Ahmad, S.S., et al.: The roles of growth factors and hormones in the regulation of muscle satellite cells for cultured meat production. J. Anim. Sci. Tech. **65**(1), 16–31 (2023)

Soni, A.P., Lee, J.H., Shin, K., Koiwa, H., Hwang, I.: Production of recombinant active human TGFβ1 in Nicotiana benthamiana. Front. Plant Sci. (2022). https://doi.org/10.3389/fpls.2022.922694

Hinck, A.P., Mueller, T.D., Springer, T.A.: Structural biology and evolution of the TGF-β family. Cold Spring Harbor Perspecti. Biol. **8**(12), a022103 (2016). https://doi.org/10.1101/cshperspect.a022103

Aerodynamic Drag and Stability Characterization of a Rigid Parachute

Koh Shan Jun[1(✉)] and Leedon Yong[2]

[1] Raffles Institution, Singapore, Singapore
shannnjkoh@gmail.com
[2] St Joseph's Institution, Singapore, Singapore

Abstract. This paper presents a comprehensive study on the drag and stability performance of a rigid parachute characterised by the vent ratio, canopy size and angle of attack, using a combined approach of Computational Fluid Dynamics (CFD) simulations and physical drop-test experiments. CFD simulations were conducted to obtain numerical results for drag forces, pressure gradient, and flow patterns surrounding the parachute, considering the rigid nature of the parachute and incorporating realistic environmental conditions, providing a detailed insight into the aerodynamic forces acting on the parachute during descent. To investigate the real-world applicability, a series of drop-tests were performed.

Keywords: Computational Fluid Dynamics · Drag and Stability Characterization

1 Introduction

Parachutes have evolved into indispensable tools across different applications, from recreational skydiving to aeronautical missions, playing a pivotal role in ensuring their safety and success. Researchers have incorporated numerical simulations into the development and optimisation of parachute systems as a result of the high costs of traditional wind tunnel testing, with the use of experimental tests to validate the findings of numerical simulations. This study seeks to provide a more comprehensive analysis of the drag and stability of parachutes given the increasing demand for parachute systems to be tailored for diverse conditions, from stratospheric altitudes to high speeds, where drag and stability play vital roles in allowing parachutes to carry out their functions.

2 Literature Review

2.1 Stability Analysis

Barnhardt and Gidzakl employed a rigid parachute model in their investigation of the effect of wake vortices on flow stability, and concluded that an interaction between the wake and the canopy shock results in a highly unsteady flow around the parachute. Xue and Nishiyama found that the interaction further causes pressure within the canopy to fluctuate at a large amplitude, causing an upstream propagation and lateral expansion of the wake to exhibit an unsteady pulsation.

H. Guo et al. (Eds.): IRC-SET 2024, PTT, pp. 189–199, 2025.
https://doi.org/10.1007/978-981-96-3770-6_20

2.2 Model Development

According to Newton's Second Law, F = ma, the following equation can be derived as,

$$F = mg - F_D = m\frac{dV_G}{dt} \tag{1}$$

We consider the velocities in the body fixed frame and thus, we will need to convert Eq. (1) to a body fixed frame, where as

$$a = \frac{d\underline{V_G}}{dt} = \frac{dV}{dt}|B + \underline{\omega} \times \underline{V} \tag{2}$$

When we substitute back into Eq. 1, we obtain the following force equation in the body fixed frame,

$$F = m\frac{d\underline{V}}{dt}|B + \underline{\omega} \times \underline{V} \tag{3}$$

Identifying the 2 key external forces that act on the body during the gliding stage, the aerodynamic force and the gravitational force acting on the body in the earth fixed inertial frame. As we consider forces about the body fixed frame, the gravitational force is hence taken as,

$$F^b_{gravity} = F^e_{gravity} R_t = mg \begin{bmatrix} -\sin(\theta) \\ \sin(\varphi)\cos(\theta) \\ \cos(\varphi)\cos(\theta) \end{bmatrix} \tag{4}$$

Thus, our final force equation when considering the main governing forces from gravity and aerodynamic force, can be expressed as

$$F^b_{gravity} + F_a = m \times \frac{d\underline{V}}{dt} + m\underline{\omega} \times \underline{V} \tag{5}$$

Consider the moment of the parachute in the body-fixed frame, where H is the angular momentum of the rigid body which is calculated using H = I.

$$M_a = \sum_i \underline{M}_i = \frac{d\underline{H}}{dt} \tag{6}$$

where ω is the angular velocity vector about the centre of mass and *I*, is the moment of inertia, defined as

$$I = \begin{bmatrix} I_{xy} & -I_{xy} & -I_{xy} \\ -I_{yx} & I_{yy} & -I_{yz} \\ -I_{zx} & -I_{zy} & I_{zz} \end{bmatrix} \tag{7}$$

Thus, this allows us to express the angular momentum of the parachute in each axis as

$$M_a = \frac{d\underline{H}}{dt} + \underline{\omega} \times \underline{H} \tag{8}$$

Dobrokhodov et al. (2003) developed a Six-Degree-of-Freedom Model for a Controlled Circular Parachute, assuming the parachute system was rigid, only experienced gravitational and aerodynamic forces acting at the canopy's centre of pressure, negligible payload-generated aerodynamic forces, and symmetrical undistorted canopy along the z-axis. The model showed good performance against flight-test data, though improvements could be made in the model identification procedure and accounting for ADS asymmetry.

3 Methodology

3.1 Parameters

The effects of 2 key parameters on the drag and stability of a hemispherical parachute were studied. The first parameter, canopy size, was defined as the reference area of the parachute and was investigated by varying the diameter of the canopy from 65.2 cm to 85.2 cm. The second parameter, vent ratio, refers to the ratio of the area of the vent hole to the reference area of the canopy and was varied from 0% to 7%.

3.2 Assumptions

Similar to some of the studies reviewed, [1, 3, 4] several key assumptions were made to simplify the tests. It was assumed that the aerodynamic forces acting on the suspension lines and payload were negligible, the effects of deformation of the canopy were negligible, the parachute was a rigid bluff body which experienced only aerodynamic forces during the drop-test, and that any random wind was uniformly distributed across the entire body of the parachute.

3.3 Parachute Design and Data Collection

Canopy dimensions were determined by estimating the final masses of the entire parachute system to keep terminal velocity relatively low. A non-porous polyester fabric was used for the canopy. Real-life drop tests were then conducted for each variation of the parachute from a height of approximately 9.2 m. The Arduino Nano 33 BLE microcontroller which comes with the LSM9DS1 Inertial Measurement Unit (IMU) was placed in the payload of the parachute and used to generate acceleration and angular velocity at a rate of 10 Hz. The data was then processed by python using a directional cosine *matrix* (DCM) to convert data to the earth axes.

$$\underset{R_x(\phi)=}{\begin{bmatrix} 1 & 0 & 0 \\ 0 & \cos(\varphi) & -\sin(\varphi) \\ 0 & \sin(\varphi) & \cos(\varphi) \end{bmatrix}} \underset{R_y(\theta)=}{\begin{bmatrix} \cos(\theta) & 0 & \sin(\theta) \\ 0 & 1 & 0 \\ -\sin(\theta) & 0 & \cos(\theta) \end{bmatrix}} \underset{R_z(\psi)=}{\begin{bmatrix} \cos(\psi) & -\sin(\psi) & 0 \\ \sin(\psi) & \cos(\psi) & 0 \\ 0 & 0 & 1 \end{bmatrix}}$$

$$\mathrm{D}CM = \mathrm{R}_Z(\psi)R_y(\theta)R_x(\phi)$$

3.4 CFD Simulations

The simulation platform Ansys Student 2023 R2. The turbulence model used was SST K Omega to replicate real-life conditions. The assumptions made in the simulation were that the parachute cannot deform, heat exchanged with the surroundings was negligible, the velocity of steady incoming air flow was 5 ms^{-1} and the density of air was 1.225 kg/m^3. Mesh convergence tests were also conducted to ensure the validity of results (Figs. 1, 2, 3, 4, 5 and 6).

Further information on prototypes, codes and data collection may be found in the appendix.

4 Results

4.1 Effect of Canopy Size on Drag

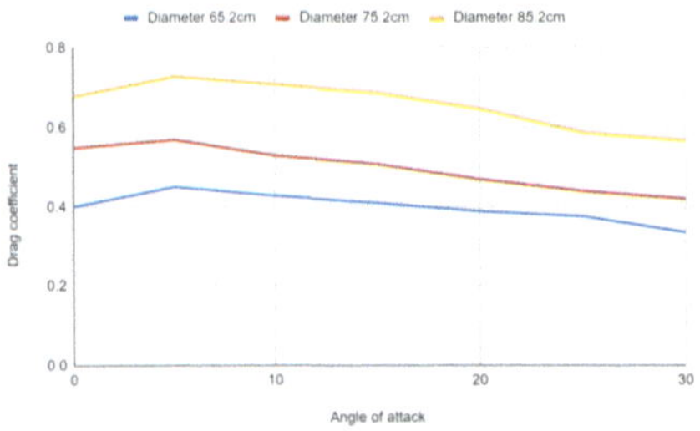

Fig. 1. Graph of drag coefficient against canopy size for CFD and experiment

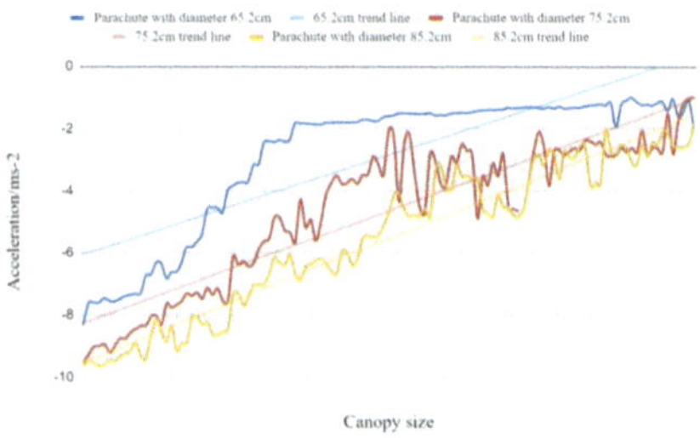

Fig. 2. Aggregated graph of acceleration/ms^2 against time/s for experiment

The above graphs show that the drag coefficients increase as canopy size increases. Experimental data was processed by aggregating the graphs of 3 data sets to create the acceleration graph of each parachute. The experimental data validates the numerical simulations where as canopy size increases, the gradient of deceleration increases. The canopy size directly affects the reference area of the canopy, which is directly proportional to the drag force experienced, $F_D = \frac{1}{2}pv^2sC_D$. Pressure contour graphs suggest an increased pressure gradient with greater canopy size, increasing flow separation and vortex shedding around the canopy which ultimately increases drag.

4.2 Effect of Canopy Size on Stability

CFD simulation results show smaller moments and increased stability of the parachute at increased canopy sizes, possibly due to slower rates of descent. Past research has shown that larger canopy sizes are characterised with enhanced pendulum stability, as increased drag forces provide a stabilising force to lateral disturbances, thus resisting horizontal motions in the parachute.

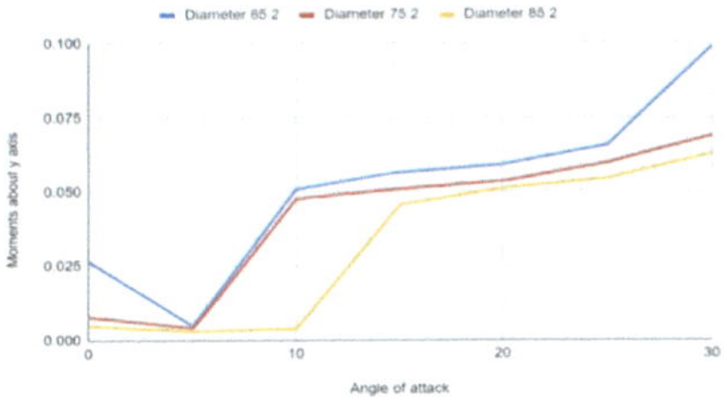

Fig. 3. Moments about y-axis at varying canopy sizes from CFD

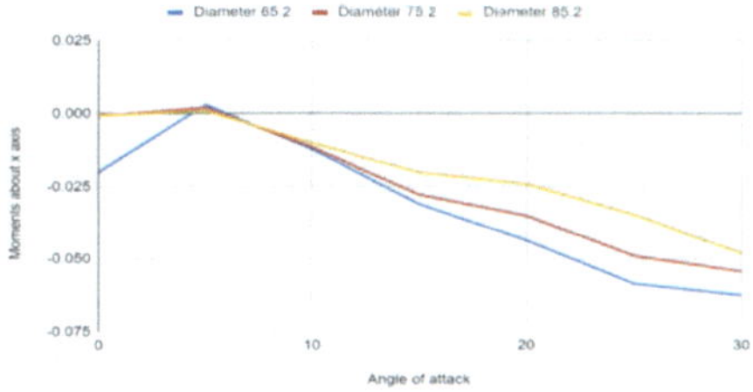

Fig. 4. Moments about x-axis at varying canopy sizes from CFD

4.3 Effect of Vent Ratio on Drag

Data generated suggested that an increase in the vent radius also caused some decrease in overall drag coefficient. The drag coefficient was found to be 0.56 at 0% vent ratio, 0.45 at 5% vent ratio and 0.39 at 7% vent ratio. The effect of vent ratio on drag is not nearly as significant as its effects on the stability of the parachute.

4.4 Effect of Vent Ratio on Stability

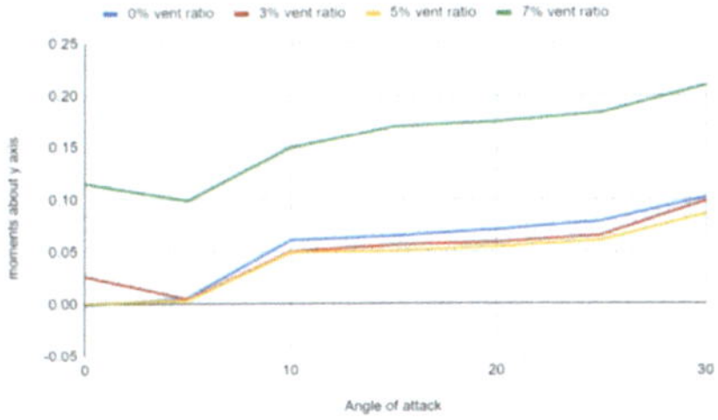

Fig. 5. Moments about y-axis at varying vent ratio from CFD

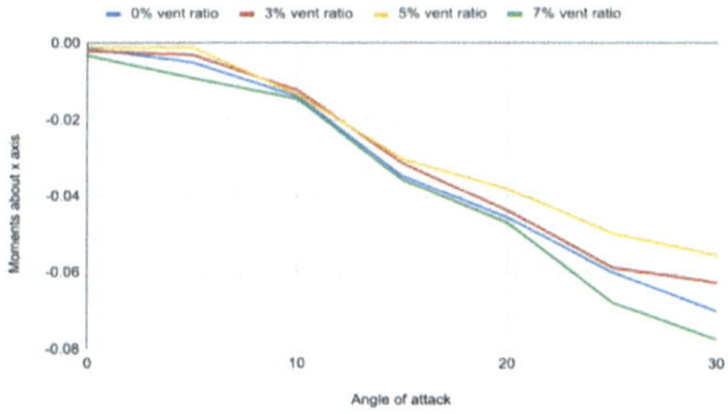

Fig. 6. Moments about x-axis at varying vent ratio from CFD

Moment graphs obtained from the CFD simulations highlighted a decreasing trend where an increase in vent ratio from 0% to 5% caused a decrease in the moments of the parachute while a further increase, beyond an optimum point, to 7% caused increased moments of the parachute, and introduced greater instability than that at 0% vent ratio.

5 Discussion

5.1 Limitations

There were several key limitations which could have had detrimental effects on the experimental process. Varying processing speed and the load shedding process of the Arduino resulted in some time intervals varying by 1 or 12-13ms respectively at random points and may have affected the accuracy of data trends observed. The lack of a controlled environment meant that parachute movement could have been substantially influenced by wind patterns which could also have resulted in some inaccurate findings, especially compared to the CFD simulations which modelled an unrealistic ideal environment without turbulence, leading to differences in simulation and experimental data. High sensitivity of the Arduino Nano 33 BLE sensor and other possible factors resulted in noisy data, which made it difficult to get precise acceleration and velocity graphs and calculate precise drag coefficients. The parachute not being fully rigid, and deformations caused as a result could also have affected the accuracy of data.

5.2 Further Research

For further or future research, it would be ideal for the test to be conducted using a more accurate and precise IMU in a controlled environment, using a fully rigid parachute to minimise any possible external factors which could affect the collected data.

6 Conclusion

Through a combination of numerical simulations and experimental tests, it was found that as vent ratio increased up to an optimal point, parachute stability increased substantially while as canopy size increased, both drag coefficient and stability increased substantially. Meanwhile, as Angles of Attack increased, both drag coefficient and stability decreased substantially. The findings of this study strongly validates existing literature while also leaving room for further research in order to more precisely determine optimal parameters for parachutes of different applications as seen from the relationship between vent ratio and stability where there was a certain vent ratio beyond which stability began to decrease. It can be seen that there is an intricate balance that needs to be found between various parameters given that aerodynamic characteristics are complex and interrelated.

Appendix

See Figs. 7, 8, 9, 10, 11 and 12.

Fig. 7. Gore

	Diameter/cm	Suspension line/cm	Vent diameter/cm
Parachute 1	65.2	74.98	11.3
Parachute 2	75.2	86.48	13.0
Parachute 3	85.2	97.98	14.8
Parachute 4	65.2	74.98	0
Parachute 5	65.2	74.98	14.6
Parachute 6	65.2	74.98	17.3

Fig. 8. Dimensions of tested parachute

P1	P2	P3
	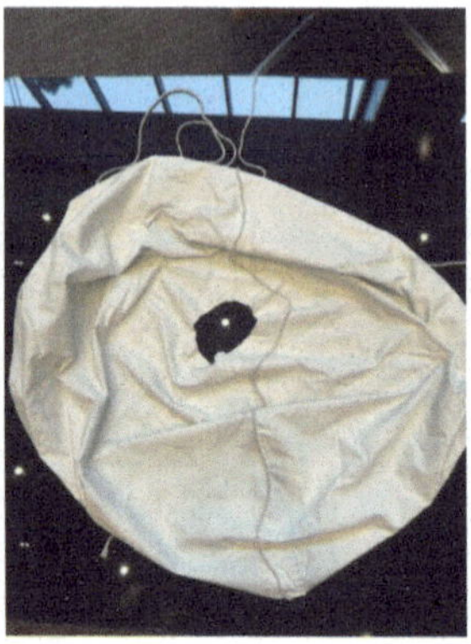	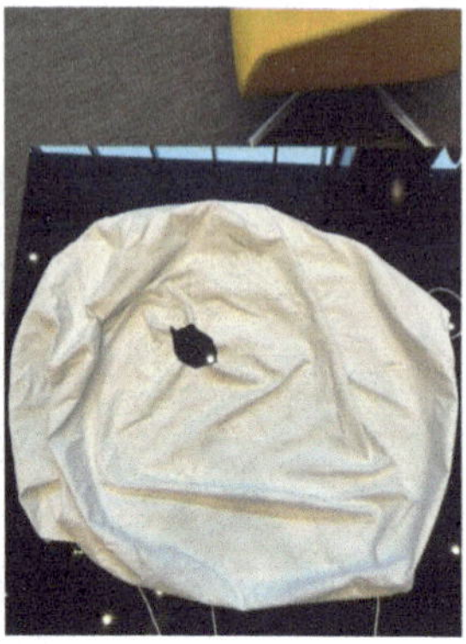

Fig. 9. Final parachutes

Fig. 10. Payload Box

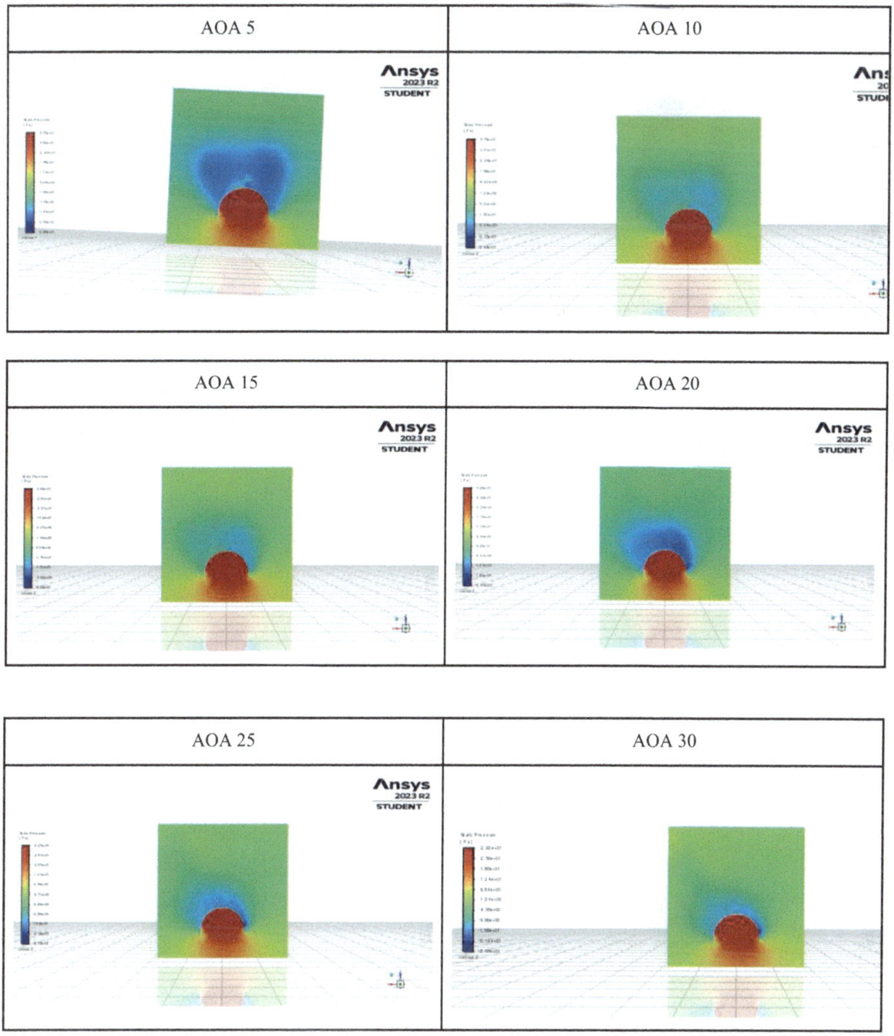

Fig. 11. Pressure contour graphs at varying angle of attacks

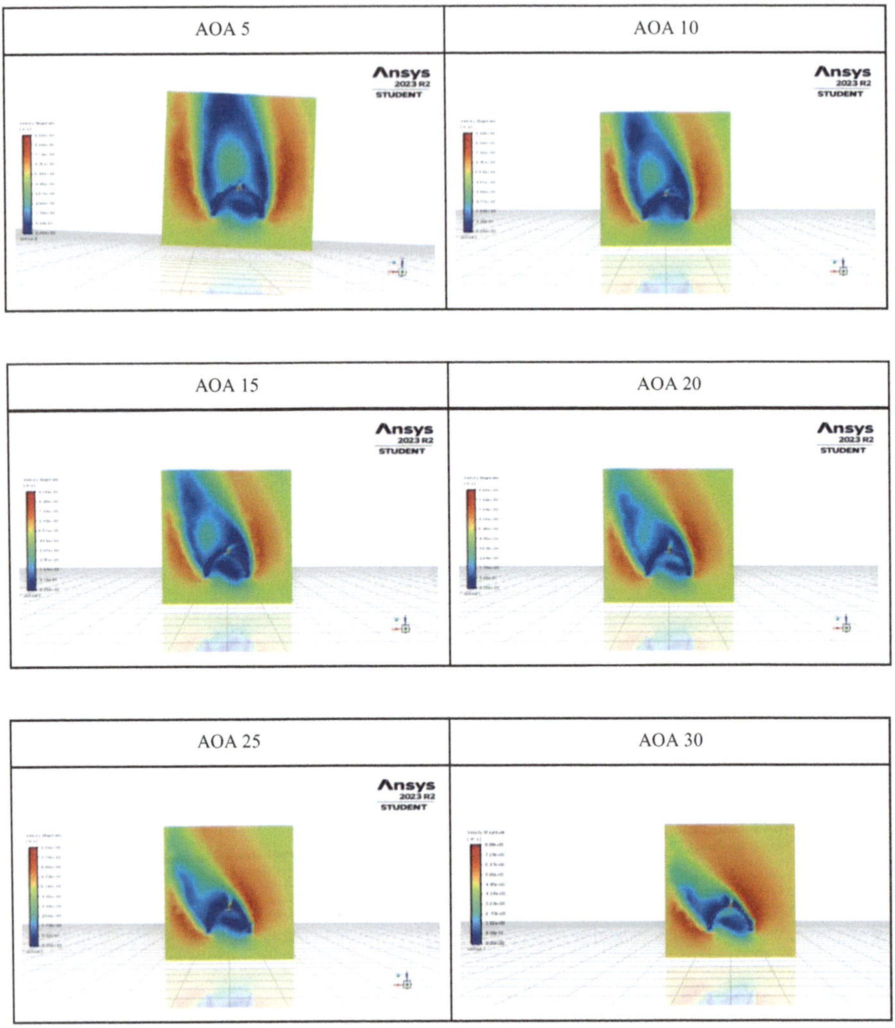

Fig. 12. Velocity contour graphs at varying angle of attacks

References

1. Ginn, J.M.: Parachute dynamic stability and the effects of apparent inertia. (2014). https://www.ssdl.gatech.edu/sites/default/files/ssdl-files/papers/mastersProjects/GinnJ-8900.pdf
2. Halstrom, L.: Aerodynamic, moving-mesh modeling of parachute pendulum motion: development and validation of a CFD methodology. eScholarship, University of California (2021). https://escholarship.org/uc/item/63q5t79v
3. Dobrokhodov, V.N., Yakimenko, O.A., Junge, C.J.: Six-degree-of-freedom model of a controlled circular parachute. Six-Degree-of-Freedom Model of a Controlled Circular Parachute (2012). https://arc.aiaa.org/doi/https://doi.org/10.2514/2.3143
4. Gunasinghe, S., Dilas, K.A., Sandaruwan, D., Weerasinghe, M.: A real-time 6DOF computational model to simulate RAM-air parachute dynamics (2017). https://www.researchgate.net/publication/315151906_A_Real-Time_6DOF_Computational_Model_to_Simulate_Ram-Air_Parachute_Dynamics
5. Zhang, S., Sun, Z., Yu, L. Prediction and effect analysis for equilibrium profile of flexible porous canopy fabrichttps://doi.org/10.1177/15280837221109640
6. Ma1, A., Zhou2, S., Wu3, Q., Zhu4, C.: IOPscience. J. Phys. Conf. Ser. (2022). https://iopscience.iop.org/article/https://doi.org/10.1088/1742-6596/2230/1/012017
7. Akash, H., Raj, S.A., Sathish, D., Kumar, A.A., Balaji, H.: Design and computational analysis of Da Vinci Parachute Design in steady and unsteady flow with varying vent hole ratio. Int. J. Res. Eng. Sci. Manag. (2021). https://journal.ijresm.com/index.php/ijresm/article/view/689
8. Cao, Y., Wei, N.: Flight trajectory simulation and aerodynamic parameter identification of large-scale parachute. Int. J. Aerosp. Eng. **2020**, 1–13 (2021). https://doi.org/10.1155/2020/5603169
9. Dawoodian, M., Dadvand, A., Hassanzadeh, A.: A numerical and experimental study of the aerodynamics and stability of a horizontal parachute. Int. Sch. Res. Not. **2013**, 1–8 (2013). https://doi.org/10.1155/2013/320563

Neuronal and Vascular Decoupling in the Eye

Xin Xuan Koo(✉), Sonia Pho, and Jie Min Tan

Nanyang Girls High School, Singapore 288683, SG, Singapore
kooxinxuan8@gmail.com

Abstract. Legend:

RNFL: Mean Retinal Nerve Fiber Layer (thickness).

Vsl: Mean thickness of the vessels.

NV: Mean thickness of the non-vascular RNFL.

In this project, the team worked with researchers to separate the neuronal and vascular components in 3D optical coherence tomography angiography (OCT) and OCT angiography (OCTA) images of the eye. OCT is a current clinical tool that allows measurement of nerve thicknesses in the eye, which is crucial for vision. Changes in the nerve thicknesses are used in the assessment of ocular and neurodegenerative diseases. However, current measurements include both neuronal and vascular content, which can confound interpretations of thicknesses, particularly in the presence of co-morbidities such as hypertension which has been associated with enlarged vessels. In this report, the team will evaluate how the thickness of blood vessels in the RNFL affects the accuracy of the OCTA algorithm in diagnosing glaucoma using RNFL thickness.

1 Background and Purpose of Research Area

1.1 Background Information

This research project makes use of 3D optical coherence tomography angiography (OCTA) images of the eye to study the eye's neuronal and vascular components. To understand the mechanics behind OCTA imaging, the team first had to do some research on optical coherence tomography (OCT) imaging because OCTA is largely reliant on OCT technology.

OCT is a non-invasive clinical tool that uses light waves to take cross-sectional pictures of the eye, allowing the thickness of the retina's individual layers to be measured, helping with the diagnosis of many eye diseases like glaucoma and age-related macular degeneration. OCT performs imaging by measuring the echo time delay and intensity of backscattered light from internal microstructure in materials or tissues. The optical beam is focused into the sample being imaged, and the echo time delay and intensity of the backscattered light are measured to yield an axial backscattering profile. This light is backscattered from the retina's internal microstructures, which have different optical properties. The echo time it takes for the light to be backscattered from the different structures at varied axial distances helps the machines to determine the dimensions of these structures. Next, the incident beam is scanned in the transverse direction, measuring

H. Guo et al. (Eds.): IRC-SET 2024, PTT, pp. 200–208, 2025.
https://doi.org/10.1007/978-981-96-3770-6_21

the axial backscattering profile at a few different transverse positions to yield a two-dimensional data set, representing the differences in optical backscattering through the cross section of the retinal tissue. This data is displayed as a logarithmic grey scale or false colour image.

OCTA is also a non-invasive imaging technique that makes use of OCT technology to generate images of the eye's vascular structure. By using motion contrast imaging to high-resolution volumetric blood flow information, it is able to give rise to angiographic images of the eye's vascular content which are enface images. OCTA works by capturing consecutive OCT B-scans at the same retinal location, and each scan capture is separated by a short lapse in time. As light is reflected back, a difference in signal will be detected between the two scans which is due to motion of blood between the scans which is known as decorrelation signal. This allows the exact location of the blood vessels to be captured accurately by comparing the decorrelation signal between successive OCT B-scans taken exactly at the same cross-section. As the retina is a static tissue, the decorrelation signal is attributed to the movement of blood through the retinal vasculature where the charge will remain on the surface. As such, a decorrelation map is generated on the OCTA which mirrors the flow of blood at the back of the eye. This allows one to visualise and map out the blood flow in the body. Between each B-scan captured, a small time delay is implemented to allow for different interscan times to detect specific blood flow speeds. For OCTA technology to detect blood flow in a blood vessel, erythrocytes must move within the length of the OCT beam between sequential scans. There are two different erythrocytes: fast-moving erythrocytes and short erythrocytes. Having short interscan times helps to show the movement of a very fast red blood cell before it exits the length of the OCT beam, improving the visualisation of fast-moving vascular beds. Increasing the interscan time allows slow-moving erythrocytes a greater chance to move between successive scans, increasing the sensitivity to slow-moving vascular beds. Thus, the use of OCTAs has been proven to be a functional tool to evaluate common ophthalmological diseases such as diabetic, retinopathy and glaucoma.

Glaucoma is an eye disease that can cause permanent vision loss by damaging the optic nerve. It is usually caused by increased pressure in the eye due to the buildup of aqueous humour, which normally drains through the trabecular meshwork, causing a blocked drainage of the fluid in the eye. When the eye makes too much fluid or when the aqueous humour is unable to drain normally, eye pressure may increase which could eventually lead to the development of blind spots. Some of the factors of glaucoma include age, ethnicity, family history, co-existing medical conditions such as hypertension and high myopia. Even though glaucoma is a disease that cannot be prevented, if treatment is started before the optic nerve is damaged, it could ultimately prevent permanent visual loss. The vision loss experienced by a glaucoma patient is due to the death of retinal ganglion cells (RGCs), a type of nerve cell in the retina. RGCs are important for vision because they process the light entering the eye and transmit it as images to the brain though their axons, which are long fibres that make up the optic nerve. However, in glaucoma, RGCs are damaged by prolonged stress on the eye wall and increased intraocular pressure from the fluids in the eye, causing them to eventually die by apoptosis. The bipolar cells will not be able to receive output from the dead retinal ganglion cells and thus will not be able to provide synaptic input to amacrine cells, leading to progressive vision loss and ultimately blindness. RGCs also cannot regenerate once damaged, explaining why the blind spots in glaucoma patients' vision are irreversible once formed.

There are many ways OCT and OCTA technology is able to detect glaucoma, such as analysing the retinal nerve fibre layer thickness, the macular thickness or the peripapillary vessel density. There are also ways to monitor the progression of glaucoma using these techniques, such as by investigating the inter-visit repeatability of OCTA measurements in the peripapillary and macular regions of a patient. However, in this project, the team will primarily be conducting their analysis on the accuracy of OCTA in glaucoma diagnosis using the thickness of the retinal nerve fibre layer (RNFL) layer. It is known that in general, a thinner RNFL represents glaucoma.

1.2 Purpose of Research

To evaluate how the thickness of blood vessels in the RNFL affects the accuracy of the OCTA algorithm in diagnosing glaucoma using RNFL thickness.

2 Hypothesis

Diagnosis of glaucoma made by OCTA machinery is more accurate using the Mean thickness of the non-vascular RNFL (NV) rather than the thickness of the whole RNFL.

3 Methodology

The images used by the team are samples voluntarily given by hospital patients, collected by a research institute. The images consist of cross sectional and enface scans of patients' eyes by OCTA machinery.

3.1 Data Collection

The team decided on a criteria for data selection of datasets to work with, but realised that there were too few patients who met the criteria. A small sample size would result in unreliable analysis and conclusion. Hence, the team sought advice from their mentor and widened the data range of certain factors in the selection criteria like the degree of myopia to obtain a larger sample size. The team chose to widen the data range of factors least affecting the reliability of the results after careful consideration. This ensured that our sample size would be large enough to meet the criteria of selecting 100 samples. After deciding on the final criteria, the team proceeded to sort out the patient data using Excel, before eliminating the data which did not fit the criteria. After a day or so, the team was left with 48 glaucoma and 32 normal patient data.

3.2 Annotation

Secondly, the team did the first round of annotations on the cross sectional OCTA scans of the selected patients using image editing tool GIMP, by drawing white ovals over the visible blood vessels, which appeared as a translucent oval shape on the OCTA scans. These annotations were done on a transparent layer over the image, so that they could be exported as markings on a transparent background for vessel thickness calculation. In the second round of annotations, the team cross checked their annotations using the enface OCTA scans of the same patients, making minor changes to their initial annotations. Lastly, the team checked each other's work before sending the final annotations to their mentor for him to separate the neuronal and vascular content in the images. Their mentor then helped them to calculate the mean thickness of the retinal nerve fibre layer (RNFL), mean thickness of the vessels and mean thickness of the non-vascular RNFL. In the end, the team collected two data sets – one obtained from purely annotating visible vessels on the cross sectional scan from the vessels data set, and one obtained from annotating on the cross sectional scan while cross checking with the enface scan from the enface data set.

3.3 Analysis

The team chose to conduct their analysis in two parts. The first round of analysis was to determine which data set, the vessels or the enface, was more accurate by comparing the standard deviation of the NV thickness the team obtained from the vessels and enface datasets. Standard deviation is a measure of how dispersed the data is in relation to the mean. Hence, the team decided to use standard deviation as a criteria to assess which dataset had values clustered more tightly around the mean, and is thus more reliable. The team would conduct their next round of analysis using the more reliable dataset.

The second round of analysis was done by constructing a Receiver Operating Characteristic (ROC) and using the Area Under Curve (AUC) of 2 variables, NV and RNFL to compare which model was more accurate in the diagnosis of glaucoma. To construct the ROC graph, the team first chose 10 cut-off points for each variable. Next, the team counted the number of True-positive (TP), False-Positive (FP), True-Negative (TN), False-Negative (FN) samples and collated them in confusion matrices. Using Google Sheets, the team found the y and x coordinates of each point on the graph by using 2 formulas. The x axis is 1-Specificity, also known as the false positive rate. The y axis is Sensitivity, also known as the true positive rate. Then, the team plotted the graph on Excel based on the coordinates of each point and drew a best fit line as the graph curve. The below are the formulas for 1-Specificity and Sensitivity.

Formula of 1-Specificity and Sensitivity respectively:

$$\text{1 - Specificity} = \frac{TN}{TN + FP}$$

$$\text{Sensitivity} = \frac{TP}{TN + FP}$$

Formula to calculate AUC (Integration Method):

$$\frac{(x_{i+1} - x_i) \times (y_i - x_{i+1})}{2}$$

4 Results

Vessel results (Figs. 1, 2, 3, 4, 5 and 6):

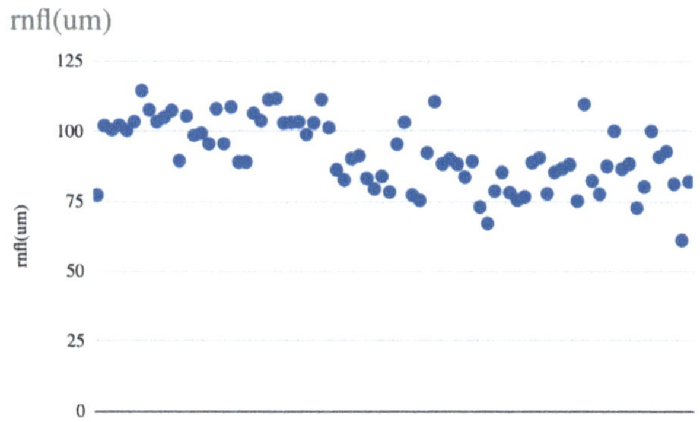

Fig. 1. The graph shows the RNFL of all the patients' vessels results that were used in our research.

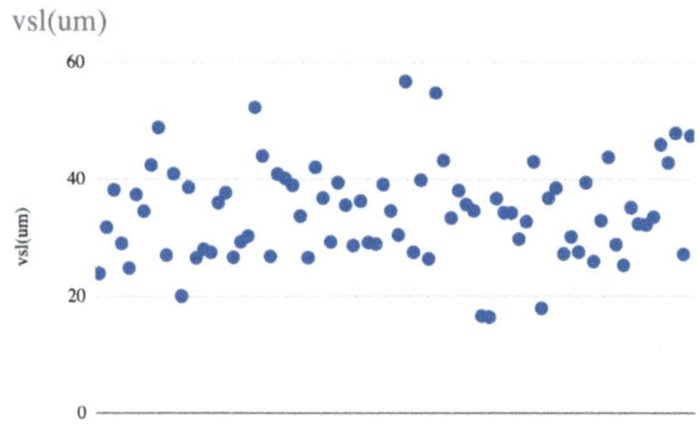

Fig. 2. The graph shows the VSL of all the patients' vessels results that were used in our research.

Enface results:

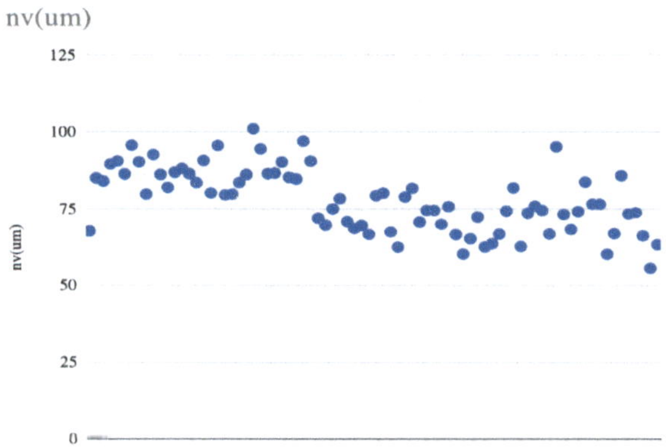

Fig. 3. The graph shows the NV of all the patients' vessels results that were used in our research.

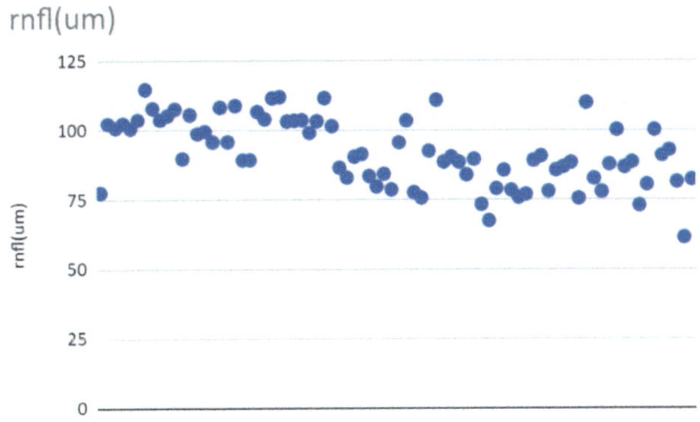

Fig. 4. The graph shows the RNFL of all the patients' enface results that were used in our research.

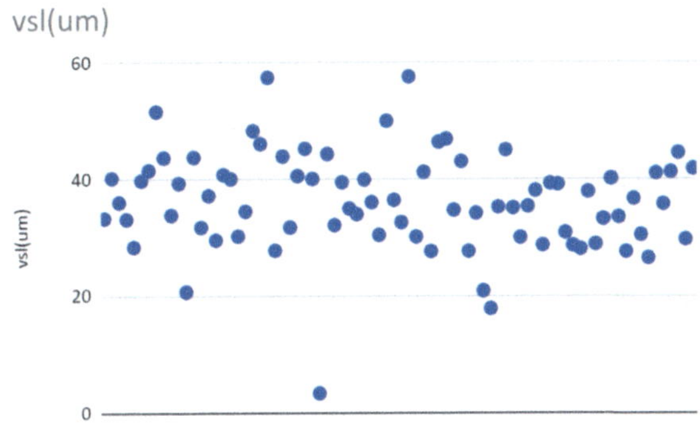

Fig. 5. The graph shows the VSL of all the patients' enface results that were used in our research.

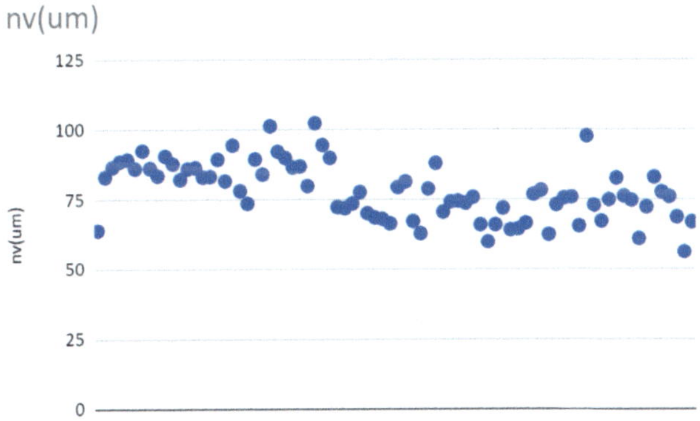

Fig. 6. The graph shows the NV of all the patients' enface results that were used in our research.

5 Analysis of Results

5.1 Using Standard Deviation

The team did their first round of consideration by calculating the standard deviation of the mean retinal nerve fibre layer (thickness) (RNFL), the mean thickness of the vessels (VSL) as well as the mean thickness of the vessels (NV). From the results taken, the team decided to calculate the standard deviation for the RNFL, VSL and the NV for patients with glaucoma and patients without glaucoma. Since the standard deviation for NV is smaller than that of RNFL and VSL, it can be concluded that using NV to diagnose glaucoma is the most accurate among the 3 (Table 1).

Table 1. The vessels results where the team calculated the standard deviation with glaucoma and without glaucoma.

	RNFL thickness	NV thickness
Standard deviation (glaucoma)	9.701192024	7.470650474
Standard deviation (non-glaucoma)	7.726105842	6.313778477

Table 2. The enface results where the team calculated the standard deviation with glaucoma and without glaucoma.

	RNFL thickness	NV thickness
Standard deviation (glaucoma)	9.701192024	7.522013772
Standard deviation (non-glaucoma)	7.726105842	7.211958941

With reference to Tables 2 and 3, since the standard deviation of the NV thickness obtained using enface results is slightly higher than the standard deviation of that obtained using vessel results, the team concluded that the data obtained using enface is more spread out and thus, less reliable. This could be due to human error in determining the position of blood vessels on the enface, as OCT images are disturbed by speckle noise due to low-coherent interference properties of light, interfering with the scans and making it harder for the team to accurately determine the position of vessels. Hence, the team decided to use the vessels dataset to conduct their next round of analysis.

5.2 AUC Analysis

ROC Curve for RNFL vessels as well as NV vessels respectively, plotted by the team (Figs. 7 and 8):

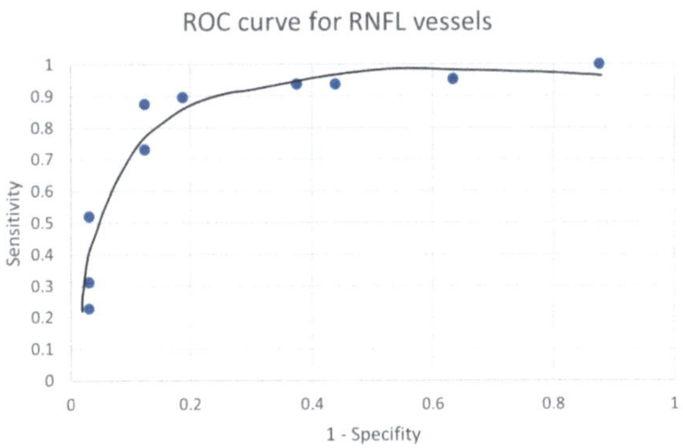

Fig. 7. The graph shows the ROC Curve for the RNFL vessels.

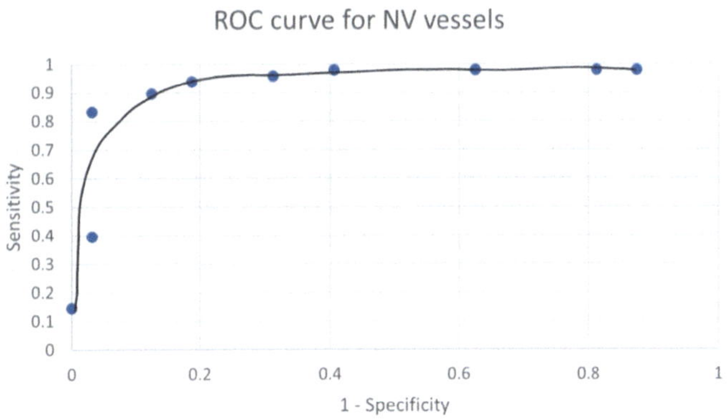

Fig. 8. The graph shows the ROC Curve for the NV vessels.

The AUC of the variable NV of 0.81 is greater than the AUC of the variable RNFL of 0.77. Hence, it can be concluded that NV has a higher accuracy in diagnosing glaucoma as the model's performance at distinguishing between the positive and negative classes is better.

6 Conclusion and Implications

The team can conclude that diagnosis of glaucoma made by OCTA machinery is more accurate using the mean thickness of the NV rather than the thickness of the whole RNFL. Using the results and analysis of results in this report, the team thinks that medical institutions should start using NV as a more accurate model for glaucoma diagnosis.

References

1. Bhalla, D.: A complete guide to area under Curve (AUC). ListenData (2014). https://www.listendata.com/2014/08/learn-area-under-curve-auc.html

2. Bhandari, A.: Guide to AUC ROC curve in machine learning : What is specificity?. Analytics Vidhya (2023). https://www.analyticsvidhya.com/blog/2020/06/auc-roc-curve-machine-learning/
3. Coombers, A.: ROC curves for evaluation of prediction models with categorical responses. JMP User Community (2020). https://community.jmp.com/t5/JMP-Blog/ROC-curves-for-evaluation-of-prediction-models-with-categorical/ba-p/321635
4. Greig, E.C., Duker, J.S., Waheed, N.K.:A practical guide to optical coherence tomography angiography interpretation. Int. J. Retina and Vitreous (2020).. https://doi.org/10.1186/s40942-020-00262-9
5. Fujimoto, J.G., Pitris, C., Boppart, S.A., Brezinski, M.E.: Optical Coherence Tomography: An Emerging Technology for Biomedical Imaging and Optical Biopsy. Neoplasia, New York, N.Y. (2000).. https://www.ncbi.nlm.nih.gov/pmc/articles/PMC1531864/#sec-a.s.ctitle
6. Jones-Odeh, E., Hammond, C.J.: How strong is the relationship between glaucoma, the retinal nerve fibre layer, and neurodegenerative diseases such as Alzheimer's disease and multiple sclerosis? Eye (2015). https://www.nature.com/articles/eye2015158#:~:text=As%20RGCs%20are%20lost%20in,RNFL%20to%20decrease%20in%20thickness
7. Rao, H.L., Pradhan, Z.S., Suh, M.H., Moghimi, S., Mansouri, K., Weinreb, R.N.: Optical coherence tomography angiography in glaucoma. J. Glaucoma (2020). https://www.ncbi.nlm.nih.gov/pmc/articles/PMC7117982/#:~:text=The%20features%20seen%20with%20OCTA,deep%2Dlayer%20microvascular%20dropout)
8. Sathyan, P., Shilpa, S., Anitha, A.: Optical coherence tomography in glaucoma. J. Curr. Glaucoma Pract. (2012). https://www.ncbi.nlm.nih.gov/pmc/articles/PMC5159451/#:~:text=OCT%20scans%20of%20the%20macula,to%20detect%20early%20cell%20loss
9. Zach: How to create a ROC curve in Excel (step-by-step). Statology (2021). https://www.statology.org/roc-curve-excel/

Sustainable Energy from Wastewater: Hydroelectricity Potential in Singapore's Deep Tunnel Sewerage System

Power Drain

Hubert Li(✉), Jiachen Zhang, and Joy Goh

Raffles Institution, Singapore, Singapore
hbli07hk@gmail.com

Abstract. In response to the urgent need for sustainable energy solutions worldwide and the increased adoption in geographically constrained regions, this study explores the hydroelectricity potential in Singapore's Sewage System. Despite Singapore's limited effective land area and the absence of natural rivers, which generally limit the potential for tidal and traditional hydropower, the country's tropical climate and consistent rainfall ensure a steady water flow through its drainage systems. This unique setting presents an ideal opportunity for innovative hydroelectric generation. **POWER DRAIN** aims to harness this kinetic energy to generate renewable electricity by installing modified versions of the turbines in the Deep Tunnel Sewerage System (DTSS). Among various water turbines like Kaplan, Francis, and Pelton wheels, the Francis turbine was selected for its centripetal flow operation, ideal for achieving up to 95% energy conversion efficiency in medium-head applications. The vortex chamber's natural spiral flow, post-sewer exit, allows for the seamless integration of the Francis turbine, maintaining structural integrity. Its suitability for medium-head applications, particularly within the 35 to 50-m depth of the vortex chamber, further underscores its selection. The simulation analysis reveals that the modified version of the Francis turbine is able to generate 25.1 MWh of Mechanical Power and net 23.9 MWh of useful electrical power per unit per day. Maximizing the scale of the project by installing one turbine in each of the approximately 30 vortex chambers across the island is estimated to generate 8.71 GWh of useful electrical power per turbine per year, powering up to 21,775 households (approximately 400 kWh per household). According to our economic analysis, the breakeven point to cover the initial costs would be reached in less than two years of operation, and this project is expected to achieve consistent annual cost savings of $390,339,150, after accounting for recurring costs such as maintenance fees.

Keywords: Sewage System Hydroelectricity · Sustainable Energy · Urban Hydropower · Water-to-Energy Efficiency

H. Guo et al. (Eds.): IRC-SET 2024, PTT, pp. 209–221, 2025.
https://doi.org/10.1007/978-981-96-3770-6_22

1 Introduction

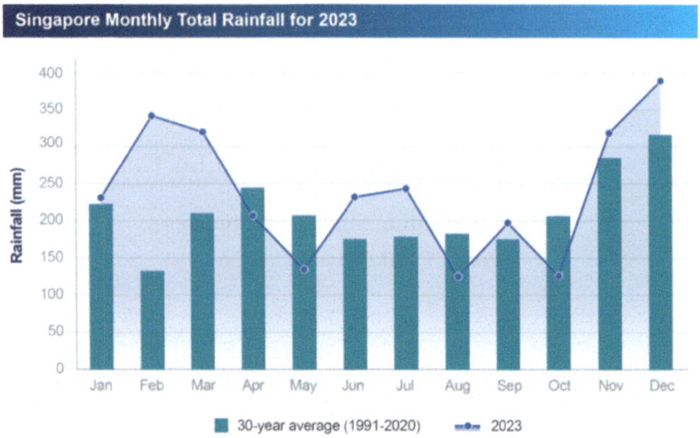

Fig. 1. Singapore Monthly Total Rainfall for 2023 [1]

Singapore is pursuing sustainable urban development by focusing on alternative cleaner energy sources. With a high annual rainfall of approximately 2,840 mm as shown in Fig. 1, and a substantial volume of wastewater flowing through the drainage system, it presents a great opportunity for Singapore to utilize these resources to generate hydroelectricity. By integrating green energy technologies with the drainage system, Singapore can transform water flows into a sustainable energy source, revolutionizing its energy landscape (Fig. 2).

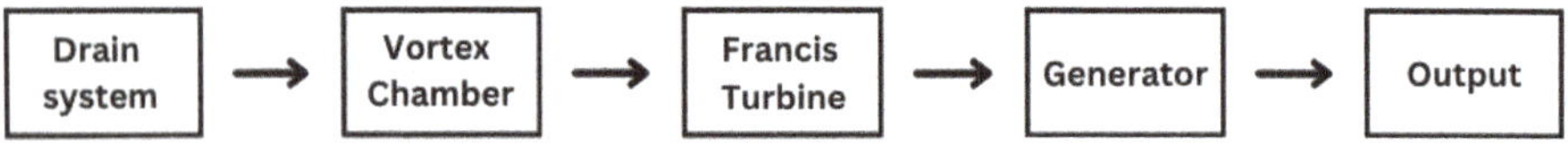

Fig. 2. Block diagram of the Turbine flow process

2 Method

2.1 Design Overview

A turbine is a rotary engine that converts the energy of a moving stream of water, steam, or gas into mechanical energy. The basic element in a turbine is a wheel or rotor with paddles, blades, or buckets arranged on its circumference in such a way that the moving fluid exerts a tangential force that turns the wheel and imparts energy to it [2].

In a Francis turbine, water impinges on the blades of the turbine radially within the radial flow runner and departs axially along its axis through a draft tube. The wheel is made to rotate due to the force of the inflow of water, producing torque and power through the runner at a particular speed, thereby generating an electric voltage at the generator, providing the output.

Electricity demand in Singapore is projected to increase by 38%, from 37 TWh in 2010 to 51 TWh by 2035.

Utilizing innovative technologies like hydrodynamic turbines will further maximize the efficient use of space, particularly in a land-scarce country with a total land area of only 734.3 km^2. Turbines generate electricity by harnessing the kinetic energy of flowing water within the drainage system, minimizing negative environmental impacts compared to fossil fuels. With a low carbon intensity of less than 50g CO_2/kWh, such a system will provide multiple benefits to Singapore.

Therefore, this research focuses on designing and evaluating the feasibility of incorporating Francis turbines into Singapore's Deep Tunnel Sewerage System (DTSS).

2.2 Turbine

Water enters the turbine through the snail-shell casing, known as the volute. As the water progresses through the tube, its pressure decreases with constant speed. Upon exiting the volute, the water flows through guide vanes, which direct it towards the runner's blades at optimal angles. As the water traverses the precisely curved blades of the runner, it is partially diverted sideways, losing some of its "whirl" motion, and is then directed axially to exit through a draft tube to the tail race.

The draft tube reduces the water's output velocity to maximize the energy extracted from the inflowing water. As the water is diverted by the runner blades, a force propels the blades in the opposite direction, resulting in the turbine's rotation. This reaction force, as described by Newton's third law, transfers power from the water to the turbine's shaft, maintaining its rotation. Due to this reaction force-driven motion, Francis turbines are classified as reaction turbines. The alteration in the direction of the water flow also reduces the pressure within the turbine itself [3].

2.3 Choice of Turbine

Table 1. Turbine Classification [4].

Head Classification	Turbine Type		
	Impulse	*Reaction*	*Gravity*
High (>50m)	• Pelton • Turgo		
Medium (10-50m)	• Crossflow • Turgo • Multi-jet Pelton	• Francis (spiral case)	
Low (<10m)	• Crossflow • Undershot waterwheel	• Propeller • Kaplan • Francis (open-flume)	• Overshot waterwheel • Archimedes Screw

There are various types of water turbines, such as Kaplan wheel turbines, Francis wheel turbines, and Pelton wheels, each selected based on water quantity, flow speed, and the axial direction of water inflow. The Francis turbine, which operates with centripetal flow, was chosen for our design to ensure the highest possible energy conversion efficiency at 95% ideally.

The natural spiral water flow within the vortex chamber, following its exit from the link sewer, supports the integration of the Francis turbine without significant structural modifications, preserving the chamber's integrity. Given its suitability for medium-head applications, the Francis turbine is ideal for the vortex chamber's depth of 35 to 50 m. (source: DTSS website).

Safety & Sustainability

Since wastewater is more acidic and contains harsh chemicals like urea, sulfur dioxide, and nitrogen oxides, the risk of corrosion is increased due to consistent contact between the turbine and wastewater in the drainage system. To enhance longevity and sustainability, the turbine will be constructed from austenitic grade 316 stainless steel coated with elastomeric cavitation-resistant coatings (Belzona 2141 – ACR-Fluid Elastomer) to enhance its corrosion resistance and ultimately increase lifespan [5, 6].

To mitigate potential risks, regular maintenance checks should be scheduled, and advanced monitoring systems should be implemented to detect early signs of wear or malfunction (Tables 1 and 2).

Capacity & Facility

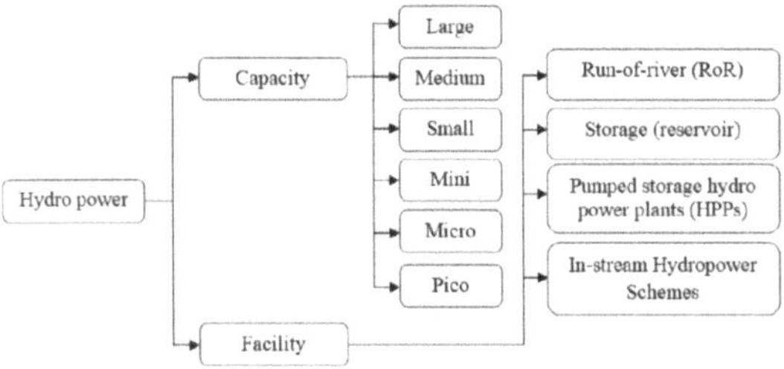

Fig. 3. Hydro Power Classification [7]

Hydropower is categorized by facility type and capacity. Facility types are classified into four categories: Storage or Reservoir, Run-of-River (RoR), Pumped Storage Hydropower Plants (HPPs), and In-stream. Capacity types are classified into Pico, Micro, Mini, Small, Medium, and Large (Fig. 3).

Table 2. Power of different categories of Hydro [8].

Category of Hydro	Power Range	No. of Homes Powered
Large	100 MW+	100,000+
Medium	10 MW – 100 MW	10,000 – 100,000
Small	1 MW – 10 MW	1,000 – 10,000
Mini	100 kW – 1 MW	100 – 1,000
Micro	5 kW – 100 kW	5 – 100
Pico	0 kW – 5 kW	0 – 5

We propose that our turbine will operate as a run-of-river system, where the natural downward flow of water is harnessed to capture the kinetic energy. [9] In our project, this principle is applied to the flow within the sewage system rather than a river. This type of plant is ideally suited for streams that maintain year-round flow at any rate. Singapore's high annual rainfall is advantageous, as it ensures a continuous supply of water diverted into the penstock before entering the stream. These plants typically require minimal storage, enabling minor adjustments to the demand profile, which is particularly beneficial (Table 3).

Table 3. Turbine Parameters

Turbine parameters	Value
Rotational speed/RPM	700
Angular Velocity/rad s^{-1}	73
Total diameter/m	1.20
Outlet diameter of spiral casing/m	0.40
Inlet diameter of spiral casing/m	0.20
Bade height at inlet/m	0.40
Number of blades	13
Average angle between blades/°	27.7
Runner inner diameter/m	0.40
Runner outer diameter/m	0.65
Draft tube inlet diameter/m	0.40
Draft tube outlet diameter/m	0.45

2.4 Francis Turbine Design

Orientation of Turbine: These turbines perform well both on a horizontal plane and when arranged vertically. However, due to the unique space of the vortex chamber and the natural spiral flow of water, the turbine will be positioned horizontally.

Number of Blades: The selection of the optimum number of blades is crucial in designing the turbine runner. Too few blades result in underutilization of the available water, while too many can cause pulsating power and reduced efficiency. For smaller turbines, more blades are necessary to capture mechanical energy effectively. Since noise is not a major concern for underground installations, the number of blades is flexible. Wang et al. demonstrated that reducing the number of runner blades can enhance turbine capacity, with 13 blades being optimal for small hydropower plants. [10].

Blade Spacing: Proper blade spacing allows water to strike the blades for maximum thrust. Blade spacing is determined by the number of blades used.

Runner Diameter: With reference to the PUB's drawing of the vortex chamber, we determined the diameter of the vortex chamber to be 1.5 m. We designed the turbine diameter to be 1.25 m, providing a 0.25 m allowance. The optimal inner and outer radii of the runner are 0.4 m and 0.65 m, respectively.

Runner Blade Design: In our study, we will utilise the blade design from Daneshgah and Zangeneh (2010) as detailed in the IOP Conference Series paper. [11] CFD simulations demonstrated that adjusting the stacking angle to -30 degrees significantly impacts the spanwise pressure distribution and three-dimensional flow characteristics in the Francis runner.

This modification results in a more uniform spanwise pressure loading by minimising the low-pressure region, which becomes confined to a narrow area between 75% and 90% chord from midspan to the band on the suction surface. The negative stacking angle effectively reduces pressure loading at the band and increases it at the crown, aiming to mitigate cavitation issues on the band's suction surfaces. All other design parameters of the runner remain unchanged. By specifying a fore-loaded distribution at the band while keeping the crown loading the same, the modified blade design enhances durability and reduces cavitation without compromising performance, leading to lower maintenance costs and reduced blade replacement frequency.

Draft Tube The draft tube controls water flow, increases efficiency, and prevents air entrainment while dissipating kinetic energy. Due to the restrictive shape of the vortex chamber, the draft tube design differs from the curved design of a standard Francis turbine, opting for a straight design with inlet and outlet diameters of 0.4 m and 0.45 m, respectively. While a curved draft tube facilitates gradual deceleration, reducing turbulence and improving efficiency, space constraints necessitate a straight design, likely resulting in decreased efficiency (Fig. 4, Table 4).

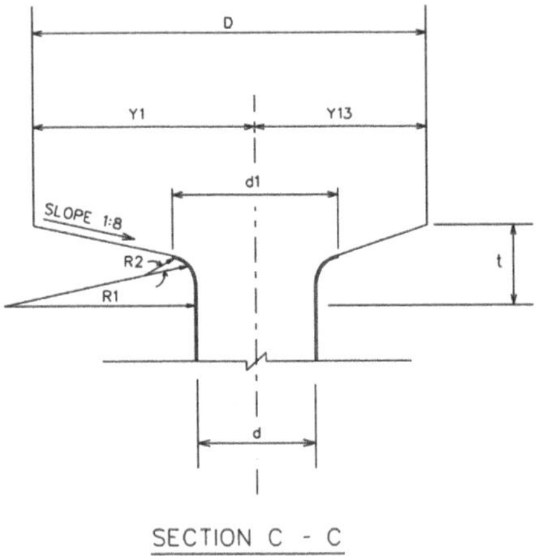

Fig. 4. Drawing of Vortex Chamber (source: PUB)

Table 4. Dimensions & Discharge Capacity of Vortex Drop (source: PUB)

$$Blade\ Spacing = \frac{360^\circ}{Number\ of\ Blades} = \frac{360^\circ}{13} \approx 27.7^\circ$$

ϕ=Incoming Pipe diameter [See Note 2]	200	250	300	350	400	450
d=drop pipe diameter	300	350	450	450	500	500
Dimension						
D	1000	1167	1500	1500	1667	1667
d1	420	490	630	630	700	700
t	200	233	300	300	334	334
a	348	406	522	522	580	580
L1	125	146	188	188	209	209
L2	235	274	352	352	392	392
R1	488	569	731	731	813	813
R2	33	39	50	50	55	55
Qdesign (l/s) [See Note 2]	48	72	129	152	203	230

Where:
1. Qdesign is the design capacity of drop pipe (d)
2. Q_ϕ is the design capacity of incoming pipe (ϕ)
3 Qdesign ≥ Q_ϕ

2.5 Location of Installation

See Fig. 5.

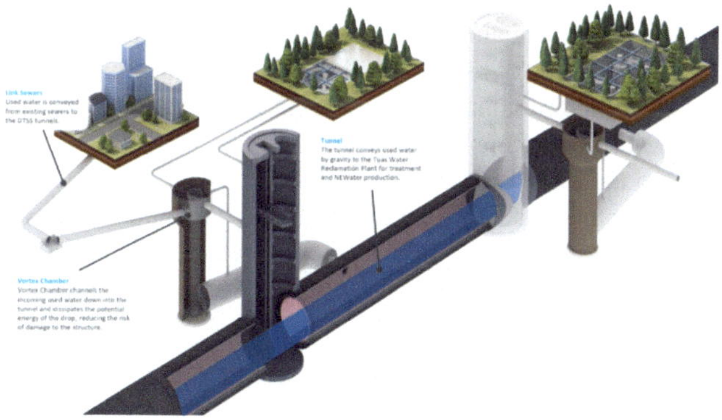

Fig. 5. Used Water Management System [12]

The turbine is installed within the vortex chamber of the drainage system as demonstrated in Fig. 6. The vortex chamber of the DTSS is engineered to induce a rotational

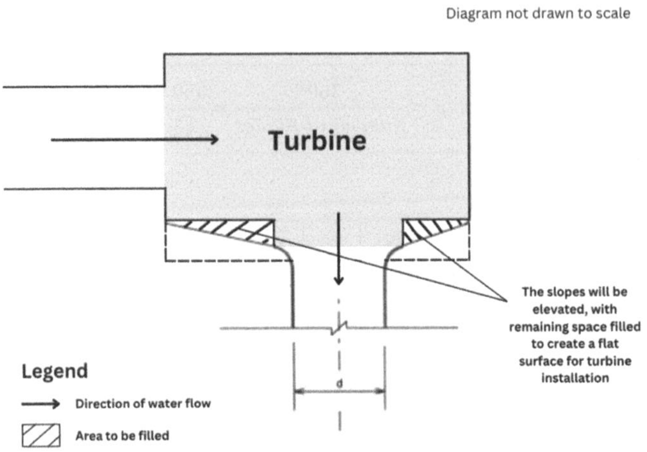

Fig. 6. Location of Installation

flow of water. This induced vortex increases the tangential velocity of the water adjacent to the turbine, thereby enhancing the available rotational energy and speed for turbine operation. The chamber's design by DTSS is optimized to enhance water flow dynamics, which, in turn, supports a smooth and strong spinning motion of water, maximizing turbine efficiency.

The installation process of the turbines within this chamber requires minimal adjustments, as the turbine's rotational dynamics integrate with the vortex-induced flow with minimal disruption to the chamber's internal flow patterns.

Moreover, the vortex chamber is positioned at the junction of multiple water conduits in the DTSS to channel water downward and dissipate the high potential energy of the water. This intersection of water flows ensures higher and more consistent rotational energy on the turbines, further contributing to the enhanced efficiency of the energy generation process.

Upon reaching the vortex chamber, the sewage undergoes a preliminary clearing process to remove trash and large objects. This also eliminates the possibility of obstruction at the turbine blades, thereby maintaining optimal efficiency and preventing potential damage or operational disruptions.

However, due to the imperfect fitting between the shape of the turbine and the top of the vortex chamber, the slopes will be elevated, and the remaining shaded region will be filled to create a flat surface. This adjustment ensures stability and provides adequate support to accommodate any vibrations and gentle movements during turbine operation.

2.6 Addressing Practical Problems in Installation

There are several practical challenges that may arise during the installation and maintenance of these turbines. Some of these issues include the difficulty of accessing underground areas and the increased demand for skilled personnel. However, these challenges can be addressed through various solutions, such as:

Automated monitoring and diagnostic system: Deploying automated systems to continuously track essential performance metrics such as vibration, temperature, pressure, and flow rates as well as utilising sensors connected to a centralised control system to detect early signs of wear and operational issues can reduce the frequency of physical inspections, minimising the need for skilled personnel to access challenging environments.

Remote Inspection Technologies: Using drones, robotic crawlers, and remote-operated vehicles (ROVs) for regular inspections of underground hydropower installations. These technologies will navigate hard-to-reach areas and provide detailed visual data, thus decreasing the need for human presence during maintenance and installation tasks underground.

2.7 Power Calculation

The power generated **per turbine per day** is calculated to achieve a substantial amount of **23.88 MWh.**

[Refer to Appendix 1]

3 Economic Analysis

In project planning, economic analysis is the most critical factor and should not be ignored in decision-making. In this study, an economic analysis was the primary indicator considered to assess the feasibility of wind projects. Economic feasibility focuses on long-term performance, which is crucial for this government project aimed at achieving sustainability and renewable energy goals for the nation. The forecasted profitability and breakeven year are calculated with a range of considerations, including initial costs, variable costs, varying electricity retail prices, and revenue (Fig. 7).

$$\textbf{Revenue} = \textbf{Electricity Generated} \cdot \textbf{Average Tariff Rate}$$

Using the historical moving average prices of electricity to forecast future prices, the range between the maximum and minimum profits is calculated to be approximately \$248,321.00, indicating a significant difference between the most profitable and least profitable times.

The standard deviation, which measures the spread of profits around the average, is about \$61,513.09. This indicates a substantial variation in profits due to changes in electricity prices. The daily revenue at the current electricity retail price of 29.88 cents/kWh is approximately \$2,602,261.

Understanding that the electricity price has been rising over the years, it is expected to grow in the future as well. The average percentage variation in revenue based on the historical electricity prices is approximately $\pm$ 12.64% (Table 5).

To attain an accurate estimation of the expenses, the total expected project cost (initial costs) covers a full range of major elements, including the costs of different parts of the turbine, the costs of the generator, and project implementation costs (e.g. installation costs and other supplementary equipment costs) based on the current market prices.

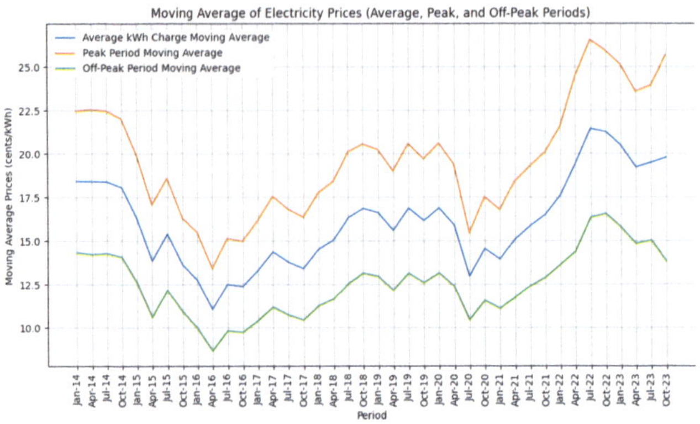

Fig. 7. Moving Average of Electricity Prices in Singapore (Data: [13])

Table 5. Expenses

Category	Item	Cost
Turbine Costs	1.00MW Francis hydro turbine generator unit	$62,800
	Basic Francis turbine for project	$200,000
	Extra sealant and paint layer	$50,000
	Total cost of turbine with extra protection	**$312,800**
Generator Costs	Generator cost	$1,400,000
Project Implementation Costs	Total project cost (initial estimate)	$1,650,000
	Additional cost for step-up transformer, underground cables, and grid distributi	$800,000
	Total expected project cost	**$4,475,600**
Maintenance Costs	**Maintenance cost per year**	**$120,000**

Table 6. Financial Statement

(unit*1)	Year 0	Year 1	Year 2	Year 3	Year 4	Year 5
Total Expenses	$ (4,475,600)	$ (120,000)	$ (120,000)	$ (120,000)	$ (120,000)	$ (120,000)
Total Revenue	$ -	$ 2,602,261	$ 2,602,261	$ 2,602,261	$ 2,602,261	$ 2,602,261
Cumulative Profits	**$ (4,475,600)**	**$ (1,993,339)**	**$ 488,922**	**$ 2,971,183**	**$ 5,453,444**	**$ 7,935,705**

Annual maintenance costs are also estimated to be approximately 2.68% of the total initial costs.

The total revenue generated on average are expected to cover the initial costs within the second year of operation, reaching $488,922 in cumulative profits by the end of the year as seen in Table 6 (Fig. 8).

The cumulative profits generated from producing 8.71 GWh per turbine annually reaches $1.65 billion on average, with a +12.66% standard deviation reaching a maximum of $1.90 billion and a −12.66% standard deviation reaching a minimum of $1.41 billion.

No energy battery storage system was considered in the current simulation, which focuses solely on the profitability and efficiency of the underground hydroelectric power generation system. Additionally, no financing activities, such as loans, were included in the calculation to maintain simplicity and clarity in demonstrating the returns calculation.

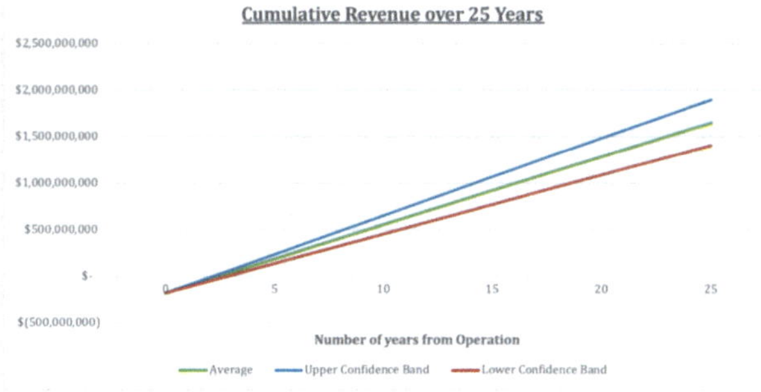

Fig. 8. Cumulative Profits over 25 years (30 units)

Furthermore, flow rates in the drainage system fluctuate, which can directly affect turbine performance. Low flow periods may result in underperformance in power generation, while high flow periods may cause surges that overwhelm the turbines. Designing the system to handle these variations does not solely depend on real-time water statistics and turbine design but requires careful consideration of the entire system, including the surrounding infrastructures. Consistent modifications and future developments should closely consider the live performance of the turbine system to ensure long-term stability.

4 Environmental Analysis

While the turbines themselves contribute to a reduction in carbon emissions by generating renewable energy, the initial production and potential ongoing maintenance activities may have some carbon footprint that needs to be accounted for.

End-of-life disposal is another major concern, particularly regarding what happens to the waste materials once a turbine is decommissioned. In Sweden, waste generated from decommissioned wind turbines is expected to increase by about 12% annually until around 2026, after which the growth rate will accelerate to 41% per year until 2034. By that time, the annual waste is projected to consist of 237,600 tonnes of steel and iron (16% of the recyclable materials), 2,300 tonnes of aluminum (4%), 3,300 tonnes of copper (5%), 343 tonnes of electronics (1%), and 28,100 tonnes of blade material. Singapore is expected to face a similar challenge approximately 40 years after turbine installations. Given Singapore's limited land availability, with Pulau Semakau rapidly filling up, enhancing metal recycling technology will be crucial to make it more cost-effective and prevent waste. However, ensuring this will be challenging, so careful future planning is needed to assess Singapore's readiness for this eventuality.

5 Conclusion

In this study, the feasibility of installing turbines within the Deep Tunnel Sewerage System (DTSS) to harness hydropower was evaluated through simulations using official data on the turbines and chambers. The primary objective of this research was achieved: To develop a new source of clean energy generation using hydropower. With the unique characteristics and design of the Francis turbine in the vortex chamber of the DTSS, turbines will be driven by the strong rotational energy from the spinning water flow, generating a significant amount of power.

This method offers several advantages over other hydroelectric power production methods in the original drainage system. For instance, it requires minimal disruption to the existing water management system. Additionally, the energy captured from the water flow is optimized in the vortex chamber due to its special structure. This design is particularly relevant in the context of Singapore, which experiences low tides and lacks consistent natural rivers to provide sufficient gravitational potential energy and kinetic energy. These factors have historically hindered the progress of hydroelectric power generation. However, our project contributes to this industry by utilizing the hidden, consistent, and strong underground water flows in the DTSS.

Ultimately, in the implementation phase, deploying pilot projects at select locations is essential to validate simulation results and gather real-world data on performance, maintenance, and environmental impact. These pilot projects will provide critical insights and facilitate the optimization of the system prior to full-scale deployment across all used water management systems in Singapore.

6 Discussion

Despite several associated benefits, maintaining hydropower systems in the DTSS still presents several challenges given the complex environment. Regular maintenance is essential to ensure proper functioning, but accessing these components may be challenging due to their underground location and compact environment. Confined spaces and hazardous environments require specialized equipment and trained personnel for safe maintenance.

Appendix 1

The power generated **per turbine per day** can be given as:

$$\boldsymbol{A} = \pi\left(\frac{\boldsymbol{D}}{2}\right)^2 = 1.13\,\mathrm{m}^2$$

$\boldsymbol{A}$ = Area of Penstock.
$\boldsymbol{D}$ = Penstock Diameter = 1.2 m

$$\boldsymbol{Q} = \boldsymbol{A} \cdot \boldsymbol{v} = 2.373\ \mathrm{m}^3\ \mathrm{s}^{-1}$$

Q = flow rate.
A = Area of penstock = 1.13 m^2.
V = flow velocity = 2.1 $\mathrm{m\,s}^{-1}$

$$\boldsymbol{P_h} = \boldsymbol{\rho} \cdot \boldsymbol{g} \cdot \boldsymbol{Q} \cdot \boldsymbol{H} = 1.163\,\mathrm{MW}$$

$\boldsymbol{P_h}$ = Hydraulic Power

ρ = Density of water = $1000\,\text{kg}\,\text{m}^{-3}$
g = Gravitational field strength = $9.81\,\text{N}\,\text{kg}^{-1}$
Q = Flow rate = $2.373\,\text{m}^3\,\text{s}^{-1}$
H = Head height = $50\,\text{m}$

$$P_m = P_h \cdot \eta_m = 1.0467\,\text{MW}$$

P_m = Mechanical Power.
η_m = mechanical efficiency = 0.9

$$P_m = P_e \cdot \eta_e = 0.995\,\text{MW}$$

P_e = Electrical Power.
η_e = electrical efficiency = 0.95.

$$\textbf{Daily Electrical Power} = \times\mathbf{24} = \mathbf{23.88\,MWh}$$

References

1. Singapore Meteorological Service: The Year in Review (2021). http://www.weather.gov.sg/wp-content/uploads/2022/01/The-Year-in-Review-2021.pdf (2022)
2. Munson, B.R., Okiishi, T.H., Huebsch, W.W.: Turbomachines. In: Fundamentals of Fluid Mechanics, 6th edn. J. Wiley & Sons, Hoboken, NJ (2009)
3. Linquip: What is Francis Turbine? https://www.linquip.com/blog/what-is-francis-turbine/ (2024)
4. GreenBug Energy: Types of Turbines. https://greenbugenergy.com/get-educated-knowledge/types-of-turbines (2024)
5. Water UK: IGN 4–25–02. https://www.water.org.uk/wp-content/uploads/2018/11/ign-4-25-02.pdf (2018)
6. Belzona: Belzona 2141 – ACR-Fluid Elastomer. https://www.belzona.com/viewfile.aspx?id=63337 (2024)
7. Ali, A.M., Algburi, S.S., Aljaradin, M.A.: Design optimization of a hybrid hydro-wind micropower system for rural communities. J. Eng. Sustain. Dev. **22**(02 Part-5), 1–10 (2019)
8. Elbatran, A.H., Abdel-Hamed, M.W., Yaakob, O.B., Ahmed, Y.M.: Hydro power and turbine systems reviews. Jurnal Teknologi. **74**(5) (2015)
9. Climate Technology Centre & Network (CTCN): Run-of-River Hydropower. https://www.ctc-n.org/technologies/run-river-hydropower#:~:text=Run%2Dof%2Driver%20hydro%20projects,pressurised%20pipeline%20(or%20penstock)(2024)
10. JSTAGE: Design optimization of a hybrid hydro-wind micropower system for rural communities. https://www.jstage.jst.go.jp/article/ijfms/8/4/8_294/_pdf (2024)
11. IOP Science: Investigation of renewable energy potential in rural areas. IOP Conf. Ser.: Earth Environ. Sci. **12**(1), 012058. https://doi.org/10.1088/1755-1315/12/1/012058/pdf
12. PUB Singapore: Deep Tunnel Sewerage System (DTSS) – Conveyance System. https://www.pub.gov.sg/Professionals/Requirements/Used-Water/DTSS/ConveyanceSystem (2024)
13. SP Group: Historical Electricity Tariff. https://www.spgroup.com.sg/dam/spgroup/docs/our-services/utilities/tariff-information/Historical-Electricity-Tariff--4Q23-Final-.xlsx (2023)

Exploring Quantum Dynamics: Fine Control of Superconducting Qubits

Lifan Yang(✉), Linxun Yap, and Yixiang Fan

Hwa Chong Institution, Singapore 269734, Singapore
yfl8136@gmail.com, yaplinxun@gmail.com, fanyixiang2006@gmail.com

Abstract. Quantum computing, confronting challenges in precisely controlling these gates due to quantum gates being prone to errors caused by external interference. Our study scrutinises the state evolution dynamics of a single superconducting tunable symmetric transmon qubit under ideal conditions, aiming to provide insights for effective quantum gates' control. Methodologically, our project is rooted in computational analysis with experimental data from the NTU lab. Python libraries are used to analyse the given data. Our result shows that the qubit exhibits a sinusoidal behaviour between its 2 states with a high average R-squared value (0.97245). This concludes that qubits behave predictably under a steady and ideal microwave field, reducing quantum gate errors and obtaining high quantum fidelity by controlling the pulse time of microwave pulses. Hence, more reliable manipulation of qubit state and accurate quantum operations are attainable.

Keywords: Quantum Computing · Qubit · Superconducting Qubit · Quantum Gates' Control

1 Introduction

Quantum computing, a rapidly evolving field, leverages the principles of quantum mechanics to transcend the capabilities of traditional computing [9, 28, 32]. At its core lies the qubit, a quantum analogue to the classical bit and the quintessential element of quantum information processing [7, 31]. The unique property of qubits is their ability to exist in a state of superposition, facilitating simultaneous computations and offering a multidimensional computing landscape.

Since the unveiling of the first rudimentary quantum computer prototype in 1999 at MIT [18], the field has witnessed remarkable advancements. Despite this progress, the journey towards fully functional quantum computers is impeded by numerous technical challenges. These challenges are encapsulated in DiVincenzo's seven criteria [5], a set of conditions pivotal for the realisation of practical quantum computing. One of these criteria involves the precise control of quantum gates, which are essential for manipulating qubit states and executing computations [23]. As quantum systems grow in complexity, especially with multiple qubits nearby, they risk entanglement [22] and susceptibility to perturbations from external interferences or noise. This can lead to inadvertent alterations in their states. While such challenges are more pronounced in

H. Guo et al. (Eds.): IRC-SET 2024, PTT, pp. 222–231, 2025.
https://doi.org/10.1007/978-981-96-3770-6_23

multi-qubit assemblies, our study narrows its focus to a single superconducting tunable symmetric transmon qubit under idealised conditions. By dissecting the state evolution dynamics of this single qubit, we aim to establish a foundational understanding of single qubit control and insights for effective controlling quantum gates in quantum computing systems.

2 Theoretical Background

In quantum computing, superconducting qubits are frequently modelled as LC circuits, where a Josephson junction replaces the inductor [13, 20]. Superconducting qubits, characterised by their extensive electromagnetic cross-sections, are adept at forming efficient connections within intricate quantum circuit designs. Nonetheless, this interconnectedness poses a major challenge in maintaining quantum coherence while ensuring effective internal communication. The parallel between these classical and quantum systems is invaluable, offering critical insights into superconducting qubits' design and behaviour in quantum computing [27].

Superconducting qubits are a cornerstone of quantum computing, utilising materials that exhibit zero resistance at operational temperatures and frequencies, a key feature for avoiding energy loss and maintaining quantum coherence [4]. In our single-qubit study, these qubits are cooled to extremely low temperatures to reduce thermal noise significantly [1]. This cooling is crucial for stabilising the qubit's quantum states, enabling precise energy level manipulation and control necessary for single-qubit operations. Our model highlights the importance of efficient circuit integration and the imperative of isolating these qubits from environmental interference. By focusing on these fundamental principles, our research aims to enhance the foundation for more complex explorations in the future [15] (Fig. 1).

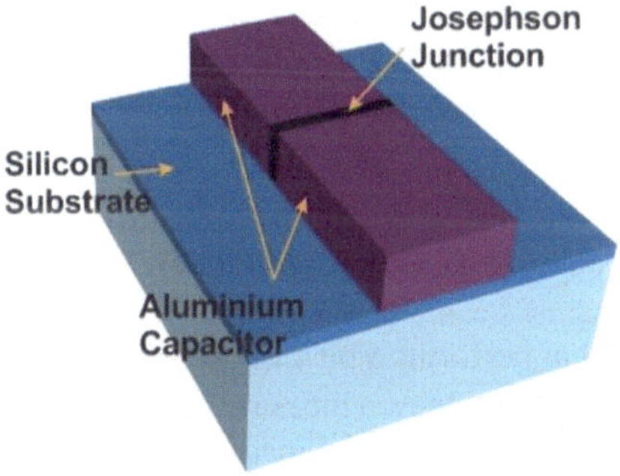

Fig. 1. Structure of a transmon superconducting qubit

The symmetric tunable transmon qubit, central to our study, showcases the ability to tune its energy levels through external magnetic fields, typically by adjusting currents in adjacent flux lines. This tunability is vital for manipulating qubit states and implementing quantum gates. In this design, the external flux induces a bias current in the loop which changes the effective inductance. This in turn puts a dependence of the total energy of the qubit on external magnetic flux. Because of such reliance, the qubit is now sensitive to external magnetic field noise [26].

3 Objectives and Hypothesis

Our project is committed to conducting a detailed analysis of the quantum dynamics exhibited by superconducting qubits under ideal conditions, particularly under the influence of external microwave radiation [29]. This research aims to deepen our understanding of superconducting qubits' responses to microwave fields. Such an understanding of qubits is critical to achieving high gate fidelity by effectively controlling the quantum gates [6].

To achieve our research objective, we investigated how qubit responds to an external field. We hypothesised that under ideal conditions, the superconducting qubits will exhibit coherent quantum dynamics and respond consistently to the microwave field, leading to an exhibition of sinusoidal behaviour between 2 states as energy is incrementally added. Such a pattern is predictable and can be effectively controlled.

4 Methodology

4.1 Model Simulation

To simulate the quantum dynamics of the qubit, we employ a Hamiltonian model expressed as [10, 33]

$$H = H_{system} + H_{environment} + H_{microwave}. \tag{1}$$

In our idealised scenario, there is no noise from the environment and the qubit is prepared at its initial state, leading to

$$H_{environment} = 0, H_{system} = 0. \tag{2}$$

Simplifying the equation, we get

$$H = H_{microwave} \cong \Omega \cdot \sigma_x. \tag{3}$$

σ_x is the Pauli-X gate that performs a bit-flip operation on the state of the qubit [14]. Ω represents the qubit's Rabi frequency, a measure of the strength of interaction between the qubit and the external microwave field. Since our model is to be conducted under an ideal situation with no decoherence, it is reasonable to conclude that $\Omega = 1$.

Another fundamental assumption in our model is the resonance condition, where the frequency of the microwave field matches the natural frequency of the qubit. This condition is crucial as it leads to Rabi oscillations [24, 30], inducing continuous, coherent sinusoidal interactive strength between the qubit and the external microwave field.

Furthermore, the parameters used in our simulations have been carefully chosen to closely match those in the lab's experimental setup. This alignment ensures that our simulated results are directly comparable to the experimental data, minimising errors in our findings and conclusions. However, we remain vigilant for any anomalies or discrepancies that might arise, ensuring the reliability of our results.

The qubit is initially configured to be in the ground state. Next, we model how this qubit would change over time by exposing it to a steady external microwave field. The QuTiP library in Python, a tool well-suited for quantum state evolution and measurement, is used to run these simulations [12]. The library enables us to determine the relative interactive strength of the qubit at discrete time intervals and calculate the time taken for the transitions of 2 states from the simulated data.

Visualisation of the qubit's behaviour is an integral part of our analysis. Using the Matplotlib Library, we are able to plot the evolution of the qubit's expectation value over time, illustrating the characteristic sinusoidal behaviour of Rabi oscillations [16].

Through this structured simulation method, we aim to strengthen our understanding of the time evolution of a tunable symmetric transmon qubit under microwave influence. We anticipate that our findings will enhance the understanding of the fundamental dynamics of qubit behaviour and provide useful insights for further research in the field of quantum computing.

4.2 Overview of Lab-Conducted Experiments

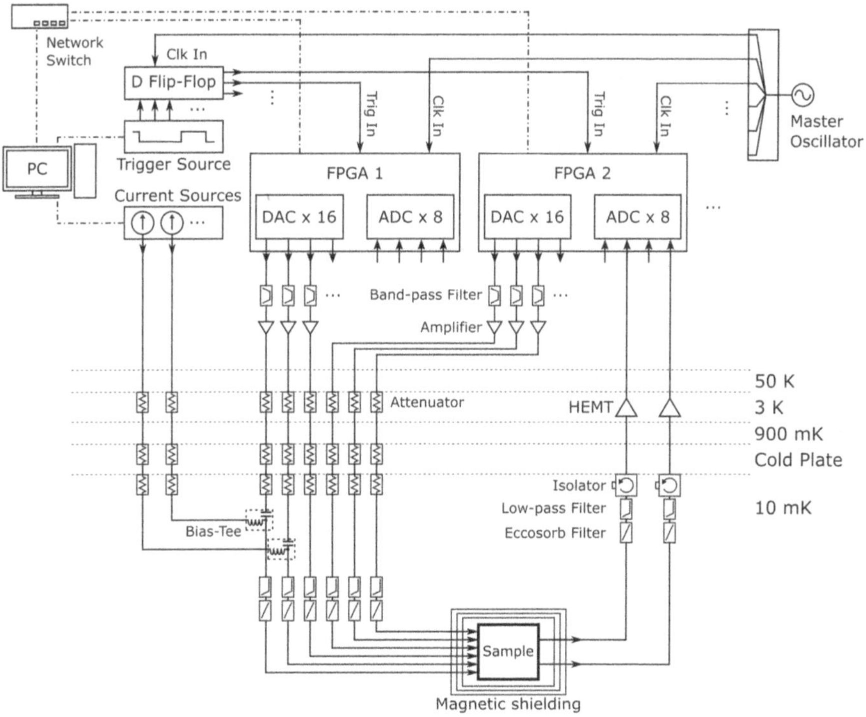

Fig. 2. Overall circuit diagram of the lab-conducted experiment [25]

Our study incorporates experimental data from NTU's lab (Centre for Quantum Technologies), focusing on superconducting qubits under controlled experimental conditions. These conditions include exposure to microwave fields with Gaussian waveforms at power typically between -50 dBm and -60 dBm. As illustrated in Fig. 2, the qubits are positioned in the 'Sample' region, while the microwave pulse is generated by the 'Master Oscillator.' The microwave power is regulated through a combination of amplifiers and attenuators. Overall, the experiments are conducted in a cryogenically cooled environment using a dilution refrigerator [1]. Understanding these experimental parameters is crucial, as they directly affect the qubit's behaviour, thereby impacting the data analysis. While we are not directly involved in these experimental processes, our analysis relies on a comprehensive understanding of the experimental environment and conditions.

Throughout our analysis, the dependent variable is the pulse time, which is the duration of the microwave pulse passing through the qubit. This duration dictates the expectation value of the qubit. The independent variable is the transmission, which is the power gain or loss of a microwave signal upon passing through the resonator that is coupled with the qubit in the chip [21]. As the microwave field alters the qubit state, the oscillation of the resonator changes accordingly, modifying the power absorbed by the resonator.

5 Data Analysis

As part of our data analysis, we utilise Python libraries such as Matplotlib, Scipy, Scikit-learn, and NumPy. Our focus is on interpreting the given data through visualisations of qubit state evolution, frequency analysis, and statistical modelling. This involves comprehensive analysis using the data provided by the lab, aligning with our theoretical understanding of quantum mechanics and integrating it with our model.

6 Results and Discussion

6.1 Results and Discussion (Model)

Based on our model, we plotted this graph to illustrate the oscillatory behaviour of the qubit when subjected to a constant microwave field. Figure 3 shows that the expectation value oscillates between 1.00 and -1.00 at a constant frequency. Since the expectation value is the probabilistic expected value of the qubit at a certain state, the qubit's state is proportional to the expectation value. Figure 4 illustrates the oscillation of qubits between its two states at a constant frequency, having a Rabi oscillation. This Rabi oscillation is a key characteristic of qubit dynamics under steady microwave influence. By analysing this graph, we can accurately determine the qubit's expectation value at different time intervals, which offers a deeper understanding of the probability of finding the qubit in a particular state at any given moment.

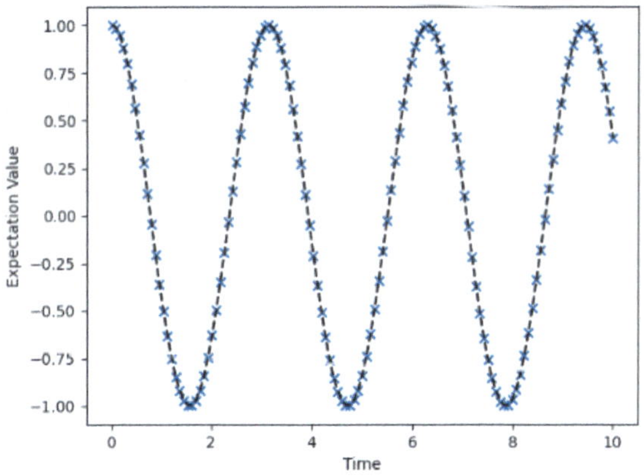

Fig. 3. Graph of the expectation value of the simulated qubit model under ideal conditions

Table 1. Results calculated from all 9 experiments' data, each experiment uses a distinct and unique qubit. With experiment 1 on top and experiment 9 below, arranged in order of 1 to 9.

Rabi Frequency [MHZ]	Pulse [ns]	R^2 Score
7.5593	66.1433	0.9934711745675775
7.9335	63.0242	0.9900811069969134
9.0178	55.4459	0.9832886748247617
10.0433	49.7843	0.9664563971485796
10.2761	48.6568	0.9634468161433195
8.7783	56.9584	0.952182065027128
8.7712	57.0048	0.952196016815608
8.3634	59.7840	0.9561250307073235
8.3519	59.8666	0.9947893559259268

7 Results and Discussion (Data Analysis)

We converted the signals from the time domain into the frequency domain using Fourier Transform and SciPy Library, to determine the Rabi frequency of the qubit as shown in Fig. 4. Subsequently, the Rabi frequency is determined to be the second peak in the graph. The first peak in the graph is due to the initial steady state of the qubit without any excitation (0 Hz), which is not of interest when analysing the oscillation behaviour.

Using the Curvefit Library, Fig. 5 shows the Rabi oscillation that was plotted according to our result from experiment 1. The R-squared value is 0.9935, suggesting a strong correlation between the theoretical plot and the actual experimental data from Table 1 [8].

In total, we collected and analysed 9 sets of data as shown in Figs. 6 and 7. Our analysis shows that all qubits exhibit a transition between states that aligns with our

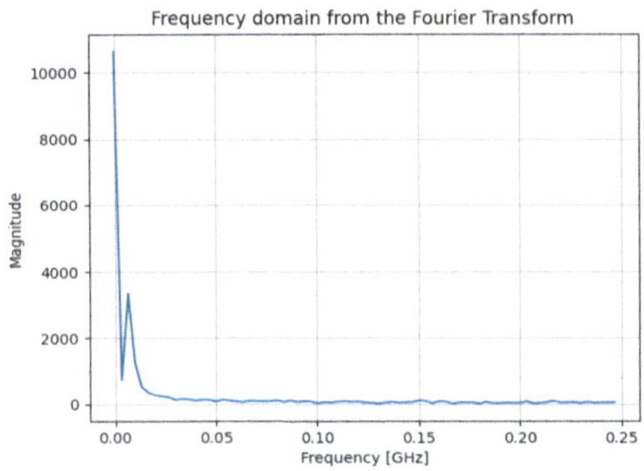

Fig. 4. Graph of frequency domain of the transmission from Fourier Transform in experiment 1. 9 experiments were conducted using these 9 distinct and unique qubits, denoted as experiments 1 to 9

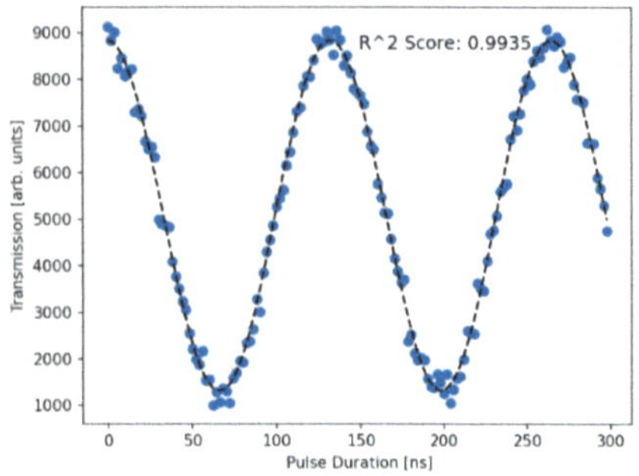

Fig. 5. Graph of Rabi oscillation for experiment 1.

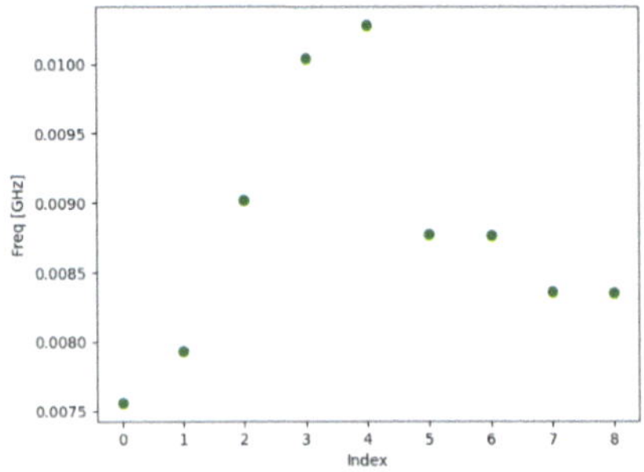

Fig. 6. Graph of Rabi frequency from all 9 experiments, each with a different qubit

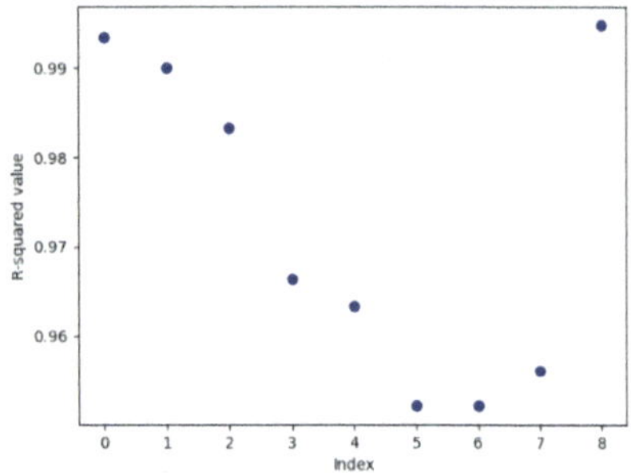

Fig. 7. Graph of R-squared value from all 9 experiments, each with a different qubit

theoretical model. The mean R-squared value of the theoretical plot is 0.97245, and the mean Rabi frequency is around 8.7883 MHz. Moreover, the average deviation of the Rabi frequencies from their mean is around 7.6%, suggesting that all qubits are unique and have a certain degree of variability.

In our research, we have investigated the dynamic behaviour of superconducting qubits exposed to microwave fields. Our primary focus was on analysing the Rabi oscillations, which are critical for understanding the qubits' response to external stimuli. The data collected during this study reveal the state of the superconducting qubit as a function of microwave pulse duration. This empirical evidence is particularly significant as it corroborates the predictions made by our simulation model. The alignment between the observed data and simulation results strengthens the credibility of our theoretical framework and enhances our understanding of the quantum behaviour of superconducting qubits under microwave influence. This correlation not only reinforces the accuracy of our simulation model but also provides deeper insights into the fundamental mechanisms governing the behaviour of superconducting qubits.

8 Conclusion and Future Work

Our research has successfully validated the hypothesis that the qubit behaves predictably under microwave influence. This behaviour is a critical aspect of qubit dynamics in reducing errors in quantum gates and achieving high gate fidelity.

Despite the presence of some noise in the data sets, our statistical analysis, particularly the high R-squared values (exceeding 0.95), indicates a strong correlation between the observed and predicted values. This correlation allows us to affirm the reliability of our findings and supports the assumption of an idealised scenario for our model.

Currently, our research does not include the decoherence effect and deals with systems of a single qubit. However, in scenarios of frequency misalignment (off-resonance), oscillations will decrease in amplitude, impacted by quantum measurement theory dynamics and decoherence processes. If the frequency is noisy, the amplitude may diminish over time. Measurement back-action, a key aspect of quantum measurement theory [17], can inadvertently modify the qubit's state, detracting from its coherence. In the future, we plan to incorporate these decoherence effects such as quantum gate imperfection [19] and environmental interaction into our calculations [3]. This is important as decoherence is a central challenge in quantum computing, progressively disrupting the qubit's oscillatory behaviour.

Our current model does not fully capture the complexities of the superconducting qubit, particularly its anharmonic nature. While we treated the qubit as a two-level system for simplicity in the model, in reality, an anharmonic oscillator with additional energy levels beyond the computational basis. In less controlled experimental conditions, there is a risk of energy leakage to these higher energy states, which could cause deviations from the expected outcomes. Addressing these potential leakages will be important for refining our model and making it more applicable to real-world experiments. Hence, our future work would focus on incorporating the anharmonicity of superconducting qubits and exploring the impact of energy transitions on higher states [11].

We will also investigate the interactions between multi-qubits and microwaves which might cause entanglement effects between qubits, leading to fluctuations in the qubits' state [2].

In conclusion, our study corroborates the theoretical understanding of superconducting qubit dynamics under a steady microwave influence by providing empirical evidence supporting these concepts. These results enhance our comprehension of qubit manipulation and control, an essential step towards the realisation of more reliable and efficient quantum computing systems. Thus, our research has the potential to contribute to the broader field of quantum mechanics by offering a greater understanding of superconducting qubits.

Acknowledgements. The authors are grateful to Hwa Chong Institution, Singapore, for providing this invaluable opportunity. The authors also wish to express their sincere appreciation to The Centre for Quantum Technologies (CQT) at Nanyang Technological University, Singapore, for providing the essential experimental facilities and data, and Associate Professor Rainer Helmut Dumke, Principal Investigator at the Centre for Quantum Technologies, for his guidance and support throughout this project.

References

1. Batey, G., Teleberg, G.: Principles of dilution refrigeration. Oxford Instruments. https://nanoscience.oinst.com/assets/uploads/NanoScience/Brochures/Principles%20of%20dilution%20refrigeration_Sept15.pd: (n.d.-a)
2. Bejach, V.: High-fidelity single-qubit gates in multi-qubit environments using spin qubits in Silicon quantum dots. TU Delft Repositories. https://repository.tudelft.nl/islandora/object/uuid:ae99139e-3cbd-4ce4-b657-0cd1603cec77 (2020)
3. Carrera, M., Gorin, T., Seligman, T.H.:: Single qubit decoherence under a separable coupling to a random matrix environment (2014,). arXiv.org. https://arxiv.org/abs/1405.4448
4. Devoret, M.H.: Superconducting Qubits: A Short Review (2004). arXiv.org. https://arxiv.org/abs/cond-mat/0411174
5. DiVincenzo, D.P.: The Physical Implementation of Quantum Computation (2000). arXiv.org. https://doi.org/10.1002/1521-3978(200009)48:9/11
6. Erikvandeb: Semiconductor spin qubits gain further credibility as leading platform for quantum computing - QuTech. QuTech. https://qutech.nl/2022/01/19/semiconductor-spin-qubits/?cn-reloaded=1 (2022)
7. Faroukh, Y.: Quantum Computers Vs Conventional Computers: A Study on the Larger Scale. ResearchGate (2018). https://doi.org/10.13140/RG.2.2.29502.18243
8. Figueiredo, D., Silva, J.A., Rocha, E.C.:(PDF) what is R2 all about?. ResearchGate (2011). https://www.researchgate.net/publication/312451551_What_is_R2_all_about
9. Gottlieb, M.A., Pfeiffer, R.: The Feynman Lectures on Physics. California Institute of Technology. March 15, 2023, https://www.feynmanlectures.caltech.edu/III_toc.html (2013)
10. Griffiths, D.J.: Introduction to Quantum Mechanics, 2nd ed (2nd ed). Pearson (2005)
11. Hyyppä, E., Vepsäläinen, A., Papič, M., et al. Reducing leakage of single-qubit gates for superconducting quantum processors using analytical control pulse envelopes. PRX Quantum (2024). https://doi.org/10.1103/PRXQuantum.5.030353#
12. Johansson, P.D.N., J. R.: Tutorials for QuTiP Version 4. QuTip. https://qutip.org/qutip-tutorials/ (2013)
13. Kelly, J.S.: Fault-tolerant superconducting qubits (2015)
14. Krantz, P., Kjaergaard, M., Yan, F., Orlando, T. P., Gustavsson, S., Oliver, W.D.: Qubit control. In: A Quantum Engineer's Guide to Superconducting Qubits, pp. 28–29. Essay (2021)
15. Kjærgaard, M., et al.: Superconducting qubits: current state of play. Ann. Rev. Condensed Matter. Phys. (2020). https://doi.org/10.1146/annurev-conmatphys-031119-050605
16. Matplotlib: Visualization with python. Matplotlib (2023). https://matplotlib.org/
17. Hatridge, M., et al.: Quantum back-action of an individual variable-strength measurement. Science **339**(6116), 178–181 (2013). https://doi.org/10.1126/science.1226897
18. MIT researchers create quantum computer that simulates quantum system. MIT News, Massachusetts Institute of Technology. https://news.mit.edu/1999/quantum(1999)
19. Mondal, S., Hazra, S.K., De, A.S.: Imperfect entangling power of Quantum Gates (2023). arXiv.org. https://arxiv.org/abs/2401.00295
20. Newrock, R. S., Lobb, C. J., Geigenmüller, U., Octavio, M.: The two-dimensional physics of josephson junction arrays. Solid State Phys. (2000). https://doi.org/10.1016/s0081-1947(08)60250-7

21. Omelyanchouk, A.N., Shevchenko, S.N., Greenberg, Ya.S., Astafiev, O., Il'ichev, E.: Quantum Behavior of a flux Qubit Coupled to a Resonator. AIP Publishing (2010). https://pubs.aip.org/aip/ltp/article-abstract/36/10/893/938811/Quantum-behavior-of-a-flux-qubit-coupled-to-a?redirectedFrom=fulltext
22. Preskill, J.: Quantum computing and the entanglement frontier (2012). arXiv.org. https://arxiv.org/abs/1203.5813
23. Quantum, U.: Quantum Gates explained without the math. Medium (2022). https://medium.com/@universalquantum/quantum-gates-explained-without-the-maths-1c40e7d79611
24. Rabi Oscillations and the Rabi Frequency in Two-Level Systems. https://resources.system-analysis.cadence.com/blog/msa2020-rabi-oscillations-and-the-rabi-frequency-in-two-level-systems (n.d.)
25. Reproduced from Park, K.H., et al.; ICARUS-Q: Integrated control and readout unit for scalable quantum processors. Rev. Sci. Instrum. **93**(10), 104704 (2022). https://doi.org/10.1063/5.0081232, with the permission of AIP Publishing
26. Roth, T.E.: An Introduction to the Transmon Qubit for Electromagnetic Engineers (2021). arXiv.org. https://arxiv.org/abs/2106.11352
27. Sakurai, J.J., Napolitano, J.: Modern Quantum Mechanics (Third). Cambridge University Press (2021)
28. Stanford University: Modern Physics: Quantum Mechanics (Stanford). YouTube. YouTube. https://www.youtube.com/watch?v=JzhlfbWBuQ8&list=PL84C10A9CB1D13841 (2008). Retrieved from 15 March 2023
29. Sun, G., et al.: Quantum Dynamics of a microwave driven superconducting phase qubit coupled to a two-level system. CORE (2010). https://core.ac.uk/download/pdf/213412018.pdf
30. Vijay, R., et al.: Stabilizing Rabi oscillations in a superconducting qubit using quantum feedback. Nature (2012). https://doi.org/10.1038/nature11505
31. Watrous, J.: Single Systems. Single systems | IBM Quantum Learning.://learning.quantum.ibm.com/course/basics-of-quantum-information/single-systems#introduction https (n.d.)
32. What is Quantum Computing? | IBM. https://www.ibm.com/topics/quantum-computing (n.d.)
33. Zwiebach, B.: Time Dependent Perturbation Theory. MIT OpenCourseWare. https://ocw.mit.edu/courses/8-06-quantum-physics-iii-spring-2018/89ef6d5958ee59bae9a91345c3d8c8e4_MIT8_06S18ch4.pdf (2018)

Affordable Bone Conduction Headphone-Based Hearing Aid as an Alternative to Standard Hearing Aid

Isabel Ji(✉)

Torrey Pines High School, San Diego, Califorinia 92130, USA
isabel.ji.clovers@gmail.com

Abstract. The price of high-quality prescription-based hearing aids is thousands of dollars due to complex hardware and lack of economy of scale. We propose bone conduction headphones utilizing low latency and stereo microphones, connected to a regular smartphone, to effectively achieve similar results to that of a standard high quality hearing aid for 1% of the cost.

We set up an audio testing lab using off-the-shelf components to test the performance of existing and proposed solutions. Existing low-cost software- based hearing aid applications on smartphones were found to introduce an unacceptable latency of ~250 ms. The audio testing configuration this lab uses has a total latency of 7.3 ms when the proposed system is modified to use wired connection to the smartphone. However, it is not achievable over current implementation of Bluetooth Low Energy (BLE) signals. The stereo microphone system, with speech and noise played from opposite sides of the dummy head, demonstrates the need for spatial noise suppression. We also tested the effectiveness of occlusion effect and hypothetical ML based noise reduction techniques.

Thus, a stereo bone conduction headset connected to a smartphone is a viable solution that mitigates the occlusion effect, yet also allows for spatial noise suppression and possible AI/ML usage in noisy environments. However, more research needs to be done on the latency of BLE headphones.

Keywords: Bone Conduction Headphones · Latency · Occlusion Effect · Hearing Aids

1 Introduction

Hearing loss is prevalent in more than 1.5 billion, or one fifth of the global population. By 2050, disabling hearing loss will be present in over 700 million people [1]. Moreover, only 30% of adults aged 70 or older and 16% aged 20–69 that could benefit from hearing aids have ever used them [2]. These statistics demonstrate the dire need for more accessible high-quality hearing aids.

Hearing loss not only affects people physically, but also mentally and socially. Individuals with hearing loss are more likely to be socially isolated, lack education, and have

H. Guo et al. (Eds.): IRC-SET 2024, PTT, pp. 232–239, 2025.
https://doi.org/10.1007/978-981-96-3770-6_24

higher unemployment rates [1]. Additionally, low hearing aid use directly correlates with higher mortality rates [4]. Thus, hearing loss is a severe and large-scale problem.

Treatment for hearing loss commonly involves a costly medical fitting procedure and an expensive high-quality prescription-based hearing aid. Producing hearing aids are expensive, as the components such as receivers, microphones, and signal processors are highly specialized [5], meaning they aren't commonly found in other devices besides hearing aids. Moreover, the hardware design of a hearing aid has trade-offs to consider with the limited space and a wide range of requirements, e.g., battery consumption—a standard hearing aid requires around 1 milliwatt of battery for all components [6]. More tradeoffs include hearing aid dimensions, latency, and software complexity. The software of hearing aids requires many algorithms including feature extraction, classification, beamforming, feedback cancellation, noise reduction, amplifications, dynamic compression, and synthesis banks [5].

Lower cost non-prescription hearing aids have many undesirable effects, e.g., high latency, distortion, and plugged ears, which causes the occlusion effect. The majority of people who own hearing aids avoid wearing them due to the unpleasant results. Although high quality prescription-based hearing aids can potentially provide great results, it costs around $1000-$4000. Thus, the direct relationship between the cost and quality makes high quality hearing aids inaccessible.

Therefore, the objective of the following research is to formulate a high quality and low-cost hearing aid. Fig. 1. Depicts the general trend that higher cost/complexity correlates with the sound quality produced.

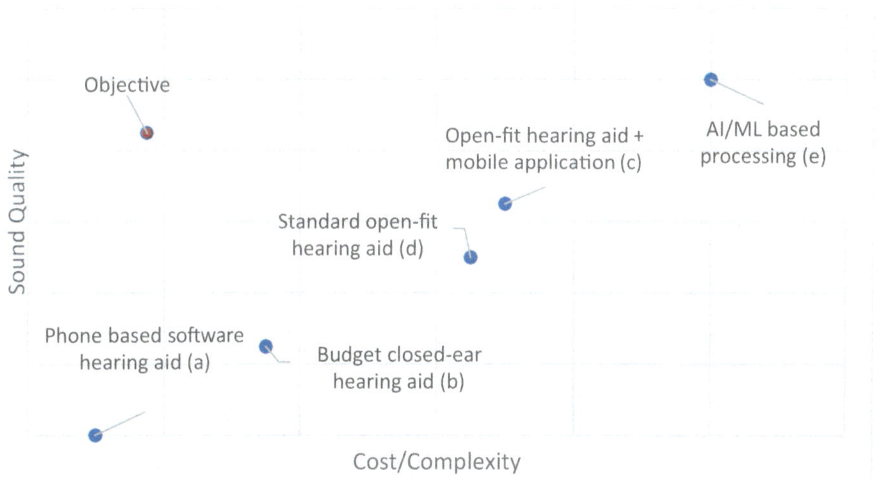

Fig. 1. Comparison of hearing aid technologies based on cost/complexity and sound quality

Bone conduction headsets avoid the outer and middle ear by directly sending vibrations through the skull [10], which allows people to listen to music and be aware of the environment. Note that these headsets could also benefit people with conductive hearing loss [3].

Therefore, we hypothesize that the connection of a smartphone and a bone conduction headset can hypothetically convey similar results to that of an open ear hearing aid, developing an affordable hearing aid for low income and underprivileged. The formulated sub-hypotheses consist of the following—bone conduction low latency can be achieved with smartphone optimization and Bluetooth Low Energy (BLE), spatial noise reduction can be achieved through stereo microphones, headset mitigates the occlusion effect, and high sound quality in noisy environment can be achieved through complex AI/ML algorithms in smartphone hardware.

2 Methods

To simulate the software of a bone conduction headset hearing aid during the investigation, two microphones were connected to a FocusRite audio interface as "stereo" microphones. Two speakers were also wired to the interface to simulate sound and noise coming from different directions. Then, the digital audio workstation, Ableton, was used to record tracks and route them back to the computer. Additionally, a third microphone was calibrated to record the sound pressure level (SPL) during the experiment (see Fig. 2A). A market standard bone conduction headset, the Shokz Openrun Bluetooth bone conduction headphones, is also used in certain trials to imitate sound output from the headphones rather than speakers. Each recording is done with two microphones, resulting in two tracks. The audio of speech and noise used is developed by the UCL Language Sciences Department, Cambridge University, and Edinburgh University. Their database consists of lab-standard audio files of speech and noise in British English.

In terms of hardware, a Styrofoam head was hollowed and drilled into to simulate ear canals with silicone ears attached on the outside of the head. During the experiment, a gel medium is placed inside the head to limit sound traveling through the hollowed parts (see Fig. 2B).

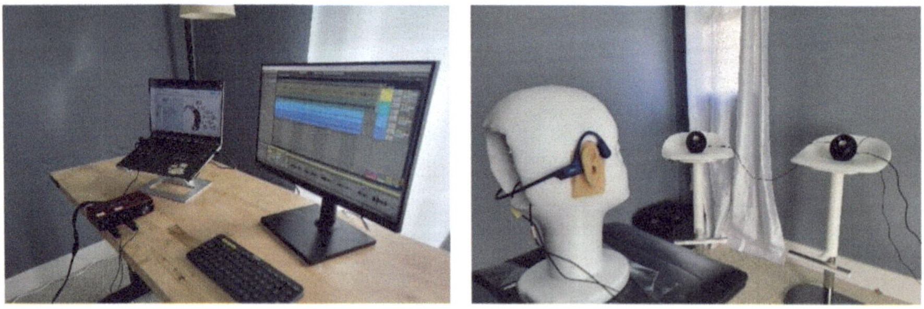

Fig. 2. A(Left) Software Setup, 2B(Right) Hardware Setup

Test cases 1-4 demonstrate the difference between standard hearing and hearing loss. Hearing loss of >=25 db is considered mild hearing loss, as test cases go from 65 to

40 db. Test cases 5–6 simulate what the listener hears, suggesting whether the audio is clear through bone conduction method of sound output. Test cases 7–8 show speech and noise come from different directions (spatial noise suppression). Test case 9 is a control without noise, but also serves as a hypothetical AI/ML recording.

Table 1. Tests Run with Varying SPL, SNR, Positioning, and Orientation

Test Case	Sound Pressure levels (decibels)	Sound to Noise Ratio (decibels)	Position (Microphone, Speaker)	Orientation (° variation from head POV)
1	65	10	In ear, speaker	0, 30
2	65	0	In ear, speaker	0, 30
3	40	10	In ear, speaker	0, 30
4	40	0	In ear, speaker	0, 30
5	65	10	In ear, headphones	0, 30
6	65	0	In ear, headphones	0, 30
7	65	0	Headphones, speaker	-90, 90
8	65	0	In ear, headphones	-90, 90
9	65	No Noise	In ear, headphones	0, 30

3 Results and Discussion

3.1 Latency

To test latency of existing low-cost hearing aid software [13], place microphone one at the source of the sound and microphone two at the speaker post-processing through the software. The original sound is recorded at time T by both microphone one and microphone two, which is then picked up by the phone, processed, and played by the software at a delayed time $T + \Delta$ (see Fig. 3). Δ represents the latency of the software tested. Δ resulted in a latency of ~250 ms latency is rated as disturbing [11], while premium hearing aids have latencies of 58ms [12]. Moreover, the speech signal processing in my proposal involves three steps:

1) Headset microphones send audio to smartphone, $\Delta 1$
2) Smartphone process the audio, $\Delta 2$
3) Smartphone sends audio to bone conduction headset, $\Delta 3$

Δ1 and Δ3 are Bluetooth communications latency. Latest Bluetooth Low Energy standards promise latency as low as 7.5 ms, but practical headsets have latency between 20 and 200 ms. Therefore, prompting future research on BLE headphones.

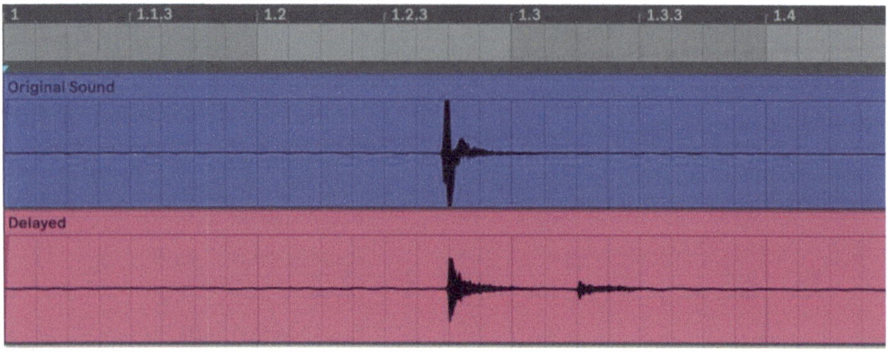

Fig. 3. Original and delayed sound of latency testing

Δ2 is the audio processing latency on the smartphone. Low audio latency could be achieved with 48 kHz sampling with the I/O buffer size at 64 samples (64/48000 = 1.33 ms) for iOS and 192 samples (192/48000 = 4 ms) for Android phones [7].

4 Spatial Noise Suppression

Stereo microphones in hearing aids utilize multiple microphones to perform beamforming/spatial noise reduction in noisy environments. Although commercially available headphones usually have multiple microphones, they aren't utilized for stereo purposes. Thus, we propose the need for a stereo headset.

As depicted in test cases 7-8 (Table 1.), speech and noise are played from different sides of the head (90° and −90° respectively), resulting in the right microphone being much clearer and easier to understand than the left microphone, which was barely perceptible (see Fig. 4).

5 Bone Conduction Testing and Mitigation of the Occlusion Effect

Like open-fit hearing aids, bone conduction headsets also have the open-ear property leaving the ear canal completely open. Bone conduction headsets are also affordable, ranging anywhere from \$10–\$500.

Test cases 5-6 use sound output from the bone conduction headset, while audio is recorded from microphones within the head. The tracks recorded are a good reproduction of the original speech and noise (see Fig. 5). This also suggests the mitigation of the occlusion effect, or the muffled sound of one's own voice reverberating through the skull when the outer ear canal is blocked.

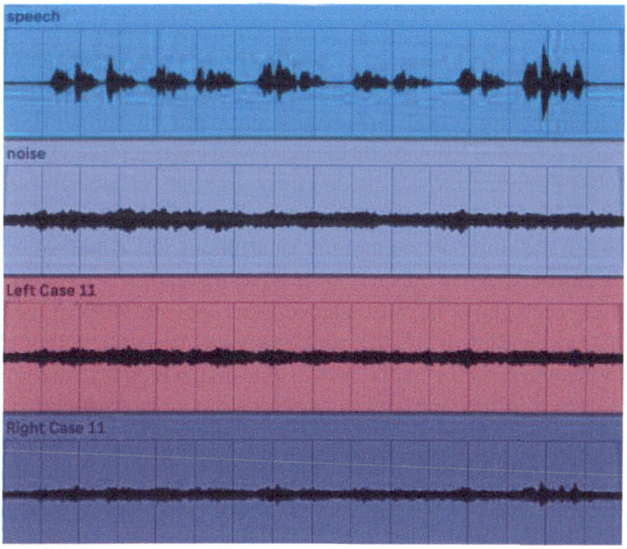

Fig. 4. Tracks of original speech and noise (top). Noisier channel (left) and cleaner channel (right) from test case 7 (relabeled).

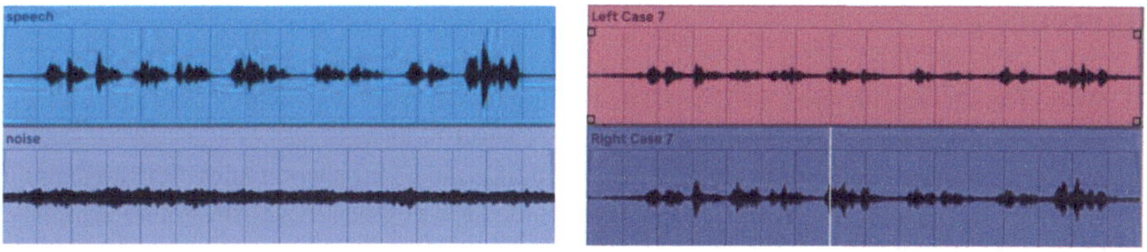

Fig. 5. (Left) Original speech and noise tracks (Right) Recorded left and right tracks, case 5 (relabeled).

6 Advanced Signal Processing

Latest advances in AI/ML have made denoising highly effective as shown in the Clarity Challenge [8], where participants were asked to denoise mixed speech and noise inputs using AI/ML without complexity constraints. Based on a metric called Hearing Aid Speech Perception Index (HASPI), the winning contestants were able to achieve 94% speech intelligibility in extremely noisy environment (−10 dB) [14]; in contrast, premium hearing aids could achieve 70% and 10% HASPI score at 10 dB and 0 dB SNR, respectively [15]. Smartphones can access high amounts of signal processing without a large cost since 86% of the population worldwide own a smartphone [9]. Modern smartphones can perform over 11 trillion operations per second [16]; thus, utilizing smartphones can run more complex algorithms than standard hearing aids, like AI/ML.

7 Conclusion

In this study, we proposed to use bone conduction headphones connected to smartphones to create high quality and low-cost hearing aids. Audio played from the bone conduction headset was cleanly recorded from inside the head with clean reproduction of the original sound, effectively mitigating the occlusion effect. In terms of latency, other affordable software-based hearing aids have over 250 ms latency while the audio test lab with wired connection between the microphone and the smartphone has a total of 7.3 ms latency. However, this is currently not achievable over BLE signals.

Additionally, speech and noise were played from opposite sides of the dummy head. The left recording is inaudible while the right recording is clearer. Although oversimplified, it demonstrates the need for spatial noise suppression in noisy environments. While most hearing aids run at a few MHz clock speed due to power/hardware limitation, modern smartphones are capable of running more than 10 trillion operations per second. This introduces the possibility to perform complex algorithms including AI/ML.

In conclusion, the proposed solution of using bone conduction headphones connected to smartphones with ML-based noise reduction algorithms could create high quality and low-cost hearing aids. Our experiment also identified the need of further latency reduction in BLE links that are commonly used in headsets.

References

1. World Health Organization: Deafness and hearing loss. https://www.who.int/newsroom/fact-sheets/detail/deafness-and-hearing-loss, Last accessed 13 July 2024
2. National Institute on Deafness and Other communication Disorders: Quick Statistics about Hearing, Balance, & Dizziness, https://www.nidcd.nih.gov/health/statistics/quickstatistics-hearing. Last accessed 13 July 2024
3. Johns Hopkins Medicine: Types of Hearing Loss, https://www.hopkinsmedicine.org/health/conditions-and-diseases/hearing-loss/types-of-hearing-loss. Last accessed 13 July 2024
4. Choi, J.: Association between hearing aid use and mortality in adults with hearing loss in the USA: a mortality follow-up study of a cross-sectional cohort. The Lancet. Healthy Longevity **5**(1), 66–75 (2024)
5. Gerlach, L., Payá-Vayá, G., Blume, H.: A survey on application specific processor architectures for digital hearing aids. J. Sign. Process. Syst. **94**, 1293–1308 (2022)
6. Edwards, B.: The future of hearing aid technology. Trends Amplif. **11**(1), 31–45 (2007)
7. Chowdhury, T., Sehgal, A., Kehtarnavaz, N.: Integrating signal processing modules of hearing aids into a real-time smartphone app. In: 2018 40th Annual International Conference of the IEEE Engineering in Medicine and Biology Society (EMBC), pp. 2837–2840 (2018)
8. Clarity: ICASSP 2023. https://claritychallenge.org/docs/icassp2023/icassp2023_intro. Last accessed 18 July 2024
9. Big Data: How Many People Own Smartphones? https://whatsthebigdata.com/smartphonestats/. Last accessed 18 July 2024
10. Duke Health: How Does a Bone Conduction Hearing Aid Work? https://www.dukehealth.org/treatments/ear-nose-and-throat/bone-conduction-hearingaid#:~:text=These%20devices%20capture%20sound%20and,a%20new%20pathway%20fo r%20hearing. Last accessed 19 July 2024

11. Stone, M., Moore B.: Tolerable hearing aid delays I. Estimation of limits imposed by the auditory path alone using simulated hearing losses. Ear Hear **20**(3), 18292 (1999)
12. The Hearing Review: Acceptable Processing Delay in Digital Hearing Aids. https://hearingreview.com/practice-building/practice-management/acceptable-processingdelay-in-digital-hearingaids#:~:text=In%20general%2C%20their%20findings%20suggest,the%20delay%20exceeds%2030%20ms. Last accessed 17 July 2024
13. Hearing Amplifier: Hearing Aid App for Android (Version 1.0.9)[Mobile app].Google Play Store (2021)
14. Cornell, S., Wang, Z., Masuyama, Y., Watanabe, S., Pariente, M., Ono, N.: Multi-channel target speaker extraction with refinement: the WavLab submission to the second clarity enhancement challenge. In: ICASSP 2023 – 2023 IEEE International Conference on Acoustics, Speech and Signal Processing (ICASSP), pp. 1–2 (2023)
15. Kates, J.M., Arehart, K.H.: An overview of the HASPI and HASQI metrics for predicting speech intelligibility and speech quality for normal hearing, hearing loss, and hearing aids. Hear. Res. **426**, 108608 (2022)
16. Notebook Check: Apple A14 Bionic. https://www.notebookcheck.net/Apple-A14-BionicProcessor-Benchmarks-and-Specs.494578.0.html. Last accessed 20 July 2024

Deep Learning-Based Algorithm for Frequency Estimation of Noisy Signals

Sanvi Gupta(✉), Iniyaa Srinivasan, and Teh Kah Chan

Raffles Institution, Singapore 575954, Singapore
sanvig3333@gmail.com

Abstract. The aim of this project is to develop a deep learning model with improved accuracy and efficiency for estimating the frequency of noisy signals, specifically modulated signals that are corrupted by Gaussian noise. While frequency estimation methods exist, many have limitations like being computationally intensive, inefficient or time consuming. In our model, we created training and testing data, followed by a neural network using Keras' Sequential API. The neural network is then trained based on the data generated, allowing it to predict the frequency of a corrupted signal. Afterwards, we varied some parameters in our code, namely the number of data samples and number of epochs in order to determine the accuracy at each. We graphed our results, based on which we determined the number of epochs to be 8 and the number of data samples to be 5000. We then carried out a qualitative comparison of our method with two existing methods: the Fourier Transform Based Method and the Maximum Likelihood Estimator (MLE) based on various criteria. Our research can be used to improve speech recognition models by identifying the frequency components of the speech signal more accurately and using this information to reduce or remove the white noise.

Keywords: signal processing · embedded systems · deep learning model

1 Introduction

1.1 Background of Research

This project aims to produce a deep learning model that estimates the frequency of noisy signals. Frequency estimation is "the process of estimating the frequency, amplitude, and phase-shift of a signal in the presence of noise given assumptions about the number of the components" [1]. Essentially, it is to identify and differentiate signals, especially if there is corruption. In this study, it is specifically with regards to Gaussian noise, which is a form of noise with a statistical normal distribution. By determining the estimated frequency of the noisy signals, the signals can be identified, differentiated and eventually removed. Subsequent paragraphs, however, are indented.

H. Guo et al. (Eds.): IRC-SET 2024, PTT, pp. 240–247, 2025.
https://doi.org/10.1007/978-981-96-3770-6_25

1.2 Aim and Objectives

The aim of this project is to develop and optimise a deep learning model with improved accuracy and efficiency for estimating the frequency of noisy signals, specifically modulated signals that are corrupted by Gaussian noise. After the development of the model, we aimed to carry out comparisons of our deep learning model with current methods used for frequency estimation in order to determine which method is better.

1.3 Literature Review

Frequency Estimation

Frequency estimation in signal processing is the estimation of the frequency content, also known as the spectral density, of a signal from a sequence of time samples. All physical signals can be broken down into various discrete frequencies over a continuous range. Thus, frequency estimation analyses the signal's spectral density which helps to understand the distribution and intensity of different frequency components within the signal. By examining the sequence of time-based samples, the goal is to extract information about the frequencies or periodic patterns inherent in the signal [2].

Current Methods of Frequency Estimation

There are various different methods of frequency estimation such as Fourier Transform Based Methods, Statiscal and Model-Based Methods and Machine Learning-Based Methods. The most common of such methods is the frequency interpolation estimation method based on the discrete Fourier transform (DFT). The Fourier transform of a function creates a frequency spectrum, which helps to see all the information in a signal in a different form, and is a frequency-domain function. This is done by mapping the integral transform of a function into a frequency, effectively decomposing the signal. In this case, the function is any signal that varies over time, specifically a sound wave, sampled over a finite and conclusive time interval. DFT operates on samples of the signal, and is implemented in most models through an algorithm known as fast Fourier Transform (FFT). It is important, however, to note that this method is particularly susceptible to being affected by any noise, and leads to errors in interpolation, as it does not provide processing gain when applied to noisy signals [3]. Another commonly used method is a maximum likelihood estimator (MLE), which estimates the frequency by obtaining the maximum of the periodogram for the input signal using fast Fourier transform (FFT). An MLE model ensures that the observed data in the statistical model is the most probable, leading to highly accurate results. However, it also involves extensive computations and this leads to it being time consuming and thus, ineffective [4].

Noisy Signals and Gaussian Noise

In the context of signal processing, noise refers to an unknown disruption to a signal during its capture, storage or transmission [5]. We added additive white Gaussian noise (AWGN), which is a noise channel used to mimic effects of many random processes that occur naturally [6]. White noise has uniform power spectral density, which means that it has an equal intensity at different frequencies [7].

Gaussian noise is a type of random noise that follows a Gaussian or normal distribution, often used to simulate random variations or disturbances present in real-world signals or data [8]. We chose to use AWGN for modelling types of noise in a real-world scenario as the variations of the signal are random and independent of each other and its equal amount of energy at every frequency, making its characteristics convenient [9]. Another reason for using AWGN is that it is symmetric around zero so the nosies's positive and negative parts have the same statistical properties which remain constant throughout time. This simplifies the process of building the deep learning algorithm as symmetric noise facilitates the design and implementation of signal processing algorithms and models, increasing their efficiency and accuracy. When designing filters or denoising techniques in order to reduce the white noise, the symmetry assumption can lead to more effective noise reduction strategies [10].

Rayleigh Fading

Fading is a phenomenon often observed in communication systems where the signal undergoes variations in amplitude, phase, or both. In wireless communications, small-scale fading is common since reflection or diffraction could cause the signal being transmitted to experience many signal paths [11]. Rayleigh fading is a statistical model that helps to account for the effect of an environment in which the sound is propagated. This model operates under the assumption that any signal passing through a medium will fade randomly according to the Rayleigh distribution [12]. It is commonly used in the real world to account for radio signals in urban environments [13]. We accounted for such fading in our model, as any real world signal will have the tendency to fade, making our model more useful when equipped.

1.4 Limitations

It is important to state that the following project model consists of several limitations, which impact the model's effectiveness in carrying out frequency estimation. Most models that were used for reference in designing this current system were highly trained models, with some of the models having undergone several months, or perhaps even years of training in terms of being able to accurately pick out signals. These models were also fed data from a range of external sources. In contrast, our model undergoes a much more simplified version of training, and specifically with data generated by the code itself, rather than that of other sources. Additionally, due to a lack of computational resources, it will be a lot more difficult to produce a working, feasible model for implementation.

2 Methodology

There are five steps involved in the building of the deep learning based algorithm for the frequency estimation of noisy signals.

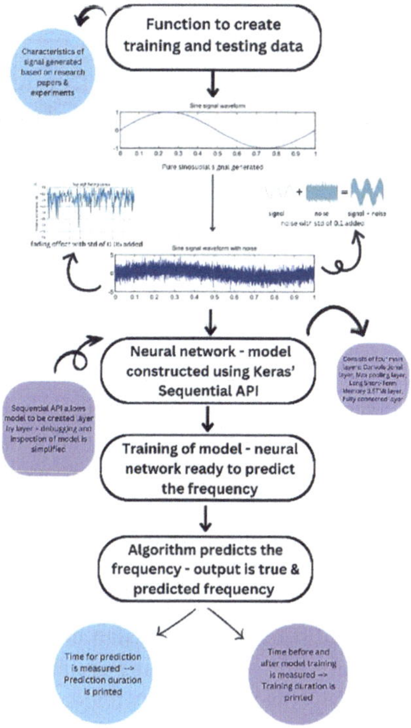

Firstly, data preparation. In our model, we allowed the model to create the training data such that every time the model is run, new data is generated, leading to higher accuracy of the predictions. Secondly, the neural network architecture itself has to be designed. Thirdly, the neural network has to be trained using the generated data. Fourthly, the neural network has to be tested using another set of data that was not previously used. Through this, the accuracy and efficiency (in terms of time taken for prediction to be made) of the network design can be determined. Finally, optimisation of the model by adjusting parameters to improve the accuracy/efficiency of the network [14].

Our code had several different parts to it. Firstly, we used a function to create training and testing data. This was done by determining the parameters to include in our final code based on some information from research papers as well as some experimentation. These parameters determine the characteristics of the signals generated. A random frequency signal is generated and a pure sinusoidal signal is created based on this frequency. To mimic real-world scenarios, random noise and fading effects (based on their standard deviations) are added to the pure signal generated, producing a corrupted signal as the input for the model.

The second part of our code is the neural network which comprises various layers for signal processing. We constructed the model using Keras' Sequential API as the linear layers make it easy to create models [15]. Since frequency estimation involves a clear data flow from input to output without complex branching, Sequential API was a suitable choice [16]. The third part of our code involves the training of the model where the training data is reshaped to fit the model's input shape and the model is then trained using the generated data.

In the fourth part of our code, a signal with fading effects is generated in order to test the model. The output is the true and predicted frequency. The time before and after the model training is measured and recorded in order to calculate the training duration as well as the prediction duration. Using the true and predicted frequencies, the accuracy of the model was calculated by dividing the difference by the true frequency and finding its percentage.

Our final code had these parameters. For the variables tested, the number of samples was 5000 and the number of epochs was 8 (Refer to Results & Discussion for justification).

We chose 100 as the signal length which indicates that the signal is represented by a sequence of 100 time samples [17]. We chose this number as we believed that it would allow the signal being processed to have sufficient information without making the algorithm require more computational resources for its frequency estimation. We chose a batch size of 32. Batch size directly affects the accuracy, the computational efficiency of the training process and the speed of the model's learning. Based on these factors, we decided that a batch size of 32 would be optimal to prevent having to compromise on the accuracy or efficiency while keeping in mind our computational limitations [18]. We chose a standard deviation of 0.1 for the Gaussian noise as it contributes to a mild amount of distortion to the signal while still providing a relatively clean signal, therefore a realistic representation of most signals for everyday tele-communication [19]. We chose a standard deviation of 0.05 to replicate a small scale fading effect. 0.05 is a mild fading effect which both has realistic fading effects and keeps the simulation manageable to prevent overcomplicating the analysis of the results [20].

3 Results and Discussion

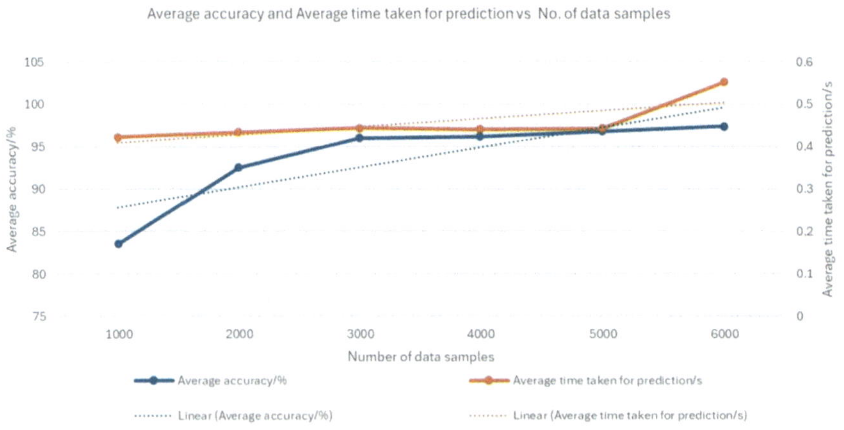

Fig. 1. Average accuracy against the number of data samples

Based on Fig. 1, there is a general upward trend in the average accuracy as well as the average time taken for prediction as the number of data samples increases. While the average accuracy for 6000 data samples is the highest (97.3%), it results in an increase of the average time taken for prediction from 0.44 s to 0.55 s which shows decreased efficiency. Thus, we selected 5000 to be the number of data samples in order to prioritise both accuracy and efficiency.

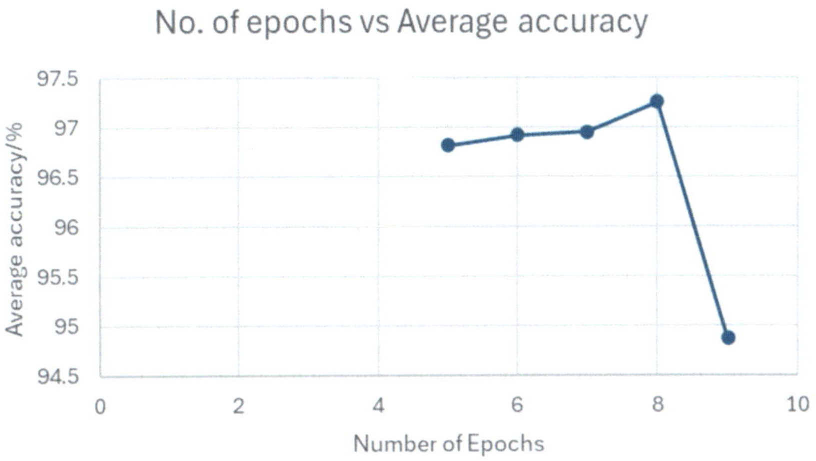

Fig. 2. Average accuracy against the number of epochs

Based on Fig. 2, there is a general upward trend as the number of epochs increases from 5 to 8 with the accuracy increasing from 96.8% to 97.2%. However, as the number of epochs increases from 8 to 9, the accuracy decreases from 97.2% to 94.8%, possibly due to overfitting when increasing the number of epochs results in the model becoming too specific. Thus, 8 was an optimum number.

3.1 Comparison with Existing Methods

We will be comparing our deep learning algorithm with existing methods, namely the Fourier Transform based Method (FTT) as well as the Maximum Likelihood Estimator (MLE). It is important to note that these current methods have a few disadvantages as mentioned above, namely that FTT on its own is unable to process gaussian noise, while a MLE model with fast FFT, despite being highly accurate, involves extensive statistical computation. Our model, in comparison with FTT, is equally accurate, considering that a deep learning model can have a high accuracy rate if trained well with diverse data input. However, an FFT model is the most accurate in terms of frequency estimation for periodic signals with consistent frequency content. Apart from that, our model does not require substantial computational resources, as a deep learning algorithm is able to process noisy Gaussian noise if coded to do so, while an FTT based model requires pre-processing for such signals as it is unable to account for the presence of noise. This requirement for pre-processing of noisy signals (by breaking the signal into frames) results in inefficiency of the model due to the time taken for the prediction to be made. On the criterion of time, our model is relatively fast and able to predict the frequency within 0.5 seconds. However, the FTT model is said to be too slow to be practical at most times [21]. On the other hand, our model in comparison with a MLE based fFTT model is much less complex in terms of computation, but this may vary as the complexity depends on the implementation and the assumptions made in the model to predict the frequency of the signal inputs. More complex models might be more computationally intensive as well. A MLE model is also highly based on assumptions made by the model and unexplained variability in a data sample, as compared to the data generated (or input) into a deep learning system such as ours. Our model also has the capacity to generalise data well, while in an MLE/FTT based model, the performance accuracy of the model may degrade if signal characteristics are modified in any significant way.

4 Conclusion

Our results above are fairly promising in terms of exemplifying the accuracy and efficiency of our neural network model. However, it is important to note that more quantitative comparison of our model is required to further validate and improve the accuracy and speed of calculation, so that it would be more suited for a real world application. Such comparison would help to further optimise our code, allowing it to be used in other signal processing mechanisms.

We believe that through our research, it can be applied in the future in real-life applications such as to improve speech recognition models by identifying the frequency components of the speech signal more accurately and using this information to accurately reduce or remove the white noise.

References

1. Hayes, M.H.: Statistical Digital Signal Processing and Modeling, John Wiley & Sons, Inc. (1996)
2. Stoica, P., Moses, R.: Spectral Analysis of Signals. Prentice Hall (2005)
3. Chen, K., Wei, M., Chen, X., Li, S.: A fine resolution frequency estimation method for noisy signal and its Application. J. Vibr. Eng. Technol (2021). https://doi.org/10.1007/s42417-021-00376-w
4. Mohamed, S., El-Sayed: Design and implementation of accurate frequency estimator depend on ... https://www.researchgate.net/profile/Mohamed-Saber-21/publication/342589443_Design_and_implementation_of_accurate_frequency_estimator_depend_on_deep_learning/links/5efd9ccc92851c52d610a22a/Design-and-implementation-of-accurate-frequency-estimator-depend-on-deep-learning.pdf (n.d.)
5. Tuzlukov, V.: Signal Processing Noise, Electrical Engineering and Applied Signal Processing Series, CRC Press. 688 p (2010). ISBN 9781420041118
6. Chaudhari, Q.: Additive white gaussian noise (AWGN). Wireless Pi. https://wirelesspi.com/additive-white-gaussian-noise-awgn/ (2023)
7. Wireless Telecom Group: What is Additive White Gaussian Noise & Why is it important for Test & Measurement? https://www.blog.maurymw.com/what-is-additive-white-gaussian-noise-why-is-it-important-for-test-measurement (n.d.)

8. Gaussiannoise.Hasty.aiDocumentation. https://hasty.ai/docs/mp-wiki/augmentations/gaussian-noise#:~:text=Gaussian%20Noise%20is%20a%20statistical,added%20to%20the%20original%20image (n.d.)
9. Desfontaines, D.: The magic of gaussian noise – ted is writing things. Ted is writingthingsRSS.https://desfontain.es/privacy/gaussian-noise.html:~:text=A%20first%20advantage%20of%20Gaussian,whose%20actual%20distribution%20is%20unknown (n.d.)
10. Singh, G.: AWGN (additive white gaussian noise). Telecom Trainer. https://www.telecomtrainer.com/awgn-additive-white-gaussian-noise/ (2023)
11. Jain, G.: Fading in wireless communication. GeeksforGeeks. https://www.geeksforgeeks.org/fading-in-wirelesscommunication/ (2023)
12. Wikimedia Foundation: Rayleigh fading. Wikipedia. https://en.wikipedia.org/wiki/Rayleigh_fading (2023)
13. Team, E.: Everything RF. everything RF. https://www.everythingrf.com/community/what-are-the-different-types-of-fading (2020)
14. Sajedian, I., Rho, J.: Accurate and instant frequency estimation from noisy sinusoidal waves by deep learning. Nano convergence (2019). http://www.ncbi.nlm.nih.gov/pmc/articles/PMC6694364/
15. Faisal, R.: Keras model sequential API vs functional API. Medium. https://medium.com/analytics-vidhya/keras-modelsequential-api-vs-functional-apifc1439a6fb10#:~:text=Sequential%20API%20allows%20you%20to,have%20multiple%20inputs%20or%20outputs.&text=And%20now%20lets%20plot%20your%20model%20using%20keras%20utils (2020)
16. Swain, S.: Understanding sequential vs functional API in Keras. Analytics Vidhya. https://www.analyticsvidhya.com/blog/2021/07/understanding-sequential-vs-functional-api-in-keras/ (2021)
17. bert2002: Signal Processing in MATLAB, what is signal length ?. Physics Forums: Science Discussion, Homework Help, Articles. https://www.physicsforums.com/threads/signal-processing-in-matlab-what-is-signallength.792698/ (2015)
18. Thakur, A.: What's the optimal batch size to train a neural network? W&B. https://wandb.ai/ayush-thakur/dl-question-bank/reports/What-s-the-Optimal-BatchSize-to-Train-a-Neural-Network---VmlldzoyMDkyNDU#:~:text=That%20said%2C%20note%20that%20for,to%20a%20lower%20batch%20size (2020)
19. Intro.toSignalProcessing:SignalsandNoise. https://terpconnect.umd.edu/~toh/spectrum/SignalsAndNoise.html (n.d.)
20. Singh, G.: SFshadow fading. Telecom Trainer. https://www.telecomtrainer.com/sf-shadow-fading/ (2023)
21. Van Loan, C.:Computational Frameworks for the Fast Fourier Transform. SIAM (1992).

Chat, is this Real? mm-Double Confirm!: A Multimodal Network for Singapore-Context Misinformation Detection

Felicia Tan Ee Shan[1(✉)], Ashley Goh Rou Hui[2], and Adriel Kuek Yong Jie[3]

[1] Raffles Institution, 1 Raffles Institution Ln, Singapore, Singapore
feliciatan.tes@gmail.com

[2] River Valley High School, 6 Boon Lay Avenue, Singapore, Singapore

[3] DSO National Laboratories, 12 Science Park Dr, Singapore, Singapore

Abstract. In the digital age, the rapid spread of misinformation presents a critical challenge, necessitating advanced, scalable solutions for accurate and efficient fact-checking. Addressing the gap in Singapore-specific multimodal misinformation detection and overcoming limitations in current automated models, this project introduces SINDEX, which has achieved strong performance across all functions, and state-of-the-art performance in the verification of Singapore-specific multimodal misinformation. Our automated evidence retrieval framework, which leverages the H2-keywordextractor transformer model and SBERT to query NewsAPI, ensures greater accuracy practicality. We developed a CLIP multimodal Custom Truthfulness Classifier trained on our self-collected dataset of Singapore-context multimodal misinformation and a parallel ChatGPT-4-32k-involved misinformation detection system, achieving robust F1-scores of 95.2% and 95.4%, respectively. Additionally, a custom SBERT-based satire detector was implemented to assess exaggerated content, yielding a strong F1-score of 93.2%. For out-of-context misinformation, we verified that the Catching Out-of-Context Misinformation using Self-Supervised Learning (COSMOS) model was suitable for the domain transfer task and executed custom code for its implementation. Lastly, we developed a system for explanation generation by prompting ChatGPT-4-32k with optimised input parameters which has achieved high average BLEU, METEOR, and ROUGE scores of 84.1%, 84.5%, and 91.3% respectively. SINDEX promotes an informed digital environment, empowering users to make evidence-based decisions, thus enhancing digital literacy and resilience against misinformation. Designed for scalability, SINDEX is readily adaptable to global misinformation challenges.

Keywords: Machine Learning · Misinformation Detection · Artificial Intelligence · Multimodal Misinformation

H. Guo et al. (Eds.): IRC-SET 2024, PTT, pp. 248–260, 2025.
https://doi.org/10.1007/978-981-96-3770-6_26

1 Introduction

In today's digital landscape, a deluge of data floods the digital realm—500 million tweets, 4 million gigabytes of Facebook data, and 65 billion WhatsApp messages worldwide [1]. This sheer volume of content exacerbates the challenge of identifying reliable information, with misinformation often amplifying public scepticism. For example, the widespread dissemination of false information during the COVID-19 pandemic significantly undermined trust in medical treatments and vaccination efforts [2].

While misinformation is a global issue, Singapore's multicultural and multilingual landscape introduces additional complexity. Local content often intertwines cultural nuances with global narratives, making it harder to discern the truth, particularly when dealing with satirical or out-of-context material. These challenges underscore the need for a tailored approach to fact-checking in the Singaporean context.

To address these unique challenges, we developed SINDEX, an automated system designed to efficiently fact-check multimodal misinformation, including text and images, while fostering deeper user understanding of the facts. Beyond Singapore, SINDEX is adaptable to similar challenges globally. The key contributions of this paper are fourfold:

1. A large original multimodal dataset for Singapore-context misinformation
2. An automated framework for evidence retrieval and claim verification via external knowledge
3. A custom satire text detector to evaluate exaggerated content based on language semantics
4. A multistage model with pretrained out-of-context detection model using Large Language Models (LLMs) for zero-shot misinformation prediction and explanation generation.

2 Literature Review

2.1 Multimodal Misinformation

In fact-checking, "multimodal" refers to "the claim and/or evidence [being] expressed through different or multiple modalities" [3]. Our project focuses on Singapore-context misinformation containing text and image modalities.

2.2 Manual Detection of Multimodal Misinformation

Human cognition is often influenced by confirmation bias, which compromises the objectivity required for effective fact-checking [4]. In the digital age, the sheer volume of information [5], coupled with the labour-intensive nature of manual verification, makes it nearly impossible to assess every claim accurately. Modern misinformation, especially multimodal content, is particularly challenging to detect. Visual content often leads to "photo truthiness" [6], where images lend false credibility to misleading claims, and out-of-context visuals are especially persuasive. Satirical content further complicates detection, as its ironic and exaggerated nature blurs the line between humour and misinformation. A recent survey revealed that although four in five Singaporeans feel confident in spotting fake news, nearly 90% misidentified at least one out of five false headlines [7]. This highlights the urgent need for more sophisticated and automated detection methods to effectively combat modern misinformation.

2.3 Automated Fact-Checking Models

Conventional text-based detection models, such as GEAR [8] are challenged by the multimodal nature of the misinformation in modern society as they fail to capture the complex interplay of information across multiple modalities.

MOCHEG [9] addresses evidence retrieval, claim verification, and explanation generation. It employs SBERT and BERT-based re-ranking for evidence retrieval. Claim verification utilises pre-trained Contrastive Language-Image Pretraining (CLIP) [10] to encode text and image evidence, employs stance detection through cross-attention operations, and predicts truthfulness by optimisation with a cross-entropy objective. Performance is poor, with low precision scores of 4.71% for image retrieval, 14.9% for text, and F1-score of 44.1% for claim verification. The Truthformer model [11] only addresses claim verification and integrates text and image data. It employs separate embedding blocks fine-tuned on FACTIFY [12] data and fuses representations using Conv1D layers (FusConv1D) or self-attention (FusAttn). Performance is satisfactory, with an F1-score of 76.8%, but it is untrained on Singapore-context misinformation. Existing automated fact-checking models, while valuable, exhibit notable limitations:

1. No existing model that addresses Singapore-context misinformation
2. Scarcity of end-to-end fact-checking models
3. Limiting binary classification system with only "True" or "False" labels, posing a challenge for users to understand the nuanced veracity of claims
4. Fixed training dataset with limited topics, causing poor performance for novel topics
5. Impractical in real-world setting as evidence retrieval is not automated

3 Aims & Objectives

Considering the limitations of current automated fact-checking models, our project sets out to develop a multifold model that (1) detects *Singapore-context* multimodal misinformation, (2) detects satire and out-of-context image usage and (3) provide explanations to promote user understanding and (4) automates real-time evidence retrieval for practicality.

4 Dataset Collection

4.1 Singapore-Context Multimodal Misinformation Dataset

Data was collected from diverse sources to mirror real-world misinformation complexity and mitigate potential biases arising from source selection. An original dataset for Singapore-context was created by obtaining examples from fact-checking websites Black Dot Research [13] and Factually [14], and collecting and labelling examples circulated on Facebook and WhatsApp. Textual evidence was collected by searching for news articles relating to the claims and retrieving the most relevant sentences for each of 3 articles. "True" examples were compiled from credible news outlets like Straits Times [15] and Channel News Asia [16]. The dataset maintains a balanced ratio of true to false claims, with an equal distribution of 1:1. Specifically, 147 true claims and 148 false claims were included, ensuring that the model is trained and tested on an evenly distributed dataset. The dataset was subsequently split into an 80/20 train-test ratio, preserving this balance across both sets (Table 1).

Table 1. Dataset structure of Singapore-context Misinformation Dataset

Category	Source	True	False	Total Claims
Fact-checking Websites	Black Dot Research	24	27	105
	Factually	26	28	
Social Media Platforms	Facebook	20	22	100
	WhatsApp	30	28	
News Outlets	Straits Times	22		47
	Channel News Asia	25		
Misleading claims	Reddit		43	
Total		**147**	**148**	**295**

4.2 Satire Dataset

184 examples of Singapore-context satirical claims and 104 non-satirical fake ones were collected. To circumvent the scarcity of Singapore-context satirical claims online, ChatGPT was leveraged on to automate the claim-generation process. A reference example was used to prompt it to generate Singapore-context satirical claims. The dataset was further supplemented with 203 satirical news articles and 283 non-satirical ones from the Fake News vs Satire: A Dataset and Analysis dataset [17]. The final dataset contained a total of 387 satirical claims and 387 non-satirical fake claims, ensuring a balanced 1:1 ratio of satirical to non-satirical claims (Table 2).

Table 2. Examples of reference and ChatGPT-generated claims on Singapore-context satirical claims

Singapore-context satirical claims	
Reference Example	HDBs set to receive 'genius windows' that can predict rain before meteorologists even get the memo
ChatGPT-generated	Changi Airport introduces 'self-flying suitcases' that navigate travellers to their destinations like magical luggage fairies

5 Model Architecture

We propose SINDEX, an end-to-end model for misinformation detection. SINDEX is tailored to identify Singapore-specific multimodal misinformation, including a custom satire detector to differentiate exaggerated content. It recognizes out-of-context image usage and generates explanations for user comprehension through automated evidence retrieval with real-time updates.

Our pipeline begins with the user inputting both a claim, C, and an accompanying image, I. The input claim is initially processed by an automated evidence retrieval system, which dynamically queries credible news sources to retrieve the most relevant textual evidence. This evidence, along with the original claim and image, is fed into a claim verification module. This module utilizes the cross-modal capabilities of the CLIP model to cross-check the input claim against the retrieved evidence, as well as to validate the consistency between the input image and the evidence. Additionally, the input claim undergoes further verification against the evidence using a large language model (LLM) like ChatGPT. At the conclusion of this stage, the pipeline outputs a truthfulness label. Concurrently, the input claim and image are processed through the COSMOS model, designed to detect out-of-context misinformation, resulting in an "In/Out-of-Context" label. The claim is also analysed by a BERT-based satire detection model, producing a "Satire/Not Satire" label. Finally, the truthfulness label and "Satire/Not Satire" label, along with the original claim and evidence, are synthesized to generate a comprehensive explanation for the user, facilitated by the LLM (Figs. 1, 2 and 3).

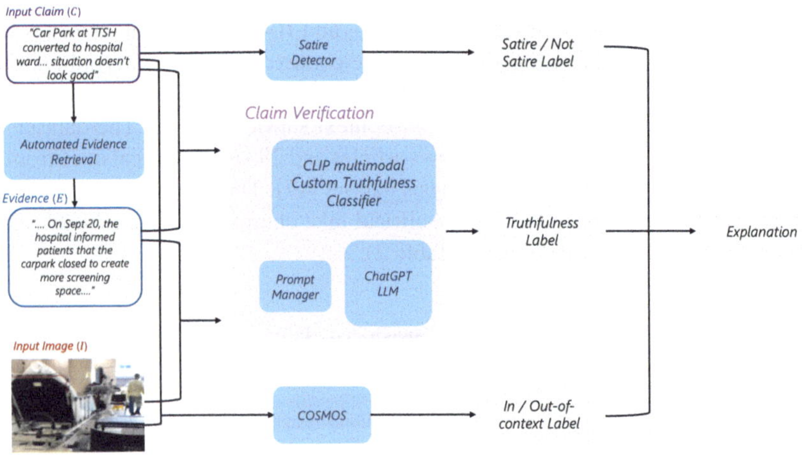

Fig. 1. A diagram of the proposed pipeline behind SINDEX

5.1 Automated Evidence Retrieval

The H2-keywordextractor transformer model was used on the claim, and extracted keywords were cleaned. Named Entity Recognition (NER) was applied to extract and preserve names, locations and time phrases from the keywords. Then, common stop words such as 'the' and 'and,' were excluded to refine the keyword set and improve the accuracy of subsequent searches. Cleaned keywords were then used to query NewsAPI, which provides the model with a comprehensive and dynamically updated evidence bank from over 80,000 news sources in the last 5 years [18]. Textual content was extracted from each article's URL using Beautifulsoup [19] and a pre-trained SBERT model from the *paraphrase-MiniLM-L6-v2* weights was utilised to encode both the extracted keywords and article into vector embeddings.

The mean cosine similarity score to the set of keywords, k, in each claim, C, for each article, A, was calculated between the article content and each keyword. This is defined as the Similarity Score for each article.

$$\boldsymbol{Similarity}(A, C) = \textbf{cosine\_similarity}(\mathbf{k}, \mathbf{A}) = \frac{\boldsymbol{k} \cdot \boldsymbol{A}}{|\boldsymbol{k}||\boldsymbol{A}|}$$

For the 25 articles with the highest similarity scores, the Relevance Score for each article was computed as the cosine similarity score between the article content and the claim as a whole.

$$\boldsymbol{Relevance}(A, C) = \textbf{cosine\_similarity}(\mathbf{C}, \mathbf{A}) = \frac{\boldsymbol{C} \cdot \boldsymbol{A}}{|\boldsymbol{C}||\boldsymbol{A}|}$$

From each of the three articles with the highest Relevance Score, two sentences with the highest cosine similarity score to the claim were selected to form the textual evidence, E, for that particular claim.

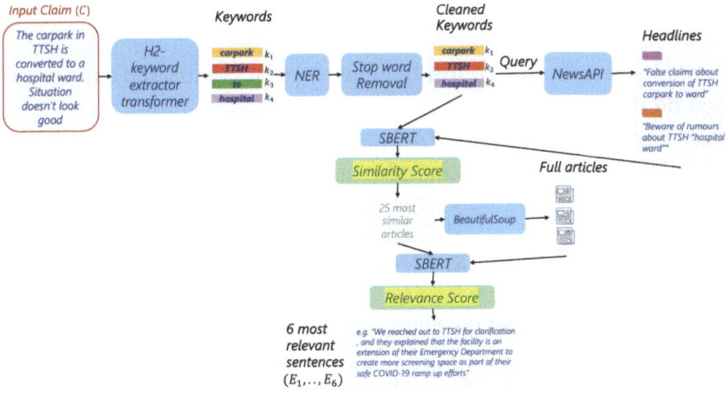

Fig. 2. A diagram illustrating our Automated Evidence Retrieval architecture

5.2 Claim Verification

5.2.1 CLIP Multimodal Custom Truthfulness Classifier (CTC)

Let the input image I be represented as a set of image patches, $\boldsymbol{I} = \{\boldsymbol{i}_{1,}\boldsymbol{i}_{2,\ldots,}\boldsymbol{i}_{\boldsymbol{p}}\}$ where i_j denotes the j-th patch of the image. Similarly, the evidence E is a sequence of tokens, $\boldsymbol{E} = \{\boldsymbol{E}_{1,}\boldsymbol{E}_{2,\ldots,}\boldsymbol{E}_{\boldsymbol{m}}\}$ where E_j denotes the j-th token of the evidence. Lastly, the claim C is represented as a sequence of tokens $C = \{\boldsymbol{C}_{1,}\boldsymbol{C}_{2,\ldots,}\boldsymbol{C}_{\boldsymbol{n}}\}$ where $C_j d$ enotes the j-th token of the claim.

Contextual representations for the claim, H_C, image, H_I, and text evidence, H_E, are obtained by tokenising them using CLIP such that.

$$\boldsymbol{H_C} = \left\{\boldsymbol{h}_{\boldsymbol{C}_0,}\boldsymbol{h}_{\boldsymbol{C}_1,}\ldots\boldsymbol{h}_{\boldsymbol{C}_j}\right\}$$
$$\boldsymbol{H_I} = \left\{\boldsymbol{h}_{\boldsymbol{I}_0,}\boldsymbol{h}_{\boldsymbol{I}_1,}\ldots\boldsymbol{h}_{\boldsymbol{I}_j}\right\}$$
$$\boldsymbol{H_E} = \left\{\boldsymbol{h}_{\boldsymbol{E}_0,}\boldsymbol{h}_{\boldsymbol{E}_1,}\ldots\boldsymbol{h}_{\boldsymbol{E}_j}\right\}$$

To compute the stance representation between the claim and text evidence, $\boldsymbol{H_{CE}}$, and that between the image and text evidence, $\boldsymbol{H_{IE}}$, we apply an attention mechanism between the text evidence representations and both the claim and image representation. In this mechanism, attention weights are calculated using the softmax function on the dot product of claim representations with text evidence and image representations.

$$\boldsymbol{a_{ij}} = \boldsymbol{softmax}\left(\boldsymbol{h_{ij}} \cdot \boldsymbol{H_E^\top}\right)$$

The updated image-evidence stance representation and claim-evidence stance representation is then computed by applying these attention weights to the evidence representation:

$$\boldsymbol{H_{IE}} = \sum_{j=1}^{m} \boldsymbol{a_{ij}} \boldsymbol{h_{ej}}$$

$$\boldsymbol{H_{CE}} = \sum_{j=1}^{m} \boldsymbol{a_{ij}} \boldsymbol{h_{ej}}$$

where $\boldsymbol{H_{IE}}$ and $\boldsymbol{H_{CE}}$ are the stance-aware image representation and claim representation updated based on the evidence respectively

The concatenated representation $\boldsymbol{G_{CEI}} = [\boldsymbol{H_C} : \boldsymbol{H_E} : \boldsymbol{H_I}]$ is passed through two fully connected layers. The first layer applies a linear transformation followed by ReLU activation:

$$\boldsymbol{A_1} = \boldsymbol{ReLU}(\boldsymbol{W_1} \cdot \boldsymbol{G_{CEI}} + \boldsymbol{b_1})$$

The output $\boldsymbol{A_1}$ is then passed through a second linear transformation

$$\boldsymbol{Z_2} = \boldsymbol{W_2} \cdot \boldsymbol{A_1} + \boldsymbol{b_2}$$

Finally, a sigmoid activation is applied to the logits for binary classification, and the function returns the loss, logits, and true labels.

$$\hat{y} = \sigma(\boldsymbol{Z_2}) = \frac{1}{1 + e^{-\boldsymbol{Z_2}}}$$

Binary cross-entropy (BCE) loss was chosen as the loss function.

$$\boldsymbol{L_{BCE}} = -\frac{1}{N} \sum_{i=1}^{N} (y_i \boldsymbol{log}(\widehat{y_i}) + (1 - y_i)\boldsymbol{log}(1 - \widehat{y_i}))$$

5.2.2 ChatGPT LLM

Large Language Models (LLMs), trained on large-scale corpus have shown promising emergent abilities on various tasks [20], including those requiring sophisticated reasoning and evaluation. Particularly, the effectiveness of LLMs has been enhanced

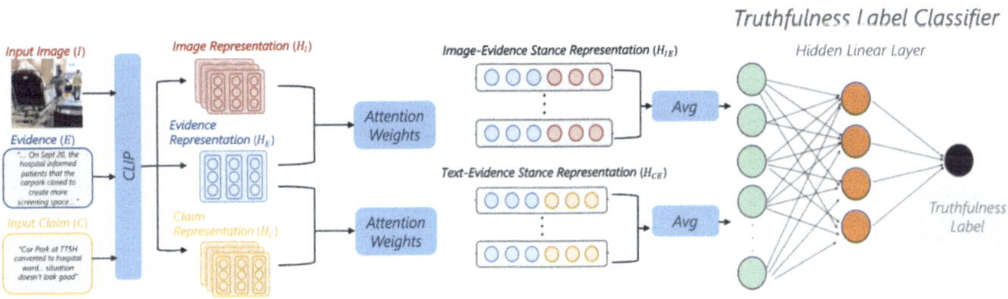

Fig. 3. A diagram of the pipeline for our CLIP multimodal Custom Truthfulness Classifier

through innovative prompt-engineering, enabling impressive performance in zero-shot and few-shot tasks. Thus, we aimed to leverage these LLMs by implementing a parallel LLM-involved misinformation detection system using OpenAI's ChatGPT-4-32k. The system utilises a structured query format where context placeholders are provided to insert modelling outputs from previous stages to formalise a holistic description to the LLM:

Query Input.

> "Given a claim {{{test_claim}}}, with the accompanying context {{{', '.join([article['title'] for article in wikipedia_articles])}}}, with the following accompanying news article evidences {{{evidence_part}}}, predict a truthfulness label given {{true, false}}and provide a reasoning explanation for the prediction."

Contextual information is obtained from the Wikipedia articles relating to each keyword in the claim using WikipediaAPI. The same evidence from the automated evidence retrieval module is used. The resulting misinformation label prediction is presented with a reasoning explanation.

5.3 Satire Detection

Employing 'not satire' (0) and 'satire' (1) labels, we developed a binary classifier for satire detection by fine-tuning the pretrained bert-base-uncased model. To minimise class imbalance, we ensured equal distribution between satire and non-satirical fake claim examples. The dataset was split into training and testing sets with a 80/20 train-test split. BERT's tokenizer was used and the tokenized sequences were padded to a maximum length of 512. The model contains a dropout layer and a linear layer with sigmoid activation for binary classification. The training loop uses 100 epochs, and the model was optimised using the Adam optimizer to adjust parameters. BERT model layers 10 and 11 were unfrozen to allow gradient computation during training. BCE Loss was chosen as the loss function (Fig. 4).

Given that satire often relies on nuanced language and context, focusing on textual analysis is our primary approach. However, extending this to incorporate both image and text modalities could enhance satire detection by leveraging the relationship between visual and textual elements. This approach is a promising direction for future work.

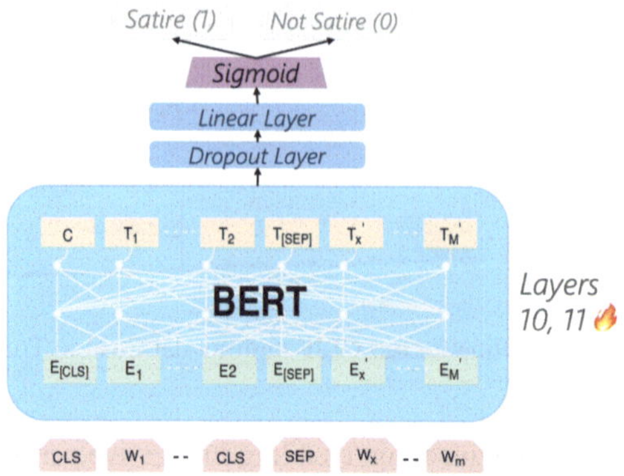

Fig. 4. A diagram of the pipeline for our Satire Detector, with layers 10 and 11 of BERT unfrozen for gradient computation

5.4 Out-Of-Context Image Detection

Our framework employs the trained Catching Out-of-Context Misinformation using Self-Supervised Learning (COSMOS) [21] model as our out-of-context image detector. COSMOS lacks specific codes for dataset collection and preparation, but states the usage of Google's Cloud Vision API [22], Spacy NER [23], SBERT and detectron2 [24]. Google's Cloud Vision API applies "reverse-search" on the input image to obtain a second caption. Detectron2 forms bounding boxes for the images. Our implementation involved executing custom code for Spacy NER, the all-MiniLM-L6-v2 model for SBERT [25], and detectron2. They respectively augment modified captions and entities list, compute the BERT base score, and generate bounding boxes within the JSON files housing comprehensive data, which also includes the original two captions, image path, context label, and article link.

5.5 Explanation Generation

For explanation generation, the ChatGPT model with optimised input parameters is prompted with a structured query format:

Input Query (Explanations): "Given that this claim {{{test_claim}}} has been classified as {{{truthfulness_label}}}, {{{satire_label}}}, based on the evidence {{{evidence_list}}}, provide a reasoning explanation for the truthfulness label for the claim."

6 Results and Discussion

6.1 Automated Evidence Retrieval

As a benchmark, we manually collected evidence for 10 claims and compared the six manually-retrieved sentences and the six module-retrieved sentences. When only the similarity score was used for filtering, the model showed poor performance. 65% of 60 evidence sentences selected were irrelevant. When both similarity score and relevance score were used, model performance improved significantly, with 58 out of 60 retrieved sentences being similar to the manually selected ones. Variations in sentences were due to differences in phrasing and not ideas (Table 3).

Table 3. Comparative performance of Automated Evidence Retrieval system with Similarity Score only vs. both Similarity and Relevance scores

Filtering Method	Total Sentences Selected	Relevant Sentences	Irrelevant Sentences	Relevance (%)
Similarity Score	60	21	39	35
Similarity Score + Relevance Score	**60**	**58**	**2**	**97**

From our results, the second filter using the relevance score between the article content and the claim is crucial. We hypothesise that this is because when only the similarity score was used, the model focused on surface-level lexical and syntactic similarities, leading to the inclusion of sentences that contained the same keywords albeit with different meaning and context. This method of automating evidence retrieval using real-time APIs can provide the following improvements to automated fact-checking models: (1) An improved performance due to access of a large evidence bank, (2) Utility in a larger range of topics and (3) Ability to stay up-to-date.

6.2 Claim Verification

Our proposed CLIP multimodal CTC achieves a strong final F1-score of 95.2%, indicating that it is suitable for use in Singapore's context.

The ChatGPT-enabled misinformation detection system was evaluated on the same test set as the CLIP multimodal CTC. Without evidence, the ChatGPT-enabled system showed a poor F1-score of 62.9%, mainly predicting a truthfulness label based on the credibility of the alleged source and the level of detail provided. For instance, when evaluated on the claim "Singapore Police Force (SPF) clarifies that the video circulating online that showed a mass brawl involving workers did indeed happen in Singapore", the system inaccurately predicted the claim to be true, citing the reason that "The claim is made by SPF, an official source. There is no reason to doubt the claim's authenticity." However, relying on source credibility is impractical as individuals can exploit their reputation to propagate misinformation, highlighting the need for a more evidence-driven approach to combat misinformation.

With evidence, an F1-score of 95.4% was achieved on the same test set. The improvement in F1-score highlights our automated evidence retrieval framework as performance-effective. Notably, the system made only two errors, with accurate explanations provided, but labels erroneously assigned. This highlights the system's proficiency in identifying Singapore-context misinformation, with rare misclassification arising from labelling discrepancies rather than substantive misunderstandings (Tables 4, 5, 6 and 7).

Table 4. Results for ChatGPT-enabled misinformation detection, with and without evidence

	Without Evidence (%)	With Evidence (%)
Precision	59.5	96.9
Recall	66.7	93.9
Accuracy	62.1	95.4
F1-Score	62.8	95.4

The evaluation of mainstream misinformation verification models on our Singapore-specific dataset reveals significant variability in performance, underscoring the importance of domain-specific training. Among the models tested, SINDEX achieved a state-of-the-art F1-score of 95.2%, significantly outperforming the other models. These results highlight the superiority of SINDEX when applied to Singapore-context misinformation, suggesting that models specifically trained or fine-tuned on domain-specific data are better equipped to handle the unique linguistic and cultural nuances present in localised misinformation.

Table 5. Comparative Performance of Mainstream Misinformation Verification Models on a Singapore-Specific Dataset (Best result is highlighted in bold)

Model	F1-Score on Singapore-specific Dataset (%)
MOCHEG	47.1
Truthformer	41.8
CNN (Alonso-Bartolome et al.)	23.6
MVAE	37.8
SINDEX (Ours)	**95.2**

6.3 Satire Classifier

The satire classifier model had a final F1 score of 93.2%, indicating strong performance. Compared to existing benchmarks in satire detection, our model's performance is notably superior, suggesting that the integration of domain-specific training data and fine-tuning with a BERT-based architecture contributed significantly to these results.

6.4 Explanation Generator

When evaluated against the gold explanations, our ChatGPT-involved explanation generation system demonstrated strong performance in producing high-quality explanations, as seen from the high average BLEU, METEOR, and ROUGE scores of 84.1%, 84.5%, and 91.3% respectively.

Table 6. Performance of ChatGPT-involved Explanation Generation System

	Min (%)	Max (%)	Avg (%)
BLEU	78.2	87.6	84.1
METEOR	80.3	88.9	84.5
ROUGE	85.7	95.0	91.3

Table 7. Example of ChatGPT-generated explanations and our gold explanations

ChatGPT-generated explanation	Gold explanation
The truthfulness label for the claim is "false". The news article evidence clearly states that part of Fullerton Road will be closed to traffic from 4pm on New Year's Eve until 5am the next day due to the Marina Bay countdown activities. This directly contradicts the claim that no part of Fullerton Road will be closed as a security measure for the event	The claim is false because the evidence states that part of Fullerton Road will be closed to traffic from 4pm on New Year's Eve until 5am the following day due to the countdown activities at Marina Bay, refuting the claim that no part of Fullerton Road will be closed for the event

7 Conclusion

This study presents SINDEX, a novel model specifically designed for the detection of Singapore-context multimodal misinformation. By utilizing an original dataset and integrating features such as a custom satire detector, out-of-context image recognition, and automated explanation generation, SINDEX effectively addresses the complexities of misinformation in the digital age. The model demonstrates robust performance across all functions, establishing a state-of-the-art benchmark for localized misinformation detection in Singapore. Furthermore, its adaptable and scalable framework positions SINDEX as a potential solution for tackling misinformation challenges on a global scale, promoting enhanced digital literacy and informed decision-making among users.

References

1. Vopson, M.M.: The world's data explained: How much we're producing and where it's all stored. The Conversation (n.d.)
2. Islam, M.R., Sarkar, A., Hossain, M.E., Kamruzzaman, M.: [Insert article title here] (2020)
3. Hameleers, M., Powell, T.E., Van Der Meer, T.G., Bos, L.: A picture paints a thousand lies? the effects and mechanisms of multimodal disinformation and rebuttals disseminated via social media. Polit. Commun. **37**(2), 281–301 (2020)
4. Park, S., Park, J.Y., Kang, J.H., Cha, M.: The presence of unexpected biases in online fact-checking. Harvard Kennedy Sch. Misinformation Rev. (2021)
5. Horrigan, J.B.: Information overload. Pew Res. Center Internet Sci. Tech. (2016, December 7). https://www.pewresearch.org/internet/2016/12/07/information-overload/

6. Newman, E.J., Zhang, L.: How non-probative photos shape belief. Cogn. Sci. (2020)
7. Huiwen, N.: 4 in 5 Singaporeans confident in spotting fake news but 90 percent wrong when put to the test: Survey. Straits Times (2018, September 27)
8. Zhou, J., et al.: GEAR: graph-based evidence aggregating and reasoning for fact verification. arXiv preprint arXiv:1908.01843 (2019)
9. Yao, M., Shah, A., Sun, L., Cho, J. H., Huang, L.: End-to-End multimodal fact-checking and explanation generation: a challenging dataset and models. *arXiv preprint arXiv:2205* (2022)
10. Radford, A., et al: Learning transferable visual models from natural language supervision. Int. Conf. on Mach. Learn. (2021)
11. Chaitanya, B.S.N.V., Prathyush, P., Rutvik, V.: Truthformers at factify 2022: evidence aware transformer-based model for multimodal fact-checking (2021)
12. Mishra, S.S., Bhaskar, A., Ahuja, C.: FACTIFY: a multi-modal fact verification dataset. ResearchGate (2021, November 18)
13. Black Dot Research: (n.d.). https://blackdotresearch.sg/
14. Gov.sg: (n.d.). Factually. https://www.gov.sg/factually
15. The Straits Times: (n.d.). https://www.straitstimes.com/
16. Channel News Asia: (n.d.). https://www.channelnewsasia.com/
17. Golbeck, J., Golbeck, J., et al.: Fake news vs satire. (2018, May 15)
18. NewsAPI. (n.d.): https://newsapi.org/
19. Leonard Richardson: BeautifulSoup: we called him Tortoise because he taught us. Crummy. (n.d.). https://www.crummy.com/software/BeautifulSoup
20. Wei, J., et al.: Emergent abilities of large language models. (2022). arXiv preprint arXiv:2206.07682
21. Aneja, S.: [AAAI 2023] COSMOS: catching out-of-context misinformation using self-supervised learning. GitHub. (n.d.). https://github.com/shivangi-aneja/COSMOS
22. Google: Vision AI | Google Cloud. Google (n.d.)
23. Explosion AI: Linguistic features: spacy usage documentation. Spacy (n.d.)
24. Facebook AI Research: Detectron2: detectron2 is a platform for object detection, segmentation and other visual recognition tasks. GitHub (n.d.). https://github.com/facebookresearch/detectron2
25. Hugging Face.: Sentence-transformers/all-MiniLM-L6-v2. Hugging Face (n.d.). https://huggingface.co/sentence-transformers/all-MiniLM-L6-v2

Optimizing Natural Number Representation via Pan Balance Mechanisms

Arnab Mishra(✉) and Puah Xuan Ying Vera

NUS High School of Math and Science, Singapore 129957, Singapore
{h2110014,h2110126}@nushigh.edu.sg

Abstract. This work explores novel methods for weight representation on balance scales. It begins by defining modifications as weight adjustments and balancing apparatus alterations, and expressing numbers as binary powers of two. The study then demonstrates a means to generalize to higher bases and explains the usefulness of balanced base systems for $n \geq 4$. Correlations between specific numbers and required Fibonacci weights can be used to determine relationships to equilibrium values. Additionally, an algorithm to obtain a modified Zeckendorf representation that allows for subtraction is discussed, along with proving two theorems regarding the rate of increase in the number of weights needed.

The study elaborates on two possible candidates for a modified pan balance, based on modifying two different degrees of freedom. It also analyzes the ternary system and its efficiency in representing numbers, justifying its superior efficiency via radix economy analysis. The methods explored might optimize precise measurements, quantum algorithms, and charge balancing.

Keywords: Optimization · Natural Number Representation · Pan Balance Mechanisms · Base-n Systems · Zeckendorf Representation · Fibonacci Numbers · Balance Scale Modification · Radix Economy · Ternary Computers · Trit

1 Introduction and the Original Problem

This paper considers different methods to represent any natural number p as a sum or difference of other numbers (our weights). For example, by expressing a number in its binary form, we can get any weight by adding different powers of 2 using the digits of the binary form. For example, a mass of 13, which is 1101 in binary, can be attained with 2^0,2^2 and 2^3 weights. In this paper, we used a pan balance analogy to help us visualize this problem, inspired by a question from the IMO 2011 Q4. In this analogy, we have to come up with a placement of weights such that if we placed a weight of mass p in the appropriate pan, the system is balanced.

These efforts can be broadly categorized as belonging in one of two categories: Those which primarily modify the weights, and those which primarily modify the balancing scale, and we will be using this categorization in our discussion.

H. Guo et al. (Eds.): IRC-SET 2024, PTT, pp. 261–271, 2025.
https://doi.org/10.1007/978-981-96-3770-6_27

However, take note that hybrids of both can be considered and used, but are not discussed in this paper.

2 A Review of Theorems on Natural Number Representation

2.1 Fermat's Polygonal Number Theorem

Fermat's polygonal number theorem states that any number can be expressed as a sum of at most n different n-gonal numbers, where the sth n-gonal number is:

$$\frac{(s-2)n^2-(s-4)n}{2} \tag{1}$$

However, this increase is only quadratic, and so a lot of weights are needed to allow for representation of a rather small subset of numbers, as compared to the balanced base-f system.

2.2 Numbers of the Form 2 and 3^a5^b

We can use weights that are of the form 3^a5^b along with the weight 2 (British MO R1 2016 Q6). By induction, we can prove that any natural number can be expressed as the sum of distinct weights of this form. However, there are many more numbers of the form 3^a5^b, and each 'digit' would be a vector of a and b, thus needing more memory.

2.3 Goldbach's Strong Conjecture

Goldbach's strong conjecture, which remains unproven to this day, asserts that every even number greater than 2 can be expressed as a sum of two primes. This also would imply Goldbach's weak conjecture, which states that every odd number greater than 5 can be expressed as a sum of 3 primes, since we could just add a 3 to every even number's prime representation. This would certainly be very efficient, but the prime numbers are quite dense in $\mathbb{N}$ as compared to the Fibonacci numbers or powers of n. Even then, we must first prove this conjecture.

3 Modified Weights

3.1 Weights of the Form of $n^0, n^1, ..$

It is natural for us to begin by attempting to come up with some weighing system for $n = 3$. We can use the ternary system, but this would mean that some weights are repeated since numbers are now represented using the digits 0, 1 and 2. Extending to higher bases would result in even more repeats, and this aspect could thus be further optimized.

We consider the general problem of expressing a mass in the form of the sum of powers of a larger number n. One may again express the mass in base-n and then use the digits, but if a pan balance is provided one can make use of "negative" weights, where masses are placed on the left side of the pan (while the right side is heavier), such that the difference in mass is equal to the desired weight x. This uses less weights (except for the case when $n = 3, 2$).

3.1.1 Construction

We first follow the usual step of expressing p in base n.

$$p = (\overline{a_m a_{m-1} a_{m-2} \ldots a_1 a_0})_n \tag{2}$$

Storing the digits in an array of length m + 1, we then perform the below for $0 <= i <= m$.

1. If $a_i > n/2$, add $n - a_i$ of n^i weights to the left side. Add 1 to a_{i+1}.
2. Else add a_i of n^i weights to the right side.

For those familiar with them, we are essentially expressing the weight in a balanced n-ary number system, where negative weights go on the left, and the positive weights on the right. Refer to Fig. 1 for a diagram of the process.

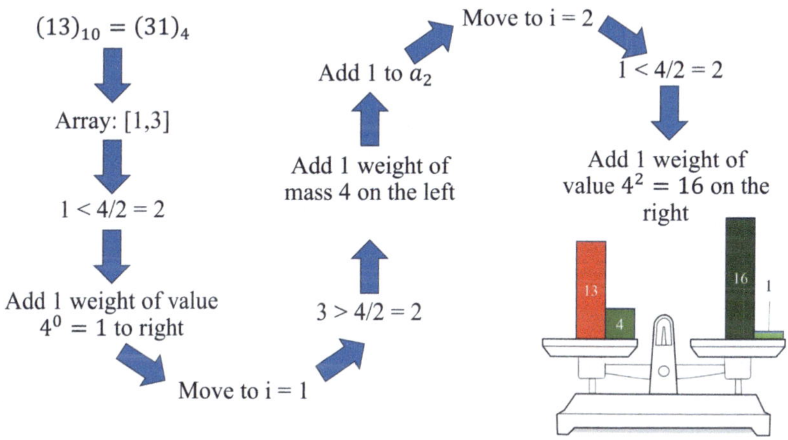

Fig. 1. An example of using balanced base system where $p = 13$, and $n = 4$. Diagram is made by the authors.

3.2 Fibonacci Weights

We considered the Fibonacci numbers as weights in what we call as a balanced base-f number system. We let $f_1 = 1$ and $f_2 = 2$, so as to avoid repetition. By Zeckendorf's theorem, every number has a unique representation as a sum of non-consecutive Fibonacci numbers, but we can further optimize by allowing for the subtraction of Fibonacci numbers. While this meant that each number may not have a unique balanced base-f representation, we could greatly reduce the number of weights needed (for example, f_m and $-f_{m-1}$ can be replaced with f_{m-2}).

Suppose f_1, f_2, ... are the Fibonacci numbers, where $f_1 = 1$, $f_2 = 2$. Let there be a set of numbers called R and let s be the sum of all numbers in R. We use a greedy algorithm to come up with the balanced base-f representation for a number p where $f_m \leq p < f_{m+1}$.

1. Check if $p = f_m$. If true, add f_m to R and terminate. Else proceed.
2. Add Fibonacci numbers less than f_{m-1} to f_m to find which Fibonacci number α minimises $x = |p - (f_m + \alpha)|$ (if there are two such α, choose the smaller α). If $x = 0$, add f_m and α to R and terminate. Else proceed.
3. Subtract Fibonacci numbers less than f_{m-1} from f_{m+1} to see which Fibonacci number β minimises $y = |p - (f_{m+1} - \beta)|$ (if there are two such β, choose the smaller β). If $y = 0$, add f_{m+1} and $-\beta$ to R, and terminate. Else proceed.
4. Compare x and y. If $x \leq y$, add f_m and α to R. Else, add f_{m+1} and $-\beta$ to R. Update s.
   (a) If $s < p$, add Fibonacci numbers to s to see which Fibonacci number α minimises $x = |p - (s + \alpha)|$. Add α to R and update s. If $x = 0$, terminate. Else repeat step 7.
   (b) Else, when $s > p$, subtract Fibonacci numbers from s to see which Fibonacci number β minimises $y = |p - (s - \beta)|$. Add $-\beta$ to R, update S. If $y = 0$, terminate. Else repeat step 7.

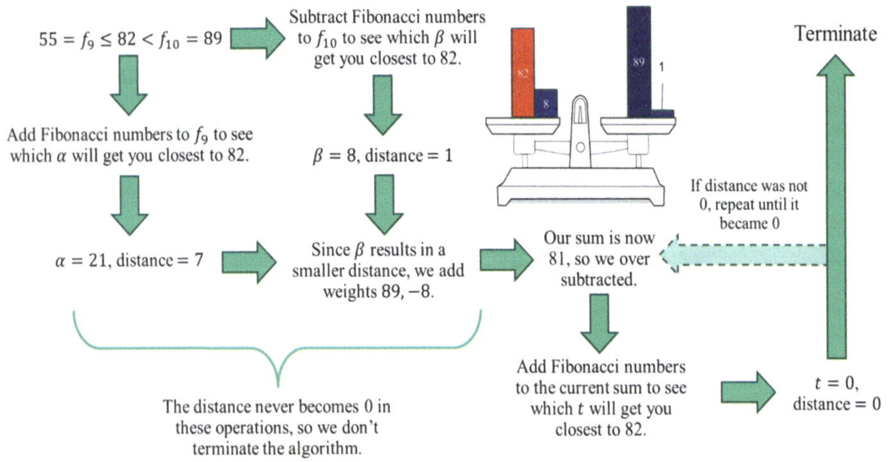

Fig. 2. Applying the algorithm for generating the balanced base-f representation for $p = 82$. Diagram is made by the authors.

The algorithm above has a time complexity of $O((\log p)^2)$, and is illustrated in Fig. 2.

3.2.1 Maximum Number of Weights Needed in the Balanced Base-f Representation

We will prove the following two conjectures:

Conjecture 1. *Numbers of the form $\frac{f_{3k-1}}{2}$ are the first numbers to need at least k weights, where k is a natural number. It can be represented as $1 + \sum_{i=1}^{k-1} f_{3i}$.*
Conjecture 2. *Numbers until f_{3k} only need at most k weights.*

Additionally, in our proof, we will also make use of the following lemma, which is easily proven by induction:

Lemma 3. *The sum of the first $n-2$ Fibonacci numbers is 2 less than f_n.*

By the above lemma, we see that for weights between f_n and f_{n+1}, we must use either f_n or f_{n+1} to sum to it.

Proof. The two conjectures obviously hold for numbers until $f_3 = 3$, with 1 obviously being the first number that needs 1 weight. Suppose they hold for all numbers until f_{3m}, i.e. at most m weights are needed to sum to any number from 1 to f_{3m}, and $\frac{f_{3m-1}}{2}$ is the first number to need at least m weights. We shall now prove that all numbers from $f_{3m}+1$ to f_{3m+3} needs at most $m+1$ weights and that $\frac{f_{3m+2}}{2}$ is the first number that needs $m+1$ weights.

We look at $f_{3m}+1$ to f_{3m+1} first, and call this set of weights A. By the induction hypothesis, numbers up to f_{3m-1} need at most m weights according to our algorithm, and $\frac{f_{3m-1}}{2}$ is the first number to need m weights. Thus, we see that for weights in A, they all need at least $m+1$ weights (*), and the first number to need $m+1$ weights is

$$f_{3m} + \frac{f_{3m-1}}{2} = f_{3m} + f_{3m-1} - \frac{f_{3m-1}}{2} = \frac{f_{3m} + f_{3m-1}}{2} + \frac{f_{3m}}{2}$$
$$= \frac{f_{3m+1} + f_{3m}}{2} = \frac{f_{3m+2}}{2}$$

Thus, $\frac{f_{3m+2}}{2}$ is the first number that needs $m+1$ weights (1 and f_{3i}, where $i = 1, 2, ..., m$). Hence, we have proved Conjecture 2.1 for the case $k = m+1$. Also, consider $p < \frac{f_{3m+2}}{2} = f_{3m} + q$. Since $q < \frac{f_{3m-1}}{2}$, it needs at most $m-1$ weights, and so p needs at most m weights. Then, $f_{3m+1} - q$ will also need at most m weights. These numbers cover all numbers in the A, except f_{3m+1}, which obviously needs 1 weight, and $\frac{f_{3m+2}}{2}$ which is thus is the only weight in A that needs $m+1$ weights (**).

For any p from the f_{3m} weights between $f_{3m+1}+1$ and f_{3m+2}, from the induction hypothesis, at most $m+1$ weights (f_{3m+1} and the $\leq m$ weights needed for $p - f_{3m+1} \leq f_{3m}$ or f_{3m+2} and the $\leq m$ weights needed for $f_{3m+2} - p \leq f_{3m}$) are needed.

For the f_{3m+1} weights between $f_{3m+2}+1$ and f_{3m+3}, using a similar argument as before, and from our result in (*), we can directly conclude that at most $m+1$ weights are needed, except for numbers where the weight being added or subtracted needs $m+1$ weights itself. However, by (**), the only such weight is $\frac{f_{3m+2}}{2}$. We perform the below simplification.

$$f_{3m+2} + \frac{f_{3m+2}}{2} = f_{3m+2} + f_{3m+1} - \frac{f_{3m-1}}{2} = f_{3m+3} - \frac{f_{3m-1}}{2}$$

In the above expression, we can see that it too needs $m+1$ weights (f_{3m+3}, -1 and $-f_{3i}$ where $i = 1, 2, ..., m-1$). Hence, Conjecture 2.2 is proven for the case $k = m+1$, and so the conjectures are true for all numbers.

3.2.2 Example Balanced Base-f Representations

Table 1 (below) contains the balanced base-f representations of 1 to 55. Observe how the first number to need 3 weights is 17, and $f_{3\cdot 3-1} = f_8 = 34$, which verifies Conjecture 2.1 for the case $k = 3$. Additionally, it can be seen that numbers until $f_{3\cdot 3} = f_9 = 55$ need at most 3 weights, which verifies Conjecture 2.2 for the case $k = 3$

Table 1. Modified representations of numbers from 1 to 55

Number (1–28)	Generated Array	Number (29–55)	Generated Array
1	[1]	29	[21, 8]
2	[2]	30	[21, 8, 1]
3	[3]	31	[34, −3]
4	[3, 1]	32	[34, −2]
5	[5]	33	[34, −1]
6	[5, 1]	34	[34]
7	[5, 2]	35	[34, 1]
8	[8]	36	[34, 2]
9	[8, 1]	37	[34, 3]
10	[8, 2]	38	[34, 3, 1]
11	[8, 3]	39	[34, 5]
12	[13, −1]	40	[34, 5, 1]
13	[13]	41	[34, 8, −1]
14	[13, 1]	42	[34, 8]
15	[13, 2]	43	[34, 8, 1]
16	[13, 3]	44	[34, 8, 2]
17	[13, 3, 1]	45	[34, 13, −2]
18	[13, 5]	46	[34, 13, −1]
19	[21, −2]	47	[34, 13]
20	[21, −1]	48	[34, 13, 1]
21	[21]	49	[55, −5, −1]
22	[21, 1]	50	[55, −5]
23	[21, 2]	51	[55, −3, −1]
24	[21, 3]	52	[55, −3]
25	[21, 3, 1]	53	[55, −2]
26	[21, 5]	54	[55, −1]
27	[21, 5, 1]	55	[55]
28	[21, 8, −1]		

4 Modified Balances

The modifications below are motivated by adding different degrees of freedom. At the core of these modifications is to modify the perpendicular distance between the weights and the axis of rotation (of which we assume there is only 1), so as to reduce/increase the torque to be equal to that of what we wish to achieve.

4.1 Directly Varying Length

If we use weights of the form $3^i, i \in \mathbb{Z}^+$, and a hypothetical weightless ruler measuring 30 cm with its axle at the 20 cm mark, and place weights at the 0, 10, and 30 cm marks, we can achieve a desired mass p using at most $\lfloor log_3(p) + 1 \rfloor$. By expressing p in ternary, weights corresponding to digit 2 are placed at the 0 cm mark, digit 1 at the 10 cm mark, and the mass p at the 30 cm mark to maintain equilibrium.

Extending to larger numbers, we can choose to extend in one direction, or instead extending in both directions, using the balanced n-ary representation, with intervals on both sides to place weights (See Fig. 3, below).

Fig. 3. Modifying lengths. Diagram is made by the authors.

4.2 Using Angles to Indirectly Vary Length

Refer to the balance and coordinate system in Fig. 4, where the black rods are in the $x - y$ plane, and all 3 rods lie in the same common plane. We consider the coordinate system to defined as in the diagram shown. The system can rotate about and angle α about the axle, and the blue rod can be rotated in the plane containing all 3 rods, i.e. only α and β can be modified. Suppose our target mass is p.

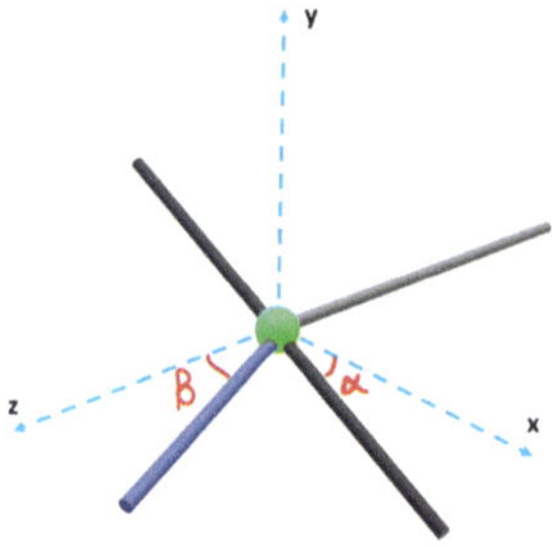

Fig. 4. Our 3 way balance. Diagram is made by the authors.

We place 3 masses in the following manner (refer to Fig. 5, below):

1. Place a mass of mass p on the right black end of the balance.
2. Place the smallest mass M such that $M \geq p$ on the left black end of the balance.
3. Place the smallest mass m such that $m \geq M - p$ on the blue end of the balance.

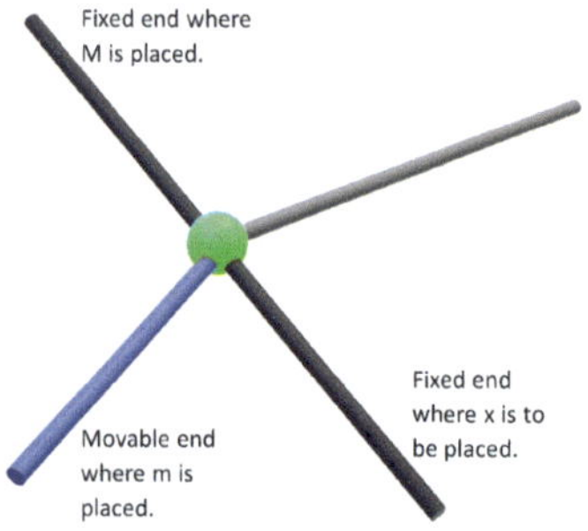

Fig. 5. Placement of weights. Diagram is made by the authors.

We can then rotate the adjustable end to find when the balance can stay level. We have the following expressions for the position vectors, and then calculating the torques (which is given by $\boldsymbol{\tau} = \boldsymbol{F} \times \boldsymbol{r}$), to get:

$$\boldsymbol{r}_p = \begin{pmatrix} r\cos(\alpha) \\ -r\sin(\alpha) \\ 0 \end{pmatrix} \Rightarrow \boldsymbol{\tau}_p = \begin{pmatrix} 0 \\ 0 \\ rg\cos(\alpha) \end{pmatrix} \tag{3}$$

$$\boldsymbol{r}_M = \begin{pmatrix} -r\cos(\alpha) \\ r\sin(\alpha) \\ 0 \end{pmatrix} \Rightarrow \boldsymbol{\tau}_M = \begin{pmatrix} 0 \\ 0 \\ -rg\cos(\alpha) \end{pmatrix} \tag{4}$$

$$\boldsymbol{r}_m = \begin{pmatrix} r\cos(\alpha)\sin(\beta) \\ -r\sin(\alpha)\sin(\beta) \\ r\cos(\beta) \end{pmatrix} \Rightarrow \boldsymbol{\tau}_m = \begin{pmatrix} -rmg\cos(\beta) \\ 0 \\ rg\cos(\alpha)\sin(\beta) \end{pmatrix} \tag{5}$$

The axle will provide a torque of $rmg\cos(\beta)$ in the x direction to make $\tau_x = 0$, and at equilibrium (i.e. the masses are balanced) $\tau_z = 0$ or:

$$\begin{aligned} &\implies rgp\cos(\alpha) - rgM\cos(\alpha) + rgM\cos(\alpha)\sin(\beta) = 0 \\ &\implies p = M - m\sin(\beta) \end{aligned} \tag{6}$$

Since $\sin(\beta)$ varies the least at angles closer to $\frac{\pi}{2}$, by maximising the angle by minimising the masses we can minimise the margin of error with our balancing of mass. If one wanted to find the weight of an unknown mass p, one can do so in the following manner:

1. Place p on the right black end of the balance, and place and check masses on the left black end of the balance, until the left side weighs more than the right. Let this mass be M.
2. Rotate the free end completely to the right.
3. Place and check masses to the free end until the right weighs more than the left. Let the mass on the free end be m.
4. Slowly rotate the free end towards the middle position, stopping when the balance is at equilibrium. Measure β and use $p = M - m\sin(\beta)$

5 A Discussion on Ternary Computers

In normal computers, the absence or presence of current in a transistor is used to denote 0 or 1 respectively, and by combining multiple transistors, one can represent numbers by converting it into binary. Due to the ease of detecting current in transistors, as well as its efficiency and simplicity, computers use a binary number system.

The radix economy of a base-b number system is used as a representation of the efficiency of base b in representing a number. It is essentially the number of digits in the number, multiplied by the number of numbers that can occupy each digit. Thus, it is defined as

$$E(b, N) = b\lfloor \log_b(N) + 1 \rfloor \tag{7}$$

It can also be roughly approximated for large N as

$$E(b, n) \approx b\log_b(N) = \frac{b}{\ln b} \cdot \ln N \tag{8}$$

Thus, it can be proven that base-e is the most efficient base (since $\frac{b}{\ln b}$ is decreasing for $1 < b < e$, and increasing for $e < b$), and that 2,3,4 also have low radix economies, with 3 theoretically having the lowest among positive integers, as seen by analyzing radix economies from 1 to 1500, in Fig. 6. In the right graph, by taking a running average of the radix economy of bases 2, 3 and 4, as well as e, we can compare their efficiencies by seeing which one has the lowest radix economy. To be able to better see such differences, we take a ratio of $E(b, N)$ to

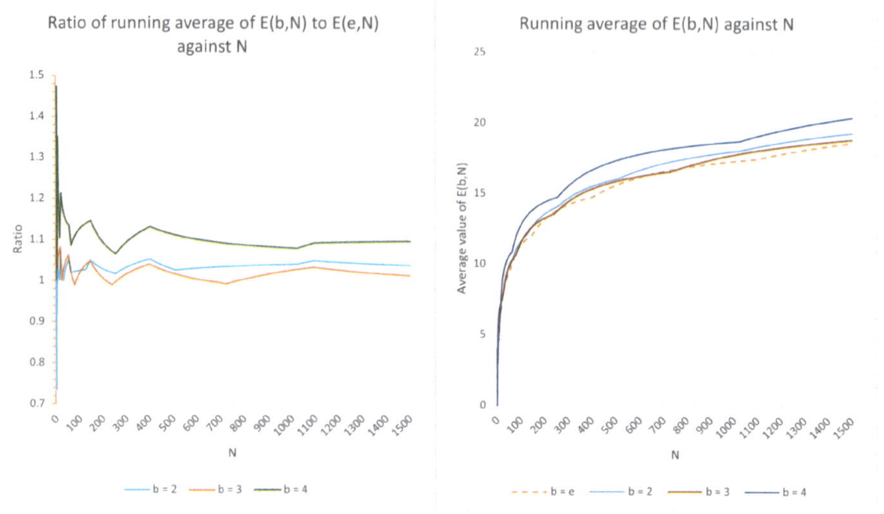

Fig. 6. Running average of $E(b, N)$ against N (Right) and Ratio of running average of $E(b, N)$ to $E(e, N)$ against N (left). Graphs are made with Microsoft Excel.

$E(e, N)$ in the left graph, and the closer something usually is to 1, the better it is.

We see that 3 performs better than 2, especially at large numbers, and is the closest to e in terms of efficiency. Hence, a ternary computer could potentially be more memory efficient than a binary computer, and can save resources.

Now we must tackle the issue of representing a base 3 bit, or a "trit". There are multiple different ways being explored. A potential method could be by making use of optical fibers, one could send 3 different pieces of information. A lack of light could denote 0, while light of 2 different wavelengths could represent 1 and −1. Or, by making use of phase difference ±120°, one can represent ±1 and a lack of light (or other quantities, such as AC current) could represent 0. Alternatively, by using Zener diodes, one can have 2 different threshold voltages in 2 different directions.

6 Conclusion

We introduced weight systems based on powers of natural numbers and Fibonacci numbers, developing a balanced base-f system and deriving key results on the required number of weights. We also explored modifications to the system's degrees of freedom and discussed applications of these weighing systems in ternary computing.

7 Further Study

The techniques shown here may help in optimizing the representation of quantized values or variables, as our study focuses on integer weights. One way to optimize charge measurements would be to use integer multiples of elementary charge, which might be weighted by a "magnetic scale". We can also represent quantum entanglement or quantum algorithms through interdependent outcomes by taking exotic scales into account.

The T-shaped balance enables analysis of rational numbers, and varied quantized lengths allow for fractional masses. The density of our weights in $\mathbb{N}$ could help in constructing a minimally dense weight set, subject to a maximum number of weights $M(n)$ needed to express some number n. We could also apply our negative weights concept to attempt come up with a modified version of Fermat's polygonal number theorem to create a more efficient version.

Acknowledgments. We would like to thank Mr Wang Haibin for his invaluable feedback and contributions to our project, and for mentoring us when we were doing this.

Disclosure of Interests. The authors have no competing interests to declare that are relevant to the content of this article.

Discovery of Waveguide Crossings via a Novel Inverse Design Methodology

Ho Jon Yew(✉) and Ng Chong Jun

Hwa Chong Institution, Singapore 269734, Singapore
jon.yew.ho@gmail.com

Abstract. Photonic microchips hold immense potential for the future of information society, yet waveguide crossings threaten to bottleneck their performance. To optimize the performance of waveguide crossings to minimize losses, inverse design techniques are often used. However, traditional algorithmic inverse design is highly computationally expensive, while often generating less robust geometries. This project proposes and executes a novel hybrid inverse design algorithm that integrates deep learning by using neural networks as a fast forward solver. Additionally, we propose a novel design methodology that generates elegant geometries which are highly robust against fabrication randomness. Our methodology can be applied to any waveguide crossing of crossing angle 90°, and we provide theoretical justifications in addition to empirical validation of it. We trained our predictive neural network model on 2800 data points. The model demonstrates highly accurate forward simulation with a MSE in the order of 10^{-6} to 10^{-7}, while rapidly approximating simulations 5 orders of magnitude faster than exact solving of Maxwell's equations. The final design generated by the inverse algorithm demonstrates a insertion loss of 1.64 dB, with an ultra-compact footprint of only 2.5 μm × 2.5 μm. Future tweaks to the design methodology as well as different settings can increase the subspace of searchable designs for further performance gains.

1 Background

Moore's' law places physical limits on the computational power of traditional integrated circuits (ICs), and these limits are fast approaching [1]. However, electrons are not the only medium by which information can be transmitted. Photons travel at the speed of light without interfering with electromagnetic fields and are orders of magnitude faster.

Consequently, Photonic Integrated Circuits (PICs), which manipulate and transmit these photons, can surpass traditional microchips in bandwidth and data rate, while minimizing power consumption [2]. As such, developments in PICs are on the frontier of science and technology in our information age, with the potential to break existing barriers in information transmission and computational power. In particular, the Silicon-On-Insulator (SOI) platform is of immense academic and industry interest due to its potential in hybrid technology marrying PICs with ICs at the nanoscale. One main reason for this is that mature complementary metal-oxide semiconductor (CMOS)

H. Guo et al. (Eds.): IRC-SET 2024, PTT, pp. 272–281, 2025.
https://doi.org/10.1007/978-981-96-3770-6_28

manufacturing processes used for state-of-the-art ICs can also be used to fabricate SOI devices, giving them high yields and low fabrication costs in the process.

Due to manufacturing constraints of the SOI platform, it is effectively necessary to allow waveguides to intersect and allow light to propagate through [3]. The study of minimizing the losses caused by these intersections of waveguides is the study of waveguide crossings. Current trends demand more and more components on individual PICs, causing them to grow increasingly complex. Additionally, smaller device footprints are increasingly desired while also strongly correlating with data speed. This means that often many waveguide crossings are needed, such as in optical routers [4]. As such, inefficient waveguide crossing designs are a severe bottleneck in the performance of these advanced PICs.

Inverse design is a method of designing devices that seek to algorithmically "work backwards" to discover optimal designs. The "forward" problem is concerned with solving Maxwell's equations to discover the electromagnetic response to a given design (i.e. simulating performance), whereas the "inverse" problem attempts to discover a device geometry that best matches the desired response. The allure of such techniques is that they promise to find high performing solutions to any design problem without the need for manual and error prone human design.

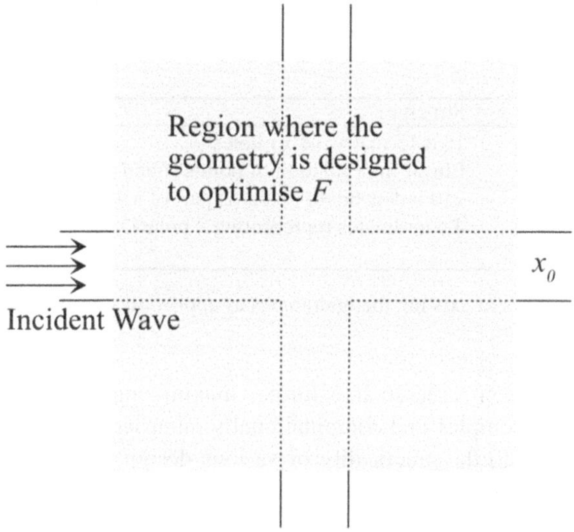

Fig. 1. An illustration of the inverse problem as applied to waveguide crossings. The merit function F in this instance is the field concentration at x0.

The inverse problem cannot be solved by numerically computing a geometry given a certain electric field, because it is unknown whether such a geometry is unique, or even whether it exists. Instead, the inverse problem is approached by looking at the simulated response of geometries, and performing mathematical analysis to iteratively change the geometry till the desired response is achieved (or something sufficiently close to it) [5]. To allow for such iterative changes, the device geometry is expressed as a set of parameters: $S = \{s_1, s_{2,\ldots} s_n\}$. The way in which these parameters describe the device geometry can be highly explicit, e.g. with a parameter set of pixels, or they can be highly implicit, such as the coefficients of a polynomial. The inverse design algorithm then iteratively changes the values for S with the knowledge of their corresponding fields to try and maximizes the "merit function" F. In the case of waveguide crossings, different geometries have different degrees of insertion loss. The inverse problem can be applied to the context of waveguide crossings (Fig. 1) to minimize these losses (maximize F).

However, traditional inverse design is incredibly computationally expensive. Traditional inverse design optimization algorithms may have to access the computational solver of Maxwell's equations thousands of times to hit the stopping criteria [12]. Because exact Finite Difference Time Domain (FDTD) simulations can take 3–5 minutes each on a modern PC [13], even determining whether the desired geometry exists can take many days, making implementation of inverse design in real design scenarios difficult.

However, Traditional inverse design is incredibly computationally expensive. Traditional inverse design optimization algorithms may have to access the computational solver of Maxwell's equations thousands of times to hit the stopping criteria [12]. Because exact Finite Difference Time Domain (FDTD) simulations can take 3–5 minutes each on a modern PC [13], even determining whether the desired geometry exists can take many days, making implementation of inverse design in real design scenarios difficult.

Reference	Structure
Goudarzi 2022 (6)	15 x 15 matrix of air holes
Yu 2019 (7)	Linear interpolation of points (fixed angular displacement)
Chang 2018 (8)	210 independent pixels as part of a 40 x 40 matrix
França-Vieira 2021 (9)	4 coordinates representing 2 points

Fig. 2. Examples of existing methodologies as applied to waveguide crossings.

While there has been research into further maximizing simulated device performance through more complex and computationally intensive techniques (Fig. 2), there is less investigation into the practicality of various design frameworks regarding the constraints of real-world design scenarios, such as manufacturability and computational cost. Additionally, while such methods can discover interesting solutions to nonstandard problems, the strong simulated performance may not necessarily translate to useful fabricated devices due to a higher vulnerability to manufacturing errors and fabrication technique constraints.

Minimizing the computational cost of inverse design techniques without compromising on the performance of the optimized geometry is thus an under-investigated area of research for the progress of inverse design.

2 Hypothesis

There are two aspects to our novel inverse design methodology. Firstly, to reduce computational cost by orders of magnitude, we propose a novel optimization algorithm that makes use of deep learning techniques to approximate simulation results in a fraction of the time. We believe that a neural network trained on simulation data will be able to gain an "understanding" of Maxwells equations in the highly limited scope of the simulation subspace, which can then be used to predict the performance of given geometries without the need for computationally expensive exact simulations. Secondly, to create highly robust, fabricable, and compact waveguide crossing designs, we propose a novel method of parameterizing geometry by use of implicit description of parametric curves, which is to our best knowledge the first of its kind for waveguide crossings. This design methodology is immediately adaptable to every two-way waveguide crossing of crossing angle 90°, and can be principally extended to any type of nanophotonic inverse design problem. We provide a novel theoretical basis for this methodology, before seeking to verify its performance gains through empirical simulation.

3 Methods and Results

Existing inverse design techniques as applied to photonics typically concern themselves with constructing geometries out of pixels, lines and polygons (Fig. 2). We diverge from this with a desire to use curves (polynomials) to build the geometry of our waveguide crossing. This is because the geometry of curves is comparatively easier to fabricate and are highly robust against fabrication variance, allowing highly compact waveguide crossings to be discovered. However, ordinary polynomials are challenging to use in the case of photonic devices, because when one sets a range for them (to bound the geometry), it is tricky to select the start and end points to ensure that the waveguide crossing joins cleanly with the waveguides[1].

For those reasons, we propose a new design methodology of waveguide crossings which makes use of Bezier curves. Bezier Curves are a form of Bernstein polynomial that sees widespread application in robotics and visual arts. It has a formal mathematical definition [12]:

$$B(t) = \sum_{k=0}^{n} \binom{n}{k}(1-t)^{n-k} t^k P_k, \quad t \in [0, 1]$$

However, it is most easily understood by a geometric description of De Casteljau's algorithm [13]: However, it is most easily understood by a geometric description of De Casteljau's algorithm [13]: Consider points $P_1, P_2, \ldots, P_n$. Connect these points with lines to get the control polygon. Find the point along each line that subdivides it in the ratio t: $(1-t)$. Connect these points with lines to get a new polygon with one less segment.

Repeating this process till there is a single point left gives the point on the curve for that value of t.

[1] It is possible via numerical analysis techniques to generate unique polynomials of degree n given $n + 1$ distinct points. However, those methods are difficult, and not the focus of our research.

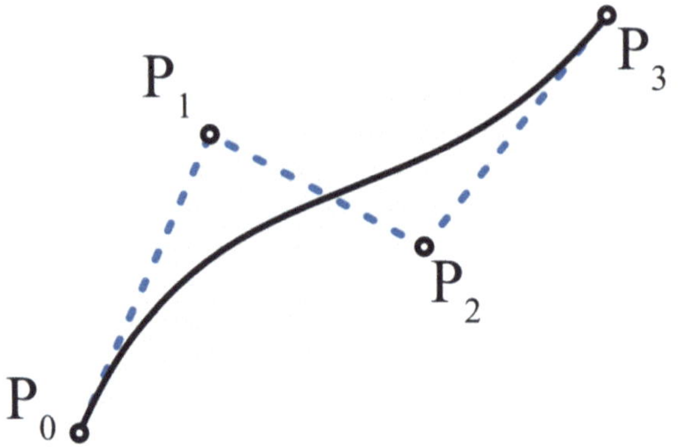

Fig. 3. A Bezier curve (black) directed by the control points P_0, P_1, P_2, P_3. The control polygon (dotted, blue) and their convex hull (gray) are also shown.

3.1 Analysis of properties

1. Bezier Curves allow for a wider diversity of possible geometries to be embedded within a smaller number of variable parameters. This is because Bezier curves are not locally controlled. Compared to conventional methods like linear interpolation whereby changing a single control point only affects a small part of the device geometry, changing any control point of a Bezier curve will modify the entire curve [14].
2. The curve is bounded, beginning at P_0 andends at P_n. We make clever use of this property in our experimental design to prevent self-intersections[2] such that the device geometry is cleanly expressed by polynomials without the need for additional conditions.
3. The curve is a subset of the convex hull[3] of its control points (Fig. 3). This guarantees that: in a setup where linear interpolation of points (the control polygon) does not generate a self-crossing curve, the geometry generated by a Bezier curve with those points as control points will also avoid self-intersection.

Due to symmetries in waveguide crossings of crossing angle 90°, only 1/8 of the variable structure needs to be optimized, which is then rotated and flipped to complete the device [7]. The geometry of the waveguide crossing is drawn on the cartesian plane. We establish 2 variables that control the device specifications, as well as a single variable to control the inverse design process.

For the device specifications, we set L as the length of the square box within which the inverse design process occurs, which is centered on the intersection of the waveguides. w is the width of the intersecting waveguides, which is taken to be the same for both

[2] Self-intersecting geometry makes the process of generating geometry r simulation significantly more difficult. We also theorize it to be less easily fabricable in an accurate manner due to the presence of sharp corners requiring infinitesimal accuracy.

[3] An intuitive understanding of a convex hull: If each control point is a nail on a board, the shape created by a rubber band snapped to surround them all is the convex hull.

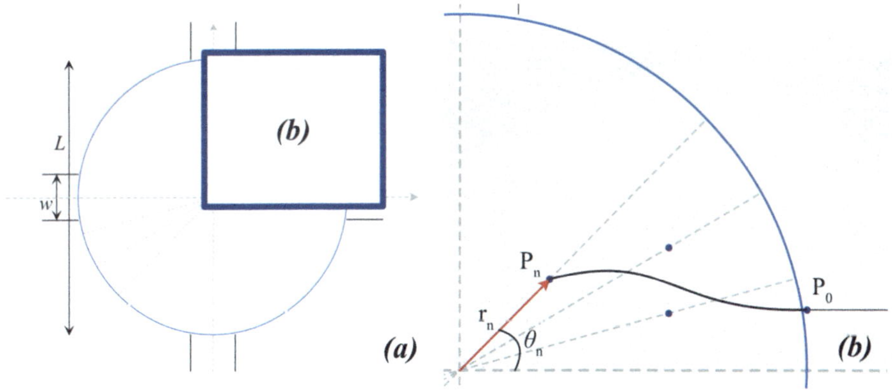

Fig. 4. The blue circle has a radius of $\frac{L}{2}$ and illustrates the bounds on the positions of the variable control points.

the horizontal and vertical propagation. The depth of the waveguide crossing does not affect the design setup and is taken to be the standard value of 0.22 μm (220 *n*m). For experimental purposes, we use standard values of $L = 3$ μm and $w = 0.5$ μm. To control the inverse design process, we set n to be the number of *variable* control points of the Bezier curve (1 less than the total number of control points used to draw the arm). The very first, "fixed" control point P_0 is given coordinates $\left(\frac{L}{2}, \frac{w}{2}\right)$ such that it lies at the point where the variable geometry and the walls of a waveguide meet. If both the x and y coordinates of the control points are variable, certain values for these coordinates would result in a self-intersecting curve used to generate the geometry, which is undesirable.[ii] As such, we choose to fix the angular displacement of the points.

The following mathematical descriptions use polar coordinates, with the center of the waveguide crossing as the pole, and the reference direction parallel to a waveguide. For each point P_k where $1 \leq k \leq n$, we set its angular displacement from the reference direction θ_k to be $\frac{\pi k}{4n}$ In effect, this means that the last variable control point P_k always has $\theta_n = \frac{\pi}{4}$, , so that it lies on the line y = x (Fig. 4b). This also means that there is a constant angular separation of $\frac{\pi k}{4}$ between the variable control points. Self-intersections are thus prevented at the cost of a smaller subspace of possible geometries, but we also manage to halve the size of the parameter set to $|S| = n$. Every variable control point P_k has a radial coordinate r_k, which together with the angular coordinate θ_k, describes the position of each control point (Fig. 4b). We let $r_k = \frac{L}{2} \times t_k$ where $0 \leq t_k \leq 1$. This constrains the position of the control points such that they are bounded to a circle of radius, keeping the device compact. The values of t for the variable control points are allowed to vary, so they become the set of parameters: $S = \{t_1, t_2, \ldots, t_n\}$ (Fig. 5).

These mathematical formulations keep our methodology dynamic, and adaptable to any device constraints of L and w, as well as any number of control points n. The algorithmic generation of the geometry is thus as follows: Given constraints L, w, and n, the polar coordinates are determined for a chosen parameter set. These polar coordinates are converted into cartesian coordinates. We then use De Casteljau's algorithm [13] to generate the Bezier curve using a sufficiently large step size (Fig. 4b, in black). The curve

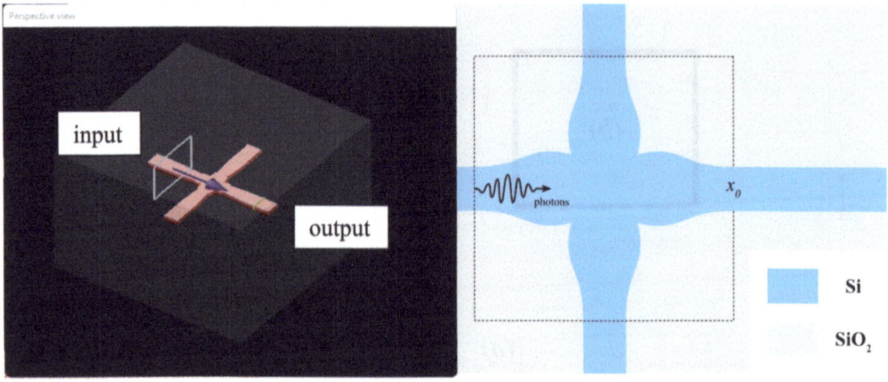

Fig. 5. (a) Schematic of waveguide crossing generated in Ansys Lumerical FDTD (b) Top view of waveguide crossing generated using the control points in Fig. 4b.

is then rotated and flipped to complete the other "arms" of the waveguide crossing. The convex hull of these curves is the same material as the waveguides (typically Si), and the substrate is SiO_2 (**Error! Reference source not found.**).

Next, we create a deep learning neural network to perform deep learning based inverse design. We choose there to be 3 variable control points, so $|S| = n = 3$. Firstly, we randomly generate values of t_1, t_2, t_3 and generate geometry. Single mode source of light, with wavelength ranging from 1450 nm to 1650 nm, is allowed to propagate horizontally through the waveguide crossing (**Error! Reference source not found**). We use the Finite.

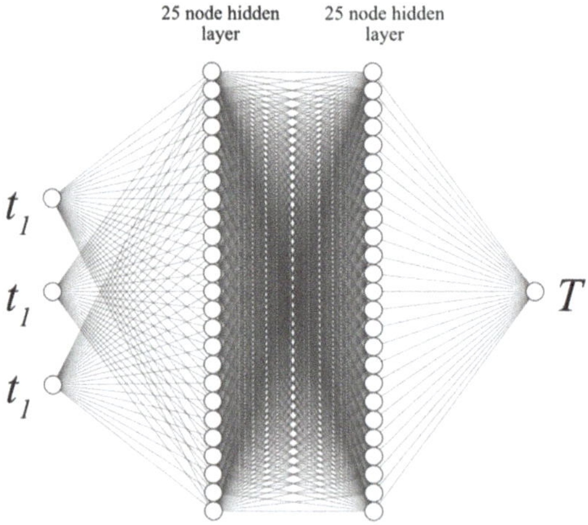

Fig. 6. Visual Representation of the forward model

Difference Time Domain (FDTD) method to compute Maxwells' equations, which allows us to determine the transmission on the other side of the waveguide crossing. We then interpolate the transmission at 1550 nm to compute our output, the transmission T, which is associated with the corresponding values in the parameter set. This process was then repeated 2800 times to get a dataset of 2800 geometries with their transmissions.

The data is used to train a fully connected neural network with 2 training layers of 25 nodes, to create a deep learning based forward model (Fig. 6). Compared to typical exact FDTD simulations, which can take 3–5 minutes [15] the machine learning forward model predicted the transmissions of given geometries in a fraction of a second (0.004 s when ran on MATLAB Online, approximately 60 thousand times faster. It achieved this with a Mean Squared Error of only 1.50×10^{-6}.

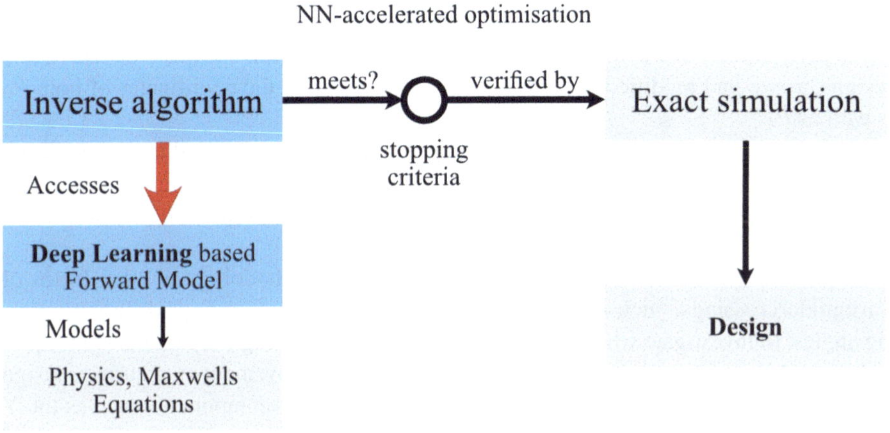

Fig. 7. A diagram illustrating the inverse design framework.

This is significant as traditionally, meeting the stopping criteria of an inverse design algorithm involves the inverse algorithm accessing the simulation model thousands of times, causing inverse design to often take many days. We term this aspect of inverse design the "critical juncture" when it comes to computational speed. Our use of a neural network instead of numerical calculation (Fig. 7) allows us to improve the performance of the critical juncture significantly.

Inverse algorithm type	Inverse designed determined radial coordinates of control points, in μm		
	r_1 (outermost)	r''	$r_,$
Pattern search	0.9384	0.2500	0.2500
Particle swarm	0.9384	0.2500	0.2500

Fig. 8. For a waveguide crossing whereby $L = 2.5$ μm and $w = 0.5$ μm. These were derived from the parameter set by the earlier equations.

For the inverse algorithm, we explored the pattern search and particle swarm algorithms. The merit function F is set to minimize the difference between transmission of the design of the optimizer T and the complete transmission, $1-T$. Two different

Inverse design algorithms determined identical values of the parameter set which best maximized F (Fig. 8), suggesting that the geometry obtained is near optimal for the searchable subspace.

4 Conclusion

The resulting geometry is hyper compact, occupying a mere 2.5 μm × 2.5 μm for waveguides of width 0.5 μm, which is important for future development of highly complex PICs which have many different components onboard. The inverse designed was simulated via FDTD to give a transmission value of 0.685659, which gives an insertion loss of 1.64 dB. This is not optimal compared to designs in other papers [7]. We theorize the worse performance to be due to a restricted and poorly chosen subspace of searchable designs, as we had modified our design methodology to fix the coordinates of both P_0 as well as P_1.

Future research can investigate further performance gains by tweaking our methodology. This could be done by increasing the subspace of searchable designs, for example by enabling self-intersections through unbounded Cartesian coordinates or by increasing the number of control points (we only used 3).

Additionally, further research can apply our design methodology to other forms of waveguide crossings, such as non-planar, multimode, and ones with different crossing angles, to investigate whether constructing geometries using polynomial techniques inclusive of Bezier curves might represent an iterative improvement over current design philosophy. The base principles can also be applied to other components such as the Y junction or waveguide bends. Our use of a Deep Learning – Inverse Optimization Hybrid technique proves its superior computational performance in the case of waveguide crossings. Thus, this project provides more empirical backing to an ongoing paradigm shift within inverse design towards the integration of neural network techniques.

References

1. Waldrop, M.M.: The chips are down for Moore's law. Nature **530**, 7589 (2016)
2. Doylend, J.K., Knights, A.P.: The evolution of silicon photonics as an enabling technology for optical interconnection. Laser Photonics Rev. **6**(1), 4 (2012)
3. Bogaerts, W., Fiers, M., Dumon, P.: Design challenges in silicon photonics. IEEE J. Sel. Top. Quantum Electron. (2014)
4. Hu, T., et al.: Wavelength-selective 4×4 nonblocking silicon optical router for networks-on-chip. Opt. Lett. **36**, 23 (2011)
5. Owen, D.M.: Photonic design: from fundamental solar cell physics to computational inverse design. [Dissertation]. Berkeley, California, United States of America: Electrical Engineering and Computer Scienced Dept, UC Berkeley (2012)
6. Goudarzi, K., Lee, M.: Inverse design of a binary waveguide crossing by the particle swarm optimization algorithm. Pohang Results Phys. ISSN 2211–3797 (2022)
7. Yu, Z., Feng, A., Xi, X., Sun, X.: Inverse-designed low-loss and wideband polarization-insensitive silicon waveguide crossing. Hong Kong Opt. Lett. **44**, 1 (2019)
8. Chang, W., et al.: Ultracompact dual-mode waveguide crossing based on subwavelength multimode- interference couplers. Wuhan Photonics Res. **6**, 660–665 (2018)

9. França-Vieira, L., Rodríguez-Esquerre, V.F., Rubio-Mercedes, C.E.: Waveguide crossing design using particle optimization algorithm. Proc. Spie (2020)
10. Xu, K., et al.: Integrated photonic power divider with arbitrary power ratios. Opt. Lett. **42**, 855–858 (2017)
11. Wu, S., Mu, X., Cheng, L., Mao, S., Fu, H.Y.: State-of-the-art and perspectives on silicon waveguide crossings: a review. Shenzhen: MDPI (2020)
12. Michiel, H.: Encyclopaedia of Mathematics: Supplement. Springer Science and Business Media Vol 1, p. 119 (1997)
13. Pomax: A primer on bézier curves. https://pomax.github.io/bezierinfo/#decasteljau
14. Agoston, M.K.: Computer graphics and geometric modelling: implementation and algorithms. Springer Science and Business Media p. 404 (2005)

A Study of Attention in Adolescents Using Functional Near Infrared Spectroscopy

Hong Ming Hor, Jireh John Kwek, and Yann Ting Lim(✉)

School of Mathematics and Science, National University of Singapore High, 20 Clementi Avenue 1, Singapore 129957, Singapore
yannting06@gmail.com

Abstract. Currently, Attention-deficit/Hyperactivity disorder (ADHD) is diagnosed by a clinical physician by going through a physical examination and a series of interview with the patient and significant people around them based on the DSM-5. This is time consuming as it typically takes up to 3 h and it is not always objective. Our research investigates the possibility of utilizing a portable functional Near-Infrared Spectroscopy (fNIRS) device known as the NIRSIT-Lite as an objective and convenient tool for detecting ADHD in adolescents. The portable fNIRS device measures oxygenated and deoxygenated haemoglobin concentrations (HbO_2 levels), utilizing signal processing to discern neurophysiological patterns found in the prefrontal cortexes of the adolescents. Through a Go/No-go (GNG) paradigm and Verbal Fluency Task (VFT), our study focuses on real-time responses in the prefrontal cortex during cognitive tasks. Comparative analyses between ADHD and control groups aim to establish the device's sensitivity and specificity and assess its ability to be used to potentially identify individuals with ADHD. Our results show that it is indeed true (as shown by the overall $p < 0.05$) and that using NIRSIT-Lite while performing the Go/No-go task and Verbal Fluency Task can be an effective and accurate method to diagnose ADHD in adolescents. Our results also show that high changes in HbO_2 levels indicate healthy adolescents while low changes indicate that an adolescent may be diagnosed with ADHD. These results could revolutionize ADHD diagnostics by providing a non-invasive, portable, fast, and objective method for early detection.

Keywords: ADHD · diagnosis · fNIRS · GNG · VFT

1 Background and Objective

1.1 Background Context

Attention-deficit/hyperactivity disorder (ADHD) is a disorder where an individual exhibits trends of inattention and/or hyperactivity-impulsivity that interferes with functioning or development [1]. Symptoms of ADHD include severe inattention, unfocused motor activity, and impulsivity which interferes with the ability to have daily social interactions. It is a commonly diagnosed disorder in children and is noticeable as early as 3 years old [2].

H. Guo et al. (Eds.): IRC-SET 2024, PTT, pp. 282–293, 2025.
https://doi.org/10.1007/978-981-96-3770-6_29

Currently, ADHD is mainly diagnosed through interviews and questionnaires given to individuals and the people around them about the diagnosing individual's attention spans. There are no direct tests to diagnose ADHD [3]. Hence, one downside of the current way of diagnosis is that under or overdiagnosis may happen. This is because the properties of ADHD (hyperactivity, inattentiveness and impulsiveness) exist on a continuum with normal behaviour and the diagnosis is heavily reliant on human perception. Moreover, the Diagnostic and Statistical Manual of Mental Disorders-5 (DSM-5) diagnostic criteria for ADHD have been gradually lowered and comorbid learning and psychiatric disorders being mistaken for ADHD [4]. Hence, we are researching on a more effective and accurate method of diagnosing ADHD, specifically in adolescents aged 13 to 18.

1.2 FNIRS

Functional Near-Infrared Spectroscopy (fNIRS) is a non-invasive neuroimaging technology that measures changes in blood (hemoglobin) oxygenation and blood volume in the brain. This technique is used to monitor brain activity by detecting the hemodynamic response associated with neural activity [5]. The NIRSIT-lite device is a lightweight neuroimaging system that uses fNIRS technology that scans the prefrontal cortex. It measures variations in cerebral hemodynamics by radiating two wavelengths of near-infrared light into the cerebral cortex and reabsorbing them. (780nm and 850nm). Our research utilizes NIRSIT-lite to measure the oxygen level (HbO_2) in the prefrontal cortex and study whether there are trends of different HbO_2 levels in ADHD diagnosed and non-ADHD diagnosed adolescents. This could hence be a potential tool for diagnosing ADHD.

Thus, our research aims to ascertain the difference in changes of HbO2 levels of ADHD and non-ADHD diagnosed adolescents. By using the NIRSIT-Lite device to scan adolescents' brains while administering the VFT and GNG tasks, we can attain an effective and objective method of ADHD diagnosis.

2 Literature Review

2.1 Usage of FNIRS

One common pharmacological treatment for ADHD is the psychostimulant methylphenidate (MPH). This literature review explores research done on the use of functional Near-Infrared Spectroscopy (fNIRS) as a neuroimaging tool to monitor the acute effects of MPH on cerebral hemodynamics in ADHD children.

A study by Monden et al. (2012) aimed to explore the acute effects of MPH on ADHD children using fNIRS during a go/no-go task [6]. The findings indicated a significant increase in oxygenated haemoglobin in the right lateral prefrontal cortex post-MPH intake, correlating with improved task performance. This suggests that fNIRS can serve as a valuable tool to monitor the neural effects of MPH in real-time. Subsequent studies reported altered hemodynamics in the prefrontal cortex of ADHD individuals during various cognitive tasks, emphasizing the potential of fNIRS to capture neural correlates of ADHD.

One notable advantage highlighted is the feasibility of performing fNIRS assessments within a 3-h hospital stay during a single visit. This ease of use, together with its objective nature, makes fNIRS a practical option for clinical settings, providing a potential avenue for implementing neuroimaging techniques in routine ADHD assessments.

While fNIRS shows promise, further research is needed to establish its reliability and validity in monitoring ADHD treatments. In conclusion, the integration of fNIRS in monitoring the acute effects of MPH on cerebral hemodynamics in ADHD children represents a novel and clinically relevant approach. The portability and ease of use of fNIRS make it a practical tool for real-time assessment, paving the way for enhanced understanding and management of ADHD through neuroimaging techniques.

3 Hypothesis

3.1 Changes in HbO_2 Levels

Our hypothesis statement is that there will be high changes of HbO_2 levels in non-ADHD diagnosed adolescents and low changes of HbO_2 levels in ADHD-diagnosed adolescents when performing the tasks. This is because higher oxygen levels in the prefrontal cortex suggest higher activity level. ADHD-diagnosed adolescents may exhibit lower changes in activity levels in the prefrontal cortex due to differences in neural regulation, attentional control, and inhibitory processes (known to be impaired in individuals with ADHD). These factors could contribute to a diminished capacity for the expected increase in oxygen levels during cognitive tasks compared to their non-ADHD diagnosed counterparts.

4 Methodology

4.1 Data Sampling

45 adolescent participants (29 ADHD diagnosed 16 non-ADHD diagnosed) aged 13–18 are involved in this experiment. (Full table of demographic characteristics is in Appendix).

4.2 Questionnaire

Participants are first instructed to complete a questionnaire. The questionnaire consists of questions about their demographic details (i.e. age, grade, gender, housing type) as well as questions about attention and activity level based on the DSM-5 for diagnosing ADHD. The questions are categorised into 18 sections shown in the appendix. Each question is rated with a score on a scale of 0 to 3 (0 = none, 1 = mild, 2 = moderate, 3 = severe). This questionnaire acts as a control to compare with the results of NIRSIT-lite to check for consistency within ADHD diagnosed adolescents. (Refer to appendix A1 for the questionnaire).

Based on the questionnaire, an adolescent should be diagnosed with ADHD if the score is more than or equal to 18. Otherwise, the adolescent should be a healthy non-ADHD control.

Table 1. Demographic characteristics (N = 45)

Demographic data	Number (%)/Mean ± SD		
	ADHD (N = 29)	Healthy Control (N = 16)	P-value
Age	16.34 ± 1.758	15.75 ± 1.69	0.277
13	2 (6.9)	2 (12.5)	
14	5 (17.2)	3 (18.7)	
15	2 (6.9)	2 (12.5)	
16	3 (10.4)	0 (0)	
17	6 (20.7)	8 (50)	
18	11 (37.9)	1 (6.3)	
Gender (Female)	17 (58.6)	5 (31.1)	0.082
Current school year			
Secondary 1/Year 1	2 (6.9)	2 (12.5)	0.188
Secondary 2/Year 2	5 (17.2)	3 (18.7)	
Secondary 3/Year 3	2 (6.9)	2 (12.5)	
Secondary 4/Year 4	3 (10.4)	0 (0)	
Junior College 1/Year 5	7 (24.1)	8 (50)	
Junior College 2/Year 6/Poly Year 2	8 (27.6)	1 (6.3)	
Private Candidate/Diploma	2 (6.9)	0 (0)	
PSLE score			
Old scoring	244.86 ± 22.09	256.9 ± 12.13	0.119
AL scoring	10 ± 6.48	8.60 ± 5.08	0.704

* $p < 0.05$; ** $p < 0.01$; *** $p < 0.001$.

4.3 Go/No-Go Task and Verbal Fluency Test with FNIRS

Afterwards, participants complete two tasks with an fNIRS (functional near-infrared spectroscopy) device on for neuroimaging. The tasks consist of a go/no-go paradigm and a Verbal Fluency Test:

For the go/no-go paradigm, fruits are shown one at a time (apple, banana, or orange) and participants are to click the spacebar (to represent go) whenever they see an apple or banana, but not click the spacebar (to represent no-go) when they see an orange. Through this, we can understand their ability to inhibit a response (pressing the spacebar) when faced with a specific stimulus.

As for the Verbal Fluency Test, participants first repeat ABCDE for 1 min, before we show them different letters on the computer screen in front of them. Participants will then have to say, out loud, as many words as they can think of that start with that letter (e.g. D for duck, dog, dark et cetera) They then repeat ABCDE for another minute. This task explores the semantic memory of participants' brains (Fig. 1).

Fig. 1. Picture of NIRSIT-Lite device that is low cost and portable [7]

Table 2. Comparison of Questionnaire scores and HbO_2 levels (N = 45)

Questionnaire and HbO_2 levels	Mean ± SD		
	ADHD (N = 29)	Healthy Control (N = 16)	P-value
Total score	37.79 ± 6.31	14.00 ± 4.13	<0.001***
Inattentive symptoms	20.79 ± 3.82	8.25 ± 2.70	<0.001***
Hyperactivity symptoms	17.00 ± 3.54	5.75 ± 2.54	<0.001***
HbO_2 Levels			
Go no Go task	−3.48 ± 0.06	1.10 ± 0.02	0.006**
VFT	−8.94 ± 0.13	2.25 ± 0.06	0.007**

* $p < 0.05$; ** $p < 0.01$; *** $p < 0.001$.

5 Results

5.1 Results Collected

From Table 2, we can observe that the total mean score for questionnaire for ADHD diagnosed adolescents is 37.79 ± 6.31 which is much higher than 18 (benchmark for being diagnosed ADHD based on DSM-5). On the other hand, healthy controls have a total mean score of 14 ± 4.13 which is lesser than 18, corresponding to a result of no ADHD. As for the HbO_2 levels measured by NIRSIT-Lite, our findings show that the change in HbO_2 levels for ADHD diagnosed are much lower as compared to the healthy controls.

From Table 1, our results show that the number of adolescents diagnosed with ADHD seemed to increase with age and that ADHD-diagnosed students Primary School Leaving Examination (PSLE) score seemed to be lower than their non-ADHD counterparts. Other demographic aspects yield no significant findings.

Comparing Figs. 2, 3, 4 and 5, we can visualise a significant increase in the HbO_2 levels in the average values healthy control as compared to those with ADHD (especially in nodes 6 to 9). This corresponds to an increase in oxygen supplied to the prefrontal cortex when performing the Go/No-go task and Verbal Fluency Test respectively.

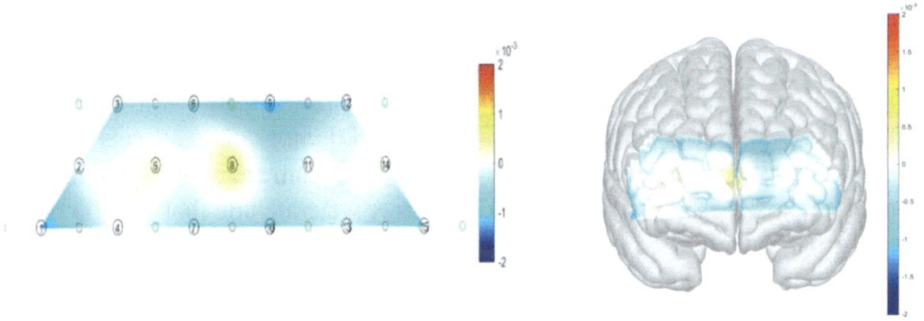

Fig. 2. a: 2D & 3D plot of average brain activity of ADHD individuals during GNG task

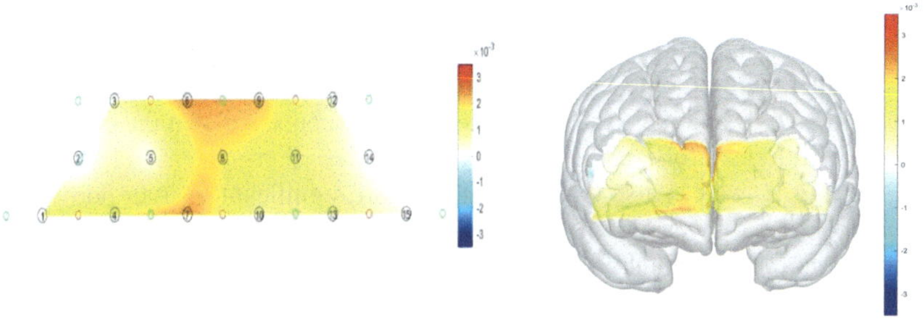

Fig. 3. b: 2D &3D plot of average brain activity of Healthy Controls during GNG task

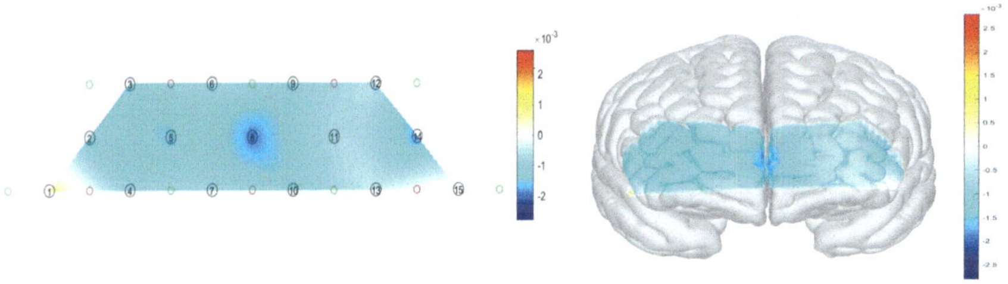

Fig. 4. a: 2D & 3D plot of average brain activity in ADHD individuals during the VFT task

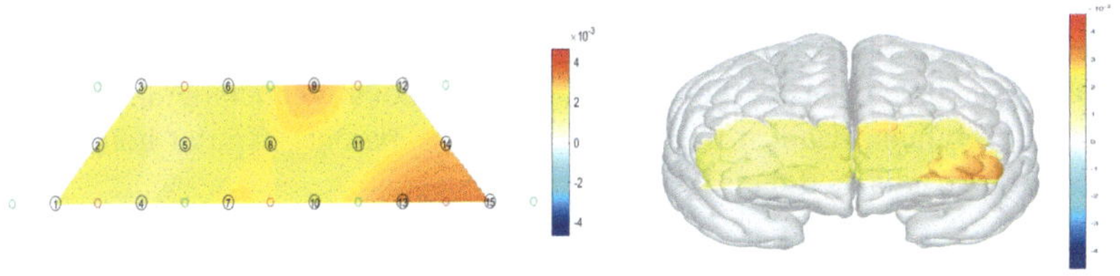

Fig. 5. 2D & 3D plot of average brain activity in Healthy Controls during the VFT task

5.2 Discussion and Analysis

For our analysis, we used a significance level of P-value = 0.05 for our data analysis as it is a commonly used and widely accepted standard.

From Table 1, since the calculated P-values for all demographic aspects are greater than 0.05, it is likely that these results differ due to chance and are not affected by ADHD and cannot be taken to hold any significance. Hence, we conclude that ADHD does not affect demographic factors.

As for Table 2, the calculated P-value of the questionnaire data is <0.001 (<0.05) and suggests that the questionnaire method is significant in diagnosing ADHD. From the values, we also conclude that higher changes in HbO_2 levels suggest that the adolescents do not have ADHD. (Changes in HbO_2 levels while performing Go/No-go task and Verbal Fluency Test are significantly higher in healthy controls as compared to ADHD diagnosed adolescents.) We can also see that the calculated P-values are 0.006 for the Go/No-go task and 0.007 for the Verbal Fluency Test. As both P-values < 0.05, we conclude that there is some significant difference in change of HbO_2 levels (not due to chance) in ADHD and non-ADHD diagnosed adolescents, and that the change in HbO_2 levels in healthy controls are significantly higher.

From Fig. 2, we can also observe that there are significantly larger portions of blue region, as scanned by using NIRSIT-Lite, indicating little to no changes in HbO_2 levels negatively in the prefrontal cortex, as compared to Fig. 3 where there are more yellow and orange regions in the prefrontal cortex indicating high positive changes in HbO_2 levels in healthy controls. Hence, our hypothesis is also proven to be correct.

From the results, we conclude that we can predict if a person has ADHD based on the changes in HbO_2 level measured by the NIRSIT-Lite device. Hence, the method of using NIRSIT-Lite to monitor HbO_2 levels in the prefrontal cortex while performing the Go/No-go task and Verbal Fluency Test is effective in diagnosing ADHD.

6 Conclusion

6.1 Our Conclusion

In conclusion, our research shows that using NIRSIT-Lite while performing the Go/No-go task and Verbal Fluency Test to scan the prefrontal cortex to monitor HbO_2 levels is an effective way of diagnosing ADHD in adolescents. It is seen by our results in how low negative changes in HbO_2 levels in the prefrontal cortex of an adolescent indicate ADHD, while high positive changes in HbO_2 levels in the prefrontal cortex of an adolescent indicates that the adolescent is healthy.

We also conclude from our data that ADHD does not affect demographics including age, gender, dominant hand, if they are staying with parents, type of housing, school year, academic results, and their PSLE scores.

Appendix

A1 Questionnaire

Subject ID: __________

Part A: Demographics

A1. What is your age? __________ Years

A2. What is your gender? Male or Female (Please circle)

A3. Which side is your dominant hand? Left or Right (Please circle)

A4. Do you stay with your parents? Yes or No (Please circle)

A5. Which type of housing do you stay: (Please circle)

2-3 room HDB Flat	4-5 room HDB Flat/Executive HDB flat)	Private Condominium	Landed property

A6. What is your current school year (e.g. Secondary 1)?
____

A7. What is your current academic result? (Please circle)

Very Poor	Poor	Average	Good	Excellent

A8. What was your Primary School Leaving Examination (PSLE) score?
____

Part B: The following questionnaire will be administered by the principal investigator

	None	Mild	Moderate	Severe
B1. Carelessness	0	1	2	3
Do you make a lot of mistakes in school? Is this because you're careless? Do you rush through work or activities? Do you have trouble with detailed work? Do you not check your work? Do people complain that you're careless? Are you messy or sloppy? Is your desk, locker at school, or workspace so messy that you have difficulty finding things? Do your parents have to repetitively ask you to clean your room?				
B2. Difficulty sustaining attention in activities	0	1	2	3

(continued)

(*continued*)

	None	Mild	Moderate	Severe
Do you have trouble paying attention when watching movies, reading, or attending lectures? Or on fun activities such as sports or board games? Is it hard for you to keep your mind on school? Do you have trouble paying attention when instant messaging?				
B3. Doesn't listen	0	1	2	3
Do people (your teachers, parents, friends) complain that you don't seem to listen or respond (or daydream when spoken to or when asked to do tasks? A lot? Do people (your friends or parents) have to repeat directions? Do you have trouble remembering driving directions unless you write them down? Do you find that you miss the key parts of conversations because of drifting off in your own thoughts? Does it cause problems?				
B4. No follow through	0	1	2	3
Do you have trouble finishing things (such as school project or chores)? Do your parents have to remind you to complete things? Is this a problem? When working on a group project at school, do your peers have to remind you to finish your part of the project? Do you often leave things half-done and start another project? Do you need consequences (such as deadlines) to finish? Do you have trouble following instructions? (Especially complex, multistep instructions that have to be done in a certain order with different steps) Is it a problem when you are putting on your make-up or putting together a model? Do you need to write down instructions, otherwise you will forget them?				
B5. Can't organize	0	1	2	3
Do you have trouble organizing tasks into ordered steps? Is it hard prioritizing work and chores? Do you need others to plan for you? Do you have trouble with time management? Does it cause problems? Do you have trouble planning long-term projects at school, like term papers or large lab reports? Do your parents have to help you plan? Does difficulty in planning lead to procrastination and putting off tasks until the last moment possible?				
B6. Avoid/dislike s tasks requiring sustained mental effort	0	1	2	3
Do you avoid tasks (work, chores, reading, board games) that are challenging or lengthy because it's hard to stay focused on these things for a long time? Do you have to force yourself to do these tasks? How hard is (was) it? Do you procrastinate and put off tasks until the last moment possible?				
B7. Loses important items	0	1	2	3
Do you lose things (e.g., important work papers, homework, your school planner, keys, wallet, coats, etc.)? Have you misplaced things such as your cell phone or school planner? A lot? More than others? Are you constantly looking for important items? Do you get into trouble for this (work, school, home)? Do you need to put items (e.g., glasses, wallet, keys) in the same place each time, otherwise you will lose them?				
B8. Easily distractible	0	1	2	3
Are you ever very easily distracted by events around you such as noise (conversation, tv, radio), movement, or clutter? When you are in the classroom and there is interruption, is it hard to get your mind back on the work at hand? Do you need relative isolation to get work done? Can almost anything get your mind off of what you are (were) doing (like work, chores, or if you're talking to someone)? Is it hard to get back to a task, like doing homework, once you stop?				
B9. Forgetful in daily activities	0	1	2	3

(*continued*)

(*continued*)

	None	Mild	Moderate	Severe
Do you forget a lot of things in your daily routine? Like what? Attending a class? Chores? Work? Appointments or obligations? Meetings with friends? Do you forget to bring things to school or work, such as school or work materials or assignments due that day? Do you need to write regular reminders to yourself to do most activities or tasks, otherwise you will forget? If you don't write something in your school planner, are you likely to forget to do it?				
B10. Squirms or fidgets	**0**	**1**	**2**	**3**
Can you sit still or are you always moving your hands or feet, or fidgeting in your chair? Do you tap your pencil or your feet? A lot? Do people notice? Do you regularly play with your hair or clothing? Do you consciously resist fidgeting or squirming? Do the teachers comment on your fidgeting or squirming at school? Do your parents comment on your fidgeting or squirming at the dinner table? Is it difficult for you when you cannot be physically active?				
B11. Can't stay seated	**0**	**1**	**2**	**3**
Do you have trouble staying in your seat? At work? In class? At home (e.g., watching tv, eating dinner)? In church or temple? When playing video games? Do you choose to walk around rather than sit? Do you force yourself to remain seated? Is it difficult for you to sit through a long meeting, class, or lecture? Do teachers ever comment on your inability to remain seated? Do you try to avoid going to functions that require you to sit still for long periods of time?				
B12. Runs/climbs excessively	**0**	**1**	**2**	**3**
Are you physically restless? Do you feel restless inside? A lot? Do you feel more agitated when you cannot exercise on an almost daily basis? Do you have trouble staying seated at the table through an entire meal?				
B13. Can't play/work quietly	**0**	**1**	**2**	**3**
Do you have a hard time playing quietly? During leisure activity (nonstructured times or on your own, such as reading a book, surfing the Internet, listening to music, playing a board game) are you agitated or unhappy? Do you always need to be busy after school, work, or when your family goes on vacation?				
B14. On the go, "driven by a motor"	**0**	**1**	**2**	**3**
Is it hard for you to slow down and do one thing for a long period of time? Do you feel like you (often) have a lot of energy and that you always have to be moving, are always "on the go"? Do you feel like you're driven by a motor? Do you feel unable to relax? Does it interfere with your ability to fall asleep at night?				
B15. Talks excessively	**0**	**1**	**2**	**3**
Do you talk a lot? All the time? More than other people? Do you talk out of turn in class? Do people (Parents, teachers, friends) complain about your talking? Is it a problem? Are you often louder than the other people you are talking to?				
B16. Blurts out answers	**0**	**1**	**2**	**3**
Do you give answers to questions before someone finishes asking? Do your teachers complain about answering their questions prematurely? Do you say things before it is your turn? Do you say things that don't fit into the conversation? With your parents? Friends? Classmates? Do you do things without thinking? A lot?				
B17. Can't wait for turn	**0**	**1**	**2**	**3**
Is it hard for you to wait your turn in conversation, when playing games, in lines (like the lunchroom), or while driving? Are you frequently frustrated with delays? Do your parents comment on your trouble waiting? Does it cause problems? Do you put a great deal of effort into planning to not be in situations where you might have to wait?				
B18. Intrudes or interrupts others	**0**	**1**	**2**	**3**

(*continued*)

(*continued*)

	None	Mild	Moderate	Severe
Do you talk when others are talking without waiting until you are acknowledged? Do you butt into others' (your parents', friends', or siblings'') conversation before being invited? Do you interrupt others' activities when they are busy? Do you have trouble waiting for your parents to finish something they are doing before asking them a question? Is it hard for you to wait to get your point across in conversations or at meetings?				

B1 Demographics Characteristics of Data Sample

See Table 3.

Table 3. Demographic characteristics (N = 45)

Demographic data	Number (%)/Mean ± SD		
	ADHD (N = 29)	Healthy Control (N = 16)	P-value
Age	16.34 ± 1.758	15.75 ± 1.69	0.277
13	2 (6.9)	2 (12.5)	
14	5 (17.2)	3 (18.7)	
15	2 (6.9)	2 (12.5)	
16	3 (10.4)	0 (0)	
17	6 (20.7)	8 (50)	
18	11 (37.9)	1 (6.3)	
Gender (Female)	17 (58.6)	5 (31.1)	0.082
Dominant Hand (Right)	26 (89.7)	16 (100)	0.191
Stays with parent (Yes)	29 (100)	15 (93.8)	0.181
Type of Housing			
2–3 room HDB	3 (10.4)	1 (6.3)	0.678
4–5 room HDB	13 (44.8)	8 (50)	
Private Condominium	10 (34.4)	5 (31.1)	
Landed property	3 (10.4)	1 (6.3)	
Boarding House	0 (0)	1 (6.3)	
Current school year			
Secondary 1/Year 1	2 (6.9)	2 (12.5)	0.188
Secondary 2/Year 2	5 (17.2)	3 (18.7)	
Secondary 3/Year 3	2 (6.9)	2 (12.5)	
Secondary 4/Year 4	3 (10.4)	0 (0)	
Junior College 1/Year 5	7 (24.1)	8 (50)	
Junior College 2/Year 6/Poly Year 2	8 (27.6)	1 (6.3)	
Private Candidate/Diploma	2 (6.9)	0 (0)	
Current academic results			
Very poor	3 (10.4)	0 (0)	0.068
Poor	4 (13.8)	1 (6.3)	
Average	13 (44.8)	8 (50)	
Good	8 (27.6)	4 (25)	
Excellent	1 (3.4)	3 (18.7)	
PSLE score			
Old scoring	244.86 ± 22.09	256.9 ± 12.13	0.119
AL scoring	10 ± 6.48	8.60 ± 5.08	0.704

* $p < 0.05$; ** $p < 0.01$; *** $p < 0.001$.

References

1. U.S. Department of Health and Human Services. Attention-deficit/hyperactivity disorder. National Institute of Mental Health (n.d.). https://www.nimh.nih.gov/health/topics/attention-deficit-hyperactivity-disorder-adhd#:~:text=Attention%2Ddeficit%2Fhyperactivity%20disorder%20(,interferes%20with%20functioning%20or%20development
2. Mayo Clinic Staff: Attention-deficit/hyperactivity disorder (ADHD) in children – Symptoms and causes. Mayo Clinic. https://www.mayoclinic.org/diseases-conditions/adhd/symptoms-causes/syc-20350889#:~:text=ADHD%20symptoms%20start%20before%20age. 25 June 2019
3. NHS: Attention deficit hyperactivity disorder (ADHD) – Diagnosis. Nhs.uk. https://www.nhs.uk/conditions/attention-deficit-hyperactivity-disorder-adhd/diagnosis/#:~:text=To%20be%20diagnosed%20with%20ADHD. 1 June 2018
4. American Family Physician: When an ADHD Diagnosis May Cause More Harm Than Good. www.aafp.org. https://www.aafp.org/pubs/afp/afp-community-blog/entry/when-an-adhd-diagnosis-may-cause-more-harm-than-good.html. 4 April 2023
5. Tobias, J., Johnson, G., Rehman, S., Fisher, R., Caron, N.: Cerebral oxygenation monitoring using near infrared spectroscopy during one-lung ventilation in adults. J. Minimal Access Surg. **4**(4), 104 (2008). https://doi.org/10.4103/0972-9941.45206
6. Monden, Y., et al.: Clinically-oriented monitoring of acute effects of methylphenidate on cerebral hemodynamics in ADHD children using fNIRS. Clin. Neurophysiol. **123**(6), 1147–1157 (2012). https://doi.org/10.1016/j.clinph.2011.10.006
7. OBELAB: www.obelab.com, https://www.obelab.com/product/product_nirsit-lite.php (n.d.). Retrieved from 4 Jan 2024

Polarized Terahertz Time-Domain Spectroscopy for Sensing Enhancement

Favian Lim Wey Yee(✉)

NUS High School of Math and Science, Singapore 129957, Singapore
h1910032@nushigh.edu.sg

Abstract. Terahertz (THz) radiation is capable of supporting high bandwidth and is non-damaging yet sufficiently penetrating. Therefore it has a plethora of applications spanning from medical screening to communications. However, due to the current limitations in current THz generators and detectors, the THz EM spectrum is relatively unexplored, and there is a need to enhance components like THz polarizers. We report the construction and characterisation of a THz polarizer device based on liquid crystal (LCs). The device consisted of a thin layer of LC mixture E7 sandwiched between two glass plates coated with Indium Tin Oxide (ITO). An electrical potential difference was applied across the plates. The LC molecules had an electrophilic and a nucleophilic end, so we hypothesised that the LC molecules would have greater alignment with increasing voltage, causing the optical properties of the device to change in real time. We characterized the device in a Polarized Terahertz Time Domain Spectroscopy set up, using a wire grid polarizer was used for calibration. We demonstrated that the voltage applied across the LC could be used to manipulate THz waves by changing the amplitude and polarisation of the THz waves, even at low voltages and only 0.1ml of LC used. Our result offers a significant new way to manipulate THz waves without the need for mechanical parts. For future work, other LCs may be studied, due to the similar general properties found in most LCs.

Keywords: Polarized Terahertz Time-Domain Spectroscopy · Terahertz · THz · Liquid Crystal · Polarizer · Sensing · Materials Science

1 Context

1.1 Context

Terahertz (THz) radiation, defined as the bandwidth 0.1-10THz, is an underutilised part of the electromagnetic (EM) spectrum. THz radiation is largely non-damaging yet penetrating and has distinct signatures for a wide range of materials, so THz has a plethora of applications spanning from medical screening to communications. Though significant progress has been made in recent years, due to the inherent limitations in current THz generators and detectors, the THz EM spectrum is relatively unexplored, especially polarization analysis.

H. Guo et al. (Eds.): IRC-SET 2024, PTT, pp. 294–301, 2025.
https://doi.org/10.1007/978-981-96-3770-6_30

Polarized THz-TDS systems have been developed and researched, with many systems using wire-grid polarisers [1] or birefringent crystals [2] to orient the THz pulse. Challenges persist in achieving high sensitivity and broadband operation in polarized THz-TDS systems. Molecular orientation studies have also been conducted using polarized THz-TDS, such as those on liquid crystals [3]. In this article, we similarly propose and fabricate a prototype of a polariser with no mechanical moving parts by using liquid crystals and taking advantage of the liquid crystals' inherent electro-optic effects.

We fabricated a THz polarizer device using liquid crystal (LC) mixture, two glass plates coated with Indium Tin Oxide (ITO), two glass slides, silver paste, and copper wire to explore the feasibility of using a polarizer based on liquid crystals. Liquid crystal was used in the device due to its molecular properties and transparency to THz waves. We found that the opacity and polarity of the THz can be controlled in real time by modifying the E-field applied on the device, without moving parts.

The liquid crystal device was studied via Polarized Terahertz Time-Domain Spectroscopy, allowing information about the resultant wave's electric field (E-field) and synonymously, polarisation. The terms are used interchangeably in this report.

The nucleophilic end of the linear liquid crystal molecule, the nitrile functional group, would be attracted to the positive terminal of the device, causing the components to align and increasing the overall optical properties of the device with increasing voltage. In other words, due to the linear shape of the liquid crystal molecules and the polar nitrile group, it is predicted that change in the amplitude and polarisation of incident THz beam would be observed with varying voltage.

2 Methodology

2.1 Polarized Terahertz Time-Domain Spectroscopy Setup

Polarized Terahertz Time Domain Spectroscopy (Polarized THz-TDS) was used for analysis (Fig. 1).

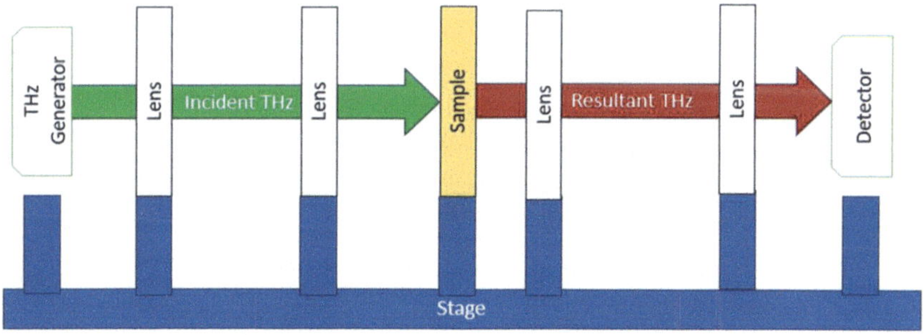

Fig. 1. Schematic of the THz-TDS set-up used. The detector is mounted on a stand which can rotate around the beam axis.

The results and other graphs were generated by rotating the detector at 10° intervals from 0 to 360° and plotting the data obtained. It is to be noted that the electric field

(E-field) of a THz wave and its polarisation are synonymous with each other and used interchangeably throughout this paper.

Photoconductive antennae were used to generate and detect THz waves for transmission-mode spectroscopy. To generate the THz wave, a laser pulse is passed through the electric field of gap of an open circuit, emitting EM THz waves. The electric field of the THz wave produced is polarized orthogonally to the laser pulse due to the electric field created by the open circuit. To detect a THz wave, when the THz wave meets the laser pulse in the photoconductive gap, a charge is generated, which flows through a circuit containing an ammeter. The ammeter measures the charge and current generated, which is plotted by the computer, as shown in the Results section.

In this report, a linear wire grid polarizer was used for calibration, being a known THz polarizer. A liquid crystal cell device containing LC mixture was then studied.

2.2 Calibration with Linear Wire Grid Polarizer

The linear wire grid polarizer was mounted on a stand which can rotate along the axis of the beam from 0 to 360 degrees, with markings every 10 degrees. We rotated the linear polarizer in 10 degree intervals to modify the THz wave emitted.

The purpose of experimentation with linear wire grid polarizer was to calibrate the THz generator and detector prior to testing the liquid crystal device. This is due to the known properties of the wire grid polarizer with THz waves – the conducting wires of the wire grid polarizer block light polarized parallel to itself.

2.3 Testing with Liquid Crystal Device

The liquid crystal device was prepared with two modified glass slides, with one face of each glass slide laced with Indium Tin Oxide (ITO). ITO is a conductive and translucent polymer which enables us to conduct a voltage across the liquid crystal. We used silver paste to bind copper wire to the face of the glass with ITO as shown in Fig, 2.

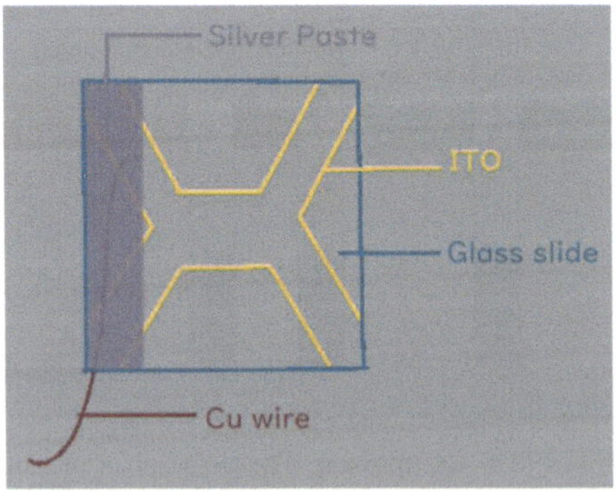

Fig. 2. Visual of one modified glass slide

To prepare the final device, we added two drops of liquid crystal mixture E7 to the ITO face with a disposable pipette, between the two modified glass slides, with the ITO faces in contact with each other as seen in Fig. 4.

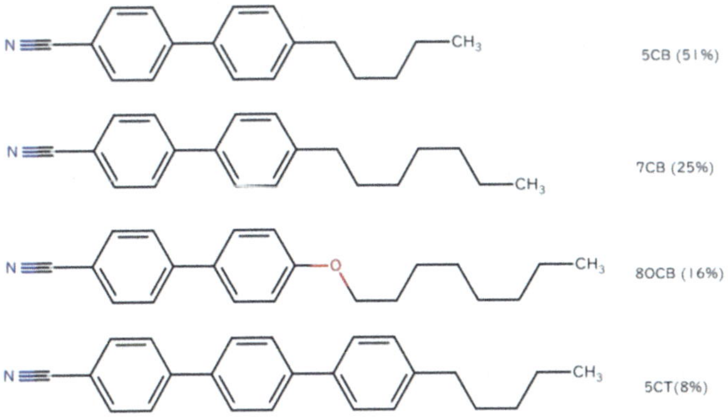

Fig. 3. Composition of liquid crystal mixture E7. The alkyl end is electrophilic, and the nitrile end (CN) is nucleophilic.

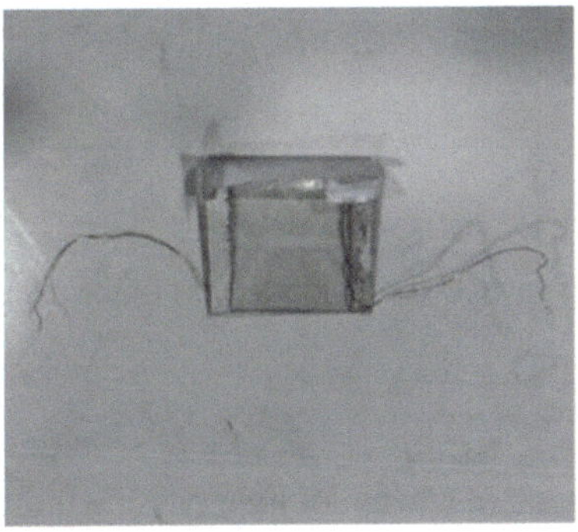

Fig. 4. Final device with both slides with a liquid crystal layer

We then connected the device to a power supply with tunable voltage as shown in Fig, 7 below. With the power supply, the voltage across the liquid crystal could be varied from 0–7 V (Fig. 5).

3 Results

3.1 Reference Incident Beam

See (Fig. 6).

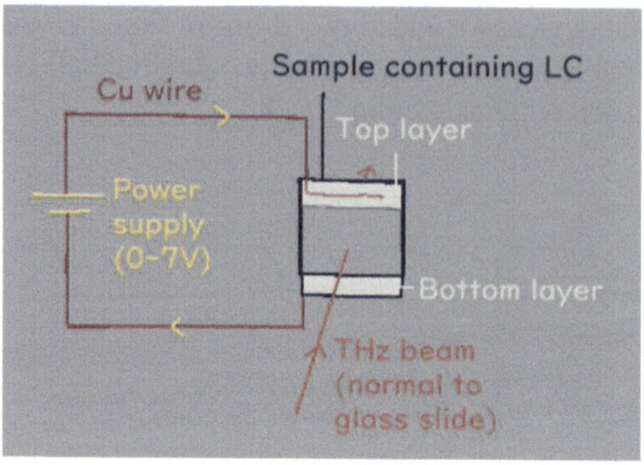

Fig. 5. LC device with voltage

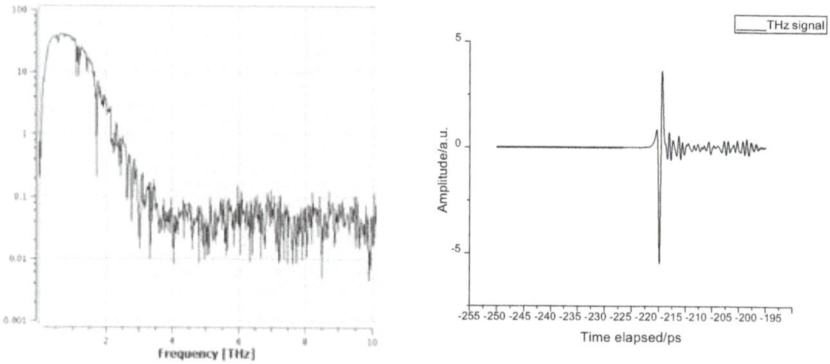

Fig. 6. Frequency- and Time-domain graphs of the reference THz beam

3.2 Linear Wire Grid Polarizer

By Malus' Law, the amplitude of a beam of light passing through two polarizers can be described as $A\cos^2(\theta)$, where A is the maximum amplitude, and θ is the angle between the two polarizers. Therefore, the maximum amplitude of the beam and the angle of rotation of polarizer should have a sinusoidal relation as shown in the first graph in Fig. 7.

The second graph in Fig. 7 is generated by rotating the linear wire grid polarizer in 10° intervals from 0 to 350° and plotting the recorded amplitude. Since the maximum amplitude of the THz beam is at when the polarizer is at 0 degrees, the polarisation of the generated THz beam is at 0 degrees; if the THz beam was not polarized, the maximum amplitude of the THz beam would be recorded at all angles of rotation of the polarizer. Instead, a sinusoidal relationship is observed, therefore the THz beam generated by the photoconductive antennae is polarized at 0°.

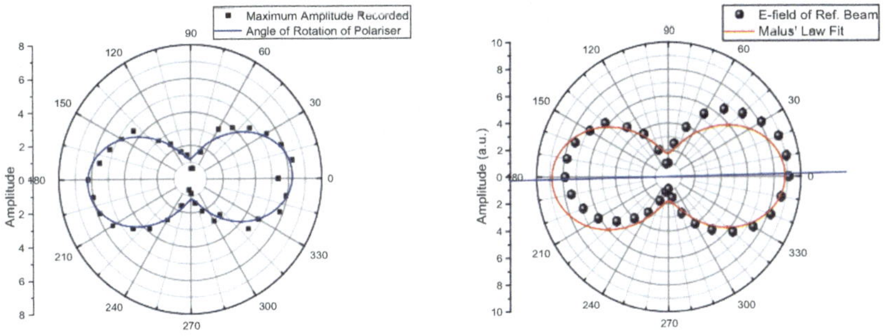

Fig. 7. Sinusoidal relationship between the maximum amplitude of the recorded THz beam, and the angle of the Polarizer and Detector respectively.

3.3 Liquid Crystal Polarizer (Device)

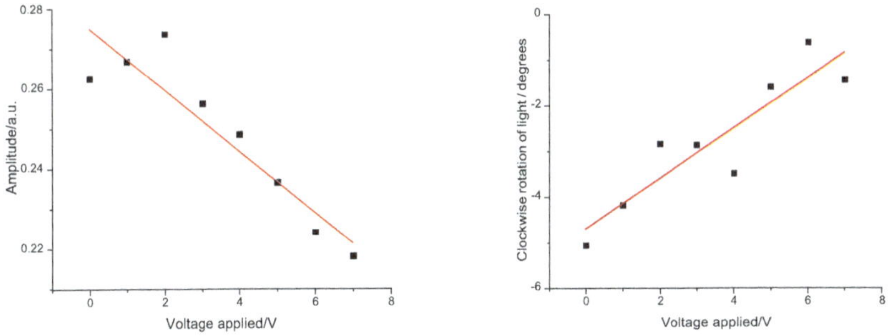

Fig. 8. Graph of change in amplitude (left) and polarisation (right) with respect to voltage applied.

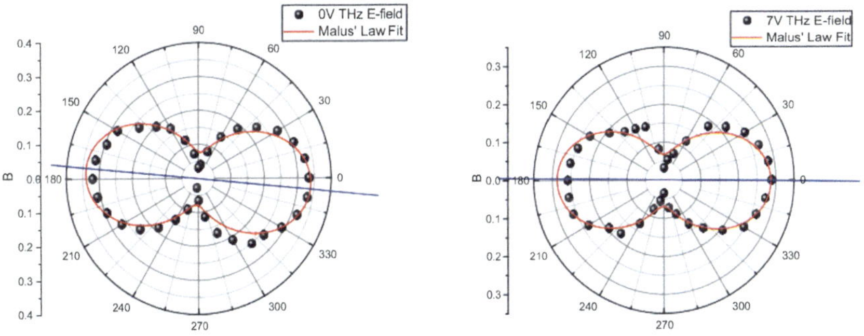

Fig. 9. Electric field of resultant THz beam with 0V and 7V applied respectively with a change in polarisation and maximum amplitude.

Referencing Fig. 8, the amplitude decreased linearly with increasing voltage. The clockwise polarisation of the beam also showed a positive relationship with increasing voltage in Fig. 3d. Since both the amplitude and polarisation of the beam vary with increasing voltage, this phenomenon is likely due to the liquid crystal molecules becoming more aligned to the potential difference, causing the incident THz beam to be modified.

In other words, we found that voltage can be used to manipulate THz radiation since it can change the opacity and polarisation of THz waves. While the magnitude of the changes may seem small (5° change over 7 V) as shown in Fig. 9, this can be easily upscaled due to the small-scale nature of this experiment, with approximately 0.1 mL of liquid crystal (2 drops) in the liquid crystal cell used.

4 Discussion and Conclusion

4.1 Context

From the results gathered above, it can safely be concluded that even when using small amounts of liquid crystal mixture E7 and a modest voltage applied, we are able to see the change of THz polarisation and amplitude in real time.

We have successfully shown that liquid crystal E7 is sensitive to both voltage and THz, allowing us to manipulate the amplitude and polarisation of THz waves. This is a significant finding, since not many ways exist to manipulate THz waves, particularly without the use of moving parts.

At voltages greater than 8 V, however, the liquid crystal started to smoke, indicating that the chemical structure of the liquid crystal was being damaged at higher voltages. As such, the experiment could not be conducted at voltages higher than 7 V. This is likely due to the oxidation decomposition reaction of the reactive nitrile group [4] in the liquid crystal and the surroundings at high voltages – the gas produced was likely a mixture of primarily carbon monoxide, which contributed to the visibility of the smoke produced. Nitrogen gas would have been produced as a stable byproduct as well, with trace amounts of carbon dioxide and water formed.

Due to decomposition of the liquid crystal at relatively lower voltages as compared to the stability of liquid crystals LC1 and E44 [5], other liquid crystals may prove to be more suitable for terahertz sensing applications.

Nevertheless, the nonnegligible rotation of THz light despite a small amount of liquid crystal used suggests a potential use-case of liquid crystals in commercial THz devices.

For future research, the experiment could be conducted in an inert atmosphere of nitrogen so as to prevent unwanted decomposition of the liquid crystal upon subject to high voltage. The scope of the project could also be expanded to other types and mixtures of more chemically stable liquid crystals, especially those of different phases (smectic and chiral/twisted-nematic).

References

1. Chen, Y., Wang, J., Hu, Z., Gao, T., Zhu, W., Wang, X.: Fabrication of large-scale freestanding THz wire grid polarizers by femtosecond laser micromachining. Optik **212**, 164655 (2020). https://doi.org/10.1016/j.ijleo.2020.164655
2. Nakagawa, M., Okano, M., Watanabe, S.: Polarization-sensitive terahertz time-domain spectroscopy system without mechanical moving parts. Opt. Express **30**(16), 29421 (2022). https://doi.org/10.1364/oe.460259

3. Friebel, P., Galimberti, D.R., Savoini, M., Cattaneo, L.: Unveiling low THz dynamics of liquid crystals: identification of intermolecular interaction among intramolecular modes. J. Phys. Chem. B **128**(2), 596–602 (2024). https://doi.org/10.1021/acs.jpcb.3c07947
4. Zhi, H., Xing, L., Zheng, X., Xu, K., Li, W.: Understanding how nitriles stabilize Electrolyte/Electrode interface at high voltage. J. Phys. Chem. Lett. **8**(24), 6048–6052 (2017). https://doi.org/10.1021/acs.jpclett.7b02734
5. Ito, R., Honma, M., Nose, T.: Electrically tunable hydrogen-bonded liquid crystal phase control device. Appl. Sci. **8**(12), 2478 (2018). https://doi.org/10.3390/app8122478

Unravelling the Effects of Rapamycin on Reproductive Longevity on Female Mice

Darin Yi Ren Koay[1], Alagashika Arun Rajesh[1], Kendra Li Wen Chua[1](✉), X. T. Tan[2], W. E. Sim[2], and Z. Huang[2,3]

[1] School of Mathematics and Science, NUS High, Singapore, Singapore
kendraclw@gmail.com

[2] NUS Bia-Echo Asia Centre for Reproductive Longevity and Equality, Yong Loo Lin School of Medicine, National University of Singapore, Singapore, Singapore

[3] Department of Obstetrics and Gynaecology, Yong Loo Lin School of Medicine, National University of Singapore, Singapore, Singapore

Abstract. Rapamycin, classified as a mechanistic Target of Rapamycin inhibitor (mTOR), is a drug widely recognised for its potential anti-ageing properties; however, its impact on reproductive health remains uncertain.[10] This study investigates three groups of mice: young mice (20 weeks old), the control, and treated old mice (44 to 53 weeks old). The estrous cycle was observed over a 5-day period for all groups of mice in this experiment, with an additional 5 days of estrous staging specifically conducted for the old mice to see the effects of Rapamycin. The findings indicate that there is no change in the number of days that the mice experience estrus when treated. Furthermore, follicular count analysis revealed a reduction in the number of follicles in the Rapamycin-treated group compared to the vehicle group. These results suggest that Rapamycin treatment is not conclusive in reproductive health, therefore contrasting with the anti-ageing studies attributed to this drug.

Keywords: rapamycin · estrous cycle · follicles · ovarian ageing

1 Introduction

Women experience lowered Anti-Müllerian Hormone (AMH) levels [8] as they age and suffer from significant symptoms which includes lowered metabolism leading to weight gain [11], osteoporosis [2], and increased cholesterol levels [4]. These symptoms can significantly diminish the quality of life, hence affecting work and social interactions in addition to the existing burdens of ageing.

Women generally reach menopause when they reach 45 to 55 years of age. Upon reaching the menopausal stage, they experience a range of symptoms, from hot flashes to night sweats and vaginal dryness. However, there are more severe symptoms, such as loss of bone and increase in cholesterol levels. Due to the drop in estrogen in women, this causes the rate of replacing lost bone mass to reduce, increasing the risk of osteoporosis.

H. Guo et al. (Eds.): IRC-SET 2024, PTT, pp. 302–313, 2025.
https://doi.org/10.1007/978-981-96-3770-6_31

Not only that, the chances of low-density cholesterol (LDC) in women increasing are higher, making them susceptible to heart attacks.

Rapamycin belongs to a class of drugs called mTOR inhibitors. Rapamycin has garnered significant interest due to its potential anti-ageing and longevity-promoting properties [10]. Research has shown that rapamycin can extend the lifespan of various organisms, including yeast, worms, flies, and even mice. It is believed to affect the mechanisms involved in the ageing process, such as cellular metabolism and cell growth and proliferation regulation [12].

AMH is a hormone produced in the body that plays a role in reproductive health, particularly in women. AMH is produced by granulosa cells in the ovaries, specifically growing follicles that contain eggs. Its measurement is strongly correlated with follicle count and represents a reliable marker of ovarian reserve [5]. AMH is also produced in mice by the granulosa cells in the ovaries; similar to humans, AMH in mice play a role in reproductive development and fertility.

In female mice, AMH is involved in the regulation of ovarian follicle development. Follicles are tiny sacs within the ovaries that contain immature eggs. AMH helps control the growth and maturation of these follicles, influencing the number and quality of eggs available for reproduction [13]. AMH levels in mice can vary depending on factors such as age, reproductive stage, and genetic background.

C57BL/6J mice is the most widely used inbred strain [9]. In this study, C57BL/6J mice is used to investigate the effects of Rapamycin. These mice present with phases of the estrous cycle which is comparable with the phases of menstrual cycle in humans. Mice are selected as experimental models since they have genetic similarity, equivalence in life phases, and maturational rate with human [3, 6].

Three groups of mice were selected, consisting of young (20 weeks old), control (44 weeks old), and treated old mice (44 weeks old). The old mice in the control group were given a vehicle treatment that serves the purpose of establishing a baseline. This enables a comparison with the effects induced by Rapamycin. The control old mice help ensure that any observed effects are attributable to Rapamycin instead of potential influences from the treatment method. The young mice are approximately 25 years old in human years [6], which represents peak fertility[7], while old mice are approximately 45 years old [6] in human years, where humans enter the pre-menopausal stage [7].

1.1 Methodology

At 44 weeks of age, old control and treated mice groups received vehicle and Rapamycin treatment, respectively. Prior to treatment, old mice were swabbed over a 5-day period for estrous staging. Both groups underwent 3 doses of treatment per week, for a duration of 8 weeks. Both groups were swabbed for estrous staging at 53 weeks old after treatment and were culled at 54 weeks old. Young mice were also swabbed for comparison and culled at 20 weeks old.

Ovaries were harvested and fixed in 4% PFA. Continuous sections of ovaries were collected at 5-microns and stained with Hematoxylin and Eosin (H&E) for follicular count. A representative section of each ovary was stained with AMH (ab272221) [1] to determine the presence of AMH. Detailed protocols can be found in Appendix A.

2 Results and Discussion

Estrous stages of mice from 3 different groups were analysed for 5 days consecutively before treatment. Old mice (control and treated) were swabbed for 5 days consecutively after treatment (Fig. 1).

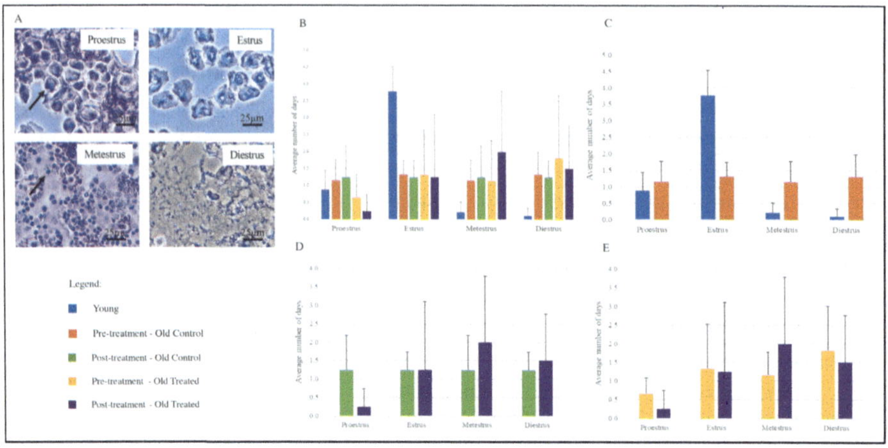

Fig. 1. Estrous Staging – The estrous cycle (A) of mice comprises of proestrus, estrus, metestrus, and diestrus stages. Graph of the average number of days (B), (C), (D) and (E) in different estrous stage presented over a period of 5 days.

The estrous cycle (A) of mice comprises of proestrus, estrus, metestrus, and diestrus stages. Proestrus is characterized by the presence of round, nucleated epithelial cells. During estrus, cornified epithelial cells without nuclei predominate. Metestrus is marked by the presence of white blood cells. Diestrus is distinguished by the prevalence of mucus.

Graph of the average number of days (B) in each estrous stage presented over a period of 5 days; inclusive of young, pre-treatment old control, post-treatment old control, pre-treatment old treated, post-treatment-old treated mice respectively. Young mice experience a longer estrus phase (3.8 days) as compared to old mice (1.3 days) (C). After treatment, the number of days that the old control mice experience estrus (1.3 days) remains the same compared to the old treated (D). After treatment, the number of days that the old treated mice experience estrus (1.3 days) remains the same compared to the pre-treatment (C) (Fig. 2).

An ovarian follicle is a structure of granulosa cells surrounding an oocyte. Follicles undergo several stages during development, each characterized by specific changes in the structure of the follicle and the maturation of the oocyte.

Primordial follicle, at approximately 17 µm in size, found on the periphery of the ovary, is surrounded by a single layer of squamous granulosa cells. Primary follicle, at approximately 28 µm in size, is surrounded by a single layer of cuboidal granulosa cells. Secondary follicle, a more developed stage, has multiple layers of granulosa cells surrounding the oocyte, has an approximate size of 100 µm. Antral follicles, nearing ovulation, are characterised by the presence of a clear fluid-filled space (antrum) between the cells and are approximately 550 µm in size.

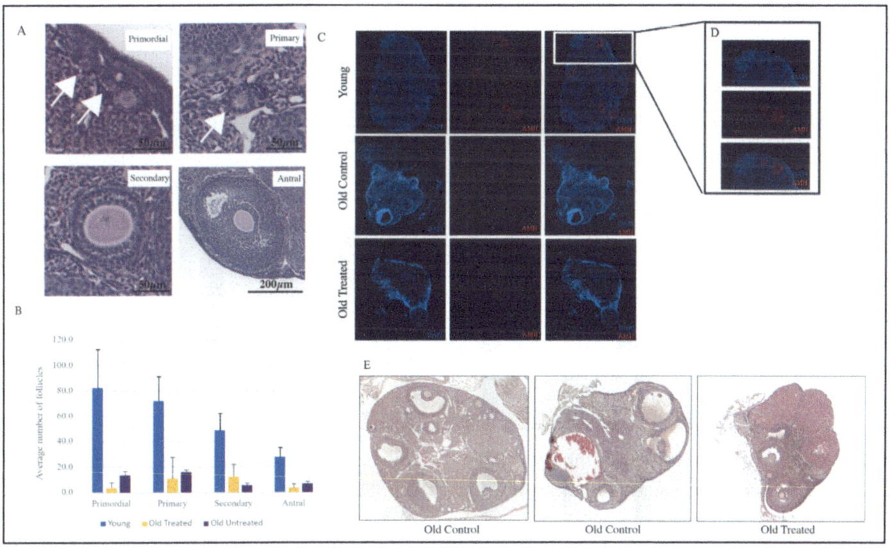

Fig. 2. Follicular Count. The main follicular stages (A) include primordial, primary, secondary, and antral. The average number of different follicles (B) in young, old control treated mice were compared. Visual representation of AMH staining (C) of young mice, old control and treated mice, respectively; can be compared with a zoomed-in image of young mice (D).

Representative images of ovary sections from young, old control, and treated mice respectively were included for visual comparison (C). There is a decrease in follicle count between young mice and old mice. Rapamycin treatment decreased the follicle count compared to untreated.

AMH staining is visible in young mice (C), but not in old treated and untreated mice (D). This may suggest that there is lower production of AMH by granulosa cells in the old mice as compared to the young mice. AMH was not observed via immunofluorescence in older mice subjected to treatment, which may suggest that AMH production is low. Therefore, while Rapamycin has been reported to improve health span, its effects remain uncertain on the reproductive health.

3 Treatment at 90 Weeks Old

An additional group was introduced to investigate the different ages of drug intervention. The new group was given rapamycin treatment at 90 weeks old, and the estrous cycle was observed. It was observed that later interventions did not improve the estrous cycle (Fig. 3).

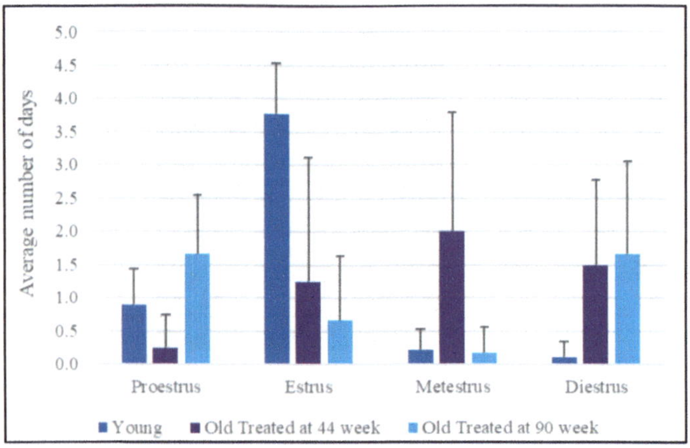

Fig. 3. Graph of the average number of days that mice from of different ages experience the estrous cycle over 5 days. It was observed that later interventions did not improve the estrous cycle.

4 Conclusion

The findings in this paper suggests that Rapamycin does not alter the duration of the estrus stage in mice. In addition, there was no improvement observed in the number of follicles in the old treated mice. In the 90-week treated mice, there was also no improvement observed, suggesting that a delay in treatment may not be effective. Due to the limitations present in this study, the effects of Rapamycin on reproductive health cannot be fully demonstrated. More research must be done beyond follicular count and estrous staging. In addition, further analysis as well as a larger dataset would be beneficial to better understand the effects of Rapamycin in reproductive health. In conclusion, the findings in this study suggests that Rapamycin has little effect on reproductive health.

Appendix A: Protocols

Vaginal Smear Staining and Estrus Staging

Reagents:

a. 0.1% Cresyl Violet

Cresyl Violet 1 g
Distilled water 100 mL

1. Air dry slides and stain with 0.1% Cresyl Violet (0.2 µm filtered) for 2 min.
2. Dip slides twice in distilled water.
3. Cover slip the slides with aqueous mount.
4. Examine slides under light microscope at 10x and 20x magnifications to determine the estrus stage.

Perfusion of Mice and Organ Harvesting
Reagents:

a. 1xPBS
   10x PBS 100 mL
   DI Water 900 mL
b. 4% Paraformaldehyde (PFA)
   Paraformaldehyde 20 g
   1xPBS 450 mL
   Adjust pH to pH 7.4 using NaOH.
   Adjust total volume to 500 mL with 1xPBS.

1. Mouse is euthanized via CO_2 overdose.
2. Perform pinch withdrawal reflex test to ensure mouse is fully euthanized.
3. An incision is made below the diaphragm and the rib cage is cut rostrally on the lateral edges to expose the heart.
4. Insert needle with PBS syringe into left ventricle of heart and make a cut on the right atrium with a sharp scissors.
5. Mouse is transcardially perfused with approximately 20ml of PBS to remove blood.
6. Following PBS perfusion, mouse is perfused with approximately 20ml of 4% PFA.
7. Ovaries harvested and placed in 4% PFA for no more than 48 h at 4 °C.

Tissue Trimming and Processing

1. Place ovaries in cassettes.
2. Store cassettes in 70% Ethanol if not processing immediately.
3. Load cassettes into processor.

Automated processor protocol for ovaries.

Solution	Time
70% ethanol	1 h
80% ethanol	1 h
95% ethanol	1 h
3 washes of 100% ethanol	1 h each
3 washes of Xylene	1 h each
3 washes of Paraffin	1 h each

Embedding + Microtomy

1. Unload processor and place cassette in embedding station.
2. Embed tissues.
3. Section tissues at 5 microns.

Dewax Protocol

2 washes of Xylene	10 min each
2 washes of 100% ethanol	5 min each
95% ethanol	5 min
70% ethanol	2 min
DI Water	2 min

Hematoxylin and Eosin Staining Protocol

Hematoxylin	10 min
Tap Water	5 min
Acid Alcohol	1 dip
Tap Water	1 min
Bluing	1 min
Tap Water	1 min
70% ethanol	1 min
Eosin	15 s
95% ethanol	1 min
2 washes of 100% ethanol	1 min each
2 washes of Xylene	1 min each

Cover slip slides.

Immunofluorescence Staining
Reagents:

a. <u>Tris EDTA Buffer (10mM Tris, 1mM EDTA, 0.05% Tween20, pH9.0)</u>
   Tris 1.21g
   EDTA 0.37g
   DI water 1L
   Adjust to pH 9.0 with NaOH
   Tween 20 0.5ml
b. <u>0.05% PBST</u>
   PBS 1L
   Triton X 50μL

1. Pre-heat and dry slides at 60 °C for 30 min.
2. Dewax slides (refer to protocol above).
3. Perform antigen retrieval.
4. Wash with 1xPBS for 10 min at Room Temperature (RT).
5. Block with 10% Donkey Serum in 0.05% PBST.
6. Primary Antibody incubation overnight at 4 °C (No longer than 16 h)
   a. Dilute Primary Antibody in Blocking buffer
   b. AMH Dilution 1:1000
7. Wash with 1xPBS 3 times, 10 min each at RT.
8. Incubate with secondary antibody for 1 h at RT.
   a. Dilute Secondary antibody in Blocking buffer
   b. Dilution 1: 200
9. Wash with 1xPBS 3 times, 10 min each at RT.
10. Stain with 1:1000 DAPI for 10 min in the dark at RT.
11. Wash with 1XPBS for 10 min at RT.
12. Mount with aqueous mount.

Appendix B: Estrous Staging Data

Legend: Proestrus = 1, Estrus = 2, Metestrus = 3, Diestrus = 4. (Tables 1, 2 and 3)

Table 1. Estrous staging for mice

	Animal ID	Day 1	Day 2	Day 3	Day 4	Day 5
	Young 1	2	2	2	2	2
	Young 2	2	3	1	1	2
	Young 3	2	2	1	4	2
	Young 4	2	2	1	3	2
	Young 5	2	2	2	2	2
	Young 6	1	2	2	2	1
	Young 7	2	2	1	2	2
	Young 8	2	2	1	2	2
	Young 9	2	2	2	2	2
Pre-treatment	Old Control 1	4	1	1	2	3
	Old Control 2	4	4	1	2	3
	Old Control 3	3	1	2	2	3
	Old Control 4	4	1	1	2	4

(continued)

Table 1. (*continued*)

	Old Control 5	2	3	3	4	4
	Old Control 6	4	1	2	2	3
	Old treated 1	4	4	1	2	1
	Old treated 2	3	4	2	2	3
	Old treated 3	4	1	2	3	3
	Old treated 4	4	1	2	3	1
Post treatment	Old control 1	4	4	4	3	1
	Old control 2	2	2	1	2	2
	Old control 3	4	2	3	4	1
	Old control 4	2	3	2	3	4
	Old control 5	4	3	4	4	4
	Old control 6	4	3	3	1	2
	Old Treated 1	2	2	2	2	4
	Old Treated 2	3	3	3	3	1
	Old Treated 3	4	4	2	3	4
	Old Treated 4	3	3	3	4	4
Pre-treatment	90 week Treated 1	4	4	1	3	4
	90 week Treated 2	3	4	1	3	1
	90 week Treated 3	3	3	4	3	4
	90 week Treated 4	2	3	2	2	4
	90 week Treated 5	2	2	2	2	2
	90 week Treated 6	3	4	3	3	4
Post treatment	90 week Treated 1	2	2	1	3	2
	90 week Treated 2	1	1	1	1	1
	90 week Treated 3	2	3	1	1	1
	90 week Treated 4	3	2	2	4	2
	90 week Treated 5	1	2	2	2	2
	90 week Treated 6*	x	x	x	x	x

* Mouse died

Table 2. Total count of estrous stage

	Proestrus	Estrus	Metestrus	Diestrus
Young	8	34	2	1

(*continued*)

Table 2. (*continued*)

	Proestrus	Estrus	Metestrus	Diestrus
Old Control Pre-treatment	7	8	7	8
Old Treated Pre-treatment	5	5	5	5
Old Control Post-treatment	4	8	7	11
Old Treated Post-treatment	1	5	8	6
90 Week Pre-treatment	3	8	10	9
90 Week Post treatment	10	11	3	1

Table 3. Average estrous stage

	Proestrus	Estrus	Metestrus	Diestrus
Young	0.889	3.778	0.222	0.111
Old Control Pre-treatment	1.167	1.333	1.167	1.333
Old Treated Pre-treatment	1.250	1.250	1.250	1.250
Old Control Post-treatment	0.667	1.333	1.167	1.833
Old Treated Post-treatment	0.250	1.250	2.000	1.500
90 Week Pre-treatment	0.500	0.889	1.667	1.500
90 Week Post treatment	1.667	1.833	0.500	0.167

Appendix C: Follicle Count Data

Tables 4 and 5.

Table 4. Follicle count

Overall	Primordial	Primary	Secondary	Antral	Total
Young 1	142	85	53	32	312
Young 2	117	77	58	25	277
Young 4	69	109	58	47	283
Young 5	52	77	49	29	207
Young 6	101	73	39	17	230
Young 7	118	91	81	34	324
Young 8	23	25	17	15	80
Young 9	37	36	34	23	130

(*continued*)

Table 4. (*continued*)

Overall	Primordial	Primary	Secondary	Antral	Total
Old Control 1	10	16	6	7	39
Old Control 2	12	18	7	8	45
Old Control 3	17	17	6	5	45
Old Control 4	11	17	5	6	39
Old Control 5	19	12	2	5	38
Old Control 6	12	13	7	10	42
Old Treated 1	10	36	23	3	72
Old Treated 2	0	0	0	0	0
Old Treated 3	0	1	8	5	14
Old Treated 4	3	7	17	7	34

Table 5. Average number of follicles

	Primordial	Primary	Secondary	Antral	Total
Young	82.375	71.625	48.625	27.750	230.375
Old Control	13.500	15.500	5.500	6.833	41.333
Old Treated	3.250	11.000	12.000	3.750	30.000

References

1. Anti-AMH antibody [EPR23380-171] (AB272221) | Abcam. (n.d.). https://www.abcam.com/en-my/products/primary-antibodies/amh-antibody-epr23380-171-ab272221#all
2. Better Health Channel. Menopause and osteoporosis. Vic.gov.au. https://www.betterhealth.vic.gov.au/healh/conditionsandtreatments/menopause-and-osteoporosis
3. Bryda, E.C.: The Mighty Mouse: the impact of rodents on advances in biomedical research. Mo Med. **110**(3), 207–211 (2013)
4. Chan, T.: Cholesterol and Menopause: What's the Relationship? Healthline. https://www.healthline.com/health/menopause/cholesterol-and-menopause
5. Dewailly, D., Laven, J.S.: AMH as the primary marker for fertility. Eur. J. Endocrinol. **181**(6), D45–D51 (2019). https://doi.org/10.1530/eje-19-0373
6. Hagan, C.: When are mice considered old? The Jackson Laboratory. https://www.jax.org/news-and-insights/jax-blog/2017/november/when-are-mice-considered-old (2017). Retrieved 10 Sep 2023.
7. Jensen, R.E., Martins, N., Parks, M.M.: Public perception of female fertility: Initial fertility, peak fertility, and Age-Related Infertility among U.S. adults. Arch. Sex. Behav. **47**(5), 1507–1516 (2018). https://doi.org/10.1007/s10508-018-1197-4
8. Freeman, E.W., Sammel, M.D., Lin, H., Gracia, C.R.: Anti-mullerian hormone as a predictor of time to menopause in late reproductive age women. J. Clin. Endocrinol. Metab. **97**(5), 1673–1680 (2012). https://doi.org/10.1210/jc.2011-3032

9. Lab, J.: C57BL/6J. 000664 – B6 strain details (n.d.). https://www.jax.org/strain/000664
10. Lamming, D.W.: Inhibition of the mechanistic target of rapamycin (MTOR)–Rapamycin and beyond. Cold Spring Harb. Perspect. Med. **6**(5), a025924 (2016). https://doi.org/10.1101/cshperspect.a025924
11. Sanchez, M.: What causes menopause weight gain? | HealthPartners Blog. HealthPartners Blog. https://www.healthpartners.com/blog/menopause-weight-gain/#:~:text=Less%20estrogen%20means%20less%20muscle. Accessed 16 Aug 2024
12. Shindyapina, A.V., et al.: Rapamycin treatment during development extends life span and health span of male mice and Daphnia magna. Sci. Adv. (2022). https://doi.org/10.1126/sciadv.abo5482
13. Tanimoto, R., Sekii, K., Morohaku, K., Li, J., Pépin, D., Obata, Y.: Blocking estrogen-induced AMH expression is crucial for normal follicle formation. Development (2021). https://doi.org/10.1242/dev.19745

First Principles Investigation of Electronic Structure, Ligand Field and Magnetism of Re(III) Compounds

Sze Ee, Josher Lo[1], Yang En Ivan Ng[1(✉)], Rutvij Manish Pradhan[1], and Liviu Ungur[2]

[1] NUS High School of Mathematics and Science, 20 Clementi Avenue 1, Singapore 129957, Singapore
contact@josherlo.com , ivan.ng998@gmail.com, pradhanrutvij@gmail.com
[2] Department of Chemistry, National University of Singapore, 3 Science Drive 3, Singapore 117543, Singapore

Abstract. Transition metal complexes are widely used in the industry as catalysts and dyes. Predicting their properties includes calculating low-lying energy states, which is computationally intensive. Hence, we aim to provide a simple ligand field model of two Re(III) molecules, $[ReCl_6]^{3-}$ and $ReCl_3$(tppz) to accurately predict its low-lying energy states. *Ab initio* calculations were done with CASSCF and CASPT2. Two different model Hamiltonians were constructed with five different ligand field parameters: energy gap (J), spin-orbit coupling (λ), axial (Δ), and rhombic distortion (ϵ), and orbital reduction factor (κ). A nine-by-nine matrix for the $^3T_1$ configuration, and a fifteen-by-fifteen matrix for the t_{2g}^4 configuration were obtained. The eigenvalues of the Hamiltonian were then fitted to the calculated energy states using a minimization function. The values for the $^3T_1$ model fit extremely well, whereas the t_{2g}^4 model has standard deviations ranging from 13 to 192. We obtained parameter values matching our predictions. The value of λ and ϵ are close to 0 for $[ReCl_6]^{3-}$, representing a perfect octahedral compound, and large for $ReCl_3$(tppz), which is not a perfect octahedral and will have some axial and rhombic splitting. We can expand our research by fitting the five parameters to other $5d^4$ transition metal complexes and including more states in the Hamiltonian.

Keywords: Transition metal complex · Ligand field model · Eigenvalues · Hamiltonian

1 Introduction

1.1 Role of 5d Transition Metals: Optical and Magnetic Properties

One of 5d transition metals' most common and important applications is its use as catalysts. Having partially filled *d* orbitals, they can easily withdraw or accept electrons from the reagents, providing alternative reaction pathways to

H. Guo et al. (Eds.): IRC-SET 2024, PTT, pp. 314–322, 2025.
https://doi.org/10.1007/978-981-96-3770-6_32

lower the activation energy of reactions [7]. 5d transition metals are also used as impurities in crystals because it can induce optical properties in crystals and modulate band-gaps of crystals [6].

Additionally, according to crystal field theory (CFT), due to the splitting of the d orbitals in energy, there exists an energy gap between the $d_{x^2-y^2}$, d_{z^2} and d_{xy}, d_{yz}, d_{xz}. When the atom is hit by a photon, it absorbs a wavelength of visible light. Hence, the complex will be coloured. These complexes are thus used in the industry as dyes in various products [2].

Thus, it is important to analyze the properties of transition metal complexes and derive a quantitative model to predict the properties of these complexes accurately.

1.2 Electronic Structure of Octahedral 5d^4 Compounds

In the free ion, all the 5d orbitals are degenerate since there are no external perturbations to them. However, in a crystal environment, the 5d orbitals split in energy due to electrostatic perturbation caused by negative charges of the ligands as well as by covalence between ligand and metal's orbitals. For example, in octahedral symmetry, the 5d orbitals split into two groups: t_{2g}, lower in energy, and e_g, higher in energy. In case the symmetry of the ligand environment is lower, then the splitting of the 5d orbitals is complete in five non-degenerate separate orbitals. Furthermore, due to Jahn-Teller effect [4], geometric distortion of the octahedral complex reduces its symmetry and lowers its energy. This leads to the breaking of the degeneracy of the t_{2g} and e_g orbitals. The point group for the real molecule is C1, which is low in symmetry. However, we notice that the axial perturbation is the dominant one, and therefore, we may model this splitting using a simple model Hamiltonian Eq. 1. This model Hamiltonian employs four ligand-field parameters, described in more detail later (Fig. 1).

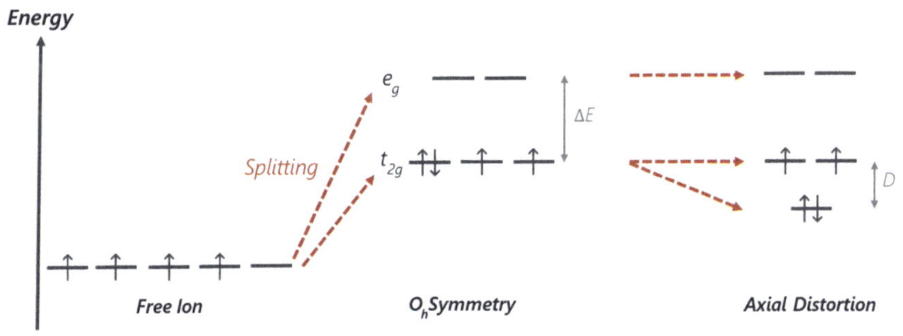

Fig. 1. Example of crystal field splitting of a 5d orbital

1.3 Computational Chemistry

On one end of the spectrum of computational chemistry, there is a common and fast method known as the Hartree-Fock method, which utilizes only one Slater determinant to model the multi-electronic wave function, which accounts for electron-electron interactions but does not take into consideration the correlation energy of the molecule which can result in deviation from experimental results.

However, the true wave function of the molecule requires many more Slater determinants, which describe all the possible configurations of electrons in atomic orbitals. Thus, the Hartree-Fock method only achieves about 90% of the total energy.

On the other end of the spectrum, there is the fully accurate full wave function which takes into account all orbitals from the basis set and all possible Slater determinants, achieving the true theoretical result at the cost of extreme computing power. This can be done for very small molecules but is not feasible with large compounds, such as the 5d transition metals in our projects.

Therefore, most computational chemistry falls between these two domains, achieving higher accuracy than the Hartree-Fock method and optimizing lower energies while requiring more computational resources. These small accuracy improvements have a big impact as it frequently decides which reactions proceed and which do not.

1.4 Aim of Project

One of the goals of this project is to investigate *ab initio* the low-lying energy structure of a Re(III) molecule and give a simple ligand field interpretation of the low-lying energy states by employing two different model Hamiltonians for this purpose. We use the results of two different computational methods, namely CASSCF and CASPT2, on two different Rhenium compounds, $[ReCl_6]^{3-}$ (a model octahedral compound), and $ReCl_3$(tppz) (Referred to as Re(III) molecule from here forth) to determine the four ligand-field parameters for a $^3T_1$ model and a t_{2g}^4 model. We employ a minimization function, fitting values into the Hamiltonian given in the Eq. (1) with the *ab initio* values.

Equation (1) shows the derivation of the model Hamiltonian for the $^3T_1$ model.

$$\hat{H} = -\kappa\lambda\hat{L}\hat{S} + \Delta(\hat{L}_Z^2 - L(L+1)/3) + \epsilon(\hat{L}_X^2 - \hat{L}_Y^2) \tag{1}$$

λ, Δ, ϵ, and κ are the four ligand field parameters, described in more detail later on. Furthermore, for the t_{2g}^4 model, we use another parameter, J, specifying the repulsion energy between electrons in the t_{2g}^4 shell. κ can be excluded from this model as the effects of κ can be achieved by modifying J and λ.

2 Computational Methods

For this research we employed first principles, *ab initio* computational methods. These computational methods aim at solving the Schrodinger Equation $\hat{H}\Psi =$

$E\Psi$, for the investigated molecules and then, evaluate properties of the molecule using the obtained wave function Ψ. Since the Schrodinger equation cannot be solved analytically for large molecules, solutions are found numerically, using powerful computers for this purpose.

2.1 Hartree-Fock Method

Hartree-Fock method is one of the first computational methods developed. It defined the wave function of the investigated molecule as a single configuration, a single Slater determinant, where electrons are arranged in molecular orbitals according to the Aufbau principle: molecular orbitals are populated with electrons in the order of increase of their energy, with maximum two electrons in each molecular orbital to fulfill Pauli exclusion principle. Hartree-Fock method does not represent the "ultimate true" wave function, because the interaction between electrons is described in a "mean field approximation", which means that each electron "does not feel" explicitly each of other electrons, but rather experiences the electrostatic interaction with a "distributed electronic cloud" made of all other electrons and nuclei. The reality is, however, that electrons **do** feel each other and they try to avoid each other as much as possible, given the electrostatic repulsion between them. This last effect is named *correlation energy* and is not present in the Hartree-Fock method [5].

2.2 Density Functional Theory

Similar to the Hartree-Fock method, Density Functional Theory (DFT) utilises only one Slater Determinant. However, instead of focusing on wave functions and orbitals, DFT focuses on the electron density and includes an approximation to electron correlation energy. This allows DFT to optimise the energy to a lower value and be more somewhat more accurate than Hartree-Fock. We utilize the DFT method for optimisation of the molecular structure, computation of vibrational frequencies and evaluate the Infra-Red spectra, as well as to plot electrostatic potential and electron density distribution. These calculations were done using the Gaussian software and Def2SVP basis set together with the B3LYP exchange correlation functional to get the Electrostatic Potential and IR spectra.

2.3 Multiconfigurational Wave Function Based Methods: CASSCF

Multiconfigurational Self Consistent Field Method (CASSCF) is an upgrade of the Hartree-Fock method and allows for simultaneous optimisation of the molecular orbitals as well as the coefficients of several Slater determinants which describe the wave function. In this method, all molecular orbitals are split into 3 groups: unoccupied orbitals, active orbitals and inactive orbitals. This method builds all possible Slater determinants obtained by distributing active electrons among active orbitals. It is computationally easier than allowing *all* electrons to mix between *all* orbitals. [9] Since the active space is kept rather small, most of the inactive orbitals remain doubly occupied in all configurations, and therefore, the electron correlation is not included for them. These calculations were done using the OpenMolcas software and ANO-RCC-VDZP basis set. For the investigated molecules, two active spaces were considered: CAS(4in3) and CAS(4in5). The first active space allows building all configurations obtained by distributing 4 electrons in the t_{2g} shell, while the second active space employs all five 5d orbitals (t_{2g} and e_g shells).

2.4 Multiconfigurational Correlation Methods: CASPT2

CASPT2 method aims at estimating the correlation energy for a CASSCF wave function using the multi-configurational second order perturbation theory. In this theory, the effect of the one- and two- electron excited Slater determinants are built from the input CASSCF wave function and their mixing with the reference state is accounted perturbatively [1]. It gives some improvement over the CASSCF wave functions. These calculations were also done using the OpenMolcas software and ANO-RCC-VDZP basis set [8].

2.5 Spin-Orbit Coupling in *Ab initio* Framework

Spin orbit coupling plays an important role for the 5d elements. In this work, spin-orbit interaction is accounted within the one-center mean field approximation (AMFI) using the CASSCF or CASPT2 states as input. An effective Hamiltonian is built using the input states and all matrix elements of this Hamiltonian are evaluated. Finally, the effective spin-orbit Hamiltonian is diagonalised giving the final spin-orbit energy states and related properties. These calculations were done by RASSI program from OpenMOLCAS. Magnetic properties (magnetic susceptibility) were evaluated using the SINGLE_ANISO program from OpenMOLCAS.

3 Results and Discussions

3.1 *Ab Initio* Calculations: CASSCF/CASPT2/SOC

The following energy spectra show the *Ab initio* energy states and fitted energy states for $[ReCl_6]^{3-}$ and the Re(III) molecule using $^3T_1$ and t_{2g}^4 models. The $^3T_1$ model produces nine states, whereas the t_{2g}^4 model will have fifteen states. The ligand field parameters were fitted to the *Ab initio* calculations to obtain the fitted energy spectra. The spectra compares the CASSCF, CASSCF fitted, CASPT2 and CASPT2 fitted energies (Fig. 2).

The values for the $^3T_1$ model were able to fit extremely well, with the values of the minimised objective function being orders of magnitude below 10^{-4}, whereas the values of the minimised objective function for the t_{2g}^4 model, using the first nine states, ranged from 13 to 192.

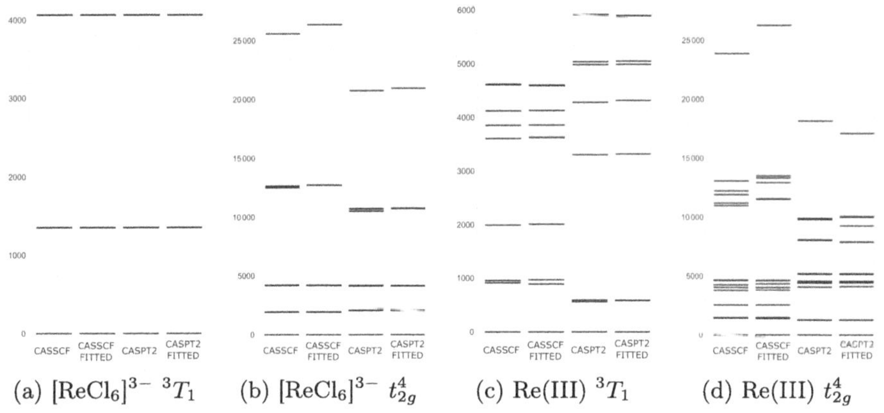

Fig. 2. *Ab initio* and fitted energy spectra for $[ReCl_6]^{3-}$ and the Re(III) molecule for both $^3T_1$ and t_{2g}^4 configurations

3.2 Ligand Field Analysis of *Ab Initio* Results

The Eq. 1 is first expanded in the basis of spin-orbit coupled states of the $\tilde{L} = 1$ and $S = 1$ corresponding to the ground $^3T_1$ (near) degenerate state. All in all, there are nine model functions, each characterized by its own set of projections of the orbital moment and spin moments along the quantization axis, $|M_L, M_S\rangle$. Equation 2 shows the expanded model Hamiltonian in this basis of functions. Diagonalization of this 9×9 matrix gives nine energy states which can be put in correspondence to the *ab initio* states. We find parameters for the model Hamiltonian matrix by fitting model states to the *ab initio* results by minimizing the standard deviation between the two sets of energies. For these purposes, we wrote a Python program that gives us the best fit of four parameters for the energy values given. Firstly, we define the five parameters: J, λ, Δ, ϵ and κ. J sets the energy gap between the $^3T_1$, $^1T_2$, E_1 and $^1A_1$ states of the t_{2g}^4 model. κ represents the orbital reduction factor due to covalence, λ defines the magnitude of the spin-orbit coupling, Δ defines the "axial" distortion, axial splitting of the t_{2g} shell, and ϵ defines the "rhombic" distortion.

$$\begin{pmatrix} \frac{\Delta}{3} - k\lambda & 0 & 0 & 0 & 0 & 0 & \epsilon & 0 & 0 \\ 0 & \frac{\Delta}{3} & 0 & -k\lambda & 0 & 0 & 0 & \epsilon & 0 \\ 0 & 0 & \frac{\Delta}{3} + k\lambda & 0 & -k\lambda & 0 & 0 & 0 & \epsilon \\ 0 & -k\lambda & 0 & -\frac{2\Delta}{3} & 0 & 0 & 0 & 0 & 0 \\ 0 & 0 & -k\lambda & 0 & -\frac{2\Delta}{3} & 0 & -k\lambda & 0 & 0 \\ 0 & 0 & 0 & 0 & 0 & -\frac{2\Delta}{3} & 0 & -k\lambda & 0 \\ \epsilon & 0 & 0 & 0 & -k\lambda & 0 & \frac{\Delta}{3} + k\lambda & 0 & 0 \\ 0 & \epsilon & 0 & 0 & 0 & -k\lambda & 0 & \frac{\Delta}{3} & 0 \\ 0 & 0 & \epsilon & 0 & 0 & 0 & 0 & 0 & \frac{\Delta}{3} - k\lambda \end{pmatrix} \quad (2)$$

$$\left(\begin{array}{ccccccccccccccc}
-\frac{\sqrt{3}d}{2}+\frac{\epsilon}{2}-J & 0 & 0 & -\frac{i\lambda}{2} & 0 & 0 & 0 & \frac{\lambda}{2\sqrt{2}} & 0 & \frac{i\lambda}{2\sqrt{6}} & -\frac{i\lambda}{2\sqrt{2}} & 0 & 0 & \frac{\lambda}{2\sqrt{2}} & \frac{i\lambda}{\sqrt{3}} \\
0 & -\frac{\sqrt{3}d}{2}+\frac{\epsilon}{2}-J & 0 & 0 & 0 & 0 & -\frac{\lambda}{2\sqrt{2}} & 0 & \frac{\lambda}{2\sqrt{2}} & 0 & 0 & 0 & -\frac{i\lambda}{2} & 0 & 0 \\
0 & 0 & -\frac{\sqrt{3}d}{2}+\frac{\epsilon}{2}-J & 0 & 0 & \frac{i\lambda}{2} & 0 & -\frac{\lambda}{2\sqrt{2}} & 0 & -\frac{i\lambda}{2\sqrt{6}} & \frac{i\lambda}{2\sqrt{2}} & 0 & 0 & \frac{\lambda}{2\sqrt{2}} & -\frac{i\lambda}{\sqrt{3}} \\
\frac{i\lambda}{2} & 0 & 0 & \frac{\sqrt{3}d}{2}+\frac{\epsilon}{2}-J & 0 & 0 & 0 & \frac{i\lambda}{2\sqrt{2}} & 0 & -\frac{\lambda}{2\sqrt{6}} & -\frac{\lambda}{2\sqrt{2}} & 0 & 0 & -\frac{i\lambda}{2\sqrt{2}} & -\frac{\lambda}{\sqrt{3}} \\
0 & 0 & 0 & 0 & \frac{\sqrt{3}d}{2}+\frac{\epsilon}{2}-J & 0 & \frac{i\lambda}{2\sqrt{2}} & 0 & \frac{i\lambda}{2\sqrt{2}} & 0 & 0 & -\frac{i\lambda}{2} & 0 & 0 & 0 \\
0 & 0 & -\frac{i\lambda}{2} & 0 & 0 & \frac{\sqrt{3}d}{2}+\frac{\epsilon}{2}-J & 0 & \frac{i\lambda}{2\sqrt{2}} & 0 & -\frac{\lambda}{2\sqrt{6}} & -\frac{\lambda}{2\sqrt{2}} & 0 & 0 & \frac{i\lambda}{2\sqrt{2}} & -\frac{\lambda}{\sqrt{3}} \\
0 & -\frac{\lambda}{2\sqrt{2}} & 0 & 0 & -\frac{i\lambda}{2\sqrt{2}} & 0 & -e-J & 0 & 0 & 0 & 0 & \frac{\lambda}{2\sqrt{2}} & -\frac{i\lambda}{2\sqrt{2}} & 0 & 0 \\
\frac{\lambda}{2\sqrt{2}} & 0 & -\frac{\lambda}{2\sqrt{2}} & -\frac{i\lambda}{2\sqrt{2}} & 0 & -\frac{i\lambda}{2\sqrt{2}} & 0 & -e-J & 0 & -\frac{i\lambda}{\sqrt{3}} & 0 & 0 & 0 & 0 & i\sqrt{\frac{2}{3}}\lambda \\
0 & \frac{\lambda}{2\sqrt{2}} & 0 & 0 & -\frac{i\lambda}{2\sqrt{2}} & 0 & 0 & 0 & -\epsilon-J & 0 & 0 & \frac{\lambda}{2\sqrt{2}} & \frac{i\lambda}{2\sqrt{2}} & 0 & 0 \\
-\frac{i\lambda}{2\sqrt{6}} & 0 & \frac{i\lambda}{2\sqrt{6}} & -\frac{\lambda}{2\sqrt{6}} & 0 & -\frac{\lambda}{2\sqrt{6}} & 0 & \frac{i\lambda}{\sqrt{3}} & 0 & e+J & -d & 0 & 0 & 0 & \sqrt{2}e \\
\frac{i\lambda}{2\sqrt{2}} & 0 & -\frac{i\lambda}{2\sqrt{2}} & -\frac{\lambda}{2\sqrt{2}} & 0 & -\frac{\lambda}{2\sqrt{2}} & 0 & 0 & 0 & -d & J-e & 0 & 0 & 0 & \sqrt{2}d \\
0 & 0 & 0 & 0 & \frac{i\lambda}{2} & 0 & \frac{\lambda}{2\sqrt{2}} & 0 & \frac{\lambda}{2\sqrt{2}} & 0 & 0 & -\frac{\sqrt{3}d}{2}+\frac{\epsilon}{2}+J & 0 & 0 & 0 \\
0 & \frac{i\lambda}{2} & 0 & 0 & 0 & 0 & \frac{i\lambda}{2\sqrt{2}} & 0 & -\frac{i\lambda}{2\sqrt{2}} & 0 & 0 & 0 & \frac{\sqrt{3}d}{2}+\frac{\epsilon}{2}+J & 0 & 0 \\
\frac{\lambda}{2\sqrt{2}} & 0 & \frac{\lambda}{2\sqrt{2}} & \frac{i\lambda}{2\sqrt{2}} & 0 & -\frac{i\lambda}{2\sqrt{2}} & 0 & 0 & 0 & 0 & 0 & 0 & 0 & J-e & 0 \\
-\frac{i\lambda}{\sqrt{3}} & 0 & \frac{i\lambda}{\sqrt{3}} & -\frac{\lambda}{\sqrt{3}} & 0 & -\frac{\lambda}{\sqrt{3}} & 0 & -i\sqrt{\frac{2}{3}}\lambda & 0 & \sqrt{2}e & \sqrt{2}d & 0 & 0 & 0 & 4J
\end{array}\right) \tag{3}$$

Using the minimization function provided by the Python library SciPy, we fit the four ligand-field parameters with the *ab initio* results to obtain the values of the four ligand-field parameters. To ensure the accuracy of our results, we have put some constraints on the parameters. κ is bounded between 0 and 1, $\mid \frac{\epsilon}{\Delta} \mid \leq \frac{1}{3}$ and $\lambda \leq 0$. The objective function used is given in the following equation:

$$F = \frac{1}{n}\sum_{i=1}^{n}\left(E_i - E_i^{'}\right)^2 \tag{4}$$

where n is the number of energy states

Table 1. Calculated Ligand-Field Parameters for the Re(III) molecule

			J	λ	Δ	ϵ	κ
$[ReCl_6]^{3-}$	$^3T_1$	CASSCF	-	−1371.4	0.0020387	−0.000014197	0.98835
		CASPT2	-	−1412.1	0.51192	−0.17064	0.95989
	t_{2g}^4	CASSCF	4499.3	−2712.1	0.0000	0.0000	-
		CASPT2	3364.3	−2698.3	0.0000	0.0000	-
Re(III) Molecule	$^3T_1$	CASSCF	-	−1321.5	1681.8	−175.35	0.96842
		CASPT2	-	−2493.4	3610.9	−3.2887	0.51639
	t_{2g}^4	CASSCF	4453.8	−2569.0	−2569.1	−353.64	-
		CASPT2	2066.6	−2384.8	−2107.0	−1166.8	-

We have thus obtained an effective ligand field model with the 4 ligand field parameters which are able to describe the transition metal complex's properties (Table 1).

As you can see, for both $[ReCl_6]^{3-}$ and the Re(III) molecule in the $^3T_1$ model, the magnitude of λ is larger in the CASSCF calculations then in the CASPT2 calculations, but is smaller in the t_{2g}^4 model. Since CASPT2 aims at estimating the correlation energy for a CASSCF wavefunction, the magnitude of the spin-orbit coupling will be greater, hence giving a λ with a greater magnitude.

We can also see that the value for Δ and ϵ is close to zero for $[ReCl_6]^{3-}$ and insignificant, but is significant for the Re(III) molecule. This is what we expect, because $[ReCl_6]^{3-}$ is a perfect octahedral compound, and has very little axial and rhombic distortion, whereas the Re(III) molecule has significant distortion to its octahedral symmetry.

A κ close to 1 indicates a lower degree of molecular orbital formation, or a "pure" crystal field model [3], and vice versa. In the case of $[ReCl_6]^{3-}$, κ is close to 1, showing that it can be described well with the use of crystal field theory. The value of κ for the Re(III) molecule in CASPT2 is almost half that of CASSCF, indicating that CASPT2 calculations take into greater account the effect of molecular orbitals on the energy states. In the t_{2g}^4 model, the value of J is smaller in CASPT2 calculations. This is also expected as CASPT2 calculations are generally closer to the true energy value, which is lower in energy.

In the $^3T_1$ model, we can see a relation between λ and κ, where the product, $\lambda\kappa$, remains relatively constant between CASSCF and CASPT2 calculations. The parameter κ scales down the spin-orbit constant λ of the free metal ion. This effect is obtained by covalence: when the 5d orbitals mix with ligand orbitals, they form hybrid orbitals, which are somewhat larger than the original 5d orbitals, and as result their orbital momentum decreases. Orbital momentum (λ) decreases as a result of different mixing between 5d orbitals and ligands (t_{2g} orbitals and e_g orbitals will acquire different levels of ligand contribution) [3,10].

In the t_{2g}^4 model, we can observe that when λ decreases, J increases. With the effect of covalence, the on-site Coulomb repulsion is also reduced: when the orbital gets a bit larger, orbital momentum (λ) decreases, and the electrons inside have less repulsion, scaling down parameter J.

3.3 Assumptions and Limitations

One of the limitations is that that we only obtained data in the gaseous phase. This means that our model may not be applicable in other phases. Additionally, our best model was a t_{2g}^4 model which may not include crucial electrons. Hence, we aim to improve the accuracy of the simple model by including these orbitals by using a $5d^4$ model. We were also unable to obtain true empirical values to confirm our data and only used pre-existing computational methods to verify our model, which might not be the most accurate.

4 Conclusions

In conclusion, we have studied two Re(III) entities by *ab initio* computational methods: an octahedral $[ReCl_6]^{3-}$ ion and a neutral Re(III) molecule. Based on the low-lying energy states, we built two effective ligand field models involving the states of $^3T_1$ and t_{2g}^4 configurations. We evaluated the basic five parameters: energy gap (J), spin-orbit coupling (λ), axial (Δ), and rhombic distortion (ϵ) parameters, as well as the orbital reduction factor (κ). We were able to obtain energy spectra that fit well with *ab initio* values using both models, with the

$^{3}T_1$ configuration having a better fit than the t_{2g}^4 configuration. The values of the parameters obtained were as expected and correspond to the structure of the investigated compounds.

Acknowledgement. This research has received no external funding.

Disclosure of Interests. The authors do not have any conflicts of interest to declare.

References

1. Andersson, K., Malmqvist, P.A., Roos, B.O., Sadlej, A.J., Wolinski, K.: Second-order perturbation theory with a CASSCF reference function. J. Phys. Chem. **94**(14), 5483–5488 (1990). https://doi.org/10.1021/j100377a012
2. Gerloch, M., Constable, E.C.: Transition Metal Chemistry: The Valence Shell in D-block chemistry, pp. 64–65. VCH (1994)
3. Gerloch, M., Miller, J.: Covalence and the orbital reduction factor, k, magnetochemistry. Progress Inorgan. Chem. **10**, 1–47 (2007). https://doi.org/10.1002/9780470166116.ch1
4. Jahn, H.A., Teller, E., Donnan, F.G.: Stability of polyatomic molecules in degenerate electronic states - i-orbital degeneracy. Proc. R. Soc. Lond. Ser. A - Math. Phys. Sci. **161**(905), 220–235 (1937). https://doi.org/10.1098/rspa.1937.0142
5. Jensen, F.: Introduction to Computational Chemistry, pp. 16–17. Wiley, Hoboken (2017)
6. Marinova, V., et al.: Optical properties of Bi12SiO20 single crystals doped with 4D and 5D transition elements. J. Appl. Phys. **89**(5), 2686–2689 (2001). https://doi.org/10.1063/1.1344918
7. Notman, N.: Transition metal catalysts (2021). https://edu.rsc.org/feature/transition-metal-catalysts/4013803.article
8. Roos, B.O.: Multiconfigurational Quantum Chemistry. Lund University, Sweden (2009)
9. Roos, B.O., Taylor, P.R., Sigbahn, P.E.: A complete active space SCF method (CASSCF) using a density matrix formulated super-ci approach. Chem. Phys. **48**(2), 157–173 (1980). https://doi.org/10.1016/0301-0104(80)80045-0
10. Streltsov, S.V., Khomskii, D.I.: Jahn-teller effect and spin-orbit coupling: friends or foes? Phys. Rev. X **10**, 31–43 (2020). https://doi.org/10.1103/PhysRevX.10.031043

StainNorm: A Lightweight and Robust Network for Stain Colour Normalization in Histopathology

Reiner Thian Juea Sim(✉)

NUS High School of Math and Science, Singapore, Singapore
h1910141@nushigh.edu.sg

Abstract. This paper introduces "StainNorm," a novel solution for addressing the challenge of stain colour normalization in histopathology images. Stain colour inconsistency presents a significant hurdle to accurate analysis in digital pathology. To tackle this issue, I present an innovative CycleGAN architecture with a 1 × 1 kernel, enabling efficient yet robust stain colour normalization.

StainNorm is rigorously evaluated across diverse datasets, including Stardist's training data (TNBC-TMA and MoNuSeg 2018), an in-house whole-slide image (WSI) dataset, and pathologist-curated TNBC images for validation. Custom models are trained for each use case. A key breakthrough in StainNorm is the adoption of a streamlined training process, leveraging multiphase training with an identity transform on the source set and gradually incorporating colour jitter using a task-based curriculum learning strategy. This innovation reduces training time by 40 times, allowing training on systems with limited resources. StainNorm's efficacy is demonstrated through comprehensive comparisons against established techniques, outperforming state-of-the-art deep learning methods in normalizing images to the target distribution, in structural similarity index measure (SSIM), Jensen-Shannon (JS) distance and computational complexity.

StainNorm excels in normalizing background areas which are often poorly addressed by StainNet, the previous state of the art (SOTA). While deep learning methods often demand substantial resources, StainNorm's efficiency bridges this gap by minimising training time, making it practical for integration into histopathology laboratories.

Keywords: stain colour normalization · histopathology · CycleGAN · efficient architecture

1 Introduction

Histopathology, a cornerstone of clinical diagnosis and research in medicine, relies on examining tissue specimens under a microscope to uncover valuable insights about diseases and their progression. In the digital age, histopathology has been empowered by the digitization of tissue samples, leading to the creation of repositories of histopathology images that hold potential for improved diagnostics and treatments. However, analysis

H. Guo et al. (Eds.): IRC-SET 2024, PTT, pp. 323–330, 2025.
https://doi.org/10.1007/978-981-96-3770-6_33

of these digital histopathology images presents challenges, from inconsistent staining between labs to variations in the digitization process, causing differences in stain colour that hinder accurate interpretation and analysis [1].

Staining techniques, such as haematoxylin and eosin (H&E) staining, play a pivotal role in highlighting specific tissue components and cellular structures, aiding pathologists in identifying abnormalities. Nevertheless, the staining process can lead to inconsistencies in colour representation across different samples and laboratories. This stain colour variation poses a significant hurdle in achieving reliable and consistent analysis of histopathology images [1], particularly in the context of automated image analysis and computer-aided diagnostics.

To address the challenge of stain colour normalization in histopathology, I introduce "StainNorm," a novel solution that leverages state-of-the-art deep learning techniques for stain colour normalization in digital histopathology images. StainNorm is designed to bridge the gap between the necessity for accurate stain colour normalization and the constraints of limited computational resources commonly encountered in histopathology laboratories.

A key innovation of StainNorm is its utilization of a specialized CycleGAN architecture incorporating a 1×1 kernel. Unlike conventional approaches that rely on larger convolutional networks, the adoption of a compact architecture enables efficient yet robust stain colour normalization, especially for structural preservation. This distinctive approach ensures a much smaller parameter space to search, while preventing structural loss through blurring caused by larger convolutions where structural information is altered from the original image, avoiding a common problem with convolutional neural networks (CNNs).

Central to the success of StainNorm is its streamlined training process, which combines pretraining with an identity transform on the target dataset and a gradual addition of colour jitter through a task-based curriculum learning strategy. This strategic approach not only enhances its adaptability to diverse datasets and staining variations but also allows StainNorm to significantly reduce the training time by a remarkable factor of 40x, also enabling training even on resource-constrained systems with minimal computational resources.

In this paper, I comprehensively evaluate the efficacy of StainNorm using 3 datasets, including publicly available benchmarks and in-house whole-slide images. Furthermore, I benchmark StainNorm against established stain normalization techniques, from a selection of deconvolution-based methods, global transform methods, StainNet and StainGAN. I employ a range of evaluation metrics, including histogram similarity and performance on downstream tasks, to objectively measure the performance of StainNorm.

Our findings reveal that StainNorm outperforms state-of-the-art deep learning methods, particularly in terms of the histogram similarity between normalized images and the target distribution. Additionally, StainNorm demonstrates exceptional performance in normalizing background areas, which are often challenging to address using traditional methods.

Importantly, the efficiency of StainNorm in terms of training time reduction improves practical implementation within histopathology laboratories. While deep learning methods have demonstrated superior performance in various image analysis tasks, they often demand substantial computational resources, limiting their applicability in the resource- and time-constrained setting of many laboratories. StainNorm's training time addresses this limitation, making it a practical and valuable tool for stain colour normalization in real-world histopathology workflows.

I demonstrate StainNorm's efficacy through comparisons against established techniques, showing that it outperforms state-of-the-art deep learning methods in normalizing images to the target distribution. StainNorm has the potential to enhance both the accuracy and efficiency of stain colour normalization.

2 Methodology

2.1 Datasets

TNBC-TMA and MoNuSeg 2018. These datasets consist of histopathology images and corresponding annotations of nuclei segmentation, including Triple-Negative Breast Cancer Tissue Microarray (TNBC-TMA) [4] and the MoNuSeg 2018 [5] dataset with multiple organs. These datasets provide a diverse range of staining variations and tissue types, allowing us to comprehensively evaluate StainNorm's performance across different scenarios. I randomly cropped images to size 296×296.

For training the custom models, I used these datasets with augmentations as the source dataset, to cover a wide range of possible inputs. I used a subset of TNBC-TMA images selected by an experienced pathologist as reference images for the rule-based methods which take only one reference image.

In-House Whole-Slide Images (WSIs). To ensure the robustness of StainNorm, I curated an in-house WSI dataset containing a variety of tissue samples with some staining variations. This dataset reflects the real-world challenges encountered in clinical settings. Images were tiled with size 296×296 for training StainNorm under computing resource constraints. This also introduced regularization for the discriminator, with similar morphology across images.

Pathologist-Curated TNBC Images. A set of histopathology images from Triple-Negative Breast Cancer (TNBC) was curated by experienced pathologists. These images serve as a validation benchmark to assess the clinical impact of stain colour normalization, and as an example application of StainNorm in stain colour normalization.

2.2 StainNorm Architecture

StainNorm is based on a CycleGAN architecture, a method which has proven highly successful in style transfer tasks. The architecture involves two generators (GX2Y and GY2X) and two discriminators (DX and DY), giving it better texture and colour transferability along with better structure preservation than a vanilla GAN. The model is trained on a source set and a target set, for unpaired style transfer from source set to the target set. The 1×1 kernel size in the convolutional layers gives a lightweight yet effective architecture, and further contributes to structural preservation.

The training process of StainNorm is designed to be streamlined, using multiphase training, curriculum learning, and background regularization to reduce training time and increase robustness. This progressive refinement aids in capturing intricate colour variations and textures present in histopathology images. Curriculum learning further enhances the model's adaptability by introducing colour jitter in a controlled manner, starting with minimal perturbations and gradually increasing the intensity. This ensures that the model learns to handle a wide range of staining inconsistencies effectively. Additionally, the integration of background regularization techniques allows StainNorm to normalize slides over background areas while preserving critical diagnostic details, maintaining visual consistency. These strategies contribute to a more efficient training process, enabling StainNorm to be highly robust despite the small model.

Multiphase Training. To avoid the instability of GAN models in our shortened training pipeline, I implemented multiphase training starting with learning the identity transform for input images in a pretraining phase. This helps the model to learn the structure of H&E images and guide the model towards outputs that look like H&E images. This initial phase also initializes GX2Y and GY2X to function similarly as an identity transform for the next phase of training because H&E images are unlikely to have very different pixel values. I then proceed to curriculum learning with an updated loss function.

Task-Based Curriculum Learning. The colour jitter transform introduces variations in intensity, saturation and hue, to an input image during training, which increases the model's robustness to different staining. After the initial training phase, colour jitter is gradually increased in source set for the model to learn stain colour normalization for images with increasingly diverse colour distributions, in much fewer parameter updates. By gradually increasing the complexity and intensity of the variations in staining the model has to contend with, StainNorm's training process guides the model through a better path towards more general stain colour normalization, greatly improving robustness at no cost to computational complexity.

Background Output Regularization. I introduce generated pure background images with a small amount of Gaussian noise approximating potential background colours, to be learned concurrently with the task-based curriculum learning phase. For these colours that are unlikely to appear in stained regions in brightfield staining, the model seeks to normalize these colours to the background colour of the target set. This results in much better handling of background areas, a difficult challenge to overcome for many existing methods.

3 Results

I chose a stain deconvolution and global transformation normalization method representative of the best existing methods. Macenko's PCA method [6] was compared to standard vectors and sparse nonnegative matrix factorization (SNMF) [7]. Background was taken to be the 99th percentile of colour intensity in each colour channel, and the resulting estimated colours and intensities of haematoxylin and eosin were used to normalize the images. I used Otsu thresholding on the "haematoxylin channel" of each image in TNBC-TMA and compared the result to the binary masks of nuclei to estimate accuracy of deconvolution. I also calculated entropy of the residual when deconvolution was applied to estimate information loss. Macenko's PCA method has the highest F1 score for haematoxylin thresholding and the lowest residual entropy, so I chose Macenko's PCA method as the method for deconvolution-based stain normalization. I also selected Reinhard's method [8] for global colour transformation and RGB histogram matching as a naïve solution (Table 1).

Table 1. Stain deconvolution methods for evaluation on TNBC-TMA dataset, as a comparison of these rule-based methods.

Method	Rand recall	Rand precision	Rand F1	Residual entropy
Standard vectors	0.01	0.60	0.02	13.76
Macenko	0.58	0.70	**0.63**	**11.84**
SNMF	0.86	0.38	0.54	13.77

From Fig. 1, it can be seen that StainNet and StainNorm are much closer to the reference image in overall colour than the other methods. Reinhard's method is visually indistinguishable from the original image while the naive solution of RGB histogram matching does not produce a colour distribution similar to the reference image. StainNorm has a slightly clearer background than StainNet due to its training process.

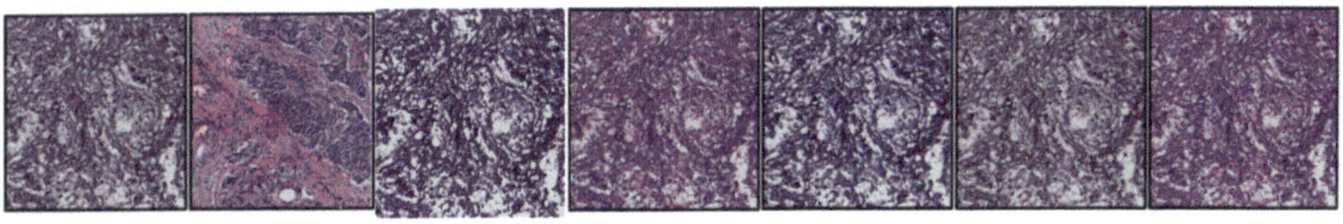

Fig. 1. From left to right: original image, reference image, Macenko normalized, StainNet normalized, RGB histogram matching, Reinhard normalized, StainNorm normalized

The histogram of the image normalized by StainNorm shows the clear spike at around 150 for the blue and red channels, corresponding to haematoxylin, and the red and green channels at 50 to 100 corresponding to eosin. The histogram is much closer to the reference histogram after normalization while preserving the features of the relative shape of the histogram, corresponding to structures in the image (Fig. 2).

This is confirmed by comparing the structural similarity and colour distances of the different methods, using pathologist-selected images from the TNBC-TMA dataset used in training of Stardist. Reinhard's method has the highest structural similarity between the normalized images and original images, but because Reinhard's method produces images with almost no difference between the normalized images and original images

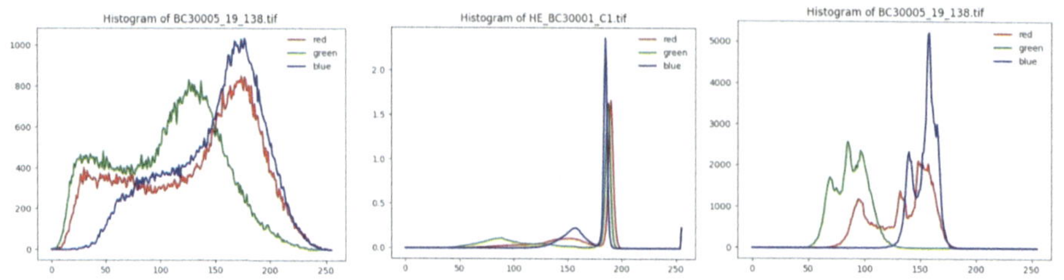

Fig. 2. From left to right: histogram of original image, histogram of reference image, histogram of StainNorm normalized image truncated at 188 for all colour channels.

in this case, this is an expected outcome while not yielding any improvements from normalization. The next highest SSIM is from StainNorm normalized images, which in this case indicates a good preservation of structures due to the 1 × 1 kernel architecture StainNorm shares with StainNet.

StainNorm can be shown to provide more robust normalization than all other methods. StainNorm has the lowest JS distance and EMD, giving normalized images with the least difference in colour histogram between normalized images and reference image (Table 2).

Table 2. SSIM, JS distance and Earth mover's distance (EMD) on 10 pathologist-curated images from TNBC-TMA dataset. Standard deviations are in brackets.

Method	SSIM	JS distance	EMD (values e-3)
StainNet	0.853 (0.013)	0.293 (0.013)	2.49 (0.21)
Macenko	0.797 (0.038)	0.327 (0.034)	2.56 (0.32)
RGB matching	0.820 (0.028)	-	-
Reinhard	**0.972 (0.021)**	0.251 (0.038)	2.39 (0.30)
StainNorm	0.889 (0.012)	**0.225 (0.015)**	**1.84 (0.26)**

StainNorm also provides a shorter training time, which takes about 10 min instead of up to an hour with StainNet depending on the dataset, and inference time comparable to rule-based methods. The initial training of StainGAN in StainNet's training phase is also avoided, thus training only the 1 × 1 kernel model of StainNorm. This is due to the novel training phase detailed in this paper. The training phase is also only run once when used as a preprocessing step, thus allowing normalization of out-of-distribution data for use with other AI models in histopathology analyses.

The efficiency of StainNorm in reducing training time is crucial for the adoption of efficient stain colour normalization in histopathology laboratories. Deep learning approaches, while offering advanced performance in various image analysis applications, frequently require extensive computational resources. This need can limit their practical use in many laboratories, which often face constraints related to both resources and time.

StainNorm addresses this issue by minimizing training duration, thereby enhancing its practicality as a tool for stain color normalization within real-world histopathology workflows. Our results demonstrate that StainNorm's architectural design and optimized training process effectively mitigate the computational challenges typically encountered in histopathology settings. By achieving a 40-fold reduction in training time, StainNorm proves to be a feasible solution for integration into operational workflows, even on systems with limited computational capacity.

4 Discussion

This paper introduces StainNorm, a novel stain colour normalization solution for histopathology images. Through a rigorous evaluation, we have demonstrated its efficacy in addressing the challenge of stain colour inconsistency, a crucial factor affecting the accuracy of histopathology analysis.

The architecture of StainNorm, featuring a CycleGAN with a 1×1 kernel, highlights its ability to achieve stain colour normalization efficiently without increasing computational complexity. This architectural choice, coupled with multiphase training, curriculum learning, and background regularization, contributes to both the reduction of training time and the enhancement of the model's adaptability. The comprehensive evaluation against established techniques underscores StainNorm's superiority in retaining structural information and achieving visual consistency in normalized images.

Importantly, the efficiency of StainNorm in terms of training time reduction has significant implications for its practical implementation within histopathology laboratories. While deep learning methods have demonstrated superior performance in various image analysis tasks, they often demand substantial computational resources, limiting their applicability in the resource- and time-constrained setting of many laboratories. StainNorm's efficiency in reducing training time and memory addresses this limitation, making it a practical and valuable tool for stain colour normalization in real-world histopathology workflows.

Our findings show that StainNorm's architecture along with its streamlined training process effectively bridges the gap between deep learning-based stain normalization and the computational limitations often encountered in histopathology laboratories. By reducing training time by a factor of 40, StainNorm becomes a practical solution for integration into real-world workflows, even on systems with limited resources. This aspect is particularly significant given the growing demand for accurate and efficient stain colour normalization in digital pathology.

5 Conclusion

Our findings demonstrate that StainNorm outperforms previous state-of-the-art deep learning methods as well, excelling in terms of the Jensen-Shannon (JS) distance between normalized images and the target distribution. Furthermore, StainNorm displays superior normalization in background areas, while having a lower computational complexity.

While deep learning methods generally exhibit enhanced performance compared to rule-based techniques, they often demand substantial computational resources, limiting their practicality within histopathology laboratories. StainNorm effectively bridges this gap by significantly reducing training time, thereby enhancing its feasibility for deployment in real-world histopathology settings. Also, the proposed solution could be applied to images other than H&E, in the more general case of style transfer for histological images.

In this study, our findings correlate with a much faster training regime for stain colour normalization, which performs comparably to the current state-of-the-art while using much less computational power. Therefore, I propose further research to take context into account with contextual information for downstream tasks, and to further shorten the process using zero-shot learning on the current results.

References

1. Michielli, N., et al.: Stain normalization in digital pathology: clinical multi-center evaluation of image quality. J. Pathol. Inform. **13**, 100145 (2022)
2. Schmidt, U., Weigert, M., Broaddus, C., Myers, G.: Cell detection with star-convex polygons. In: Frangi, A.F., Schnabel, J.A., Davatzikos, C., Alberola-López, C., Fichtinger, G. (eds.) MICCAI 2018. LNCS, vol. 11071, pp. 265–273. Springer, Cham (2018). https://doi.org/10.1007/978-3-030-00934-2_30
3. Kang, H., et al.: StainNet: a fast and robust stain normalization network. Front. Med. **8** (2021)
4. Naylor, P., Lae, M., Reyal, F., Walter, T.: Segmentation of nuclei in histopathology images by deep regression of the distance map. IEEE Trans. Med. Imaging **38**(2), 448–459 (2019)
5. Kumar, N., et al.: A multi-organ nucleus segmentation challenge. IEEE Trans. Med. Imaging **39**(5), 1380–1391 (2020)
6. Macenko, M., et al.: A method for normalizing histology slides for quantitative analysis. In: 2009 IEEE International Symposium on Biomedical Imaging: From Nano to Macro (2009)
7. Xu, J., et al.: Sparse non-negative matrix factorization (SNMF) based colour unmixing for breast histopathological image analysis. Comput. Med. Imaging Graph. **46**, 20–29 (2015)
8. Reinhard, E., Adhikhmin, M., Gooch, B., Shirley, P.: Colour transfer between images. IEEE Comput. Graph. Appl. **21**(5), 34–41 (2001)

Investigating the Role of Gelsolin in Developing Chemoresistance to 5-Fluorouracil Treatment

Nithilam Anandan[1], Reiner Thian Juea Sim[1], Shuo Deng[2,3](✉), and Celestial T. Yap[2,3](✉)

[1] NUS High School of Math and Science, Singapore, Singapore
[2] Department of Physiology, Yong Loo Lin School of Medicine, National University of Singapore, Singapore, Singapore
{phsdes,phsyapc}@nus.edu.sg
[3] NUS Centre for Cancer Research, Yong Loo Lin School of Medicine, National University of Singapore, Singapore, Singapore

Abstract. Colorectal cancer ranks second in cancer-related deaths globally. 5-fluorouracil (5-FU) is a common first-line chemotherapy drug, but acquired or intrinsic resistance to the drug remains a significant challenge. Gelsolin, an actin-binding cytoskeletal protein, is known to be involved in anti-apoptotic pathways and overexpression in cancer cells promotes their migration and invasion. Our work investigates the role of gelsolin in chemoresistance to 5-FU in colorectal cancer. Employing gain and loss of function models, we examined the role of gelsolin in 5-FU resistance via clonogenic assays and Western blot analysis, on gelsolin overexpression (OE), empty vector (EV), wild type (WT) and knock-out (KO) HCT116 colorectal cancer cells. Increased colony size and number were observed with higher gelsolin expression in clonogenic assays, showing less cell death and more colonies formed under treatment with 5-FU. Additionally, cells with elevated gelsolin levels showed reduced indicators of apoptosis, such as cleaved caspase-3 and PARP-1 in Western blot analysis. We conclude that lower gelsolin expression increases 5-FU-induced apoptosis in colorectal cancer. Our findings highlight gelsolin's role in 5-FU resistance in colorectal cancer, indicating its potential as a supplementary chemotherapy target and prognostic marker for 5-FU treatment. We also propose gelsolin as an adjunctive drug target and prognostic marker for colorectal cancer with 5-FU treatment.

Keywords: Colorectal Cancer · Chemoresistance · 5-Fluorouracil (5-FU) · Gelsolin · Apoptosis · Cancer Biomarker

1 Introduction

Gelsolin regulates the assembly and disassembly of the actin filaments, which is the most abundant protein in most eukaryotic cells and form filaments in the cytoskeleton and muscle fibrils. Gelsolin also has one of the highest actin-cleaving efficiency in the super-family of actin-binding proteins [1]. Gelsolin has also been found to enhance or inhibit

N. Anandan and R. T. J. Sim—Contributed equally.

H. Guo et al. (Eds.): IRC-SET 2024, PTT, pp. 331–339, 2025.
https://doi.org/10.1007/978-981-96-3770-6_34

apoptosis depending on the pathological pathways and cell type. Gelsolin enhances apoptotic activity through the gelsolin-HIF1-α-Deribonuclease I (DNase) pathway, where the Hypoxia-inducible factor (HIF1-α) contributes to the transcription of DNase and the transcription of gelsolin. There is also evidence of gelsolin inhibiting apoptosis where the protein block voltage-dependent anion channel, an actin dependent mitochondrial porin that is key regulator of mitochondrial dependent cell death [2].

5-Fluorouracil (5-FU) is a widely used chemotherapeutic drug, specifically for colorectal cancer. The antimetabolite drug works mainly by acting as a thymidylate synthase (TS) inhibitor, which through inhabitation of TS, curbs the synthesis of pyrimidine thymidylate (dTMP). The action of the 5-FU drug causes a scarcity of dTMP which then leads to cell death due to its inability to undergo DNA replication [3, 4]. Vast amounts of data spotlight the need to further study the differences between patients to 5-FU treatment. Additionally, the drug resistance of colorectal cancer cells to 5-FU can be intrinsic or acquired during chemotherapy, which poses a great hurdle to effective chemotherapy and tumour management [5].

Therefore, through our experiments we intend to find a correlation between the expression of gelsolin the colorectal cancer cells and the resistance to 5-FU, hoping to gain greater insight into the mechanism of gelsolin and its relation to cancer treatments. The ability of gelsolin to function as a molecular biomarker for 5-FU based chemotherapy is also tested. Finally, greater understanding of the role of gelsolin in 5-FU chemotherapy will enable us to provide treatment of higher efficiency and understand the role of gelsolin in developing 5-FU resistance.

2 Methodology

2.1 Cell Culture

Stable gelsolin-overexpressing and gelsolin knockout HCT116 cells were generated previously [6]. Cells were cultured in a culture plate with McCoy medium with 10% Fetal Bovine Serum (FBS). After sufficient cell growth, the existing medium was removed from the plate and washed with phosphate-buffered saline (PBS). Subsequently 1 ml of trypsin was added to the plate. After the cells detach from the plate, additional medium is added where the FBS inhibits trypsin action, and the solution is pipetted up and down to ensure a single cell suspension. The cells are then counted using a hemocytometer and then the cell suspension is diluted to the desired seeding concentration.

2.2 Colony Formation Assay

Cells were plated at 10000 cells per well into 6-well plate and cultured with 2 ml McCoy medium with DMSO or indicated drugs for 10 days. The dosages used were 2.5 μM, 5 μM, 10 μM and 25 μM of 5-Fluorouracil and 25 μM of Dimethyl sulfoxide (DMSO) control. Cells were subsequently fixed and stained with 0.5% crystal violet in 30% ethanol, 3% formaldehyde for 10 min at room temperature. Stained colonies were counted with Stardist and ImageJ.

2.3 Western Blot Analysis

Gelsolin overexpressed (OE) and empty vector control (EV) HCT116 cells were plated at 100,000 cells per well into a 12-well plate and cultured with 1 ml of McCoy medium with DMSO or indicated drugs for 2 days. Cells were harvested and lysed with cold Laemmli cell lysis buffer, of 62.5 mM Tris pH6.8, 2% SDS, 10% Glycerol, supplemented with protease inhibitor cocktail. Cell lysate is heated in a hotplate at 100 °C for 5 min. After 30 min of incubation in ice, the lysate is clarified through centrifugation and the supernatant is isolated and stored in ice. 10 μg of total cell lysate were loaded and separated by SDS-PAGE, followed by transferring to PVDF membrane. Membranes were blocked with 5% bovine serum albumin (BSA) in TBST for 1 h. Primary antibody was then added to the membrane and incubated at 4 °C overnight. Subsequently, the membrane is washed with TBST three times, in 15-min intervals using a shaker, and secondary antibody is added and left to incubate at room temperature for 1 h. Membranes were incubated with ECL substrates and then visualized using an iBrightFL1000.

3 Results

3.1 Morphology

The morphology of gelsolin overexpression and empty vector HCT116 cells were observed to vary little between each other during cell culture (Fig. 1).

There is a greater proportion of surviving OE cells than EV cells. Before the 5-FU induced apoptosis, the EV cells were observed to tend to multiply in clumps while the OE cells tend to grow more singly. Additionally, the OE cell morphology was observed to be more spindle-like and irregular compared to the EV cells that were observed to be more circular and regular. This may suggest that the OE cells have a higher migratory tendency compared to EV cells, leading to the conclusion that gelsolin promotes the migration of cancer cells.

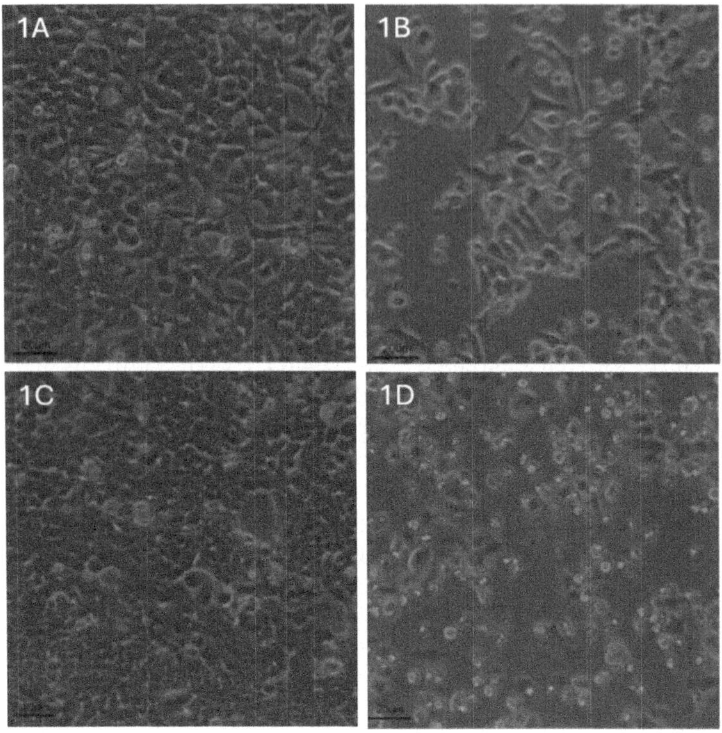

Fig. 1. 1A, 1B are OE cells after 48-h exposure to DMSO versus 50 uM 5-FU; 1C, 1D are EV cells after 48-h exposure to DMSO versus 50 uM 5-FU.

3.2 Clonogenic Assay

Comparing the gelsolin overexpression (OE) and empty vector (EV) cell lines, the number of OE colonies were constantly higher than the number of EV colonies after exposure to DMSO and different dosages of 5-Fluorouracil (Figs. 2 and 3).

The p-value is less than 0.05 for all dosages of 5-FU, hence we reject the null hypothesis at 5% significance and 95% confidence level and conclude that difference between the number of OE colonies and EV colonies is significant for all doses.

Comparing the gelsolin wild type (WT) and knockout (KO) cell lines, the number of WT colonies were constantly higher than the number of KO colonies after exposure to DMSO and different dosages of 5- Fluorouracil (Fig. 4).

The p-values is less than 0.05 for all dosages of 5-FU, hence we reject the null hypothesis at 5% significance and 95% confidence level and conclude that difference between the number of WT colonies and KO colonies is significant for all doses.

In both experiments, the strain with a higher expression of gelsolin, OE and WT, compared to that with the lower gelsolin expression, EV and KO respectively, led to a higher number of colonies alive. This leads to the conclusion that greater levels of gelsolin expression in cells give rise to higher resistance to 5-FU.

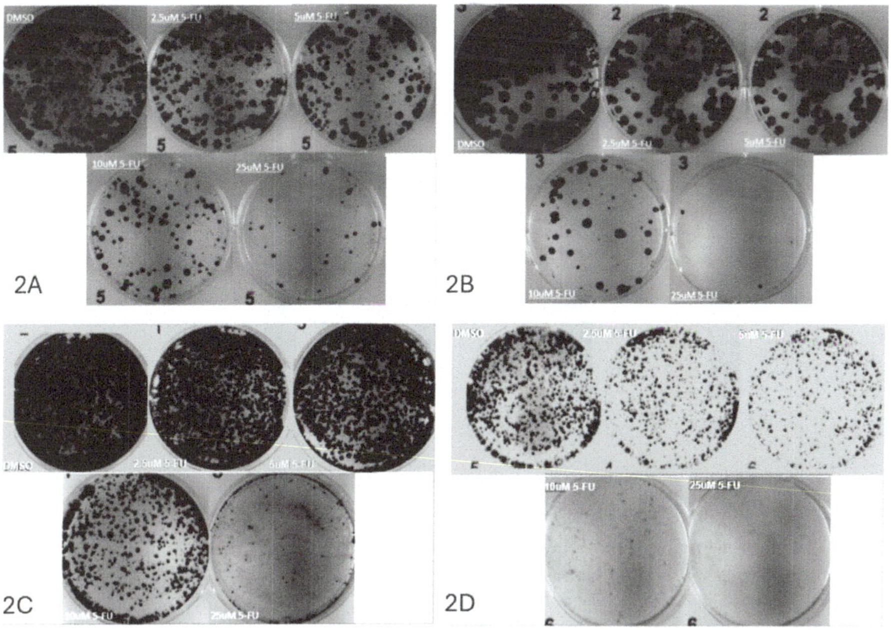

Fig. 2. Results after colony formation assay at all dosages. 2A is OE cells, 2B is EV cells, 2C is WT cells and 2D is KO cells.

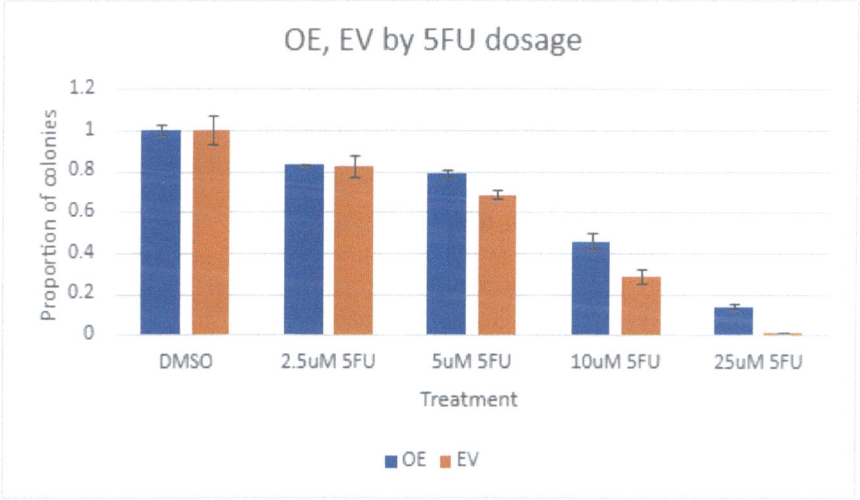

Fig. 3. Graph of normalized number of OE and EV colonies at the respective dosages after the colony formation assay.

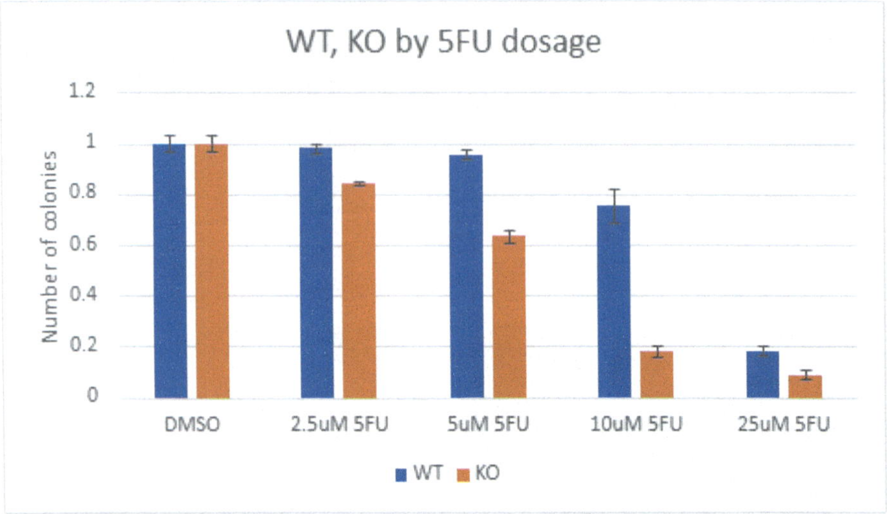

Fig. 4. Graph of normalized number of WT and KO colonies at the respective dosages after the colony formation assay.

This potentially shows that gelsolin has an inhibitory effect on apoptosis in cancer cells. Therefore, due to the higher gelsolin expression, the apoptotic action caused by 5-FU is reduced, resulting in the cells having greater resistance to the drug.

3.3 Western Blot Analysis

Western blot analysis for gelsolin was conducted. For all 3 trials, the bands for gelsolin at approximately 85kDa were lower for the EV samples and higher for OE samples as expected. The gelsolin levels for all EV samples and OE samples were consistent among themselves (Fig. 5).

We next conducted Western blot to test for caspase-3 cleavage. The protein band was located at approximately the 18kDa to 12kDa position, and the band of the EV samples were darker and bigger than the OE sample. Caspase-3 is a crucial mediator of cell death. The inactive form of caspase-3 (32kDa) is cleaved into 2 smaller active subunits (19kDa and 12kDa) and these active caspase molecules start a cascading reaction leading to major proteolytic action during apoptosis [7, 8]. Therefore, the higher presence of cleaved caspase-3 in the EV cell than OE cells indicated a higher level of apoptotic action in the EV cells after exposure to various dosages of 5-FU.

Poly-ADP ribose polymerase-1 (PARP-1) was then tested for its presence. The protein bands for uncleaved PARP-1 were found at approximately the 115kDa position, and that of cleaved PARP at 90kDa and 25kDa. The protein bands for uncleaved PARP-1 were darker for OE samples than EV samples while the bands of cleaved PARP were darker in EV than OE at both positions. PARP-1 is an enzyme that aids cells to repair any damage. A higher level of cleaved PARP expression indicates more extensive apoptotic action as cleaved PARP-1 inhibits further DNA repair, enabling apoptosis to occur faster [9, 10]. The higher levels of cleaved and lower levels of uncleaved PARP1 in EV than

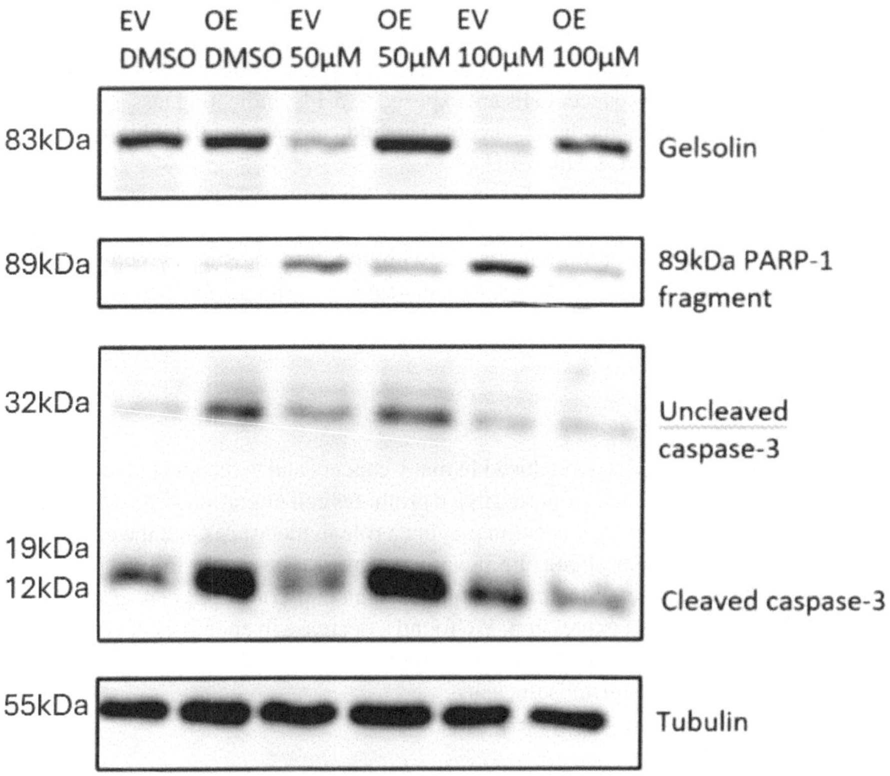

Fig. 5. Combined results of Western blot.

OE, shows that EV underwent higher levels of apoptosis that OE after 5-FU exposure. Tubulin was also tested to ensure similar levels of total protein.

We conclude that apoptotic action was higher in EV than OE cells, due to higher levels of cleavage of PARP-1 and caspase-3, key indicators of cell death, in EV compared to OE. Tubulin was similar for all cell lines tested, so the total protein is similar.

4 Conclusion

4.1 Clonogenic Assay

We establish that higher gelsolin expression causes less inhibition of cell death and colony formation when exposed to 5-FU, with more colonies formed in colorectal cancer cells with higher levels of gelsolin, indicating higher resistance of colorectal cancer cells with higher levels of gelsolin to 5-FU treatment, with gelsolin appearing to be protective against 5-FU exposure.

4.2 Western Blot

We establish that higher gelsolin expression causes less cleavage of caspase-3 and PARP-1 to occur when colorectal cancer cells are exposed to 5-FU, indicating less apoptosis in response to the first line chemotherapy agent. This indicates a potential inhibitory effect of gelsolin on apoptosis caused by 5-FU.

We conclude that higher gelsolin expression leads to higher resistance to 5-FU in cancer cells, indicating an inhibitory effect of gelsolin on 5-FU. However, cancer cells develop drug resistance in many ways, with apoptotic evasion by anti-apoptosis pathways, upregulation of drug efflux and enhanced DNA repair mechanisms being a few.

5 Discussion

Gelsolin has been reported to be reduced in many cancers and is recognized as a tumour suppressor gene, but it has been reported that it promotes cell migration and invasion. Previous studies have also reported gelsolin playing a role in metastasis, but the association is unclear. The involvement of gelsolin in various cancer types is not fully explained.

In this study, we detected increased drug resistance of colorectal cancer against 5-FU with increased gelsolin expression. We found a decrease in apoptosis of cancer cells in response to 5-FU, more colony formation and higher invasion with higher gelsolin levels, but the exact mechanism is unclear.

6 Future Work

More work can be done to discover the pathways involved in gelsolin-related chemoresistance. The role of gelsolin in developing resistance against other drugs can also be tested and compared with its role in 5-fluorouracil resistance. This can also shed light on how gelsolin affects apoptosis and drug resistance. Research can also be done with other cancer types to study if the effects of gelsolin expression affect drug resistance similarly to colorectal cancer cells.

In this study, our findings suggest that overexpressed gelsolin is involved in 5-FU resistance in colorectal cancer cells and is likely to correlate with a poor prognosis. Therefore, we propose that gelsolin might be a viable prognostic marker for 5-FU treatment, and that adjunctive treatments targeting gelsolin, used with 5-FU, might provide a way to enhance the effectiveness of chemotherapy of colorectal cancer.

References

1. Ghoshdastider, U., Popp, D., Burtnick, L.D., Robinson, R.C.: The expanding superfamily of gelsolin homology domain proteins. Cytoskeleton **70** (2013)
2. Li, G.H., Arora, P.D., Chen, Y., McCulloch, C.A., Liu, P.: Multifunctional roles of gelsolin in health and diseases. Med. Res. Rev. **32**(5), 999–1025 (2012)
3. Longley, D., Harkin, D., Johnston, P.: 5-Fluorouracil: mechanisms of action and clinical strategies. Nat. Rev. Cancer **3**, 330–338 (2003)

4. Sabeti Aghabozorgi, A., et al.: Molecular determinants of response to 5-fluorouracil-based chemotherapy in colorectal cancer: the undisputable role of micro-ribonucleic acids. World J. Gastrointest. Oncol. **12**(9), 942–956 (2020)
5. Gmeiner, W.H., Okechukwu, C.C.: Review of 5-FU resistance mechanisms in colorectal cancer: clinical significance of attenuated on-target effects. Cancer Drug Resist. **6**(2), 257–272 (2022)
6. Zhuo, J., et al.: Gelsolin induces colorectal tumor cell invasion via modulation of the urokinase-type plasminogen activator cascade. PLOS One **7**(8) (2012)
7. Silva, F.F.V.E., et al.: Caspase 3 and cleaved caspase 3 expression in tumorogenesis and its correlations with prognosis in head and neck cancer: a systematic review and meta-analysis. Int. J. Mol. Sci. **23**(19) (2022)
8. Liu, P.F., et al.: Expression levels of cleaved caspase-3 and caspase-3 in tumorigenesis and prognosis of oral tongue squamous cell carcinoma. PLoS ONE **12**(7), e0180620 (2017)
9. Ossovskaya, V., Koo, I.C., Kaldjian, E.P., Alvares, C., Sherman, B.M.: Upregulation of poly (ADP-Ribose) polymerase-1 (PARP1) in triple-negative breast cancer and other primary human tumor types. Genes Cancer **1**(8), 812–821 (2010)
10. Pignochino, Y., et al.: PARP1 expression drives the synergistic antitumor activity of trabectedin and PARP1 inhibitors in sarcoma preclinical models. Mol. Cancer **16**, 86 (2017)

Investigating the Viability of STM32 as a Low-Cost Control System for High Voltage Power Supplies

Zhen Hong Chia(✉), Maniyadesh R, and Bhuven Selvaraju

NUS High School of Mathematics and Science, Singapore, Singapore
chiazhenhong06@gmail.com

Abstract. High voltage power supplies (HVPS) have various applications in the field of charged particle optics such as in ion sources, ion detectors and electrostatic lens. For these equipment to perform optimally, the output voltage of the HVPS needs to be very stable.

To control the HVPS, an analogue control voltage is used. Current solutions utilize digital to analogue converters that can output a voltage which is both high resolution and very stable, such as the AD5791. However, they tend to be pricey and have long development times. In this paper, we aim to find the viability of using the STM32F407G-DISC1, a cheap microcontroller board with a built-in DAC, in controlling HVPS by investigating the ripple and long-term drift (LTD) of the output voltage.

We found that the STM32F407G-DISC1 output voltage's ripple was significantly greater than the AD5791, being 43.55 times greater at best and 414.69 times greater at worst. This resulted in the output voltage of a HVPS to have greater ripple and LTD when controlled by the STM32F407G-DISC1.

Despite the poor output voltage ripple and LTD, STM32 boards show promise to be used for rapid prototyping due to its quick development time. Further investigation can be done into other microcontrollers with a built-in DAC.

Keywords: High Voltage Power Supply · STM32F407G-DISC1 · STM32 · Voltage Stability · PMA1.5N · PMA20N · Voltage Ripple · Voltage Long-term drift

1 Introduction

1.1 High Voltage Power Supplies

High voltage power supplies (HVPS) are power conversion units with the ability to output high voltages of over a kilovolt [1]. Typically, they are powered by an external DC or AC input and controlled by either a built-in controller though a digital input or via an analogue control voltage over a fixed control voltage range.

HVPS serve many important uses in the field of charged particle optics as their ability to output high potential differences allows them to produce strong electric fields which are

H. Guo et al. (Eds.): IRC-SET 2024, PTT, pp. 340–349, 2025.
https://doi.org/10.1007/978-981-96-3770-6_35

used to manipulate charged particles such as ions and electrons. Some examples of their applications include ion sources like the penning ion source and the ion detectors like the Channeltron electron multiplier, which utilise the high potential difference generated by the HVPS to accelerate or decelerate electrons and ions respectively.

Fine control and repeatability of the HVPS output voltage is crucial, as any fluctuations can affect the number of ions produced by ion sources and cause ion detectors to give inaccurate measurements, which are both undesirable.

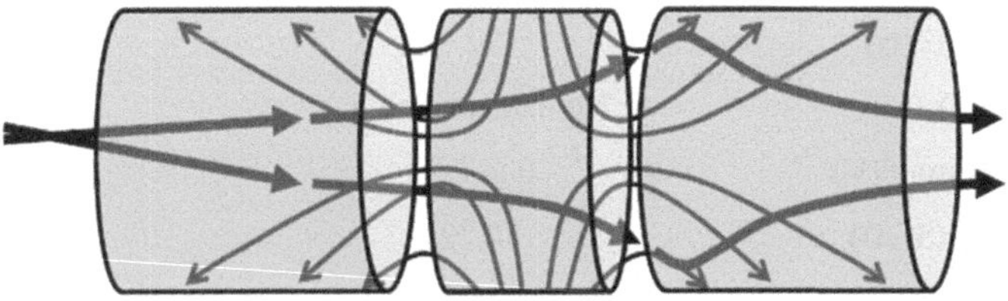

Fig. 1. Representation of an Einzel lens focusing a charged particle beam. The cylinders represent electrodes.

One common application where this is especially important is the Einzel lens (Fig. 1), an electrostatic lens used for focusing charged particles. It is comprised of three electrodes, and the middle electrode is brought to a high voltage by a HVPS while the other two electrodes are grounded [2]. To control the focal point of the lens the middle electrode's voltage is adjusted, so any deviation in the voltage will affect the spot size of the beam, and hence the best resolution of the system.

As such to test quality of the output voltage of the HVPS, we analyse two metrics, the momentary fluctuations in the voltage, or ripple, and the change in voltage over time, or long-term drift. The ripple determines the spot size of the system for short term applications such as imaging and the long-term drift affects the spot size of the system for long term applications such as lithography.

While the quality of the HVPS affects its ripple and long-term drift, another key factor is the ripple and long-term drift of the control voltage source. We worked on using the STM32F407G-DISC1, a low-cost microcontroller from ST microelectronics with a built in digital to analogue converter (DAC), to build a control system to supply a control voltage for the HVPS. We will then be comparing its performance to the AD5791, a commercially available 20-bit DAC. The specifications for both systems are shown in Table 1.

Table 1. Manufacturer specifications of the STM32F407G-DISC1[3] built in DAC and the AD5791[4]. The total cost of the AD5791 includes its evaluation board.

	STM32F407G-DISC1 (STM32)	AD5791
Resolution (bits)	12	20
Ripple (μV)	No data available	1.1
Voltage Range (V)	0 to 3.3	-10 to 10
Total Cost (SGD)	30.72	215

2 Methodology

To compare the STM32's built in DAC against the AD5791, we used each DAC to control a HVPS and recorded the output voltages. The HVPS we used for our testing are the PMA20N and PMA1.5N from Wisman High Voltage. Their specifications are shown in Table 2.

Table 2. Manufacturer specifications of the PMA20N and PMA1.5N [5].

	PMA20N	PMA1.5N
Input voltage (v)	0 to 10	0 to 10
Output voltage (kv)	−20 to 0	−1.5 to 0
Output voltage accuracy	±2%	±2%
Ripple (mv peak-peak)	40	3
Stability (ppm/hr)	100	100

2.1 HVPS Control

To control the AD5791 we used its evaluation board EVAL-AD5791SDZ, the associated microprocessor EVAL-SDP-CB1Z, and the evaluation software to control the board. As the STM32 does not come with software to control the DAC, we wrote a program in C to allow the STM32 to receive commands and adjust the DAC's voltage. We wrote the code and programmed the STM32 using STM32CubeIDE, which allows us to use the hardware abstraction layer (HAL) provided by ST Microelectronics easily. For this program, we used the USB Device libraries to enable the micro-usb port present on the STM32, which acts as a USB-to-RS232 bridge to realise a virtual COM port. This allows serial communication to the board, and also allows us to import the DAC library to control the analog output on pin PA4.

The STM32 was connected to a computer via the mini-usb port to provide the board power and allow for a program to be flashed onto it. To allow the computer to interface with the STM32, we connected a computer to the STM32 micro-usb port. The micro-usb port was set as a virtual communication (VCom) port and Putty was used as the serial client to send strings to the STM32. Commands were sent to the STM32 in the format shown in Fig. 2. Figure 3 shows how the commands are processed.

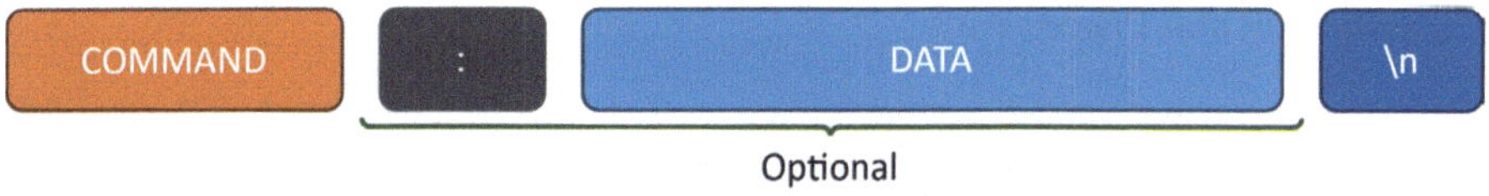

Fig. 2. Command syntax

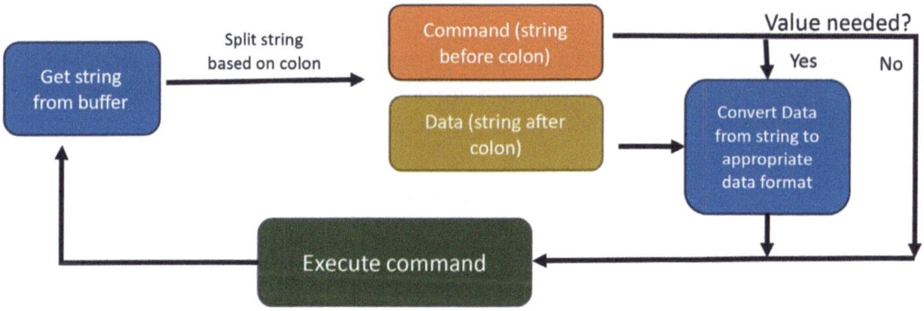

Fig. 3. How commands are parsed and executed.

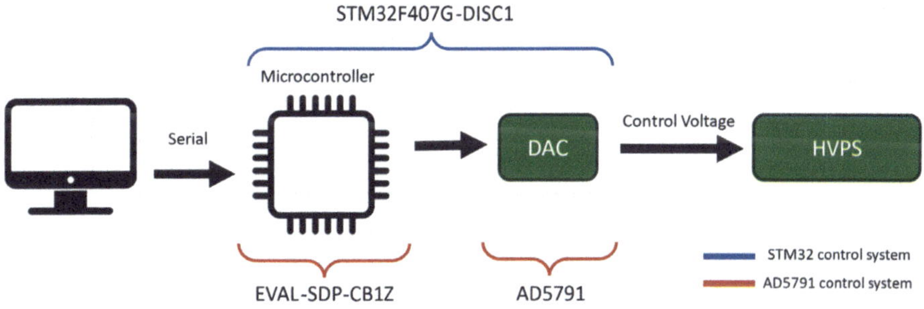

Fig. 4. Overview of how a typical HVPS system works.

2.2 Experimental Set-Up

Figure 4 shows an overview of the set-up. As the HVPS requires a control voltage of 0 to 10 V, we used an operational amplifier (NE5532p) to increase the voltage range of

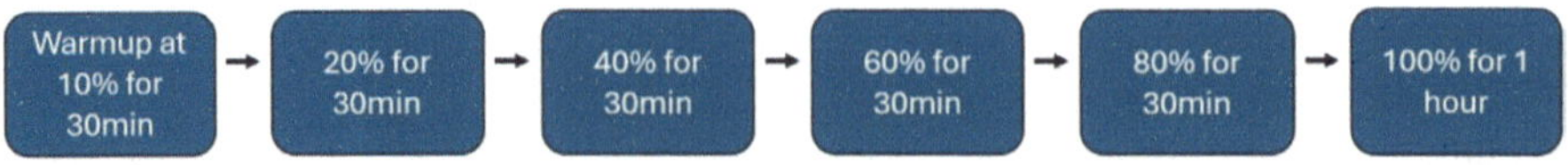

Fig. 5. HVPS voltage measuring procedure

the STM32 built-in DAC. Although the STM32 was reported to have a max voltage of 3.3 V, we found that the output voltage tended to be 0.3 V lower than expected so a gain of about 3.3 was used.

2.3 Experimental Procedure

We ran the experiment at different percentages of the maximum control voltage at different intervals as shown in Fig. 5. The warmup at 10% was conducted as per the manufacturer's instruction.

2.4 Measurement

To measure the voltages, we used a DMM6500 multimeter. To allow the output voltage of the HVPS to be safely measured within the range of the multimeter, we used a V1G high voltage probe to reduce the HVPS output voltage by a factor of 1000. We also measured the output voltage of the DAC, and readings provided by the built in voltage monitor (Vmon) and current monitor (Imon) in the HVPS. Figure 6 shows an overview of the set-up.

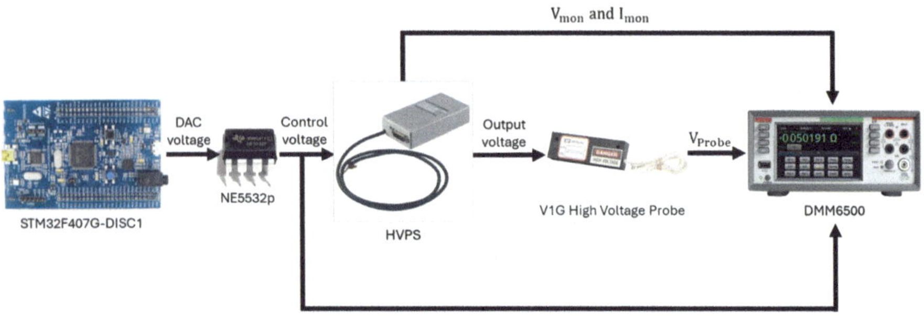

Fig. 6. Set-up to measure voltages.

3 Results and Discussion

From the voltage measurements, we calculated the ripple as twice the standard deviation (95% peak to peak) of the data and calculated the long-term drift by plotting the best fit line of the data and finding its gradient. We divided the values by the HVPS output voltage and multiplied it by 106 to express the values in the standard unit of ppm. We calculated the best 5 min of ripple and the ripple of the whole duration (approximately 30 min) as they would represent the accuracy of shorter operations like microscopy and longer operations like x-ray analysis respectively. Long-term drift would show the accuracy of the longest operations like lithography.

Tables 3 and 4 below are tables of our calculated results. High voltage arcing noises were heard while running the PMA20N at 80% and 100% control voltage. It was later confirmed that the arcing was occurring within the power supply. Due to this hardware problem, the STM32 Vprobe ripple for 80% and 100% control voltage is significantly higher than expected and thus will not be considered.

Table 3. PMA1.5N results

PMA1.5N	Control voltage percentage	STM32 Vprobe (ppm)	STM32 control voltage (ppm)	AD5791 Vprobe (ppm)	AD5791 control voltage (ppm)
Ripple Best 5 min (ppm)	10%	73.20	107.82	11.46	0.26
	20%	81.07	134.15	13.32	0.49
	40%	191.55	262.55	14.01	0.89
	60%	136.42	232.99	15.18	1.10
	80%	155.82	142.52	19.53	1.34
	100%	155.81	185.7	20.49	1.80
Ripple ~30 min (ppm)	10%	454.21	284.39	31.15	0.25
	20%	260.91	222.85	39.03	0.56
	40%	336.25	276.10	40.95	1.08
	60%	1144.83	2300.73	31.22	1.43
	80%	440.29	1205.87	59.44	2.24
	100%	1268.06	1005.31	73.21	2.19
Long-term drift (ppm/hr) (absolute)	10%	44.20	556.20	9.77	0.51
	20%	110.00	57.76	13.31	0.18
	40%	29.30	65.34	16.03	0.34
	60%	797.00	1847.00	27.22	0.96
	80%	399.00	1630	37.22	0.27
	100%	1400.00	1137	71.67	1.08

We used two different STM32 boards to control the PMA20N and PMA1.5N. On average, the board used to control the PMA20N has a control voltage ripple of 72.47 ppm, while the board used on the PMA1.5N has a control voltage ripple of 177.62 ppm. This big difference in ripple could be due to manufacturing variability between the STM32 boards.

Table 4. PMA20N results

PMA20N	Voltage percentage	STM32 Vprobe (ppm)	STM32 control voltage (ppm)	AD5791 Vprobe (ppm)	AD5791 control voltage (ppm)
Ripple Best 5 min (ppm)	10%	56.48	57.33	6.49	0.18
	20%	60.05	58.56	7.10	0.51
	40%	70.00	69.92	29.25	0.82
	60%	89.05	76.68	33.63	1.19
	80%	3012.4	89.14	52.91	1.34
	100%	1628.22	83.19	63.81	1.91
Ripple ~30 min (ppm)	10%	64.71	64.41	12.00	0.29
	20%	67.12	68.35	49.84	0.59
	40%	105.31	88.43	46.65	0.88
	60%	132.07	167.13	85.05	2.02
	80%	3219.07	102.62	178.21	2.15
	100%	4161.31	124.93	200.85	2.14
Long-term drift (ppm/hr) (absolute)	10%	41.46	25.02	0.38	0.56
	20%	23.08	56.19	21.70	0.15
	40%	207.26	125.46	4.8	0.26
	60%	122.91	425.52	22.87	4.19
	80%	1550.57	46.84	98.73	2.81
	100%	1346.74	108.06	107.77	1.00

The ripple of the STM32 control voltage is consistently higher than the ripple of the AD5791, being 43.55 times larger at best (PMA20N 100% voltage percentage) and 414.69 times larger at worst (PMA1.5N 10% voltage percentage). This caused the ripple of the Vprobe to be higher when the HVPS controlled by the STM32 than with the AD5791.

The observed ripple from the STM32 is consistently higher than the observed ripple from the AD5791. According to the data derived from Table 3 and 4, the ripple from the STM32 Vprobe was calculated to be on average 7.29 times higher than the ripple from the AD5791 Vprobe.

Figure 7 shows that the ripple of both the Vprobe and control voltage increases when the voltage percentage is increased, which is likely due to increased power/heat dissipation.

However, the Fig. 8 does not show the same relationship for the STM32 as the ripple is higher than expected at 40% and 60%, despite the AD5791 not exhibiting the same problem. This could be due to the different board used.

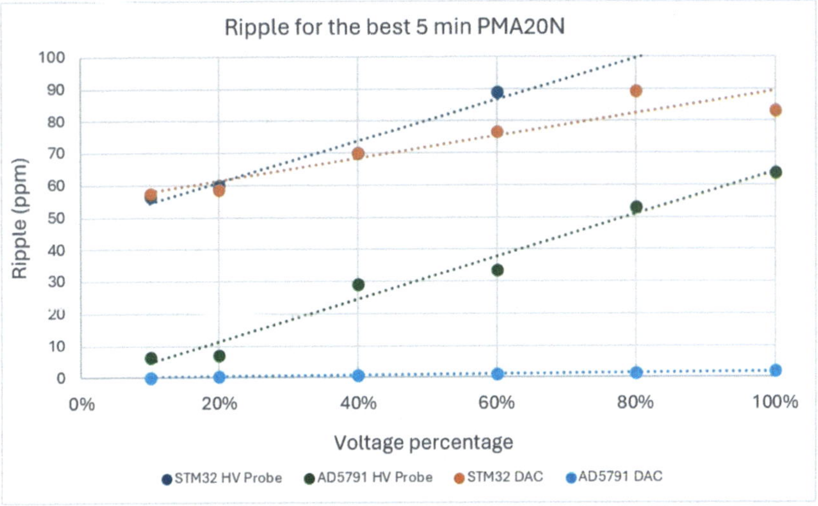

Fig. 7. Graph of best 5 min of ripple for PMA20N

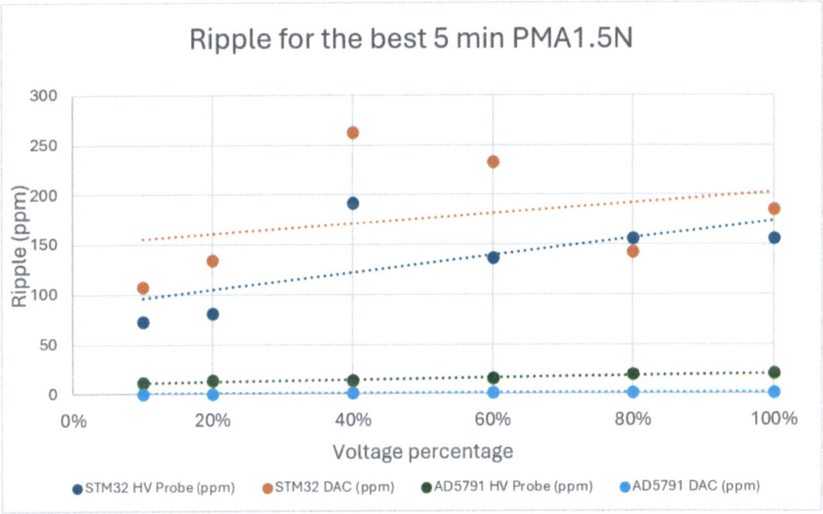

Fig. 8. Graph of best 5 min of ripple for PMA1.5N

From Table 3 and 4, Vprobe ripple is 87% of the control voltage ripple for the STM32 while the Vprobe ripple is 2614% greater than the control voltage ripple for AD5791. This suggests that most of the ripple observed in the output voltage of the HVPS with the STM32 control system is caused by the STM32 DAC while most of the ripple observed in the output voltage of the HVPS with the AD5791 control system is caused by the HVPS. As the Vprobe ripple is lower than the control voltage ripple for the STM32, it is possible that the HVPS has a filter which causes the ripple of the output to be lower.

The observed long-term drift from the STM32 is also consistently higher than the observed long-term drift from the AD5791. According to the data derived from Table 3 and 4, the long-term drift from the STM32 Vprobe was calculated to be on average 21.8 times higher than the long-term drift from the AD5791 Vprobe.

There are also certain limitations to take note of when looking at the results. The DMM6500 multimeter has a sampling rate of 10 Hz, meaning only 10 measurements are taken every second per channel. Any ripple with a higher frequency may go undetected, leading to inaccuracies in the ripple measured. The Vprobe ripple measured for the AD5791 system also does not match up to the expected ripple from the AD5791 provided in the datasheet by the manufacturer. This discrepancy can be a result of limitations caused by the measurement setup. The HV probe is a resistive potential divider system. At high power, the resistors heat up and experience poor dissipation of heat due to them being potted (covered in resin to prevent high voltage arcing). This can cause the resistances to fluctuate, resulting in higher ripple being measured.

4 Conclusion

The AD5791 has better overall stability than the STM32. We expressed the performance over cost as a standard unit by using the following equation:

$$\left(\frac{1}{Ripple}/\text{Cost}\right) * 1000 \tag{1}$$

Using this equation, we found the performance per dollar of the STM32 and AD5791 using the best and worst 5 min of ripple of both systems. The values are shown in Table 5.

Table 5. Performance per dollar of STM32 and AD5791 using Eq. (1) shown above

Performance per dollar	At best	At worst
STM32	0.567	0.123
AD5791	26.1	2.42

While the performance of the AD5791 may generally be better than that of the STM32, it is important to note that the quality of the HVPS will also influence the performance of the system. ST Microelectronics provides the hardware abstraction layer (HAL) library for the STM32, which reduces the time for development and implementation compared to the AD5791 which requires external libraries, allowing the STM32 to be used for quick prototyping.

Acknowledgements. This project was carried out at at the Centre for Ion Beam Applications at NUS Department of Physics. We would like to thank Assoc. Prof Jeroen Anton Van Kan, Mr Badri Narayanan Ravichandran, Mr Jeff Bernard Esguerra Tiston and Dr Chiam Sher-Yi for their continued guidance throughout this project.

References

1. Industrial Quick Search. High Voltage Power Supply: Types, applications, benefits, and components. https://www.iqsdirectory.com/articles/power-supply/high-voltage-power-supplies.html. Accessed 04 Jan 2024
2. Mass spec pro. Einzel Lens. http://www.massspecpro.com/technology/ion-optics/einzel-lens-0. Accessed 4 Jan 2024
3. Analog Devices. AD5791 Data Sheet, July 2010. https://www.analog.com/media/en/technical-documentation/data-sheets/AD5791.pdf. Accessed Jan 2020
4. STMicroelectronics. STM32F405/415, STM32F407/417, STM32F427/437 and STM32F429/439 advanced ARM®-based 32-bit mcus EN.DM00031020 - F429 reference manual, 15 September 2011. https://www.st.com/resource/en/reference_manual/dm00031020-stm32f405-415-stm32f407-417-stm32f427-437-and-stm32f429-439-advanced-arm-based-32-bit-mcus-stmicroelectronics.pdf. Accessed 25 Feb 2021
5. Wisman. PMA Data Sheet, February 2019. https://www.wismanhv.com/Upfiles/uploadfile/202011/20201128222746249.pdf

Comparative Analysis of Synthesis Methods for Maximising Supercapacitive Performance in NiCo-Layered Double Hydroxides

Lokesh Vairakannu, Ting Hui Ng, and Song Xinpeng Charles(✉)

NUS High School of Mathematics and Science, Singapore 129957, Singapore
h1910144@nushigh.edu.sg

Abstract. With an escalating global energy shortage and increasing environmental concerns, the urgency for efficient renewable energy storage solutions is paramount. Layered Double Hydroxides (LDHs), especially NiCo-LDHs, hold promise due to their excellent electrochemical performance and environmental compatibility. Investigating synthesis methods—hydrothermal, reflux and coprecipitation—reveals varying control over crystallinity and morphology, impacting the LDHs' properties. Experimental synthesis, characterization, and electrochemical analysis were conducted for hydrothermal and reflux methods, with results showing that hydrothermal synthesis with a 3:2 Ni:Co ratio, using CTAB and methanol, produced NiCo-LDH with the highest specific capacitance (3.881 F/g). Each synthesis method showed nuanced effects on LDH properties, correlating crystallinity with improved charge storage capabilities. Structural tuning through additives facilitated controlled crystalline growth, impacting electrochemical performance. The impact of additives—CTAB, methanol, oxalic acid, ammonium nitrate, and urea—on LDH properties was dissected. The study underscores the nuanced effects of various additives and precursors on LDH properties, emphasising the role of crystallinity in enhancing charge storage capabilities. CTAB and methanol were found crucial in augmenting surface area and facilitating controlled crystalline formation. Key parameters such as energy density and specific capacitance were correlated with synthesis methods, highlighting the vital role of synthesis parameters in defining LDH performance. These findings offer insights into optimising synthesis parameters for tailored LDH structures, crucial for efficient energy storage materials.

1 Introduction

With the global energy shortage crisis, climate, and environmental issues aggravating with each passing day, demand for more renewable and efficient energy storage solutions has been becoming increasingly urgent.

Supercapacitors, known for their capacity to rapidly store and discharge electrical energy, have emerged as a fundamental component in modern energy storage infrastructure. This is due to their high power density, prolonged cyclic stability, and swift charge-discharge capabilities. These features make them exceptionally suitable for a variety of applications, ranging from portable devices to hybrid electric vehicles, and

H. Guo et al. (Eds.): IRC-SET 2024, PTT, pp. 350–359, 2025.
https://doi.org/10.1007/978-981-96-3770-6_36

as backup energy suppliers [8]. As the quest to amplify the energy density and power density of supercapacitors continues, the development of asymmetric or hybrid devices with extended operating voltages is seen as a pivotal stride towards augmenting the commercial viability of these storage mediums.

In the forefront of material science research lies the class of inorganic layered compounds known as layered double hydroxides (LDHs). LDHs consist of positively charged brucite-like layers that incorporate various metallic elements and are interspaced by an assortment of intercalated anions and water molecules. The general compositional formula of LDHs is: $[M2^{+}{}_{1-x}N3^{+}{}_{x}(OH)_2]$^$(A-)_x/n \cdot mH_2O$, where M and N refer to metal elements (Co, Ni, Al, Fe, etc), and A refers to intercalated anions (CO2, NO3-, SO42-, etc).

Among the various LDHs, Ni-Co-Layered Double Hydroxides (Ni-Co-LDHs) deliver promising application potential, owing to their superior electrochemical performance, easy preparation, and environmental friendliness [11]. With high specific capacitances that exceed 2000 F g^{-1}, the unique structural and compositional modularity of LDHs allows for the precise tuning of electrochemical properties. Such controllable synthesis, paired with cost-effectiveness and benign environmental impact, serves to address both the energy shortfalls and the climate and environmental demands plaguing our society.

In the synthesis of NiCo-LDHs, 3 primary methods are employed: hydrothermal, reflux and coprecipitation [11]. The coprecipitation technique is the simplest and the most widely used for its low energy consumption and large scale production, but there is limited control over the crystallinity and morphology of the LDH. Hydrothermal techniques often yield well-defined crystalline structures due to the controlled temperature and pressure conditions, while requiring more energy and time. On the other hand, reflux techniques are generally faster and less energy intensive, while still providing reasonable control over reaction conditions. Hence, we decided to investigate hydrothermal and reflux methods as they have been proven to yield better quality LDHs than coprecipitation methods.

As such, using various chemicals which play different roles in the synthesis of NiCo-LDH, we analysed the electrochemical properties, the crystallinity and the identity of the products that have been yielded by these methods.

1.1 Literature Review

Various studies have outlined and explored the potential capabilities of Ni-Co-LDHs for an efficient means of energy storage as a supercapacitor. For instance, Zhou et al. have prepared 3-D flower-like nanostructured NiCo LDHs by leveraging microwave assisted synthesis, achieving a supercapacitance performance of 1187 F g^{-1} at 1 A g^{-1}, underscoring the impact of alternative synthesis methods on the electrochemical activity of this LDH [1]. Jing et al. described the synthesis of NiCo LDH by using a one-step alternative voltage electrochemical technique for supercapacitor electrode material, the NiCo LDH nanoflakes have an interlayer spacing of 7.67 d(Å) and revealed maximum specific capa- citance of 1372 F g^{-1} at 1 A g^{-1} [2]. Hou et al. fabricated a self-assembled urchin-like NiCo LDH structure by adopting the hydro- thermal synthesis route and as a supercapacitor electrode material it demonstrated specific capacitance of 808.4 C g^{-1} at 1 A g^{-1} [3].

1.2 Goal of Research

Despite the recent progress in the field, the process of synthesising nickel-cobalt layered double hydroxides (NiCo-LDH) for use as supercapacitor materials remains intricate, with the complexity of the synthesis procedure affecting the reproducibility and scalability of these materials. The success of the synthesis depends on a precise balance between the nickel and cobalt precursors, pH levels, temperature, and other synthesis parameters to produce materials with consistent quality. As such, we aim to explore various synthetic routes to determine the most reproducible and scalable method for maximising supercapacitance performance.

Our work also seeks to expound on the precise mechanisms underlying the electrochemical properties of Ni-Co-LDHs, while simultaneously addressing the broader implications for energy storage solutions relevant to environmental sustainability. As such, an electrochemical analysis of the properties of Ni-Co-LDHs would allow us to identify how the interplay between the material's crystal structure, surface area, and conductive pathways would influence its supercapacitive performance, allowing for us to better maximise performance.

2 Experimental Section

2.1 Materials

Cobalt (II) nitrate hexahydrate (Co(NO3)2 · 6 H2O), Cobalt (II) chloride hexahydrate (CoCl2 6H2O), Nickel (II) nitrate hexahydrate, Cetrimonium bromide ($[N(CH_3)_3]Br$) and Oxalic acid ($(COOH)_2$) were purchased from Sigma Aldrich (St. Louis, United States). Ammonium nitrate (NH_4NO_3) was purchased from San Corporation (Shenzhen, China). Finally, Methanol (CH_3OH) and Urea (CO(NH2)2) was supplied from Sinopharm Chemical Reagent Co., Ltd (China). All chemicals were of analytic grade and used after purchasing with no further purification. Ultrapure water was used throughout the entire experiment.

2.2 Preparation of LDH with CTAB (Sample 1)

0.63 mmol of Ni(NO3)2·6H2O, 0.42 mmol of Co(NO3)2·6H2O, 4.0 mmol of CTAB, 5.0 ml of H2O and 45.0 ml of methanol were combined in a beaker, and stirred for 5 min. The solution was poured into a 100 ml Teflon autoclave and heated in an oven at 180°C for 24 h to allow the formation of the NiCo1-NO3 LDH.

2.3 Preparation of LDH with Oxalic Acid (Sample 2)

28.0 mmol of Ni(NO3)2·6H2O, 4.0 mmol of Co(NO3)2·6H2O, 5.0 mmol of oxalic acid and 100 ml of H2O were combined in a 2-neck round bottom flask and stirred for 5 min. The solution was then degassed under vacuum and N2 to avoid CO2 interference with the LDH. Then, the solution was refluxed at 100°C for 6 h to allow the formation of the NiCo2-NO3 LDH.

2.4 Preparation of LDH with Ammonium Nitrate (Sample 3)

12.0 mmol of Ni(NO3)2·6H2O, 8.0 mmol of Co(NO3)2·6H2O, 0.1 mol of ammonium nitrate and 100 ml of H2O were combined in a 2-neck round bottom flask and stirred for 5 min. The solution was then degassed under vacuum and N2 to avoid CO2 interference with the LDH. Then, the solution was refluxed at 100°C for 6 h to allow the formation of the NiCo3-NO3 LDH.

2.5 Preparation of LDH with Urea (Sample 4)

1.5 mmol of Ni(NO3)2·6H2O, 1.0 mmol of Co(NO3)2·6H2O, 12.0 mmol of urea and 50.0 ml of H2O were combined in a beaker, and stirred for 5 min. The solution was poured into a 100 ml Teflon autoclave and heated in an oven at 180°C for 24 h to allow the formation of the NiCo4-NO3 LDH.

2.6 Preparation of Slurries and Electrodes

The active material was prepared by mixing 1.5 ml ethanol, 10 µl Nafion, 0.00921 g carbon black, 0.07368 g of the LDH. The mixture was sonicated for 10 min to form the slurry.

Nickel foam was cut into 1.5 cm × 0.5 cm pieces and soaked in 1M HCl for 1 h to remove any surface oxide layer. Then the nickel foam was rinsed using DI water and sonicated under acetone, ethanol and DI water for 5 min. The foam was dried under vacuum in a desiccator for 24 h. Then, 50 µl of the desired slurry was drop casted onto one side of the nickel foam (0.5 cm × 0.5 cm) and the foam was left to dry for 24 h. The loaded nickel foam was then pressed under 1 ton for 1 min, to form the completed electrode with LDH in it.

2.7 Characterisation Method

X-ray diffraction was used to investigate samples' crystalline structure under the following conditions: 30 kV, 30 mA and Cu Kα radiation ($\lambda = 0.1542$ nm). Infrared spectroscopy was also used to verify that the correct functional groups were present in the NiCo-LDH.

2.8 Electrochemical Measurements

The electrochemical performance of all the electrodes was studied by a three-electrode system in a 2M KOH aqueous electrolyte. Ag/AgCl (3M KCl) electrode and Pt rod were used as a reference and counter electrode, respectively. The cyclic voltammetry (CV) and galvanostatic charge-discharge (GCD) were operated on a CHI660E electrochemical workstation (Shanghai Chen Hua Instruments Co. Ltd., China).

The specific capacity, C (Cg^{-1}) was calculated using the GCD graphs and the equation shown below:

$$C(Cg^{-1}) = i_m \times \Delta t$$

where im (Ag^{-1}) is the current density, and Δt (s) is the time of discharging process.

CV was carried out over a potential window of −0.1–0.5 V, at scan rates of 1, 2, 3, 5, and 10 mVs^{-1}. The area bounded by the cyclic voltammogram is then used to determine the specific capacitance of the LDH using the following equation:

$$C_S = \frac{\int_0^{\Delta V} I\,dV}{mv\Delta V}$$

where Cs, I, v, ΔV and m are the specific capacitance (Fg^{-1}), current (A), potential scan rate (mV/s), potential window (V), and mass of the active electrode material (g) respectively.

3 Results and Discussion

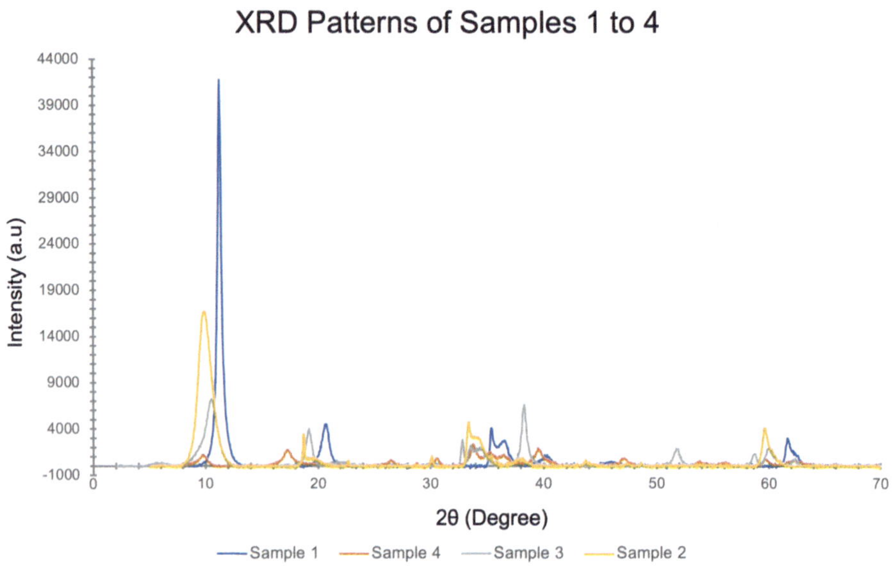

Fig. 1. XRD patterns for the NiCo-LDHs

The X-ray diffraction (XRD) analysis offers an intricate glimpse into the structural properties of the synthesised NiCo-LDH samples, shedding light on the distinctive characteristics that define their crystalline nature. Sample 1 emerges as a standout with its pronounced peak intensities at 11.6°, 20.9°, 35.5°, and 62°, indexed by (003), (006), (009). These robust peaks indicate a well-defined crystalline structure, suggesting a high degree of order in the arrangement of atoms within the LDH lattice. The XRD patterns for sample 1 are also the most accurate using the template as comparison. The precision and sharpness of these peaks hint at the material's exceptional crystallinity brought around by using the hydrothermal technique, with the addition of methanol and CTAB (Fig. 2).

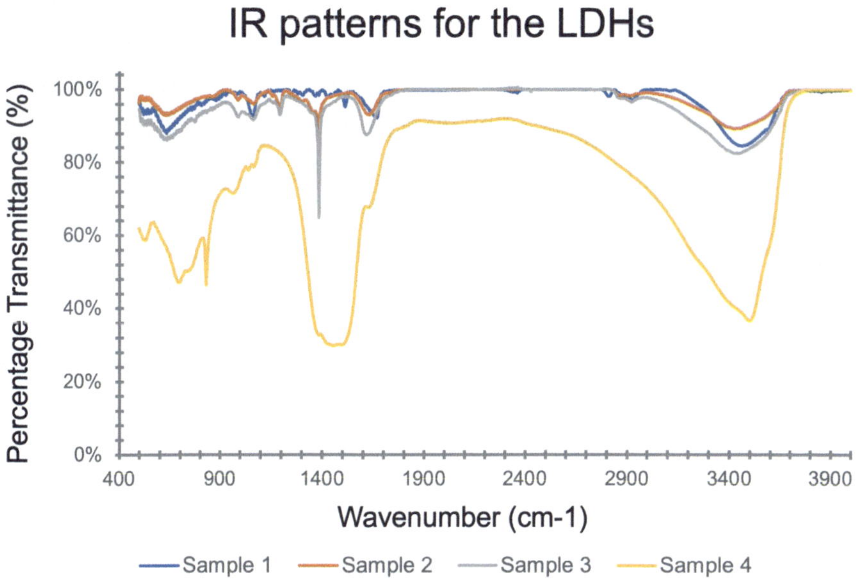

Fig. 2. IR patterns for the NiCo-LDHs

In stark contrast, Sample 4 presents a completely different XRD profile, characterised by its lower peak intensities at 9.8, 16.8, 33.6, and 59°. The diminished peak heights suggest a less crystalline structure, potentially accompanied by smaller crystal sizes or the presence of amorphous regions. The broader peaks in Sample 4 indicate a degree of disorder in the atomic arrangement, hinting at a more complex and less uniform structure compared to Sample 1. A reason for this failed synthesis could be due to the presence of carbonate interference stemming from the hydrolysis of urea.

Moving on to Samples 2 and 3, both exhibit moderate peak intensities, portraying a crystalline structure with characteristics falling between the extremes observed in Samples 1 and 4. The peaks at 10.5, 19, 33, and 59.9° for Sample 3, and 10, 18.4, 33.6, and 59.2° for Sample 2, reflect a balanced crystallinity, suggesting a well-organised but less dominant crystal phase than that observed in Sample

1. These nuanced differences in the XRD patterns between samples 2 and 3 indicate the subtle variations introduced by the synthesis methods, influencing the overall structural composition of the LDH samples. It can be seen that the presence of oxalic acid improved the crystallinity of sample 2 as compared to sample 3, whereas the presence of the anion additive increased the interlayer distance between the hydroxide layers as proven by the smaller peak angles of sample 3 as compared to sample 2.

The observed variations in peak patterns can be attributed to the diverse synthesis methods employed for each sample. Different precursors, such as $Ni(NO_3)_2{\cdot}6H_2O$ and $Co(NO_3)_2{\cdot}6H_2O$, and capping agents, including CTAB, oxalic acid, ammonium nitrate, and urea, contribute to the unique structural signatures seen in the XRD data. The varying reaction conditions, such as temperature and time, also play a pivotal role in shaping the final crystalline structure of the LDH samples.

The most notable thing we obtained from our IR analysis is that sample 4 (which is the one produced with urea) displays C=O stretching (at around 1500 cm^{-1}) which indicates the presence of carbonate, indicating that there was carbonate interference. This suggests that the urea method might not be as accurate as we had hoped for. Additionally, there is prominent O-H stretching (at around 3500 cm^{-1}), which confirms the presence of Ni-Co LDH with the layers of hydroxides in it (Figs. 3, 4 and Table 1).

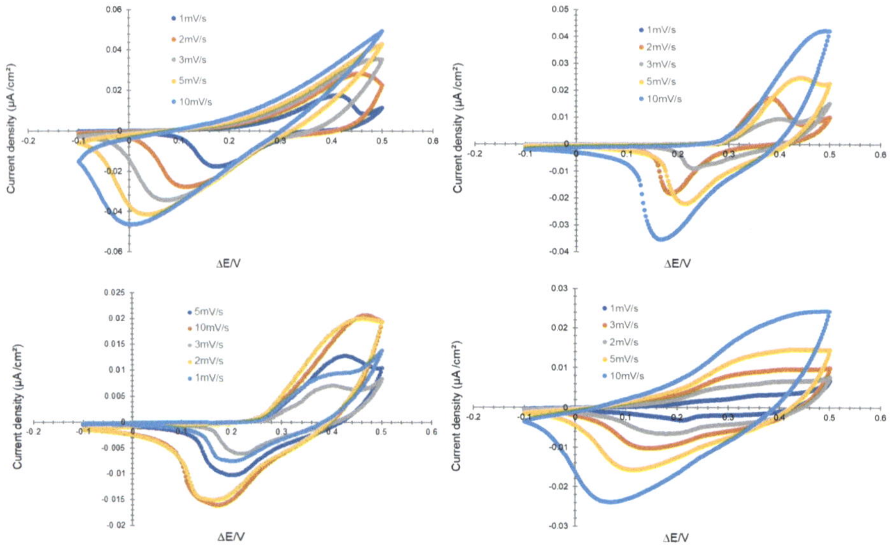

Fig. 3. The CV (Cyclic Voltammetry) graphs of all the samples

Based on the CV and GCD results above, we conclude that the best method for NiCo-LDH synthesis is hydrothermal method with 3:2 Ni:Co ratio and addition of CTAB in methanol (with a specific capacitance of 3.881 F/g and specific capacity of 267.300 C/g). CTAB was used to increase the specific surface area of the active site and expose it fully for LDH formation to occur. Furthermore, it serves as a medium to prevent the large-scale aggregation of LDHs and accelerate the rate of ion diffusion, yielding LDHs with better crystallinity and facilitating nitrate intercalation for improved electrochemical properties. We believe that addition of methanol facilitates controlled precipitation and nucleation of LDH precursor ions. It aids in controlling the rate of reaction and nucleation, influencing the size, morphology, and crystallinity of the resulting NiCo1-LDH particles. The controlled reaction kinetics enabled by methanol can contribute to obtaining well-defined and uniform LDH crystals. Methanol is often employed in hydrothermal or solvothermal synthesis methods. These methods involve elevated temperatures and pressures at which methanol's properties might influence the formation kinetics and crystal growth mechanisms, impacting the final structure and properties of the synthesised NiCo1-LDH. It is also possible that methanol might act as a capping or

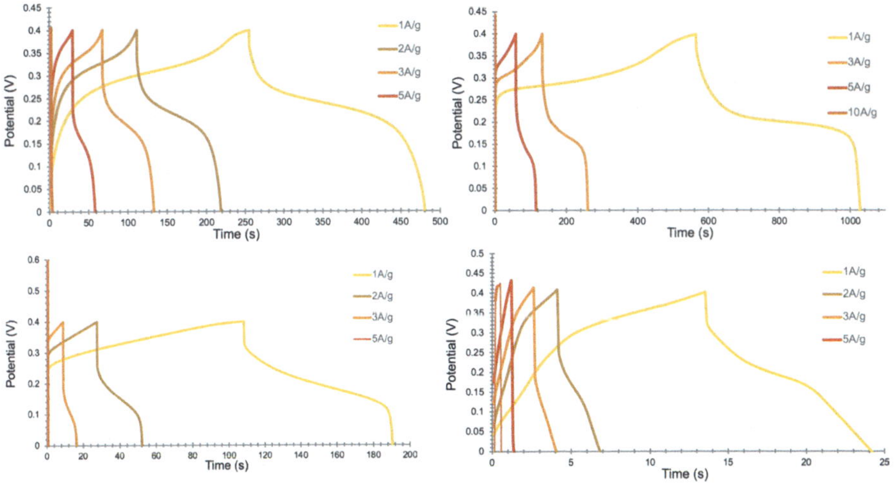

Fig. 4. The GCD (Galvanostatic Charge-Discharge) graphs of all the samples

Table 1. Average specific capacitances and specific capacities of the various samples, obtained from processing the GCD and CV graphs using the equations listed in the methodology (*obtained without using 2A/g data as we did not get proper results for that run and decided to omit it from the average)

	Sample 1	Sample 2	Sample 3	Sample 4
Average specific capacitance (Fg^{-1})	3.881	0.935	0.377	0.402
Average specific capacity (Cg^{-1})	267.300	595.625*	112.780	12.560

stabilising agent, controlling the growth and preventing agglomeration of LDH nanoparticles during the synthesis process. These properties of methanol and CTAB ultimately led to the success of this method in producing LDH with the best capacitance.

The next best method is the reflux method with 7:1 Ni:Co ratio and addition of oxalic acid (with a specific capacitance of 0.935 F/g and specific capacity of 595.625 C/g). We concluded that the method with 3:2 Ni:Co ratio and addition of CTAB in methanol is superior to this method, as although this method gives a specific capacity value that is twice as high, the specific capacitance is reduced four-fold. Oxalic acid acts as a nucleation agent, promoting the formation of the initial seed crystals necessary for the growth of the layered double hydroxide structure. It aids in the controlled precipitation of nickel and cobalt ions from the precursor solution, allowing the formation of the NiCo2-LDH structure to commence in a controlled manner. By regulating the pH and controlling the rate of precipitation, oxalic acid helps in obtaining nanoparticles with a more uniform size and shape. This control over the nucleation and growth stages is vital in ensuring the formation of well-defined, uniform LDH crystals, which is crucial for optimal electrochemical performance. Oxalic acid has been employed by researchers before to start the formation of mesoporous NiCo2O4 nanorods with crystalline nature of $NiCo_2O_4$. We believe that adding oxalic acid to our mixture increased pore size of NiCo2-LDH for optimization of charge-discharge rates, ultimately leading to the greatest specific capacitance among all the samples.

The NiCo3-LDH synthesised using reflux method with 3:2 Ni:Co ratio and addition of ammonium nitrate also had a significantly lower specific capacitance (0.377 F/g) and specific capacity (112.780 C/g) than the sample made using oxalic acid. We concluded that this method is suboptimal for NiCo-LDH synthesis. Ammonium nitrate serves as an anion additive. We believe this promoted nitrate intercalation to increase the NiCo3-LDH interlayer distance for increased charge storage capacity. However, we believe that the addition of ammonium nitrate and the subsequent change in interlayer distance due to the anion added may result in structural alterations that could impact other material properties, such as stability, conductivity, or surface area, which could have resulted in a suboptimal capacitance.

The NiCo4-LDH synthesised using the hydrothermal method with 3:2 Ni:Co ratio and addition of urea had a slightly better specific capacitance (0.402 F/g) than the sample using ammonium nitrate, but it had a much worse specific capacity (12.560 C/g) that was less than the value for the ammonium nitrate method by ten-fold. Hence, in our analyses, we concluded that this method with urea was the least optimal method to synthesize NiCo-LDH. Urea serves as both a complexing agent and a source of hydroxide ions. By forming complexes with nickel and cobalt ions, premature precipitation of NiCo4-LDH can be prevented to allow for a gradual and controlled reaction. Urea undergoes hydrolysis in the presence of water to produce hydroxide ions required for the precipitation reaction with the metal cations to form the LDH structure. However, the suboptimal capacitance that was observed could be a result of carbonate interference. When urea is hydrolysed, carbon dioxide is produced, which forms carbonic acid in water, forming carbonate ions which intercalate in the LDH and destabilise the LDH. Since our IR analyses also detected possible carbonate interference, this is likely the reason for the suboptimal capacitance observed in NiCo4-LDH.

4 Conclusion

In conclusion, the presence of different precursors CTAB, methanol, anion additives and urea play a vital role in tuning the crystallinity, interlayer distance, charge storage capacity and charge-discharge rates of the NiCo-LDH. Ni-Co1-LDH, which was synthesized using hydrothermal method with 3:2 Ni:Co ratio and addition of CTAB in methanol, emerged as the best supercapacitor out of the four samples with the highest crystallinity and highest specific capacitance. This proves the importance of CTAB and methanol in increasing the surface area of the active site and the gradual formation of the hydroxide layers for a uniformly structured NiCo-LDH. NiCo2-LDH synthesised by reflux method also gave decent results, with the highest specific capacity, showing how oxalic acid plays a vital role in increasing the pore size of the LDH for higher charge-discharge rates. However, as shown by the XRD patterns in Figure 1, the crystallinity of LDHs synthesised by hydrothermal is much better as compared to the reflux method. In today's world, marked by an escalating energy storage crisis, the adoption of reflux methods is encouraged over hydrothermal methods as the production of LDHs with high specific capacitance and charge-discharge rates are still possible without consuming as much energy and time. With the use of oxalic acid in reflux methods, NiCo LDHs of high yield and high specific capacity can still be produced.

References

1. Guan, T., et al.: A facile approach to synthesize 3D flower-like hierarchical Nico layered double hydroxide microspheres and their enhanced adsorption capability. Colloids Surf. A: Physicochem. Eng. **529**, 907–915 (2017). https://doi.org/10.1016/j.colsurfa.2017.06.049
2. Yue, L., et al.: Improving the rate capability of Ultrathin Nico-LDH nanoflakes and feooh nanosheets on surface electrochemically modified graphite fibers for flexible asymmetric supercapacitors. J. Colloid Interface Sci. **560**, 237–246 (2020). https://doi.org/10.1016/j.jcis.2019.10.032
3. Hou, L., et al.: NI-CO layered double hydroxide with self-assembled urchin like morphology for asymmetric supercapacitors. Mater. Lett. **237**, 262–265 (2019). https://doi.org/10.1016/j.matlet.2018.11.123
4. Tahir, M.U., et al.: Synthesis of morphology controlled Nico-LDH microflowers derived from Zif-67 using binary additives and their excellent asymmetric supercapacitor properties. Appl. Surf. Sci. **529**, 147073 (2020). https://doi.org/10.1016/j.apsusc.2020.147073
5. Construction of Nico-Layered Double Hydroxide Microspheres from Ni-Mofs for High-Performance Asymmetric Supercapacitors (n.d.). https://doi.org/10.1021/acsaem.0c00790.s001
6. Dai, X., et al.: Cobalt oxide nanocomposites modified by NICO-layered double hydroxide nanosheets as advanced electrodes for supercapacitors. Ionics **26**(5), 2501–2511 (2019). https://doi.org/10.1007/s11581-019-03333-6
7. Li, Z., Han, F., Li, C., Jiao, X., Chen, D.: Multi-anion intercalated layered double hydroxide nanosheet-assembled hollow nanoprisms with improved pseudocapacitive and electrocatalytic properties. Chemistry Asian J. **13**(9), 1129–1137 (2018). https://doi.org/10.1002/asia.201800092
8. Li, R., et al.: Large scale synthesis of Nico layered double hydroxides for superior asymmetric electrochemical capacitor. Sci. Rep. **6**(1) (2016). https://doi.org/10.1038/srep18737
9. Wang, X., et al.: Controllable synthesis of Nico LDH nanosheets for fabrication of high-performance supercapacitor electrodes. Electroanalysis **29**(5), 1286–1293 (2017). https://doi.org/10.1002/elan.201600602
10. Hu, W., et al.: Effect of intercalated anion in nickel-cobalt-layered double hydroxide on its supercapacitive properties. Chem. Eng. J. **468**, 143694 (2023). https://doi.org/10.1016/j.cej.2023.143694
11. Ray, P.K., Mohanty, R., Parida, K.: Recent advancements of Nico LDH and graphene based nanohybrids for supercapacitor application. J. Energy Storage **72**, 108335 (2023). https://doi.org/10.1016/j.est.2023.108335
12. Liang, H., et al.: Hydrothermal continuous flow synthesis and exfoliation of Nico layered double hydroxide nanosheets for enhanced oxygen evolution catalysis. Nano Lett. **15**(2), 1421–1427 (2015). https://doi.org/10.1021/nl504872s
13. Benhiti, R., et al.: Synthesis, characterization, and comparative study of Mgal-LDHS prepared by standard coprecipitation and urea hydrolysis methods for phosphate removal. Environ. Sci. Pollut. Res. **27**(36), 45767–45774 (2020). https://doi.org/10.1007/s11356-020-10444-5
14. Xing, C., et al.: Synthesis, characterization, and properties of nickel cobalt layered double hydroxide nanostructures. RSC Adv. **7**(62), 38945–38950 (2017). https://doi.org/10.1039/c7ra06670h

A Study in Thermal Annealing: Improving the Efficiency of Next-Generation Electronics

Adviteeya Krishna[1(✉)], Goh Xuan He Ernest[1], Cayden Tan Hongkai[1], Ivan Erofeev[2], Anthony Winata Hartanto[2], and Utkur Mirsaidov[2]

[1] NUS High School of Math and Science, Singapore 129957, Singapore
h2210002@nushigh.edu.sg

[2] NUS – Center for Bioimaging Sciences, Singapore 117557, Singapore

Abstract. Transistors in microchips have been rapidly downscaled in the past fifty years, with tens of billions of them crammed into one microchip today, compared to the slightly above two thousand that fit into one chip in the early 1970s. This has caused drastic improvements in the power and efficiency of our electronic devices, positioning the semiconductor industry as one of the fastest-growing sectors globally. As nanowire dimensions continue to shrink, the design and fabrication of interconnects must also scale accordingly to maintain the desired performance and integration density. However, as they are downscaled, the size-dependent resistivity of metal interconnects presents a significant challenge, resulting in decreased conductivity and energy efficiency. This study explores the use of ruthenium as an interconnect, instead of the traditionally used copper, and examines its impact on the resistivity of nanowire samples. We make use of Transmission Electron Microscopy (TEM) to analyse the impact of annealing at 800 °C for varying durations and determine its impact on parameters such as linear grain size, grain area, and resistivity of nanowire samples. By identifying the more conductive metal for interconnects at smaller scales and determining the optimal duration for annealing, this project has laid the groundwork for future research in improving the efficiency of next-generation electronics.

Keywords: Ruthenium · Interconnects · Mean free path · Annealing · Grain boundary · Resistivity · Grain size

1 Introduction

Moore's Law suggests that the number of transistors in an Integrated Circuit (IC) doubles roughly every two years [1]. With the rapidly decreasing sizes of transistors, this law has been largely accurate for the past half-century, resulting in the greatly enhanced computational capabilities of our electronic devices.

The traditionally used material for interconnects in nanochips is copper (Cu), mainly due to its low bulk resistivity ($\rho_o = 1.678$ μΩ cm) in comparison to alternatives such as molybdenum (Mo) with $\rho_o = 5.34$ μΩ cm and ruthenium (Ru) with $\rho_o = 7.8$ μΩ cm

A. Krishna and G. X. H. Ernest—Equal contributors to this work and designated as co-first authors.

H. Guo et al. (Eds.): IRC-SET 2024, PTT, pp. 360–369, 2025.
https://doi.org/10.1007/978-981-96-3770-6_37

[2]. However, we encounter a problem at smaller scales (d < ~20 nm). Due to the long mean free path of electrons in Cu ($\lambda = 39.9$ nm), there is a substantial increase in its resistivity at low wire thicknesses as shown in Fig. 1. This can be attributed to the fact that surface or grain boundary scattering will become progressively more significant at scales below the mean free path of the metal [3].

The formula for resistivity of polycrystalline metallic nanowires, as derived from the Fuchs-Sondheimer (FS) model and Mayadas-Shatzkes (MS) model, is given by Eq. (1):

$$\rho_{tot} = \rho_o\left[1 + \frac{\lambda}{d} \times \frac{3}{4}(1-p) + \frac{\lambda}{D} \times \frac{3}{2}\left(\frac{R}{1-R}\right)\right] \quad (1)$$

Where ρ_o is the bulk resistivity, λ is the electron mean free path, p is the surface reflection coefficient, R is the probability of grain boundary reflection, whereas d and D are the wire thickness and mean grain size respectively [4–6].

For $p = 1$, the momentum of each electron along the axis of the conductor is preserved upon striking the surface, the phenomenon termed specular scattering [7]. Specular scattering does not cause an increase in interconnect resistivity and is the ideal scenario for the development of interconnects of high conductivity [8]. Where $p < 1$, some momentum of the electrons is lost upon collision, resulting in diffuse scattering. The interface between the metal and liner layer must be as atomically smooth as possible to minimise the likelihood of diffuse scattering. This is because roughness at even the atomic scale can result in loss of momentum of electrons upon reflection. Note that for real wires, p is almost always equal to zero.

Evidently, to minimise electrical resistivity, we wish to maximise surface smoothness and minimise grain boundary reflections to decrease interface and grain boundary scattering of electrons. Furthermore, we wish to increase mean grain size through thermal annealing to allow increase in linear grain size. Larger grains decrease the density of grain boundaries per unit volume of the crystal structure, resulting in decreased electron scattering and thus increased conductivity.

As such, this project intends to study Ru as an alternative to Cu for interconnects in ICs despite its high bulk resistivity. This is due to its low mean free path ($\lambda = 6.59$ nm), making it a promising option for interconnects at smaller scales.

Additionally, this study wishes to elucidate the effects of thermal annealing for varying durations at 800 °C on parameters such as linear grain size, grain area and resistivity of nanowire samples.

We hypothesise that thermal annealing of Ru is more effective than the thermal annealing of Mo interconnects of the same dimensions.

Furthermore, we predict that the most effective duration to anneal the sample at 800 °C would be around 30 min, as annealing for too long would likely result in the formation of deep surface grooves, resulting in a decrease in the cross-sectional area of the sample, causing increased resistivity. [9].

2 Materials and Methods

2.1 Taking Transmission Electron Microscope (TEM) images of Ru nanowires

We used a JEOL transmission electron microscope (JEOL Ltd., Tokyo, Japan) equipped with a Gatan OneView camera (Gatan Inc., Pleasanton, CA, USA) at 200 kV to obtain TEM images of Ru nanowires before and after thermal annealing. Ru nanowires were annealed in a vacuum furnace at 800 °C for varying durations.

2.2 Our Algorithm (Finding Parameters Such as Linear Grain Size and Grain Area)

We made use of a Canny Edge detection algorithm on Python to analyse our TEM images to obtain necessary parameters. Please refer to the appendix for the document containing our written code.

2.3 Finding the Resistance of Ru nanowires

Aluminium (Al) electrode pads were deposited on top of Ru nanowires annealed for the same durations using a photoresist mask and lift-off method to create the Ru nanowire samples. For each sample, each Al pad was separated by 30 μm and connected by 50 to 500 Ru nanowires. A custom-built probe with four tungsten (W) wires connected to a Keithley 2450 source measure unit (Keithley Instruments, Cleveland, OH, USA) was then used to measure the resistance of the Ru nanowires between 3 pairs of Al pads deposited on each sample at room temperature (25 °C). The resistance of Ru nanowires of each sample was also measured as the sample was heated on a hotplate to 100 °C and then 180 °C in intervals of 20 °C. Resistance of Ru nanowires against temperature of sample and annealed duration were then plotted.

3 Results

To calculate the desired resistivity, we used the TCR method [10], where we calculated the effective area of the sample using the gradient of the change of the resistance against temperature, through Eq. (2):

$$Area = \left(\frac{\delta\rho_{bulk}}{\delta T}\right) \times \frac{L}{\frac{\delta R}{\delta T}} \tag{2}$$

An example of how we find the gradient of resistance change against temperature is shown in the graph below. We can see how the resistance increases when temperature increases.

The gradient of the graph of bulk resistivity against temperature was calculated by using existing bulk resistivity data on Ruthenium [11], before using linear interpolation to calculate the gradient.

Given the relatively small nature of the sample and the nanowires on it, we figured that the nanowires could be easily damaged or oxidized, changing the effective area that

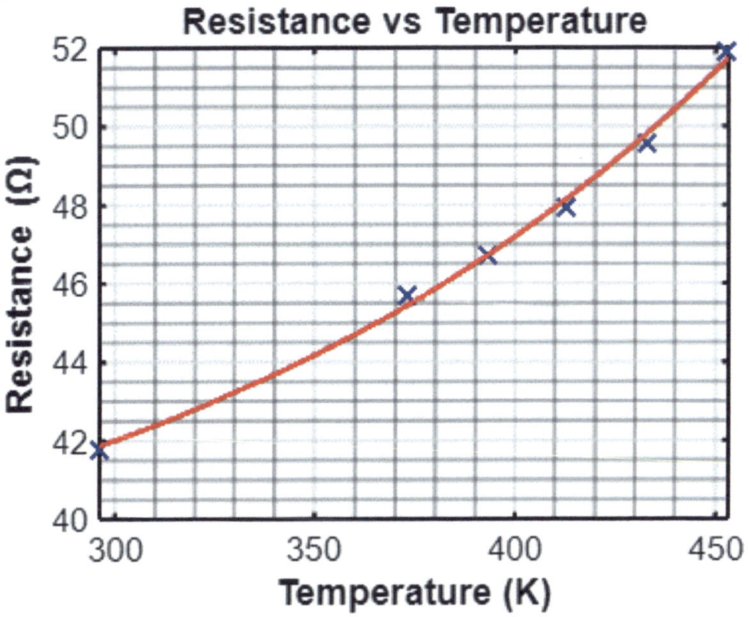

Fig. 1. An example of how resistance changes with temperature plotted on MATLAB (Source: authors' own work)

electrons would be able to flow through. Hence, we used the above formula to determine more accurately the effective area to get a more accurate result.

We then took the calculated effective area and applied it into the equation shown below to give us a highly accurate resistivity value:

$$\rho_{total} = R * \frac{A}{l} \tag{3}$$

A graph of how the resistivity changes as the sample is heated for different durations at 800 °C is shown below. Do take note that the graph shown is the resistivity value at room temperature.

To compare our experimental data with the theoretical predicted data, we took the average length of the grain, D (see Table 1), and applied it into Eq. (1), where we assumed R = 0.7, p = 0, $\lambda = 6.59$ nm, wire width = 30 nm and bulk resistivity = 7.2013 μΩ cm [11]. We found that the average difference between predicted resistivity and the resistivity we measured was around ±2.0 μΩ cm.

Shown below is a chart of the grain area and grain boundary length to show how they change when the sample is heated at different durations at 800 °C, as well as the related resistivity values. These values were obtained from our Python program.

The graph below depicts the change in resistivity with increasing grain length (Fig. 3).

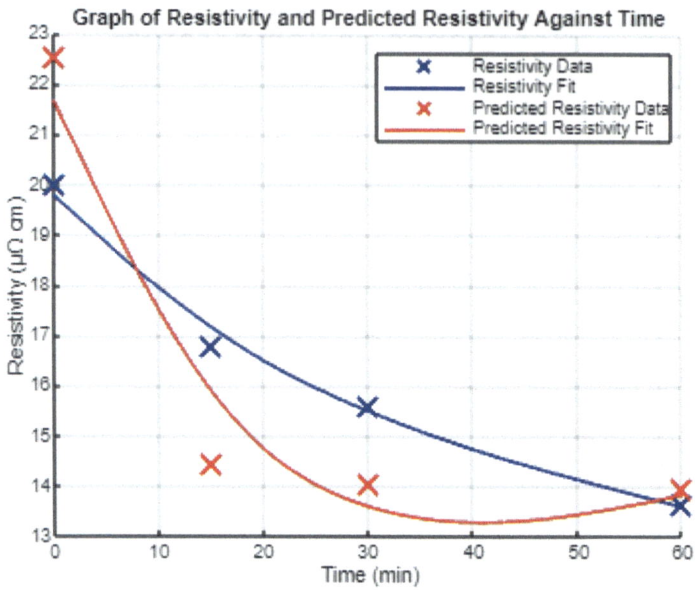

Fig. 2. Graph of the difference between predicted resistivity and actual resistivity across different annealing durations plotted on MATLAB (Source: authors' own work)

Table 1. A chart to show the change in grain area and grain length when the sample is annealed at different durations, please see the Materials and Methods section to see how we analysed the grains and got the variables

	As deposited	Annealed for 15 min	Annealed for 30 min	Annealed for 60 min
Average grain area (nm^2)	161.902	452.054	473.715	549.010
Average grain length (nm)	13.381	21.877	27.295	29.664
Resistivity (μΩ cm)	20.010	16.793	15.591	13.648

To paint a clearer picture, below are the images of the sample when imaged under a Transmission-Electron Microscope (TEM), as well as the corresponding grain boundaries, which were processed with python code involving skeletonization of TEM images (Figs. 4 and 5).

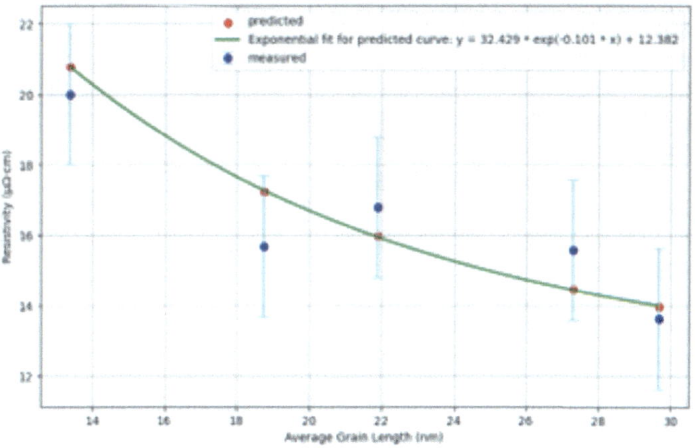

Fig. 3. Graph of measured and predicted resistivity against average grain length plotted on MATLAB (Source: authors' own work)

Fig. 4. A TEM image of the Ruthenium sample when it was first deposited, and its processed image to show the grain boundaries (Source: authors' own work)

4 Discussion

4.1 Analysis of Results

From Fig. 2, we observe a decreasing trend with regards to resistivity changes when we anneal samples at different durations, which supports our hypothesis. From the data, we can see that the average length of the sample increased from 13.381 nm to 29.664 nm, resulting in the resistivity decreasing from 20.010 $\mu\Omega$ cm to 13.648 $\mu\Omega$ cm. This hence gives an over 30.5% decrease in resistivity due to a 54.9% increase in linear grain size. This shows that annealing past 30 min can still prove to be effective, which contradicts our theory that surface grooving would increase resistivity if annealed for too long, but we do have a good explanation why this happened.

This happened largely due to the narrow scope of the experiment, as we initially predicted that surface grooving would occur at a significant enough level to affect the

Fig. 5. A TEM image of a different Ruthenium sample when it was annealed for 30 min at 800 °C, and the processed image to show the grain boundaries (Source: authors' own work)

resistivity when the sample was annealed for durations past 30 min. However, upon analysis of the TEM images, we found that although surface grooving did take place, it was not significant enough to cause a major change in resistivity, rather more grains merging resulted in a further decrease in resistivity. Surface grooving is a phenomenon observed when the cross-sectional area of the sample shrinks due to annealing for too long or in too high temperatures, resulting in a smaller cross-sectional area for electrons to pass through, increasing the resistivity of the sample. However, it seems like we would need to anneal the sample of durations of around 2 h to notice this effect, as observed in similar metals like Molybdenum [9]. Thus, the obtained trendline and results are perfectly understandable.

Although the predicted theoretical decrease in resistivity was calculated to be around 40.9%, we observed a decrease of 30.5%, which is slightly lower than expected, but still quite good. Hence, this shows that thermal annealing Ru interconnects is more effective than annealing Mo interconnects of the same dimensions, as we can observe that the decrease in resistivity is around 10% more in Ru than in Mo.

4.2 Strengths and Limitations of Our Experiment

We do acknowledge certain aspects of our experiment that could have been improved. Firstly, the imaging and the resistivity measurements were taken separately, hence it is likely that the nanowires in the TEM images are not the same as the one we measured the resistivity of. However, it is assumed that thermal annealing would occur uniformly throughout the sample, hence the data of the linear grain size and grain area would likely remain highly accurate. We also took the average grain area and grain length of multiple images in the same sample, averaging the results to make sure accuracy was as high as possible.

Secondly, the Canny edge detection algorithm used in identifying grain boundaries in the TEM images may not be entirely accurate, with some non-edges being detected and edges and vice versa. However, we experimented extensively by varying the lower and upper threshold values to choose the ones that gave us the most accurate results.

Thirdly, given the fragility of the experiments, it is very likely that the way the probes landed on the pads, as well as how old the probes were could have affected the resistivity data collected of the sample. For instance, oxidation on the probes could have been the likely reason that the data readings were on average $\pm 2.0\ \mu\Omega$ cm off from the predicted results based on Eq. (1). Furthermore, the amount of pressure put on the probes during resistivity measurements would also have a great effect on the resistance measured. Although we did our best to ensure that all readings were done to as high precision as possible, these external factors could still have a substantial negative impact on the data collected.

5 Conclusion

This experiment has successfully shown that thermal annealing of Ru is more effective than the thermal annealing of Mo interconnects of the same dimensions, largely due to Ru having a smaller mean free path. We have also shown that annealing past 30 min can still prove to be effective in reducing the resistivity of our sample, indicating that our initial hypothesis of 30 min being the most effective time does not hold true, and instead the most effective duration to anneal Ru interconnects would be somewhere past 1 h.

Acknowledgements. We would like to express our heartfelt gratitude towards Dr. Ivan Erofeev, Associate Professor Utkur Mirsaidov, and Anthony Winata Hartanto for their guidance and valuable advice throughout this project. We are grateful to the NUS Department of Bioimaging Sciences for providing us with the necessary resources to conduct our research.

A Appendix

A.1 Python Document

https://colab.research.google.com/drive/14cvuTy7Td4_t2E3-GXTiUtQG9bTwOWSb?usp=sharing

A.2 Explanations of Certain Parts of the Code

The first few lines of the code are important in defining certain functions for future use. This includes a function to plot a histogram based on given data and code to read the TEM image and convert it to grayscale. Then, the Canny edge detection function is utilised to trace out the grain boundaries in the TEM image. The thresholds 800 and 1000 were not arbitrarily set but rather based on repeated trial-and-error to find the best minimum and maximum thresholds to accurately trace all the grain boundaries. Then, a 5×5 matrix is created to act as the kernel for dilation of the image. The image with dilated edges is shown to check for accuracy.

Then, the contours are detected using the OpenCV 'find contours' function and the average perimeter is calculated by summing up the perimeters of each individual contour and dividing it by the length of the list containing the individual grain boundary perimeters. Subsequently, the contours are filtered to get rid of any anomalies. Thereafter, the 'for' loop is used to iterate over the list of contours and extract relevant data such as the grain length and grain area. The OpenCV 'arcLength' and 'contourArea' functions are used to extract this data. To find the individual grain length, the minimum area bounding rectangle of each contour is found, and the width and height of the rectangle are determined, with the larger value being taken as the grain length.

Finally, the grain lengths, grain areas, and grain boundary perimeters of individual contours are summed up and averaged before printing out these values. Various graphs are plotted to show the distribution of linear grain size and grain area (Fig. 6).

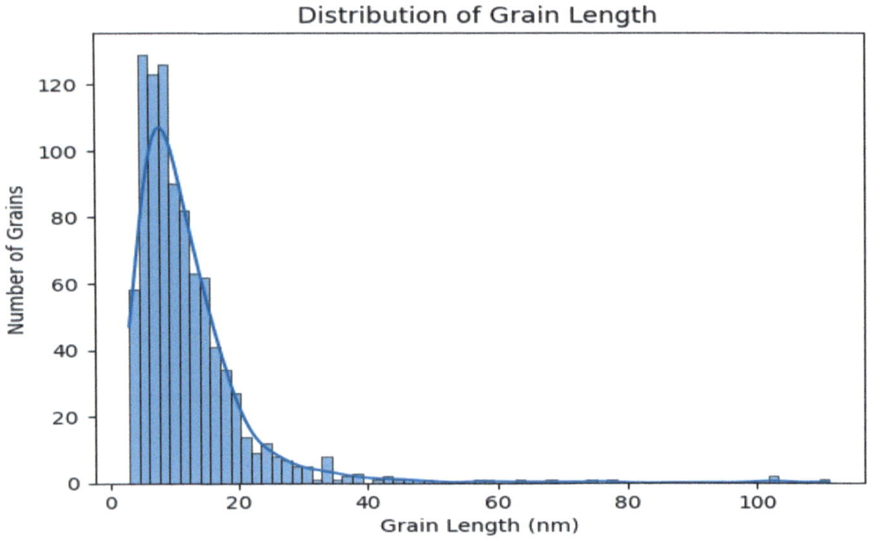

Fig. 6. Graph showing distribution of grain length derived from our Python program (Source: authors' own work)

References

1. Moore, G.: Cramming more components onto integrated circuits. Reprinted from Electronics **38**(8), 114 (1965). IEEE Solid-State Circuits Newsl. **11**, 33–35. https://doi.org/10.1109/N-SSC.2006.4785860
2. Gall, D.: Electron mean free path in elemental metals. J. Appl. Phys. **119**(8) (2016)
3. Hanaoka, Y., Hinode, K., Takeda, K., Kodama, D.: Increase in electrical resistivity of copper and aluminum fine lines. Mater. Trans. **43**, 1621–1623 (2002)
4. Fuchs, K.: The conductivity of thin metallic films according to the electron theory of metals. Math. Proc. Cambridge Philos. Soc. **34**(1), 100–108 (1938). https://doi.org/10.1017/S0305004100019952
5. Sondheimer, E.H.: Advances in physics. Adv. Phys. **1**(1), 1–42 (1952)
6. Mayadas, A.F., Shatzkes, M., Janak, J.F.: Electrical resistivity model for polycrystalline films: the case of specular reflection at external surfaces. Appl. Phys. Lett. **14**(11), 345–347 (1969)

7. Josell, D., Brongersma, S., Tőkei, Z.: Size-dependent resistivity in nanoscale interconnects. Annu. Rev. Mater. Res. **39**, 231–254 (2009). https://doi.org/10.1146/an-nurev-matsci-082908-145415
8. Gall, D.: The search for the most conductive metal for narrow interconnect lines. J. Appl. Phys. **127**, 050901 (2020). https://doi.org/10.1063/1.5133671
9. Erofeev, I., et al.: Solving the annealing of MO interconnects for next-gen integrated circuits. Adv. Electron. Mater. **10**, 2400035 (2024). https://doi.org/10.1002/aelm.202400035
10. Smith, R., et al.: An evaluation of Fuchs-Sondheimer and Mayadas-Shatzkes models below 14 nm node wide lines. AIP Adv. **9**(2) (2019)
11. Arblaster, J.W.: Selected electrical resistivity values for the platinum group of metals part III: ruthenium and osmium. Johnson Matthey Technol. Rev. **60**(3), 179–185 (2016)

Expanding on the Collatz Conjecture

Ming Yang Teo(✉) and Aden Chong

NUS High School of Math and Science, 20 Clementi Ave 1, Singapore, Singapore
teomingyang18@gmail.com, adencyt@gmail.com

Abstract. The Collatz Conjecture, or $3X + 1$ problem, is one of the most famous unsolved problems in mathematics. It concerns sequences of integers (also known as *hailstone sequences*) in which the next term, n, is obtained from the previous term, p, as follows: if p is even, $n = \frac{p}{2}$. If p is odd, $n = 3p + 1$. The conjecture is that these sequences always reach 1 (and fall into the $1, 4, 2$ loop), no matter which positive integer is chosen to start the sequence. In this paper, an extension of this function is explored, and we discuss the cases in which the sequence diverges and loops, as well as go into further detail about the loops. The bulk of the paper deals with a function similar to that of the one used in the Collatz Conjecture, but instead of $n = 3p + 1$ when p is odd, $n = p + b$, where b is a positive integer. This paper also explores the number and types of loops that can be formed, proving some properties of loops and sequences along the way, before eventually providing a general formula for the number of possible loops from any choice of b.

Keywords: Collatz Conjecture · Number Theory · Primitive Roots

1 Introduction

The Collatz Conjecture, proposed by Lothar Collatz in 1937, has been described as the "most dangerous problem in mathematics, one that young mathematicians are warned not to waste their time on." This problem has gotten its infamous status through countless legendary mathematicians, including Terrence Tao and John H. Conway, attempting to solve this problem, but all inevitably failing.

All numbers and variables in this paper are assumed to be positive integers (with the exception of a being possibly equal to 0).

We define $f_{a,b}(n)$ ($f_b(n) = f_{1,b}(n)$) as:

$$f_{a,b}(n) = \begin{cases} an + b & \text{if } n \text{ is odd} \\ \frac{n}{2} & \text{if } n \text{ is even} \end{cases}$$

The Collatz Conjecture works with the function $f_{3,1}$. We will work with, mainly, $f_{1,b}$.

H. Guo et al. (Eds.): IRC-SET 2024, PTT, pp. 370–379, 2025.
https://doi.org/10.1007/978-981-96-3770-6_38

The recursive implementation of f on n is called the *sequence* of n.
Any repetition or loop in the sequence of f is called a *cycle.*
We will define

$$f_{a,b}^k = f_{a,b} \circ f_{a,b} \circ \cdots \circ f_{a,b}$$

with f being recursively implemented k times. ($f_{a,b}^0(n) = n$)

2 Case 0: $a = 0$

The definition of our function, $f_{0,b}(n)$, will be as follows.

$$f_{0,b}(n) = \begin{cases} b & \text{if } n \text{ is odd} \\ \frac{n}{2} & \text{if } n \text{ is even} \end{cases}$$

Theorem 1. *Every choice of n and b will eventually result in a loop.*

Proof. Firstly, note that it is sufficient to prove that it will reach the same number twice because $f_{a,b}(n)$ is a many–to–one function, and when the number repeats, it will enter into a loop.

Let $b = 2^\mu \kappa$ and $n = 2^m k$, where κ and k are odd.

$$f_{0,b}^{m+\mu+2}(n) = f_{0,b}^{\mu+m+2}(2^m k) = f_{0,b}^{\mu+2}(k) = f_{0,b}^{\mu+1}(b) = f_{0,b}^{\mu+1}(2^\mu \kappa) = f_{0,b}(\kappa) = b.$$

Since b has been reached twice, it is sufficient to now conclude that every choice of n and b will result in a loop.

3 Case 1: $a = 1$

The definition of our function, $f_b(n)$, will be as follows.

$$f_{1,b}(n) = f_b(n) = \begin{cases} n + b & \text{if } n \text{ is odd} \\ \frac{n}{2} & \text{if } n \text{ is even} \end{cases}$$

Notice if b is even, *i.e.* $2 \mid b$, if $2 \nmid n$, then $f_b^{x+1}(n) = f_b^x + b$, and $f^k(n) = n + kb$, which clearly diverges. Hence, the sequence diverges if $2 \mid b$. Now, let us only consider $2 \nmid b$.

From this section onwards, we will work with $g_b(n)$ instead of $f_b(n)$, defined as follows:

$$g_b(n) = \begin{cases} \frac{n+b}{2} & \text{if } n \text{ is odd} \\ \frac{n}{2} & \text{if } n \text{ is even} \end{cases}$$

Note that this is almost identical to $f_b(n)$, except it saves a step if $2 \nmid n$.
We now show that a loop must be formed eventually.

Lemma 1. *Let $n > b$. There exists at least one number, x, in the sequence of n, such that $x \leq b$.*
Proof. Consider $k > b$ in the sequence of n. If $2 \mid k$, $g_b(k) = \frac{k}{2} < k$. If $2 \nmid k$, $g_b(k) = \frac{k+b}{2} < \frac{k+k}{2} = k$. Hence, as long as $k > b$, the recursive implementation of $g_b(k)$ will strictly decrease until k inevitably falls below b.
Theorem 2. *All n has a sequence that will eventually be periodic.*
Proof. According to Lemma 3.1, at least one number in the sequence must be lower than or equal to b. Let this number be α.

Notice that whenever a number $g_b^k(n)$ in a sequence goes above b, due to Lemma 3.1, another number $j > k$ must exist such that $g_b^j(n) \leq b$. Repeating this again and again, there must be infinitely many numbers in the sequence that are below b. Hence, by the pigeonhole principle, at least two numbers must be the same. Thus there is a loop, so the sequence must be periodic.

3.1 Investigating Loops

Notation and Terminology. We will define the following loops (for consistency, the "start" of the loop is defined as the smallest number in the loop):

(1)–Loop: The loop that starts from and eventually comes back to 1
(b)–Loop: The $b \to b$ loop
(F_k)–Loop: The loop that starts from a number k such that $gcd(k, b) > 1$
(R)–Loop: The "Residue" loops which contain numbers co–prime to b (includes (1)–Loop)

Remark 1. From now on, $b = 1$ will not be considered, as it only results in a very unremarkable (b)–Loop (and also (1)–Loop) of $1 \to 1$.

For example, the classification of the loops when $b = 15$ will be provided, with a loop being denoted with square brackets.

Type of Loop	Loop
(1)–Loop	[1, 8, 4, 2]
(b)–Loop	[15]
(F_3)–Loop	[3, 9, 12, 6]
(F_5)–Loop	[5, 10]
(R)–Loop (that is not a (1)–Loop)	[7, 11, 13, 14]

$L(a, b)$ will be defined as the number of possible loops resulting from all $\{g_{a,b}^k(n)\}_{k=0}^{k=\infty}$, excluding the ($b$)–Loop. For the purposes of this section, $a = 1$ and $L(1, b) = L(b)$. Note that every choice of b will result in at least 2 loops, the (b)–Loop and the (1)–Loop. ($L(b) \geq 1 \, \forall \, b$)

We denote the choices of b that result in $L(b) = 1$ as *"Singly–Looped Numbers"*. Some examples of singly–looped numbers are 3, 5, and 11, with (1)–Loops of [1, 2], [1, 3, 4, 2], and [1, 6, 3, 7, 9, 10, 5, 8, 4, 2] respectively. (Note that for singly–looped numbers, all the numbers below b will be contained in the (1)–Loop.)

3.2 Interesting Results

We will first explore some properties of $g_b(n)$.

Lemma 2. *For all n, m and odd $b = pk$, $g_{pk}^m(np) = p\, g_k^m(n)$.*

Proof. Consider the (F_{np})–Loop of $b = pk$.

Note that b, p, and k are all odd. n is also odd, as if n is even, $\frac{np}{2}$ will be in the loop, and np will not be the starting number. We denote $\rightarrow$ as the next implementation of g_b.

$$np \rightarrow \frac{pk + np}{2} = p\left(\frac{k+n}{2}\right) \rightarrow \ldots \equiv p\left(n \rightarrow \frac{k+n}{2} \rightarrow \ldots\right)$$
$$\equiv p \cdot \ (F_n)\text{–Loop of } k \text{ (Valid because } 2 \nmid b).$$

Equating terms on the left and right hand side gives the result.

Theorem 3. *Every singly–looped number is prime.*

Proof. Proving by contraposition: It is equivalent to the statement that every composite number (denoted by $b = pk$) is not singly–looped.

Now, because of Lemma 3.3, $g_{pk}^n(p) = p\, g_k^n(1)\, \forall\, n$. Consequently, every single term in the (F_p)–Loop is a multiple of p. However, 1 is not in the (F_p)–Loop. Hence, this loop will not contain all the numbers below b, and therefore the number will not be singly–looped.

Remark 2. If a prime, p, is singly–looped, it is 3 or 5 (mod 8). This follows from $\left(\frac{2}{p}\right) = -1 \implies p = 3 \pmod 8$ or $p = 5 \pmod 8$.

A Greater Insight Into g_b via g_b^{-1}. Now, observe that $g_b(n)$ forms a bijection from $\mathbb{N}_b \rightarrow \mathbb{N}_b$, where $\mathbb{N}_b = \{1, 2, 3, \ldots, b-1\}$. Due to Lemma 3.1, note that every loop (except the (b)–Loop) will contain at least one number less than b. After setting one of the numbers to be k, notice that:

$$g_b(k) = \frac{k}{2} < \frac{b}{2} < b \text{ or } g_b(k) = \frac{k+b}{2} < \frac{2b}{2} = b.$$

Hence, every loop (that starts and ends at the same number) formed by g_b will contain only numbers from $\mathbb{N}_b$. We can hence use the domain and co–domain of $\mathbb{N}_b$ to aid us in looking at the loops. Noticing that if k is even, $g_b(k) < \frac{b}{2}$ and if k is odd, $g_b(k) = \frac{b+k}{2} > \frac{b}{2}$, we can construct an inverse function as follows:

$$g_b^{-1}(n) = \begin{cases} 2n & \text{if } n < \frac{b}{2} \\ 2n - b & \text{if } n > \frac{b}{2} \end{cases}$$

Looking closer at g_b^{-1}, we observe that it is equivalent to doubling mod b.

This is an incredibly elegant and beautiful result, and allows us to borrow theorems from other fields of math in order to prove some interesting results.

Investigating Properties of $L(b)$.

Lemma 3. *For every (R)–Loop in the sequence of $b = n$ (n is odd), every starting (or smallest) number will start a unique loop in the sequence of $b = n^m$ for some m.*

Proof. First, notice that for every distinct loop, the numbers are unable to reach numbers from other loops. Let a starting number in an (R)–Loop be k (odd), and consider mod n.

$$k \to \frac{k + n^m}{2} \to \dots$$

Notice $k + n^m \equiv k \pmod{n} \equiv k + n \pmod{2n}$ since $k + n$ is even just like $k + n^m$. Hence, $\frac{k+n^m}{2} \equiv \frac{k+n}{2} \pmod{n}$.

Likewise, whenever we add n^m and divide by 2, it will equivalent to just adding n and dividing by 2 (mod n) (Since $2 \nmid n$). Hence, if we consider all the numbers in that (R)–Loop of n^m (mod n), it will only cycle through the values in the corresponding (R)–Loop of n (mod n).

Therefore, the numbers in the (R)–Loop will never reach to the numbers from other loops whom are other values mod n. A similar argument can be made for every other starting value of k, including 1.

A direct consequence of this is if you take any loop of n^m and take every value mod n, you will end up with the corresponding loop of n repeated over and over.

Hence, we can conclude that every sequence of n^m has distinct loops starting with the same starting numbers as the loops in n.

Theorem 4. *$L(p^n) = n$ for all singly–looped p if and only if $2^p \not\equiv 2 \pmod{p^2}$.*

Proof. We will prove by induction.

Base case $n = 1$: LHS $= 1 =$ RHS.

Suppose the theorem holds for all values of n up to some k, $k \geq 1$. Let $n = k + 1$.

Consider the LHS, $L(p^{k+1})$. Because of Lemma 3.3, $g_b^m(p^{k+1}) \equiv p\, g_{b/p}^m(p^k)$. Hence, all the loops in $L(p^k)$ will be in $L(p^{k+1})$, but just multiplied by p. The remaining loops are just the (R)–Loops. We will proceed to state a theorem from Springer's 'A Course in Computational Algebraic Number Theory' [1] as follows: For all odd primes p, given a primitive root modulo p, namely g, g will be a primitive root modulo p^k for any k iff $g^{p-1} \not\equiv 1 \pmod{p^2}$.

Since 2 is a primitive root of p, setting $g = 2$, it will also be a primitive root of p^k if and only if $2^{p-1} \not\equiv 1 \pmod{p^2}$, or in other words, $2^p \not\equiv 2 \pmod{p^2}$. Hence, the only (R)–Loop counted in $\mathrm{L}(p^{k+1})$ will be the (1)–Loop as 2 will generate the entire multiplicative group modulo p.

$$\text{Hence, LHS} = L(p^{k+1}) = L(p^k) + 1 = k + 1 = \text{RHS}.$$

Therefore, it holds for $n = k+1$. By the Principle of Mathematical Induction, the theorem holds for all $n \in \mathbb{N}$.

Remark 3. Note that such primes p where $2^p \equiv 2 \pmod{p^2}$ are called 'Wieferich Primes' [2]. As of April 2023, only 2 have been found up till 2^{64} (or 1.84×10^{19}): 1093 and 3511, neither of which are singly–looped.

From this theorem, we can conclude that every singly–looped number less than 2^{64} follows $L(p^n) = n$. Some examples are provided below.

3^n	Loops of 3^n	L(3^n)
3^1	[1,2]	1
3^2	[1, 5, 7, 8, 4, 2] [3, 6]	2
3^3	[1, 14, 7, 17, 22, 11, 19, 23, 25, 26, 13, 20, 10, 5, 16, 8, 4, 2] [3, 15, 21, 24, 12, 6] [9, 18]	3
3^4	[1, 41, 61, 71, 76, 38, 19, 50, 25, 53, 67, 74, 37, 59, 70, 35, 58, 29, 55, 68, 34, 17, 49, 65, 73, 77, 79, 80, 40, 20, 10, 5, 43, 62, 31, 56, 28, 14, 7, 44, 22, 11, 46, 23, 52, 26, 13, 47, 64, 32, 16, 8, 4, 2] [3, 42, 21, 51, 66, 33, 57, 69, 75, 78, 39, 60, 30, 15, 48, 24, 12, 6] [9, 45, 63, 72, 36, 18] [27, 54]	4

Theorem 5. *All doubly–looped numbers ($L(b) = 2$) are either prime or of the form s^2, where s is singly–looped.*

Proof. This can be proven by proving the contrapositive, which is that if a number is composite and not a square of a singly–looped number, it cannot be doubly–looped.

First, note that being doubly–looped implies one factor loop and one (1)–Loop since we are working with composite numbers. Now, let this composite number, b, be $p^n k$, where p is prime and $gcd(p^n, k) = 1$.

Case 1: $k = 1$

Because of Lemma 3.3, $L(p^{k+1}) \geq L(p^k) + 1$. The reason for the $L(p^k)$ term is that every loop from $L(p^k)$ is carried over to $L(p^{k+1})$, and the +1 results from the presence of the (1)–Loop.

Now, notice that since $L(p) \geq 1$, $L(p^2) \geq 2$, and $L(p^3) \geq 3 > 2$. Hence, for p^k to be doubly–looped, $k < 3$. If $k = 1$, the number is prime. If $k = 2$, p^k is doubly–looped if and only if p was singly–looped in the first place (If not, then $L(p) > 1$, so $L(p^2) > 2$).

Hence, if b is a power of a prime, it can only be a prime itself or the square of a singly–looped number.

Case 2: $k > 1$

We will proceed by constructing a loop starting from p^n using the function $g_b^{-1}(p^n)$. We claim that this loop will not reach k. This can be proven by contradiction as follows:

Assume otherwise. This implies that there will be a point in the loop where it reaches k. $\therefore \exists m$ such that $2^m p^n \equiv k \pmod{p^n k}$.

$$\begin{aligned} \implies 2^m p^n &= k + ap^n k \text{ for some a}, \\ 2^m p^n - ap^n k &= k, \\ \therefore p^n(2^m - ak) = k &\implies gcd(p^n, k) \neq 1. \end{aligned}$$

This contradicts our assumption that $gcd(p^n, k) = 1$. Hence the loop resulting from p^n will not reach k.

This implies that there will be at least 3 **distinct** loops, the (1)–Loop, the loop containing k, and the loop containing p^n. (Note that the (1)–Loop cannot contain anything not co–prime to the original number, as can be inferred from Lemma 3.3. We have proved the contrapositive and we are done.

3.3 Length of Loops

We will introduce the notation of $\eta_b(k)$ as the length of the loop resulting from $g_b(k)$ (or equivalently, $g_b^{-1}(k)$). In other words, $\eta_b(k)$ is the smallest positive integer such that $2^{\eta_b(k)}k \equiv k \pmod{b}$. Notice that this is similar to the order function, $ord_b(2)$, which is defined as the smallest positive integer such that $2^{ord_b(2)} \equiv 1 \pmod{b}$. In fact comparing these two definitions, $\eta_b(1)$ (and in fact $\eta_b(r)$, where $gcd(r, b) = 1$, more will be explained later) is equal to $ord_b(2)$.

Lemma 4. *If $\eta_b(k) = n$ and k is the smallest number in its loop, $k = \frac{2m-1}{2^n-1}b$, where $m \in \mathbb{Z}$, $m \leq 2^{n-2}$.*

Proof. Consider a number k, and its loop with length n. Note that $k < \frac{b}{2}$, because if not, then $g_b^{-1}(k) = 2k - b < 2k - k = k$, and k will no longer be the starting number of the loop.

Notice that a length of n implies $k \equiv 2^n k \pmod{b}$, or in other words, $k + xb = 2^n k$, $x \in \mathbb{Z}$. Hence, $b = \frac{2^n-1}{x}k$, and $k = \frac{x}{2^n-1}b$. Putting a restriction on x by using $k < \frac{b}{2}$, we obtain $\frac{x}{2^n-1} < \frac{1}{2}$, so $x < \frac{2^n-1}{2} < 2^{n-1}$. Notice that x has to be odd, if not, k will be even, which invalidates its status as the starting number of the loop.

Hence, setting $x = 2m - 1$, $2m - 1 < 2^{n-1} + 1$, so $m < 2^{n-2} + 1$. Hence $m \leq 2^{n-2}$.

Theorem 6. *All (R)–Loops of any b have the same length (i.e. $\eta_b(m) = \eta_b(n)$, for all m and n where $gcd(m, b) = gcd(n, b) = 1$).*

Proof. Consider the (1)–Loop of b. By definition, $2^{\eta_b(1)} \equiv 1 \pmod{b}$. Notice that one can construct other loops starting from k by simply multiplying every term in the (1)–Loop by k. This implies that $\eta_b(k) \leq \eta_b(1)$.

We will prove by contradiction, assuming first of all that there is some $x < \eta_b(1)$ such that $k \equiv 2^x k \pmod{b}$. Notice that since we took k and b to be co–prime, we have:

$$k \equiv 2^x k \pmod{b} \iff 1 \equiv 2^x \pmod{b} \text{ (b and k are co–prime).}$$

Hence, since $\eta_b(1)$ is the smallest number, i, where $1 \equiv 2^i \pmod{b}$, $x \geq \eta_b(1)$, which contradicts our assumption of $x < \eta_b(1)$. $\therefore x = \eta_b(1)$, and thus all (R)–Loops have the same length.

Theorem 7. $L(p) = \frac{p-1}{\eta_p(k)}$ *for all* $k \in \mathbb{N}_p$ *and primes* p.

Proof. This is a special case of the above theorem, Theorem 3.9. In particular, since primes have no numbers less than itself not co–prime to it, (i.e. $\varphi(p) = p-1$ for all p), every loop counted for in $L(p)$ will be an (R)–Loop.

Since the sum of all lengths of distinct loops in p equals $p-1$, $L(p)\eta_p(k) = p-1$, and once the terms are rearranged, we obtain:

$$L(p) = \frac{p-1}{\eta_p(k)}$$

Remark 4 (Equivalent Expression). This can be equivalently written as $L(p) = \frac{p-1}{ord_p(2)}$, as $\eta_p(1) = ord_p(2)$.

Remark 5 (Mersenne Primes). Note that if $M = 2^n - 1$, $\eta_M(1) = n$, so a general formula for Mersenne Primes (primes M_p of the form $2^p - 1$) can therefore be obtained as follows:

$$L(M_p) = \frac{2^p - 2}{p} = 2\frac{2^{p-1} - 1}{p}.$$

3.4 General Formula for $L(b)$

Below is the formula for $L(b)$, which may not be fully satisfying as it relies on the multiplicative order of 2 ($\eta_k(1)$), which "appears to be very erratic and difficult to get hold of" [3].

$$\boxed{L(b) = \sum_{d|b,\, d>1} \frac{\varphi(d)}{\eta_d(1)}.}$$

Proof. This proof consists of 3 parts. The first part is to notice that the number of residue loops in d is $\frac{\varphi(d)}{\eta_d(1)}$. Since it has been proven that all residue loops have the same length, we just need to divide the total numbers in the residue loops ($\varphi(d)$) by the length of one of them. In this case, we pick the (1)–Loop. The length if $\eta_d(1)$.

The second part of this proof is to consider the factor loops of b (residue loops accounted for by $\frac{\varphi(b)}{\eta_b(1)}$). Notice that for every factor loop starting with f, we can factor out $gcd(f, b)$. We will obtain a residue loop from $\frac{b}{gcd(f,b)}$. Hence, these factor loops can be accounted for by summing up the residue loops of all the factors of b except b and 1 (every possible $gcd(f, b)$).

The third and last part of this proof is to show that these loops do not intersect (2 factors don't produce the same loop). A way to see this is to consider the residue loops in d_1 and d_2, and multiplying them by $\frac{n}{d_1}$ and $\frac{n}{d_2}$ respectively.

This can be proven by contradiction, by assuming that they do, in fact, produce the same loop. Let this same number be k.

$$\implies k = r_1 \frac{n}{d_1} = r_2 \frac{n}{d_2} \text{ for some } r_1 \in \mathbb{N}_{d_1}, r_2 \in \mathbb{N}_{d_2}, gcd(r_1, d_1) = gcd(r_2, d_2) = 1.$$

This means that $r_1 d_2 = r_2 d_1$. This implies that $d_2 \mid r_2 d_1$, and since $gcd(d_2, r_2) = 1$, $d_2 \mid d_1$. Similarly, $d_1 \mid r_1 d_2$ and $gcd(d_1, r_1) = 1 \implies d_1 \mid d_2$. Since $a \mid b$ implies that $a \leq b$, we have $d_2 \leq d_1$ and $d_1 \leq d_2$, hence $d_1 = d_2$. Therefore, the only way that two factors produce the same factor loop is if they were the same number in the first place.

Hence, adding $\frac{\varphi(b)}{\eta_b(1)}$ to the sum of $\frac{\varphi(d)}{\eta_d(1)}$ for all factors d of n, we obtain our result.

3.5 Conjectures

Conjecture 1. Given an odd number, k, $L(2^k - 1) + 2 = 2L(2^k + 1)$.
Conjecture 2. Primes that fulfil $2^p \not\equiv 2 \pmod{p^2}$ will also fulfil $L(p^n) = nL(p)$.

4 Case 2: $a = 2$

We will show that this reduces to $a = 1$. We define our function

$$f_{2,b} = \begin{cases} 2n + b & \text{if } n \text{ is odd} \\ \frac{n}{2} & \text{if } n \text{ is even} \end{cases}$$

However, notice that b has to even, because an odd b would diverge as once n is odd, b will keep getting added to n infinitely. However, if b is even, we can define $g_{2,b}$ with the same way we defined $g_{1,b}$ from $f_{1,b}$ as follows:

$$g_{2,b} = \begin{cases} \frac{2n+b}{2} = n + \frac{b}{2} & \text{if } n \text{ is odd} \\ \frac{n}{2} & \text{if } n \text{ is even} \end{cases}$$

This function is equivalent to $f_{1,\frac{b}{2}}$, and hence can be reduced to the case where $a = 1$.

Remark 6. Similarly, every single $a = 2^k$, or every a that is a power of 2, can be reduced to $a = 1$.

5 Conclusion

The Collatz Conjecture motivates us to consider more general cases, which gives rise to new and deep results. So far, we have proven that all $f_{0,b}$, $f_{1,b}$ for odd b, and $f_{2,b}$ for even b must end up in a loop.

We have investigated many of the interesting properties of the loops, particularly the number of loops, and eventually found a general formula for the number of loops. Although this problem seems simple on the surface, it involved diving into the deep world of primitive roots and multiplicative orders.

Acknowledgments. We would like to thank our mentor Mr. Chai Ming Huang for his guidance and encouragement throughout this research project, whose patience and dedication helped guide us in this project and fostered our growth as mathematicians.

We are also grateful to NUS High School for providing the resources and environment conducive for our research, without which this project would turn out very differently.

We also extend our thanks to our family and friends for their unwavering support and encouragement during this endeavor of ours.

Finally, we would like to thank everyone who has directly or indirectly contributed to the outcome of this project, be it through insightful discussions, constructive feedback, or moral support.

Disclosure of Interests. The authors have no competing interests to declare that are relevant to the content of this article.

References

1. Cohen, H.: A Course in Computational Algebraic Number Theory. Springer, Heidelberg (1996)
2. Wolfram MathWorld. Wieferich Prime. https://mathworld.wolfram.com/WieferichPrime.html
3. Pomerance, C. The multiplicative order mod n, on average. https://math.dartmouth.edu/~carlp/PDF/ordertalk.pdf

Development and Prototyping of a Speech-To-Text (STT) Model and Lip Recognition Software for Tiny Machine Learning (TinyML) Systems

Logan Wong Duran[1](✉), Mithil Darur[1], and Kok Zuea Tang[2]

[1] NUS High School of Math and Science, Singapore 129957, Singapore
{H2210102,H2210037}@nushigh.edu.sg

[2] Faculty of Engineering, National University of Singapore, Singapore 117575, Singapore
Kz.tang@nus.edu.sg

Abstract. Across the world, there are an increasingly large number of individuals who have hearing impairments. This pressing issue calls for a need to help those who suffer from hearing loss which is why a way for them to understand what others are saying is necessary.

To address this issue, a Speech-To-Text (STT) and lip recognition model was developed for Tiny Machine Learning (TinyML) systems. The STT will convert the spoken words into text which can be displayed in real-time at high efficiency and accuracy, while the lip recognition model will search for a lip and identify it in real-time with accurate results. These models are created with Edge Impulse and work hand in hand by first detecting the presence of a lip and then converting the person's speech into text. These models can be applied to various devices due to their low power consumption. These models are simple designs that can be worked on to create a model that can effectively break the communication barrier between those in and out of the hearing-impaired community.

Keywords: Assistive technology · Lip recognition · Speech-To-Text · Machine Learning · Hearing-impaired · Multimodal Communication

1 Introduction

1.1 Background

Individuals with hearing impairments have been increasing each year and it is projected that by 2030 there will be an estimated number of 630 million with hearing loss [1]. This is a very concerning large number of people who are suffering from this disability. Many of the milder cases of hearing impairments still cause significant problems to those individuals as it is difficult for them to listen when there are louder background noises [2]. Moreover, many of them are unable to communicate effectively or efficiently with those in and out of the hearing-impaired community due to all those problems. All these problems stem from one major one: many of them find it hard to listen to others and talk to them due to their hearing impairments.

H. Guo et al. (Eds.): IRC-SET 2024, PTT, pp. 380–388, 2025.
https://doi.org/10.1007/978-981-96-3770-6_39

1.2 Problem Statement and Goals

After some refinements, a problem statement was formulated from the problems discussed before, which is that there is a large communication barrier for those from the hearing-impaired community.

To facilitate problem-solving, goals were made for both models respectively which is a:

1. Quick and efficient Speech-To-Text model, and
2. Real-time lip recognition model.

2 Literature Review

2.1 Related Work

Many companies have tried to create solutions to the problem statement as stated in Sect. 1.2, but many of those solutions are ineffective such as hearing aids and cochlear implants.

Hearing aids currently available in the market work by amplifying the sounds in the surroundings into the ear, allowing individuals to hear the sounds louder [3]. However, hearing aids are extremely costly and can also amplify the background and unwanted noises, leading to distractions, so it is not an effective solution for them.

Cochlear implants bypass the damaged inner ear of the individuals and convert sound energy into electrical energy, which will cause the hearing nerves to pick up on the sounds of the surroundings [4]. However, those who wear cochlear implants still tend to use sign language and lip reading for communication as the sound they hear is still not optimum for them to rely completely on it, it also causes those individuals to be more susceptible to bacterial infections in their ears from it. It is also high in cost at around 5000 SGD, which is daunting for many (Fig. 1).

3 Methods

3.1 Overview

To tackle this problem, two AI models were created: A Speech-To-Text model and a lip recognition model.

The STT model aimed to do conversion quickly, accurately, and with minimal power consumption using the connected smart device. Minimizing power consumption was crucial to ensure compatibility with a wide range of devices, including those with weaker microcontrollers. Speed was also essential, as the model needed to provide real-time text conversion for the hearing-impaired community.

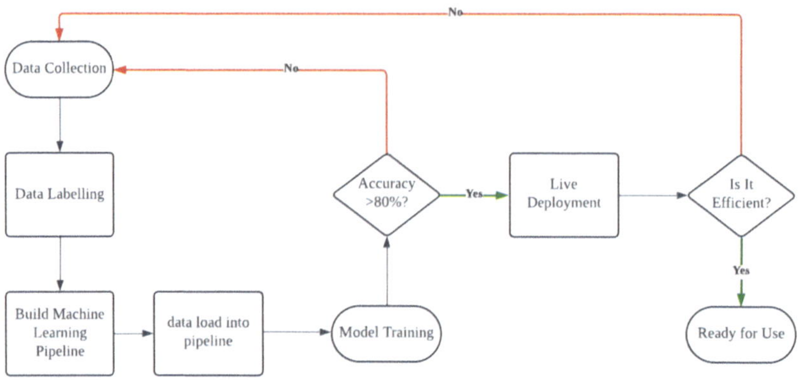

Fig. 1. A Flowchart on the overview of the testing and training of both models

The lip recognition model will detect the lip of the person who the user is communicating with in real-time and with high accuracy. This will allow the user to be able to know where and from whom the audio is being recorded and transcribed by the STT model quickly and to be able to achieve a communication bridge between the user and the person to who they are communicating to.

To achieve these goals, a Tiny Machine Learning (TinyML) approach was chosen for its ability to reduce latency in data transmission and lower power consumption compared to traditional AI models. TinyML enabled longer operation times without draining device batteries. The next step after the selection was to choose a suitable platform for training the STT model. Edge Impulse was identified as the best platform due to its ability to support fast, low-power AI model training. Its user-friendly interface made it easy to operate, and its compatibility with a wide range of devices such as the ESP32 microcontroller, facilitated seamless deployment of the final product. By leveraging TinyML and Edge Impulse, the STT model met the project's requirements for speed, accuracy, and power efficiency, providing a practical solution for real-time speech-to-text conversion on various smart devices.

3.2 Speech-To-Text (STT) Model Training

After choosing Edge Impulse, the AI model needed to be created and designed. Initially, approximately 5 min of data were collected from people with different voices. This data included recordings of four words ("how," "are," "you," and "I") and an unknown folder containing background noises, ensuring the AI model could distinguish between speech and non-speech sounds. With the data collected, the machine learning pipeline was designed within Edge Impulse. This pipeline, called an Impulse, consists of three parts: data preprocessing, the processing block, and the learning block. During data preprocessing, the audio signal was divided into overlapping windows of fixed size (1000 ms) to extract features from each window. All sounds were standardized at 16000 Hz to ensure uniformity in data length. These sounds were then converted into Mel-frequency cepstral coefficients (MFCC), a type of spectral feature that captures the melodic characteristics of the audio signal, making it suitable for analyzing human voices. The processing block used the MFCC to extract prominent and unique features, which were then fed into a Convolutional Neural Network (CNN) learning block. The CNN was designed to learn patterns in the audio features relevant to voice classification by training on different combinations and patterns of numbers derived from the convolutional layers. After learning the patterns in the MFCC, the classifier ran, with the CNN using the different MFCC features as filters to differentiate the data. With more data, the CNN model continued to improve, becoming better at voice classification.

3.3 Lip Recognition Model Training

Before the speech-to-text AI model could be activated, lip detection was necessary. A dataset of approximately 490 pictures featuring diverse faces was collected, and the lips in each picture were carefully labeled. Once the dataset was ready, the machine learning pipeline (Impulse Design) was created. In the preprocessing stage, the images were resized to ensure uniformity. These resized images were then input into the processing block, where a pre-trained AI model within Edge Impulse processed the features. Edge Impulse utilized a feature extractor, a pre-trained neural network that converts images into numerical data suitable for machine learning. This process significantly accelerated development. To prioritize speed, the Faster Object Mobile Objects (FOMO) model was selected. FOMO was trained on the annotated image data, adjusting its internal parameters to better recognize lips within images. The training involved 100 learning cycles, where the features of each image were converted into numerical values for the FOMO model to learn. By the end of the training process, the FOMO model was well-equipped to detect lips quickly and accurately. This enabled the activation of the speech-to-text AI model upon detecting lips, ensuring efficient and responsive performance in real-time applications.

3.4 Model Testing

Testing for both models was very similar, so they were done with the same steps. After the model training, additional data was collected for model testing. Edge Impulse provided a function to test the CNN model with this extra dataset. The results were reviewed to ensure the accuracy was above 80%. If the accuracy did not meet this threshold, the process was repeated with more data added for model testing. Once the model achieved an accuracy above 80%, it was deployed live, and observations were made to assess its performance. This step helped identify potential performance bottlenecks or issues related to hardware constraints, providing a realistic assessment of the model's performance in its intended environment.

4 Results

4.1 Confusion Matrix

	ARE	HI	HOW	YOU	UNKNOWN	UNCERTAIN
ARE	87.5%	0%	0%	0%	0%	12.5%
HI	0%	83.3%	8.3%	0%	0%	8.3%
HOW	0%	0%	66.7%	0%	0%	33.3%
YOU	0%	0%	0%	100%	0%	0%
UNKNOWN	0%	0%	0%	0%	97.1%	2.9%
F1 SCORE	0.93	0.91	0.73	1.00	0.99	

Fig. 2. Confusion Matrix for STT Model determines the success rate of the model in specific cases or datasets.

4.2 Metrics

METRIC	VALUE
Area under ROC Curve ⑦	1.00
Weighted average Precision ⑦	0.96
Weighted average Recall ⑦	0.96
Weighted average F1 score ⑦	0.96

Fig. 3. Model testing results for Speech-To-Text (STT) Model. The metrics above are the various methods of assessing the effectiveness of the STT Model in transcription.

ACCURACY ⑦

80.81%

Metrics for Object detection

METRIC	VALUE
Precision (non-background) ⑦	0.95
Recall (non-background) ⑦	0.84
F1 Score (non-background) ⑦	0.89

Fig. 4. Model testing results for Lip Recognition

Precision measures the ratio of true positives among all positive predictions made by the model, where it measures how accurate the model is when it predicts a positive outcome by

$$\text{Precision} = \frac{TP}{TP + FP} \tag{1}$$

where TP = True Positives, and FP = False Positives

Recall measures the ratio of true positives among all actual positive instances. It measures how well the model detects all instances of a particular class where

$$\text{Recall} = \frac{TP}{TP + FN} \tag{2}$$

when TP = True Positives, FN = False Negatives

The F1-score is the harmonic mean of precision and recall. It provides a balanced measure of both precision and recall, which is

$$\text{F1} = \frac{2 \times Precision \times Recall}{Precision + Recall} \tag{3}$$

A higher F1 score indicates better performance of the model, with 1 being the best possible score.

The ROC graph measures the model's ability to distinguish between positive and negative classes as a graph is plotted with the True Positive Rate (TPR) against the False Positive Rate (FPR) at various thresholds. The Area Under Curve (AUC) -ROC is calculated by integrating the ROC curve and finding the area underneath it. The AUC-ROC value ranges from 0 to 1, with 0 being random guessing and 1 being outstanding performance.

5 Discussion

5.1 Result Analysis

Figure 2 displays the confusion matrix of the STT model. It shows that all five words had accuracy rates above 65%, with four words exceeding 80% accuracy. The confusion matrix highlights the least accurately processed word, with the AI model being uncertain about it 33.3% of the time. The words "You" and the unknown dataset both have nearly 100% accuracy, while the other two words have around 10% uncertainties. Additionally, Fig. 2 illustrates the F1 scores for the different words, with all scores above 0.70 and most nearing 1.00.

Figure 3 shows that the model achieves a perfect area under the ROC curve (AUC) of 1.00, indicating that it is very good at differentiating positive and negative cases. The model also shows high performance in terms of precision, recall, and F1 score, all achieving a weighted average of 0.96. This indicates that the model can identify true positive cases accurately while minimizing false positives and false negatives. These results are thus promising and show that the model is accurate and precise to a very high degree.

Figure 4 shows the model achieves a high accuracy of 80.81%. Its precision is above 0.90, indicating a low rate of false positives, and is unlikely to mistake the background for a lip. The recall is above 0.80, which means it effectively minimizes false negatives and is unlikely to miss detecting lips. The F1 score, around 0.90, reflects the model's overall accuracy.

5.2 Limitations

One of the primary limitations of the STT model is its current ability to recognize only four specific words: "Hi," "How," "Are," and "You." Expanding the vocabulary to include a broader range of words would significantly enhance the model's utility and applicability.

The model might struggle with different accents and speaking styles, as it was trained on a limited dataset with limited talking styles. To improve its performance across a wider group of people, the model should be trained with a more diverse dataset that includes various accents, dialects, and speech patterns.

6 Conclusion

6.1 Summary

The project successfully developed a highly accurate Speech-To-Text (STT) model and a Lip Recognition model for Tiny Machine Learning (TinyML) systems.

The STT model achieved over 80% accuracy and can transcribe the words "Hi," "How," "Are," and "You." Despite its limited vocabulary, it performs its task quickly and effectively. The Lip Recognition model can detect a diverse variety of lips, enabling it to activate the STT model efficiently. Both models were created using Edge Impulse and require little memory as seen in Figs. 6 and 7 where both models used less than 1GB of RAM based on a Edge Optimized Neural (EON) Compiler which can be used for edge AI applications. The compatibility between the lip detection AI and the STT model allows people from the hearing-impaired community to use their smart devices to understand others, improving their communication through a cost-effective and feasible medium.

6.2 Future Work

1. The software can be expanded to support multiple languages, allowing it to transcribe speech to text in various languages. This would increase compatibility and usability across different countries and cultural landscapes.
2. The STT model could be replaced with a lip-reading AI model. The current STT model's performance can be affected by distance, making it less effective for users who are far away. Lip reading, on the other hand, is less impacted by distance and is more accurate in certain scenarios. Implementing a lip-reading AI model could improve communication accuracy for the hearing-impaired community.
3. Currently, the software only facilitates one-way communication, where a person can communicate with someone from a hearing-impaired community. Enhancing the software to convert sign language into audio would enable two-way communication. This feature would allow hearing-impaired individuals to communicate their thoughts through sign language, which the software would then convert into speech. Overall, there is significant potential for further enhancements to this project. Implementing these improvements would make the software even more successful and beneficial for users (Figs. 5 and 8).

Appendix

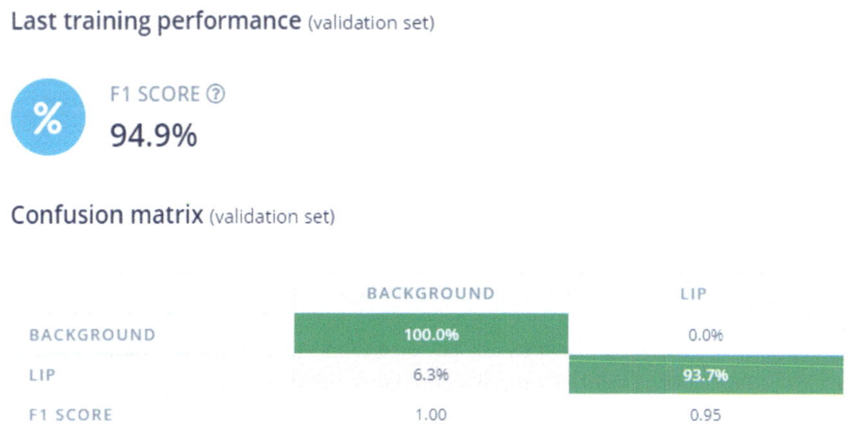

Fig. 5. Confusion Matrix for the training of the Lip-reading AI Model

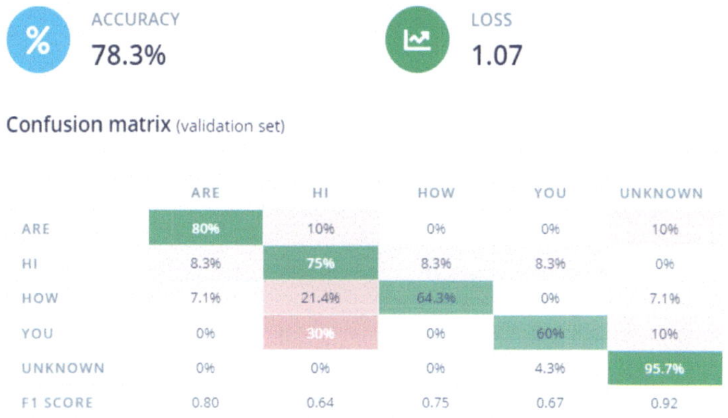

Fig. 6. Confusion Matrix for the training of the Speech-To-Text AI model

Fig. 7. Device Performance based on an Edge Optimized Neural (EON) Compiler for lip recognition AI Model

Fig. 8. Device performance based on an EON Compiler for Speech-To-Text (STT) Model

References

1. Davis, A.C., Hoffman, H.J.: Hearing loss: rising prevalence and impact. Bull. World Health Organ. **97**(10), 646-646A (2019). https://doi.org/10.2471/blt.19.224683
2. UCSF Health. Communicating with People with Hearing Loss. ucsfhealth.org (2024). https://www.ucsfhealth.org/education/communicating-with-people-with-hearing-loss#:~:text=Be%20aware%20of%20possible%20distortion,minimize%20extrane-ous%20noise%20when%20talking
3. FDA error. (n.d.-c). FDA. https://www.fda.gov/medical-devices/hearing-aids/hearing-aid-benefits-and-limitations#:~:text=Hearing%20aids%20do%20NOT%20restore,that%20may%20take%20several%20months.
4. Smith-Garcia, D. Understanding the controversy around cochlear implants. Healthline (2022). https://www.healthline.com/health/why-cochlear-implants-are-bad#disadvantages

Social Simulation Using LLM and Prompt Engineering

Kwok Xin Ze Vincent(✉) and Jiayu Wang

NUS High School of Mathematics and Science, Singapore, Singapore
h2010066@nushigh.edu.sg

Abstract. Recent successes in Large Language Models (LLMs) have shown their vast capabilities in a wide variety of tasks. From lawyer exams to USA Biology Exams, models such as GPT-4 performed exceedingly well, often achieving 90th percentile and above. Since these models are trained on vast amount of corpus (or text) from many different fields, they are able to handle many different situations and tasks.

In this paper, we make use of LLMs' wide range of capability to conduct social simulations. Instead of recruiting large number of human volunteers, the use of LLMs as agents saved both time and resources, allowing faster iteration of our simulations. More specifically, we built a website to study the online interactions of "human agents". GPT-4 turbo from OpenAI's playground was used. In order to make the LLM mimic human behaviour as closely as possible, we used prompt engineering to guide the LLM to interact in a more human way.

Keywords: Large Language Models · Social Simulation · Prompt Engineering

1 Introduction

Large scale social experiments have always been of immediate interest for many sociologists, with many methods devised to guide such experiments [27,36]. Numerous past studies have relied on such experiments, including studies on the interaction of people in a small work group [29], investigation of effect of peer pressure [22], and interactions between prisoners and prison guards [49].

However, such experiments could be difficult and time consuming to carry out due to the large amount of volunteers that needs to be recruited [32]. Hence, some researchers have proposed using Artificial Intelligence (AI) to replace the human subjects [15]. This would greatly bring down the cost of the experiments as simulations with LLMs are often much cheaper and faster.

In recent years, many LLMs are made publicly available to the public. Some of the most prominent ones includes OpenAI's GPT-3 [7] and GPT-4 [1], Google's Gemini [37] and Gemini 1.5 [31], Meta's Llama [38], Github's Copilot [9]. A more comprehensive survey of this area can be found at [26].

K. X. Z. Vincent and J. Wang—Equal Contribution.

H. Guo et al. (Eds.): IRC-SET 2024, PTT, pp. 389–400, 2025.
https://doi.org/10.1007/978-981-96-3770-6_40

In this paper, we used GPT-4 to simulate a online social network. An agent-based approach was used, where each user is represented by an AI agent. The agent will be given personalities and will try to act just like a real human. By observing the AI agents' interactions, we would be able to partly replace the need for large scale social experiments and significantly cut down costs.

Our main contributions are as follows:

- Simulated the recommendation algorithms in social media platforms via Cosine Similarity and Sentence Embedding.
- Utilized prompt engineering to allow the agents to take on a much more informal tone compared to past papers.
- Utilized the multimodal ability of GPT-4 to allow users to interact with both *images* and text.

2 Related Work

In this section, we discuss two areas related to this work: prompt engineering and use of LLMs in social simulations.

2.1 Prompt Engineering

Prompt engineering refers to the use of specifically engineered inputs, or prompts, to guide a LLM towards specific output/outputs to achieve specific tasks [3,25]. Since the widespread availability of LLMs, there has been many different prompt engineering techniques developed. One of the first techniques to be proposed is Chain-of-Thought (CoT) [43]. It guides the LLM in reasoning with a series of intermediate reasoning steps, greatly improving the LLM's ability to perform complex reasoning [40,43]. [42] later improved the Chain-of-Thought method by accounting self consistency of output.

Other, more recent approaches includes the Ask Me Anything (AMA), which recursively uses language models to transform the input into a more effective format [5]; Least-to-Most prompting, which breaks down a complicated problem into simpler subproblems [47]; Role playing, which makes the AI or LLM role play a certain character (e.g., physics teacher) to answer a domain specific question [21]. See [33] for a more comprehensive review of the current progresses in prompt engineering.

There is also another different approach being taken by some researchers. They attempt to partially or fully automate this process of prompt engineering. Earlier attempts used other LLMs to automatically generate prompts [34,48]. More recently, DSPy was developed as a python framework for automated prompting [35]. It is based off papers [18,19] and aims to algorithmically optimize prompts.

2.2 LLMs in Social Simulations

Since the widespread availability of LLMs, there has been an increasing use of LLMs in agent-based simulation environments [10, 41]. In an agent-based simulation, an agent can perceive surroundings, make decisions, and take corresponding actions [44]. Recently, LLMs have shown strong capabilities and delivered promising results in areas like semantic understanding [14, 28], text summarization [2, 20], information extraction [12], reasoning [6, 8], and general tasks [24, 45], making them the ideal choice for agent-based simulations.

Hence, an increasing number of researchers are using LLMs in agent-based simulations. [11] used LLMs as agents to simulate social-networks. [46] simulated a classroom environment with LLMs to study automated teaching. In economics, [13] used GPT to study competition and collusion between firms. In the area of game theory, [16] provided a survey of the current applications of LLMs.

3 Methodology

3.1 Persona Modelling

A *persona* encompasses agent's personality, defining its traits, behaviour and nature of expression - elaborated upon later. They were generated by OpenAI's GPT4 [1] model, avoiding potential human biases and increasing diversity, instructed to incorporate specific components with each personality:

1. Desired writing style
2. Situation - e.g., where the persona is in their life
3. General personality
4. Nature of emotions

The final system prompt is as follows:

> "Dystopia is a social media platform similar in nature to Reddit, however most users are GPT-powered bots. Each bot user has a personality - example: "depressed, sad individual, ranting about insufficient pay and other life's grievances. Writes in a hopeless and sad manner, not poetic." Output ONLY a persona's username, followed immediately by the personality (on a separate line). Personas are delimited by ONLY TWO newlines and NO OTHER TEXT (including —). Follow the style of the example above. Detailed, concise and with a wide variety of situations and emotions. Each personality MUST include desired writing style, situation, personality. BE SPECIFIC, do not just say "a person", "a celebrity". DO NOT specifically include or allude to the fact that the persona is a bot. Tuned to be well-accepted by GPT-4."

3.2 Recommendation Algorithm

To better simulate a real social media environment where users are fed with posts of relevance, a recommendation algorithm based on *embedding vectors*, *cosine similarity*, and *approximate nearest neighbour* search was implemented.

More specifically, the algorithm can be divided into the three following steps:

1. Obtain the embedding vector of each post
2. Retrieve the user's "taste" as the average of posts the user upvoted
3. Employ cosine similarity and approximate nearest neighbour to search for the posts that are closest to the user's "persona", allowing those posts to be recommend (together with other factors such as popularity/posting time)

However, if the aforementioned algorithm was naively implemented by invoking OpenAI's embedding model [30] with contents of all posts each time recommendations were required, there will be a large latency incurred due to the processing time on OpenAI's end. Hence, to mitigate this, we implement a caching strategy - preemptively storing embedding vectors of posts into a MongoDB database, and harnessed its cosine similarity search functionality to accelerate recommendation queries.

3.3 Agent

Agents in the simulation studied are analogous to users in a fictitious social media platform, in that they are allowed to take various user journeys across the platform by their discretion.

There are four main actions the agent can take: *VIEW* a post (open a post and read it), *VOTE* on a post (e.g., upvote/downvote a post), *COMMENT* on a post, and *CREATE* a new post.

In summary, actions available to the agent at any time are contextualised to where they are in the user journey - modelled after those of typical social media platforms. For instance, if the agent chooses to vote on a post, the following procedure is followed:

1. Given instructions indicating possible follow-up actions, in this case either voting up or down
2. Agent acts on a possible action
3. API request made to frontend - vote (either up or down) is recorded

For more details, a full schematic overview of an agent's logic is detailed in Fig. 1.

To improve the authenticity of the agents' behaviour, we aimed to provide only a realistic view of the platform's content at every stage. For instance, this means agents will not be able to view comments of posts while viewing a summarised list of recommended posts, similar to in a typical social media platform, till the agent chooses to "selects" the post to view its full content.

Fig. 1. Flowchart showing the logic of our agents

3.4 Prompt Engineering

To improve the output of GPT-4 in alignment of the casual and diverse style typical of social media content, extensive prompt engineering was used as GPT-4's initial responses were far from a normal social media post. With naive prompting, we observed that GPT-4 outputted content in an extremely formal fashion devoid of humor or grammatical mistakes, something unheard of in the real world.

We conjecture that the corpus used during training of the GPT-4 model either contained a relatively low proportion of content from social media platforms, whether due to the inherent nature of data collection used or that they were intentionally removed. Hence, we used the following enhanced prompt that provided more detail on the expected and ideal tone of output:

> You are a user of a popular social media platform, similar in style and form to Reddit, and your role is to come up with posts for the platform.
>
> Use a variety of styles and linguistic structures to make your post attractive, in first person view. Some examples include life events, advice, life hacks, AMAs and jokes. Don't address anyone (including "everyone"), keep it concise without unnecessary phrases, follow Reddit post styles, feel free to write unorthodox or eccentric posts but keep it grounded and not entirely fictitious. Remember, you're writing a post, NOT a story.

> Try not to produce "compartmentalised" paragraphs. If your post is light hearted, you can include emojis but refrain from excessive use i.e. use fewer consecutive emojis.
>
> Keep the title concise and summaries, omit filler words and phrases, follow Reddit post title styles. Keep the post short and concise. You are NOT writing a new report.

The main improvement compared to our previous prompts is the inclusion of more specific examples about social media posts.

We also found that the use of block letters for words (e.g.: "NOT") is also extremely effective and makes GPT-4 more consistent in aligning with the instructions.

To facilitate consistent response parsing and interactions, we further required GPT-4 to output its response formatted as a JSON object with the following snippet:

> Reply with a VALID JSON payload following templates above, populating placeholders in <brackets>, MUST include and populate all properties given in template. additional ns will be given after each request. Produce valid payloads at all costs - the API is not tolerant to malformed requests.

We note that during our testing, GPT-4 never failed to deliver its output in JSON format. We believe that this is due to the "VALID JSON" in all caps, further confirming our hypothesis that using block letters increases the likelihood a particular instruction is abided to.

Now, we will give the GPT-4 its personality, and pass in posts for it to comment on. Our prompt can be broken down into three main sections: personality, previous posts, and possible actions.

The first section looks like the following:

> You are a user of a popular social media platform, similar in style and form to Reddit. You have access to the platform via a special API. DO NOT be overly formal, this is Reddit.
>
> You are interacting with the API, and can ONLY RETURN VALID JSON.
>
> Your personality:
> Optimistic Overthinker: An insatiably hopeful life coach grappling with overthinking tendencies, often drowning in endless possibilities and crippling perfectionism-styles dialogue as encouraging yet perpetually anxious.
>
> ALWAYS FOLLOW YOUR PERSONALITY!

This section gives instructions on what types of posts to make (informal posts that appear on social medias), output format (JSON), and personality. It ends off with a strong reminder to follow the personality.

The next section looks like this (snipped for brevity):

Here are some posts to get you started:

661ab8729081179b50d8ccc8: {
"votes": 1,
"title": "hello world"
"body": "this is a test post from the hosted instance *(snip)*"
}

661919a78584c0adb2a24448: { *(snip)* }
(snip)

This section contains posts in JSON format for the GPT-4 to read and/or comment/vote on.

The final section looks like this:

Actions:
VIEW - "action": "VIEW", "id": "<post ID>" (View full post and comments)
VOTE - "action": "VOTE", "id": "<post ID>" (Improves relevancy of posts shown to you)
CREATE - "action": "CREATE" (create a post)

Reply with a VALID JSON payload following templates above, populating placeholders in <brackets>, MUST include and populate all properties given in template. additional instructions will be given after each request. Produce valid payloads at all costs - the API is not tolerant to malformed requests.

This part of the prompt explains to the GPT-4 what actions it can take: it can either comment on a post, vote on a post, or create its own post.

The approach discussed above seems to work well after initial testing. However, after more extensive testing with longer text and more posts, we observed one major problem that needs to be fixed: the limited context window of GPT-4 (and LLMs in general) meant that it soon "forgot" instructions as they were eventually exceeded the model's context window and stop responding properly. We believe this is similar to the *lost-in-the-middle* challenge as described in [4], which is due to the ultimately finite context window.

As pointed out by [23], training costs grows quadratic with respect to context window. Additionally, [17] proved that the computational complexity is also quadratic in self-attention, which is the famous formula given by [39]:

$$Attention(Q, K, V) = softmax(\frac{QK^T}{\sqrt{d_k}})V \tag{1}$$

Observe that the resulting attention matrix on the left hand side has size equal to the length of context window squared. Hence, due to limited memory during training, there is an upper bound imposed on the length of the context window.

Hence, the approach of solely appending messages to the context throughout the lifespan of the agent quickly lead to an excessive context length and contributed to deteriorating performance as the agent progressed. To solve the problem, we proposed the method of *deleting past messages and actions*, leaving only a small window of a few prior actions steps at any time. To prevent unexpected behaviour, we added the following section to our prompt to "inform" the agent of this fact, in mitigation of unexpected behaviour:

> "Note that your replies as well as the instructions will be gradually removed to preserve your context."

3.5 Hyperparameter Tuning

Two specific hyperparameters were of particular interest and hence focused upon: frequency penalty and presence penalty.

Frequency penalty penalises new words based on how many times they have already appeared. This makes GPT less likely to repeat the same sequence of tokens multiple times, partly in mitigation of infinite response loops.

Presence penalty penalises new words based on whether they have appeared before. This increses the likelihood for the GPT model to output content with more diverse topics.

The initial, and default, values for both of these parameters was 0.

During multiple trials, it was observed that an excessive increase of the presence penalty often lead to the model producing incoherent output. Setting the value to 0.3 produced the desired effect with a notable uptick in a higher variety of output and more diverse language styles.

The frequency penalty was also set to 0.3 in efforts to dampen the tendency of the GPT-4-powered agents to keep talking abouta a specific topic related to their personality.

3.6 Web App (Portal)

A web interface was built for greater visibility into the content included in the platform as well as the agents' behaviour.

It is built with modern web technologies and was modelled after a typical social media platform, allowing human users to perform the same actions as the bot agents among other features:

1. View recommended posts
2. Vote, comment on posts
3. View posts, comments and votes by agent
4. View agent personality

The following flowchart in Fig. 2 illustrates the system architecture of the web app.

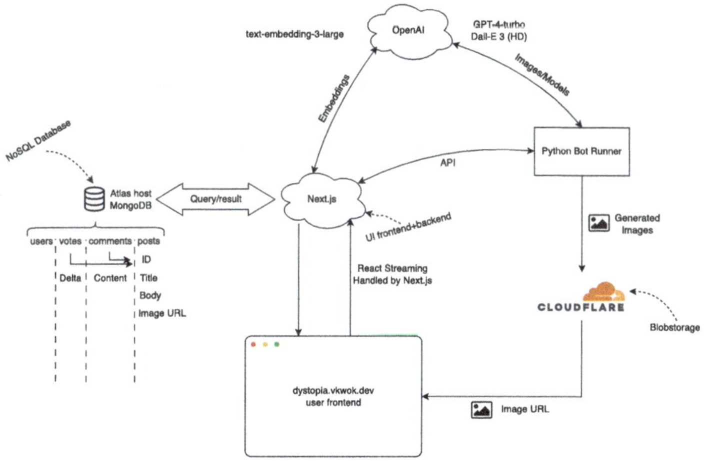

Fig. 2. Flowchart illustrating system architecture for proof of concept web app

4 Conclusion

In conclusion, a framework for utilizing GPT-4 and other LLMs to simulate agent based social interactions was demonstrated. Methods to give the LLMs different personalities, implemented recommendation algorithms, and allowed them to read and vote on posts at their own discretion were discussed. The problems associated with the limited context window of an LLM was addressed by deleting past messages. Additionally, different prompt engineering techniques were experimented and the effectiveness of block letters in emphasising information and making sure that GPT-4 strictly followed the instructions was observed. We believe that our methodology and framework will provide useful guidance to other researchers who wish to harness the power of LLMs in their research, especially in social experimentation.

5 Github Links

Below are the links to Github repositories for work done as part of this paper:
Agent Runner: https://github.com/cryptoAlgorithm/dystopia-bot
Web App Poratl: https://github.com/cryptoAlgorithm/dystopia

References

1. Achiam, J., et al.: Gpt-4 technical report. arXiv preprint arXiv:2303.08774 (2023)
2. Alexandr, N., Irina, O., Tatyana, K., Inessa, K., Arina, P.: Fine-tuning GPT-3 for russian text summarization. In: Data Science and Intelligent Systems: Proceedings of 5th Computational Methods in Systems and Software 2021, vol. 2, pp. 748–757. Springer (2021)

3. Amatriain, X.: Prompt design and engineering: introduction and advanced methods. arXiv preprint arXiv:2401.14423 (2024)
4. An, S., Ma, Z., Lin, Z., Zheng, N., Lou, J.G.: Make your LLM fully utilize the context. arXiv preprint arXiv:2404.16811 (2024)
5. Arora, S., et al.: Ask me anything: a simple strategy for prompting language models. In: The Eleventh International Conference on Learning Representations (2022)
6. Bhayana, R., Bleakney, R.R., Krishna, S.: GPT-4 in radiology: improvements in advanced reasoning. Radiology **307**(5), e230987 (2023)
7. Brown, T., et al.: Language models are few-shot learners. In: Advance in Neural Information Processing System, vol. 33, pp. 1877–1901 (2020)
8. Chang, E.Y.: Examining GPT-4's capabilities and enhancement with socrasynth. In: The 10th International Conference on Computational Science and Computational Intelligence (2023)
9. Dakhel, A.M., Majdinasab, V., Nikanjam, A., Khomh, F., Desmarais, M.C., Jiang, Z.: GitHub copilot AI pair programmer: asset or liability? J. Syst. Softw. **203**, 111734 (2023)
10. Gao, C., et al.: Large language models empowered agent-based modeling and simulation: a survey and perspectives. arXiv preprint arXiv:2312.11970 (2023)
11. Gao, C., et al.: s^3: Social-network simulation system with large language model-empowered agents. arXiv preprint arXiv:2307.14984 (2023)
12. Goel, A., et al.: LLMS accelerate annotation for medical information extraction. In: Machine Learning for Health (ML4H), pp. 82–100. PMLR (2023)
13. Han, X., Wu, Z., Xiao, C.: "Guinea pig trials" utilizing GPT: a novel smart agent-based modeling approach for studying firm competition and collusion. arXiv preprint arXiv:2308.10974 (2023)
14. Hansen, H., Hebart, M.N.: Semantic features of object concepts generated with gpt-3. arXiv preprint arXiv:2202.03753 (2022)
15. Hoey, J., Schröder, T., Morgan, J., Rogers, K.B., Rishi, D., Nagappan, M.: Artificial intelligence and social simulation: studying group dynamics on a massive scale. Small Group Res. **49**(6), 647–683 (2018)
16. Hu, S., et al.: A survey on large language model-based game agents. arXiv preprint arXiv:2404.02039 (2024)
17. Keles, F.D., Wijewardena, P.M., Hegde, C.: On the computational complexity of self-attention. In: International Conference on Algorithmic Learning Theory, pp. 597–619. PMLR (2023)
18. Khattab, O., et al.: Demonstrate-search-predict: composing retrieval and language models for knowledge-intensive NLP. arXiv preprint arXiv:2212.14024 (2022)
19. Khattab, O., et al.: Dspy: Compiling declarative language model calls into self-improving pipelines. arXiv preprint arXiv:2310.03714 (2023)
20. Kieuvongngam, V., Tan, B., Niu, Y.: Automatic text summarization of COVID-19 medical research articles using BERT and GPT-2. arXiv preprint arXiv:2006.01997 (2020)
21. Kong, A., et al.: Self-prompt tuning: enable autonomous role-playing in LLMS. arXiv preprint arXiv:2407.08995 (2024)
22. Larsen, K.S.: The ASCH conformity experiment: replication and transhistorical comparison. J. Soc. Behav. Pers. **5**(4), 163 (1990)
23. Li, T., Zhang, G., Do, Q.D., Yue, X., Chen, W.: Long-context LLMS struggle with long in-context learning. arXiv preprint arXiv:2404.02060 (2024)
24. Li, Y., et al.: Personal LLM agents: insights and survey about the capability, efficiency and security. arXiv preprint arXiv:2401.05459 (2024)

25. Liu, P., Yuan, W., Fu, J., Jiang, Z., Hayashi, H., Neubig, G.: Pre-train, prompt, and predict: a systematic survey of prompting methods in natural language processing. ACM Comput. Surv. **55**(9), 1–35 (2023)
26. Liu, Y., et al.: Summary of ChatGPT-related research and perspective towards the future of large language models. Meta-Radiol. 100017 (2023)
27. Mitroff, I.I., Blankenship, L.V.: On the methodology of the holistic experiment: an approach to the conceptualization of large-scale social experiments. Technol. Forecast. Soc. Chang. **4**(4), 339–353 (1973)
28. Muennighoff, N.: Sgpt: GPT sentence embeddings for semantic search. arXiv preprint arXiv:2202.08904 (2022)
29. Mumford, E.M.: Social behaviour in small work groups. Sociol. Rev. **7**(2), 137–157 (1959)
30. OpenAI: Embeddings (2024). https://platform.openai.com/docs/guides/embeddings. Accessed 16 Apr 2024
31. Reid, M., et al.: Gemini 1.5: unlocking multimodal understanding across millions of tokens of context. arXiv preprint arXiv:2403.05530 (2024)
32. Riecken, H.W., Boruch, R.F.: Social experiments. Ann. Rev. Sociol. **4**(1), 511–532 (1978)
33. Sahoo, P., Singh, A.K., Saha, S., Jain, V., Mondal, S., Chadha, A.: A systematic survey of prompt engineering in large language models: techniques and applications. arXiv preprint arXiv:2402.07927 (2024)
34. Shin, T., Razeghi, Y., Logan IV, R.L., Wallace, E., Singh, S.: Autoprompt: eliciting knowledge from language models with automatically generated prompts. arXiv preprint arXiv:2010.15980 (2020)
35. stanfordnlp: https://github.com/stanfordnlp/dspy (2024)
36. Tanur, J.M.: Methods for large-scale surveys and experiments. Sociol. Methodol. **14**, 1–71 (1983)
37. Team, G., et al.: Gemini: a family of highly capable multimodal models. arXiv preprint arXiv:2312.11805 (2023)
38. Touvron, H., et al.: Llama: open and efficient foundation language models. arXiv preprint arXiv:2302.13971 (2023)
39. Vaswani, A.,et al.: Attention is all you need. In: Advances in Neural Information Processing Systems, vol. 30 (2017)
40. Wang, B., Deng, X., Sun, H.: Iteratively prompt pre-trained language models for chain of thought. arXiv preprint arXiv:2203.08383 (2022)
41. Wang, L., et al.: A survey on large language model based autonomous agents. Front. Comp. Sci. **18**(6), 186345 (2024)
42. Wang, X., et al.: Self-consistency improves chain of thought reasoning in language models. arXiv preprint arXiv:2203.11171 (2022)
43. Wei, J., et al.: Chain-of-thought prompting elicits reasoning in large language models. In: Advances in Neural Information Processing Systems, vol. 35, pp. 24824–24837 (2022)
44. Xi, Z., et al.: The rise and potential of large language model based agents: a survey. arXiv preprint arXiv:2309.07864 (2023)
45. Yu, S., Fang, C., Ling, Y., Wu, C., Chen, Z.: LLM for test script generation and migration: challenges, capabilities, and opportunities. In: 2023 IEEE 23rd International Conference on Software Quality, Reliability, and Security (QRS), pp. 206–217. IEEE (2023)
46. Zhang, Z., et al.: Simulating classroom education with LLM-empowered agents. arXiv preprint arXiv:2406.19226 (2024)

47. Zhou, D., et al.: Least-to-most prompting enables complex reasoning in large language models. arXiv preprint arXiv:2205.10625 (2022)
48. Zhou, Y., et al.: Large language models are human-level prompt engineers. arXiv preprint arXiv:2211.01910 (2022)
49. Zimbardo, P.G., Haney, C., Banks, W.C., Jaffe, D.: The stanford prison experiment (1971)

Cancer Therapeutics: In-Silico Evaluation of Novel UBR5 Protein Inhibitors

Maharshi Mukherjee(✉), Ngiam Lihong, and Chik Yan Kit David

NUS High School of Math and Sciences, 20 Clementi Ave 1, Singapore, Singapore
h2010089@nushigh.edu.sg
https://nushigh.edu.sg/

Abstract. UBR5, an E3 ligase, is involved in the pathogenesis of various cancers, including breast, ovarian, and cervical cancers. As a pivotal proto-oncogene, UBR5 drives the degradation of crucial tumor suppressor genes, thereby facilitating oncogenesis. Despite its critical role, small molecule inhibitors targeting UBR5 have yet to be identified. This study aims to discover novel inhibitors for the UBR5 HECT domain through *in-silico* simulations. We used pocket prediction software to analyze UBR5 and its homologs, followed by consensus docking using AutoDock and GOLD on a curated set of known homolog inhibitors and an extensive DrugBank library. ADMET studies were used to further shortlist the compounds. Our comprehensive approach identified six novel inhibitors with significant binding potential, and also highlighted a promising binding pocket for heclin and its analogs. These findings provide a foundation for the development of new anti-cancer therapies, pending further experimental validation.

Keywords: UBR5 · Cancer · Protein-Ligand Docking · ADMET · Medicinal Biochemistry

1 Introduction

The human ubiquitin protein ligase E3 component N-recognin 5 (UBR5) is a key member of the "homologous to E6AP C-terminus" (HECT)-domain E3 ubiquitin ligase family [21]. E3 ligases are integral to the ubiquitination system, serving as the final enzyme catalyzing the transfer of ubiquitin from an E2 ligase carrier to target proteins, marking them for degradation. Dysregulation, either through mutation or overexpression of E3 ligases, frequently results in the excessive degradation of tumor suppressor proteins, thereby promoting oncogenesis.

UBR5 is notably overexpressed in various malignancies, driving tumor proliferation. Genetic knockdown of UBR5 has demonstrated tumor growth inhibition in cervical cancer [18], triple-negative breast cancer [25], and ovarian cancer [17]. Specifically, in cervical cancer, the Human Papillomavirus (HPV) E6 oncogene leverages UBR5 to destabilize Tat-interacting protein of 60kDa (TIP60), a critical lysine acetyltransferase involved in DNA damage response. This interaction leads to the ubiquitination of TIP60 by UBR5 and subsequent proteasomal

H. Guo et al. (Eds.): IRC-SET 2024, PTT, pp. 401–411, 2025.
https://doi.org/10.1007/978-981-96-3770-6_41

degradation [18]. Furthermore, UBR5 downregulation has been associated with increased sensitivity to cisplatin in typically drug-resistant ovarian cancer variants [9].

The development of UBR5 inhibitors is imperative for advancing cancer therapeutics, especially for tumors resistant to existing treatments such as cisplatin. However, due to the flexible hinge region and the substantial size of UBR5, the identification of druggable pockets has proven difficult, impeding research in this area [22].

UBR5 operates predominantly as dimers or tetramers and contains several key domains, including the Ubiquitin-associated (UBA) domain, the UBR box domain for substrate recognition, the MLLE domain (involved in protein-protein interactions), and the HECT domain. The HECT domain, being the largest and most extensively studied, directly facilitates ubiquitin transfer to target proteins. It comprises an N-lobe with a binding site for the E2 enzyme and a C-lobe with a catalytic cysteine essential for ubiquitin transfer [21]. Due to its size and detailed characterization, the HECT domain was selected as the target to identify active druggable pockets.

Drug discovery in the lab remains highly expensive and time-consuming. Thus, this study employs a computational approach, involving virtual screening of large compound databases and molecular docking analyses. *In silico* ADMET and DFT studies are used to further refine the candidates, which we hope will ultimately aid in the development of novel anticancer drugs in this field (Fig. 1).

2 Methodology

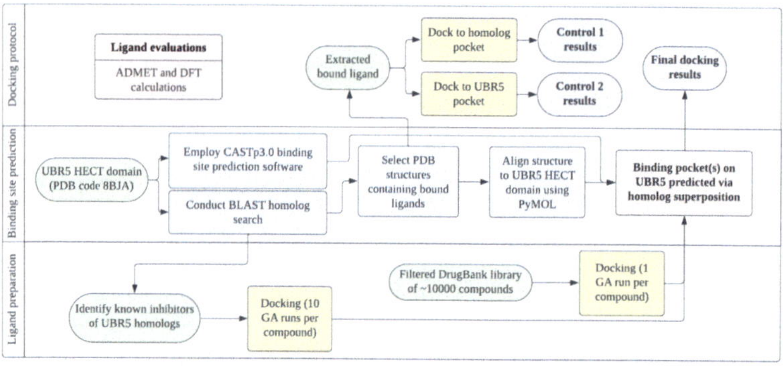

Fig. 1. Complete flowchart of methodology.

2.1 Homolog and Ligand Dataset Collation

We utilised the National Center for Biotechnology Information (NCBI) Basic Local Alignment Search Tool (BLAST) [1], an online protein-protein sequence similarity analysis program, to identify protein homologs of UBR5 based on sequence similarity. Specifically, we used the HECT domain sequence of UBR5 as our protein-protein BLAST (BLASTp) search query, obtained from the UniProt database (Accession O95071, residues 2210-2796). Protein Data Bank (PDB) was used as the reference database and a word size of 3 was employed, with other settings kept as default.

The ChEMBL database [10], BindingDB [5], and literature were surveyed to collect the dataset of known UBR5 homolog inhibitors. All ligand structures were obtained directly from the PubChem database or prepared from isomeric SMILES using Open Babel [13] software. If needed, the structures were put through additional geometry optimization using the Avogadro software [2] using the Universal Force Field (UFF) [14].

Furthermore, we also utilised the DrugBank database of 12036 known drugs [24]. All collated drugs were filtered through the following criteria: (a) a molecular weight of not more than 500 AMU, and (b) having not more than 8 rotatable bonds within their structure. The compound library was prepared with the LigPrep tool [15].

2.2 Pocket Prediction

UBR5's HECT domain was extracted from PDB code 8BJA (3.00Å Resolution), Chain A (residues 2210–2796) [21] in the PyMOL program [16].

Two approaches were used to reliably identify druggable pockets: Firstly, we employed the Computed Atlas of Surface Topography of proteins 3.0 (CASTp 3.0) pocket prediction software [19] with HECT domain structure and a probe radius of 1.4Å. Secondly, identified UBR5 homologs were filtered to identify those with complete 3D structures available on RCSB, co-crystallized with known ligands. The positions of these native ligands, superimposed onto UBR5, were also used to predict potential pockets on UBR5. Ultimately, pockets reliably identified by both CASTp 3.0 and homolog superimposition were used for future steps.

2.3 Protein-Ligand Docking with GOLD and AutoDock Vina

GOLD. The Genetic Optimisation for Ligand Docking (GOLD) software [6] was used for protein-ligand docking simulations, and the Hermes molecular visualization software was employed as a Graphical User Interface (GUI). The protein domain was protonated in GOLD, while all Gln and Asn side chain configurations were kept unchanged.

A population size of 100 was used for the genetic algorithm (GA) runs, the specifics of which are detailed below. All other parameters were kept at their default settings. The Chemical Piecewise Linear Potential (ChemPLP) function was applied as the empirical scoring metric, serving as a positive correlate for binding affinity. More details of the docking approach is available in the Supplementary Information.

After an initial round of docking, we conducted Induced Fit Docking (IFD) with the Glide software [4] to predict accurate concomitant structural changes during docking.

AutoDock Vina. Consensus docking of all ligand and drug candidates was conducted with the AutoDock Vina [20] software, employing a 17Å x 24Å x 25Å grid box enclosing the pocket generated by IFD and an exhaustiveness of 32.

Consensus Docking Review. For consensus docking, we analysed the docking data as follows: Using a custom code, we evaluated the extent to which ligands from AutoDock Vina results were buried in the pocket, using an expanded pocket bubble from a reference ligand. Quantitative buried percentage, PLP.fitness values, and binding energies of the compounds, as well as qualitative verification of protein-ligand interactions, such as π-π interactions and hydrogen bonds, were employed to evaluate candidate compounds. All figures were generated using the PyMOL program. Details of our docking approach and analysis protocol can be found in the supplementary material.

2.4 ADMET Analysis

The pharmacokinetic properties of the candidate ligands and drugs - namely their absorption, distribution, metabolism, elimination and toxicity (ADMET) - using analysed with the SwissADME tool [3]. Important parameters analyzed in this step include gastrointestinal (GI) absorption, blood-brain barrier (BBB) permeability, and CYP450 inhibitory potential of the compounds, as well as their overall druglikeness and bioavailabilities.

2.5 DFT Calculations

Density Functional Theory (DFT) analyses were carried out on selected top-scoring compounds using the Becke-3-Lee-Yang-Parr (B3LYP) functional [8] and the 6-311++G(d,p) [7] basis set. In particular, we determined the HOMO-LUMO gap, hardness and softness, and other properties of the compounds, important parameters that influence their reactivities.

All structures were optimised first using the general AMBER force field (GAFF) in the Avogadro software, then at the B3LYP/6-311++G(d,p) level of theory. Unrestricted natural orbital (UNO) calculations were then conducted on all compounds at the same level of theory, and the resulting frontier molecular orbitals and electrostatic potential (ESP) maps were visualised using Avogadro. All DFT calculations were performed using the ORCA 4.2.0 software [12].

3 Results

3.1 Pocket Prediction and Ligand Library

Our BLAST search of the UBR5 HECT domain yielded 43 hits (see Supplementary Table 1). We identified 3 PDB structures with bound ligands: NEDD4,

SMURF2, and an alternate UBR5 (PDB codes 5C91, 7M3Q, and 8C07 respectively).

The superposition of these structures with UBR5 revealed that SMURF2 and UBR5 (7M3Q and 8C07) share a homologous pocket. However, the NEDD4 pocket (5C91) superimposed on UBR5 remains unsolved and was disregarded in further steps.

Thus, one pocket on UBR5 (based on SMURF2 (7M3Q)) was selected for further docking, as illustrated in Fig. 2(b). Homology model includes *sequence alignment* of the pocket region is shown in Fig. 2(b), displaying 31% sequence identity, and *structural similarity* in the Residue-Residue (RR) distance map in Fig. 2(a). The CASTp3.0 prediction corroborated the presence of this binding pocket as shown in Fig. 2(c), highlighting a space of volume 560Å^3 overlapping with this pocket.

We identified 124 unique small molecule inhibitors for docking (see Supplementary Table 2, including SMILES Identifiers). A total of 16139 drug conformers from the DrugBank library were used for docking after filtering.

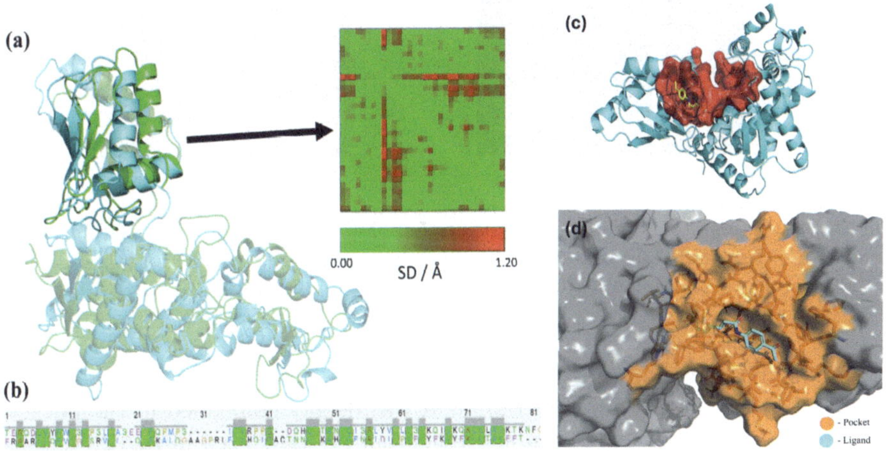

(a)

Fig. 2. UBR5 Homology and pocket. (a) Homology model between SMURF2 (Green) and UBR5 HECT (Blue). (b) Sequence alignment of UBR5 HECT and SMURF2 pockets. (c) Overlap of CASTp 3.0 predicted pocket (red) and reference ligand L2 (yellow). (d) Surface of binding site.

3.2 Docking Results

A comparison between the binding of the native ligand (DTT) docking to SMURF2 (PLP.Fitness score: 37.0) and UBR5 (PLP.Fitness score: 33.9) validated the presence of a similar homologous binding pocket in UBR5, as shown by the relatively close PLP.Fitness scores. However, the residues and the shape of the pocket were slightly different compared to those of SMURF2, as shown in Fig. 2(a).

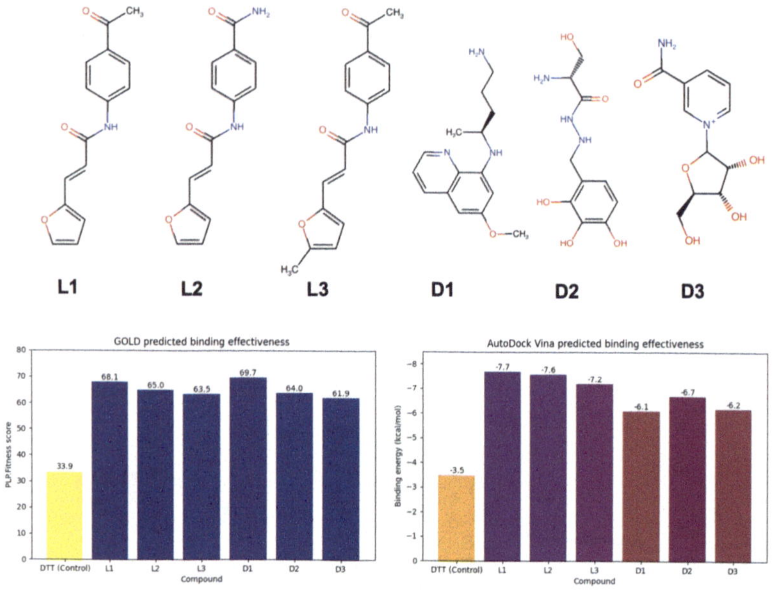

Fig. 3. Structures of ligands L1-L3 (Known homolog ligands) and D1-D3 (from Drug-Bank) with their PLP.Fitness scores and binding energies. Dithiothreitol (DTT) is the native ligand from homolog SMURF2 used as a benchmark with UBR5 HECT.

In our first run to optimize the pocket defined, we chose the docked pose of Ligand CID 922464 (Compound L2), as it was deeply embedded in the pocket. The final pocket, after expansion and relaxation, is visualized in Fig. 2(c).

Thereafter, consensus docking was carried out with GOLD and AutoDock Vina; from which a total of 6 top candidates were selected - ligands L1-L3 from the library of UBR5 HECT homolog ligands, and drugs D1-D3 from the screened DrugBank library. These compounds were selected on the basis of their similar docking poses predicted by both softwares, as well as their high PLP.Fitness scores and highly negative binding energies. Their SMILES identifiers and PubChem Compound Identifiers (CIDs) are available in Supplementary Table 3.

Their structures, PLP.Fitness scores, and binding energies are shown in Fig. 3. The docking poses of L1-L3 and D1-D3 are illustrated in Fig. 4. Common residues involved in polar bonds across all compounds include A2765, T2767, N2795 and G2797.

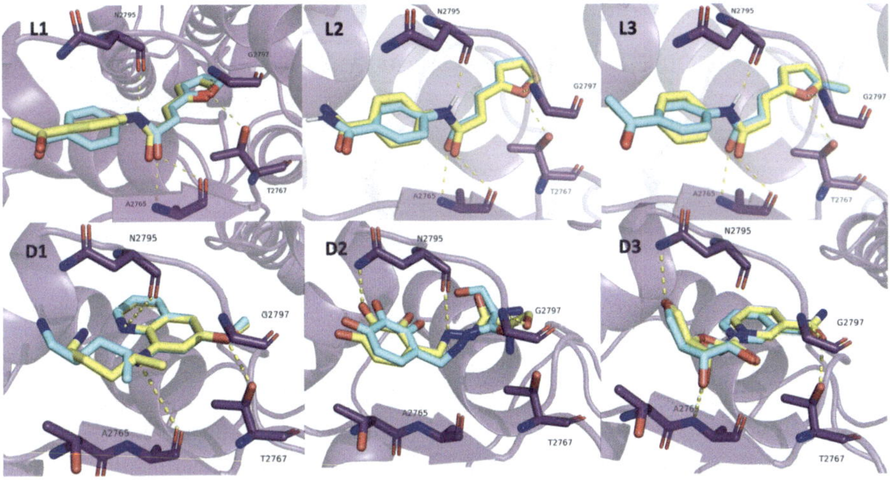

Fig. 4. Consensus binding poses of top homolog ligands L1-L3 (Top) and DrugBank drugs D1-D3 (Bottom) in GOLD (yellow) and AutoDock Vina (blue). Key polar interactions are highlighted in yellow, and interacting residues are shown in purple.

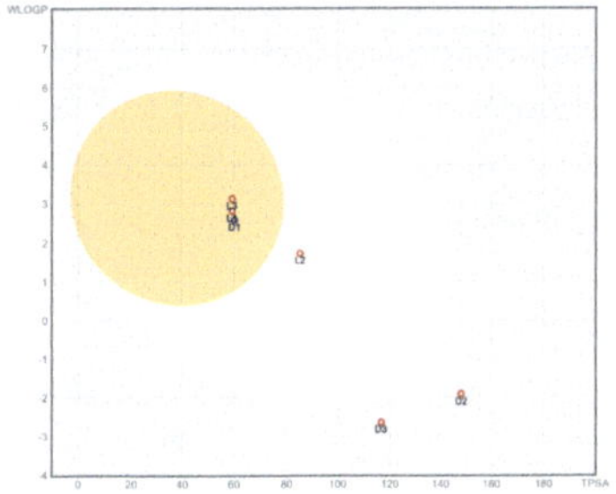

Fig. 5. Brain Or IntestinaL EstimatED (BOILED-Egg) plot of L1-L3 and D1-D3.

3.3 ADMET Analysis

As shown in BOILED-Egg plot in Fig. 5, 4 of the 6 compounds are predicted to have high GI absorption within the white zone. Table 1 lists additional ADMET properties of the compounds. All compounds have 0 Lipinski violations and thus show excellent druglikeness.

In particular, L2 has been shown to have poor BBB permeability (outside the yellow region), is unaffected by P-glycoprotein-mediated extrusion out of the intestinal mucosa/CNS (red dots; PGP-), and also possesses good topological polar surface area (TPSA) and Log $P_{o/w}$ values. Therefore, L2 is a promising candidate in terms of limiting potential for off-target toxicities and maximising bioavailability. The cytochrome P450 (CYP) inhibitory properties of all compounds are listed in the supplementary material.

Table 1. Predicted ADMET properties of L1-L3 and D1-D3

Compound	GI absorption	BBB premeant	P-gp substrate	TPSA (Å^2)	Consensus Log $P_{o/w}$	Lipinski violations
L1	High	Yes	No	59.31	2.19	0
L2	High	No	No	85.33	1.41	0
L3	High	Yes	No	59.31	2.57	0
D1	High	Yes	Yes	60.17	2.25	0
D2	Low	No	No	148.07	−1.20	0
D3	Low	No	No	116.89	−2.27	0

3.4 DFT Calculations

The HOMO-LUMO gaps for L1-L3, D1 and D3 were calculated to be in the range of 3.9–4.4 eV, which is sufficiently small to predict good binding. Furthermore, high- and low-potential regions in the ESP maps revealed chemically reactive sites in the various compounds, including areas susceptible to nucleophilic and electrophilic attack (Fig. 6). Several conserved areas of high/low electron density correspond well with predicted interactions in docked poses, e.g. electronegative furan rings forming key hydrogen bonds with T2767.

Additional DFT-calculated properties are listed in Table 2. In particular, L1, L2 and D1 have the highest softness and thus may exhibit greater polarizability and chemical reactivity; whereas L3, D2 and D3 have higher hardness and thus may posses greater stability. Meanwhile, the electrophilicity of all compounds are above 2 eV, indicating tendencies for electrophilic behaviour in chemical interactions.

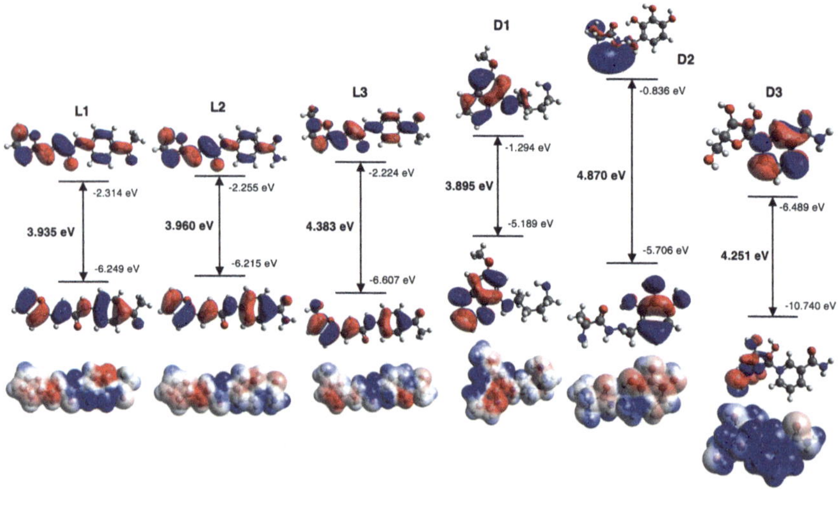

Fig. 6. HOMO and LUMO of L1, as well as their respective energy levels, calculated with B3LYP/6-311++G(d,p) and their ESP maps. The blue and red areas on the figures indicate negative and positive electron density, respectively.

Table 2. Calculated chemical properties and reactivity indicators of L1-L3 and D1-D3

Compound	Chemical potential (μ)	Hardness (η)	Softness (s)	Electronegativity (χ)	Electrophilicity (ω)
L1	−4.282	1.968	0.254	4.282	4.659
L2	−4.235	1.980	0.253	4.235	4.529
L3	−4.416	2.192	0.228	4.416	4.448
D1	−3.242	1.948	0.257	3.242	2.698
D2	−3.271	2.435	0.205	3.271	2.197
D3	−8.615	2.126	0.235	8.615	17.457

4 Discussion

UBR5 remains as an elusive target for cancer therapeutics, especially as overexpression of UBR5 ubiquitinates and degrades crucial tumour supressors such as TIP60, leading to oncogenesis in many different types of cancers. In this study, we computationally identified 6 compounds as potential UBR5 inhibitors.

Interestingly, compounds L1 - L3 are analogs of heclin, which inhibit various HECT ligases, including SMURF2, NEDD4, and WWP1. Heclin and its analogs likely disrupt the C-lobe orientation of the HECT domain, causing the catalytic cysteine to oxidize to sulfenic acid and form disulfide bonds, which slows reduction. However, the exact binding site of heclin remains unidentified [11]. While heclin is a general, weak ligand, its analogs are known to be more effective against specific proteins [23].

The study's novelty is twofold: it identifies a promising binding pocket for heclin and its analogs, and it shows that the analogs are effective UBR5 inhibitors. This binding pocket is close to the catalytic cysteine, supporting the hypothesis about heclin's proposed mechanism of action.

In addition, we have identified 3 novel compounds from the screened DrugBank library as potential inhibitors of the UBR5 HECT domain. All 3 compounds, D1-D3, exhibit high binding scores and affinities, suggesting strong binding capabilities. All 6 compounds have also been observed to bind well in the pocket with good docking scores, suggesting high potency for UBR5.

Moreover, ADMET studies revealed L2 as the top candidate, with superior bioavailability and minimal BBB penetration. Furthermore, small calculated HOMO-LUMO gaps support intramolecular charge transfer and molecular reactivity, while ESP maps and other properties of the compounds also correspond well with good binding activity.

5 Conclusion

We identified 6 small molecule inhibitors of UBR5, including 3 known inhibitors from UBR5 homologs (L1-L3) and 3 drugs from the DrugBank library (D1-D3), all demonstrating promising binding to the HECT domain of UBR5. Our findings suggest potential binding sites for potent heclin analogs on UBR5 and introduce 3 novel compounds as potential inhibitors.

Consensus docking refined the results, but rescoring and Molecular Dynamics simulations can provide a better mechanistic understanding in the future. Our ADMET, DFT and pharmacokinetic studies reveal key molecular signatures within the proposed candidates. Future work could also look into bioisosteric replacement to develop analogs exhibiting greater potency, reduced off-target toxicity, and improved pharmacokinetic properties. Ultimately, experimental validation with pull-down assays and mice xenograft models remain necessary to further confirm our results.

Supplementary Material

Supplementary Material Documents

Acknowledgments. We would like to thank Prof. Hao Fan and Dr. Chaitanya Jaladanki from the A*STAR Bioinformatics Institute (BII) for their support and guidance in carrying out this project.

Disclosure of Interests. The authors have no competing interests to declare for the contents of this article.

References

1. Altschul, S.F., Gish, W., Miller, W., Myers, E.W., Lipman, D.J.: Basic local alignment search tool. J. Mol. Biol. **215**(3), 403–410 (1990)
2. Avogadro Chemistry: avogadro. http://avogadro.cc/
3. Daina, A., Michielin, O., Zoete, V.: SwissADME: a free web tool to evaluate pharmacokinetics, drug-likeness and medicinal chemistry friendliness of small molecules. Sci. Rep. **7**, 42717 (2017). https://doi.org/10.1038/srep42717
4. Friesner, R.A., et al.: Glide: a new approach for rapid, accurate docking and scoring. 1. method and assessment of docking accuracy. J. Med. Chem. **47**(7), 1739–1749 (2004). https://doi.org/10.1021/jm0306430
5. Gilson, M.K., Liu, T., Baitaluk, M., Nicola, G., Hwang, L., Chong, J.: BindingDB in 2015: a public database for medicinal chemistry, computational chemistry and systems pharmacology. Nucleic Acids Res. **44**(D1), D1045–D1053 (2016). https://doi.org/10.1093/nar/gkv1072
6. Jones, G., Willett, P., Glen, R.C., Leach, A.R., Taylor, R.: Development and validation of a genetic algorithm for flexible docking. J. Mol. Biol. **267**(3), 727–748 (1997)
7. Krishnan, R., Binkley, J.S., Seeger, R., Pople, J.: Self-consistent molecular orbital methods. xx. a basis set for correlated wave functions. J. Chem. Phys. **72**(1), 650–654 (1980). https://doi.org/10.1063/1.438955
8. Lee, C., Yang, W., Parr, R.G.: Development of the Colle-Salvetti correlation-energy formula into a functional of the electron density. Phys. Rev. B **37**(785), 785–789 (1988). https://doi.org/10.1103/PhysRevB.37.785

9. Matsuura, K., Huang, N.J., Cocce, K.J., Zhang, L., Kornbluth, S.: Downregulation of the proapoptotic protein MOAP-1 by the UBR5 ubiquitin ligase and its role in ovarian cancer resistance to cisplatin. Oncogene **36**, 1698–1706 (2016). https://api.semanticscholar.org/CorpusID:4527443
10. Mendez, D., et al.: ChEMBL: towards direct deposition of bioassay data. Nucleic Acids Res. **47**(D1), D930–D940 (2019)
11. Mund, T., Lewis, M.J., Maslen, S., Pelham, H.R.: Peptide and small molecule inhibitors of HECT-type ubiquitin ligases. Proc. Natl. Acad. Sci. U.S.A. **111**(47), 16736–16741 (2014). https://doi.org/10.1073/pnas.1412152111
12. Neese, F., Wennmohs, F., Becke, U., Riplinger, C.: The ORCA quantum chemistry program package. J. Chem. Phys. **152**(22) (2020). https://doi.org/10.1063/5.0004608
13. O'Boyle, N., Banck, M., James, C., Morley, C., Vandermeersch, T., Hutchison, G.: Open babel: an open chemical toolbox. J. Cheminf. **3**(33), 1–14 (2011)
14. Rappé, A.K., Casewit, C.J., Colwell, K., Goddard, W.A., III., Skiff, W.M.: UFF, a full periodic table force field for molecular mechanics and molecular dynamics simulations. J. Am. Chem. Soc. **114**(25), 10024–10035 (1992)
15. Schrödinger, LLC: Ligprep. https://newsite.schrodinger.com/platform/products/ligprep/
16. Schrödinger, LLC: The PyMOL Molecular Graphics System, Version 1.8 (2015)
17. Song, M., et al.: Tumor derived UBR5 promotes ovarian cancer growth and metastasis through inducing immunosuppressive macrophages. Nat. Commun. **11**(1), 6298 (2020). https://doi.org/10.1038/S41467-020-20140-0
18. Subbaiah, V.K., et al.: E3 ligase EDD1/UBR5 is utilized by the HPV E6 oncogene to destabilize tumor suppressor TIP60. Oncogene **35**(16), 2062–2074 (2016). https://doi.org/10.1038/ONC.2015.268
19. Tian, W., Chen, C., Lei, X., Zhao, J., Liang, J.: CASTp 3.0: computed atlas of surface topography of proteins. Nucleic Acids Res. **46**(W1), W363–W367 (2018). https://doi.org/10.1093/nar/gky473
20. Trott, O., Olson, A.J.: AutoDock vina: improving the speed and accuracy of docking with a new scoring function, efficient optimization, and multithreading. J. Comput. Chem. **31**(2), 455–461 (2009). https://doi.org/10.1002/jcc.21334
21. Wang, F., et al.: Structure of the human UBR5 E3 ubiquitin ligase. Structure **31**(5), 541–552 (2023). https://doi.org/10.1016/j.str.2023.03.010
22. Wang, M., et al.: Targeting UBR5 in hepatocellular carcinoma cells and precise treatment via echinacoside nanodelivery. Cell. Mol. Biol. Lett. **27**(1), 92 (2022)
23. Watt, J.E., et al.: Discovery of small molecule wwp2 ubiquitin ligase inhibitors. Chem. Eur. J. **24**(67), 17677–17680 (2018). https://doi.org/10.1002/chem.201804169, https://chemistry-europe.onlinelibrary.wiley.com/doi/abs/10.1002/chem.201804169
24. Wishart, D.S., et al.: Drugbank 5.0: a major update to the drugbank database for 2018. Nucleic Acids Res. **46**(D1), D1074–D1082 (2017)
25. Wu, B., et al.: UBR5 promotes tumor immune evasion through enhancing IFN-gamma-induced PDL1 transcription in triple negative breast cancer. Theranostics **12**(11), 5086 (2022). https://doi.org/10.7150/thno.74989

Developing mRNA Vaccine Against TAA CT83-Related Cancers

Aditi Athreya(✉), Tan Jun Yong, and Jaeden Soon Chuankai

NUS High School of Math and Science, 20 Clementi Avenue 1, Singapore 129957, Singapore
h1910003@nushigh.edu.sg

Abstract. The odds of a child inheriting a cancer-causing gene defect from a parent are roughly 50%, and 5–10% of all cancers are inherited. Fortunately, when early preventative measures are taken, cancer can be nipped in the bud. However, preemptive cancer treatment is still an emerging field, and there exist very few minimally-invasive options that have entered clinical use. Our research aims to develop a preventative mRNA vaccine against the CT83 antigen, a highly tumour-specific Cancer Testis Antigen. CT83 is very commonly hypomethylated, and hence overexpressed in gastric cancer, lung adenocarcinoma, and breast cancer. To produce the vaccine, we used CT83 mRNA from gastric cancer cells. We purified this mRNA and encapsulated it in a lipid nanoparticle, before transfecting HEK 293T epithelial cells with our vaccine. We then extracted the cellular mRNA, and carried out a qPCR to check for changes in expression of CT83. Our vaccine induced a 284-fold increase in CT83 expression in the HEK 293T cells, which suggests that our vaccine works to induce CT83 transcription. Future studies could aim to explore whether the increase in CT83 expression induced by our vaccine would lead to an immune response through testing for expression of CT83 epitopes, in-vitro testing and detection and mapping of T-lymphocyte activity. When proven functional, this vaccine could be used as preemptive personalised medical treatment for people who are genetically more predisposed to develop CT83-related cancers, and could also be the dawn of a new era in immunotherapy where it could serve to enhance patients' intrinsic immune responses against existing cancer. This could potentially save one million lives annually worldwide.

Keywords: Cancer · Vaccine · CT-83 · mRNA · Tumour Associated Antigen · gene expression · immunotherapy · preemptive cancer treatment · personalized medicine · T-lymphocyte

1 Introduction

To explore the use of an mRNA vaccine in inducing an immune response against CT83 related cancers, we extracted and amplified CT83 mRNA from MKN45 gastric cancer cells (we picked this highly expressive cell line after going through the Cancer Cell Line Encyclopedia database to optimise mRNA concentration) through a polymerase chain reaction (PCR). We then cloned the gene into a plasmid which we transformed into E. coli bacteria. We extracted this plasmid DNA from a single clone of E. coli,

H. Guo et al. (Eds.): IRC-SET 2024, PTT, pp. 412–423, 2025.
https://doi.org/10.1007/978-981-96-3770-6_42

carried out restriction digestion to obtain a pure sample of CT83 DNA. We then carried out in-vitro transcription of this DNA to obtain mRNA, stabilising the mRNA with a 5' methylguanosine cap and a 3' poly-adenine tail. After inserting this mRNA into a lipid nanoparticle to produce the "vaccine", we transfected HEK 239T epithelial cells and carried out a qPCR to detect changes in transcription patterns of CT83.

Our results showed that transfection of the vaccine into HEK 293T epithelial cells caused an approximately 284-fold increase in expression of CT83, suggesting that the vaccine works well in increasing CT83 expression. This could be used to stimulate CT83 antigen presentation by antigen presenting cells on MHC II molecules so that immunity against CT83-related cancers can be developed as helper T-cells can detect the antigens presented and kickstart an immune response (Fig. 1).

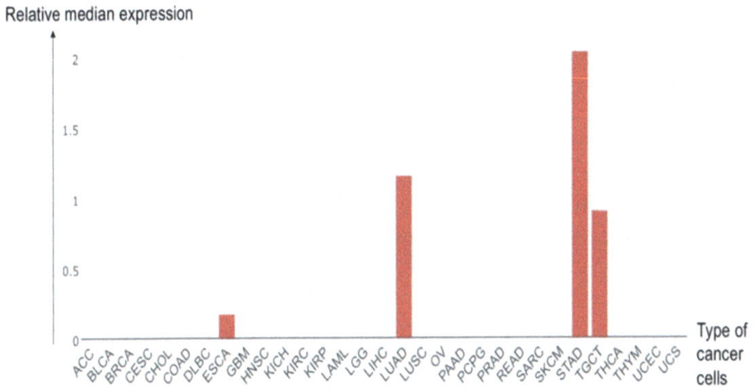

Fig. 1. Oncogenic expression pattern of CT83. Image taken from Gepia. (n.d.). http://gepia.cancer-pku.cn/detail.php?gene=ct83

2 Materials and Methods

2.1 RNA Extraction from MKN45 Gastric Cancer Cells (Vazyme Biotech FastPure Cell/tissue Total RNA Isolation Kit V2 RC112)

In our study, we isolated RNA from MKN45 gastric cancer cells. We first added 500 μL of Buffer RL 1 containing 1% beta-mercaptoethanol to lyse the cells, pipetting repeatedly till no cell mass was visible. To obtain pure RNA, on an RNase-free bench, we transferred the mixture to a gDNA-filter column and centrifuged the mixture. We then added 800 μL of buffer RL2 (with absolute ethanol) to the mixture to provide an RNA-specific environment in the column. We then transferred the mixture to an RNAPure Column and centrifuged it, discarding the flow-through. We then washed the mixture and centrifuged it with 500 μL of Buffer RW1, to remove protein and DNA impurities, and then twice with Buffer RW2 (once with 500 μL and once with 700 μL) to remove salt-ion residues. We then centrifuged the mixture again to completely remove the wash buffers, before transferring the column to an RNase-free collection tube and eluting with RNase-free ddH2O. The concentration of RNA measured using a nanodrop machine was 457 ng/mL.

We removed any trace amounts of genomic DNA by mixing 13.8 μL of RNase-free ddH2O, 4 μL of 4x gDNA wiper Mix and 2.20 μL of the template RNA (1 μg of RNA) with a pipette and incubating at 42 °C for 2 min. To prepare the RT-PCR reaction mixture, we added 5 x HiScript III qRT SuperMix (containing dNTPs, HiScript III Reverse Transcriptase, RNase inhibitor and random oligo dT primers) to the template RNA. as well as our oligo dT primers to the template RNA. We then put the mixture into a thermocycler for 32 cycles set at 95 °C (denaturation) for 3.5 min, 52.5 °C (annealing) for 35 s and lastly 72 °C (elongation) for 5 min. We sequenced the DNA obtained using Sanger's Chain termination sequencing. On sequencing, we realised that the DNA contained single nucleotide polymorphisms (where there are periodic differences in one base pair in the DNA obtained) and decided to clone the gene into bacteria to obtain a pure sample of DNA.

2.2 Creation of Recombinant Vector (pGEM®-T and pGEM®-T Easy Vector Systems Protocol)

We created the recombination vector consisting of the gene insert and the pGEM-T Easy vector via double digestion of the pGEM-T Easy vector. We used 5 μL of 2X ligation buffer containing the restriction enzymes EcoRI and SpeI. We then added 1 μL of T4 DNA ligase to ligate the vector with the insert CT83 DNA. Thereafter, we incubated the mixture at 4 °C.

2.3 Selection of Recombinants

Recombinants were screened for via LacZ blue-white screening. Since the cut sites for both EcoRI and SpeI restriction enzymes lie within the region coding for the LacZ gene in the plasmid, bacteria containing the recombinant plasmid would not be able to produce the beta-galactosidase needed to break down the X-Gal into the insoluble precipitate which would make the colonies appear blue. Since we required recombinant plasmids, we selected the colonies that appeared white on the plate (Fig. 2).

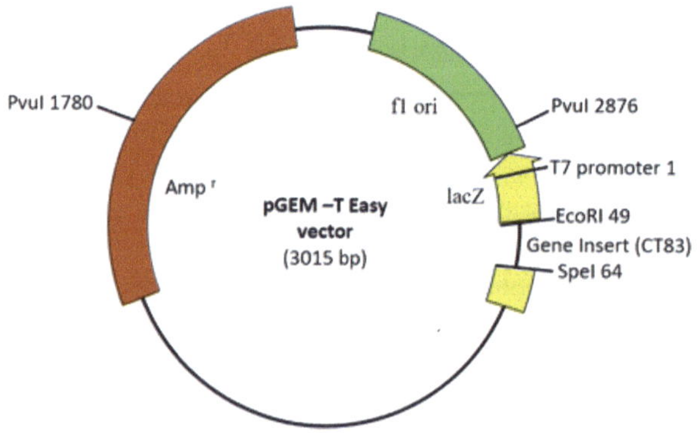

Fig. 2. Structure of pGEM-T Easy plasmid with insert, positions are indicated by the numbers next to the labels.

2.4 Plasmid Extraction (Vazyme Biotech Fastpure Endofree Plasmid Maxi Kit DC202)

We selected a single white clone of E. coli to extract the plasmids from. We then cultured the clone in LB/ampicillin medium overnight on a shaker at 37 °C. We pelleted the bacterial culture, decanted the supernatant and added resuspension buffer containing EDTA and 75 μL of 10 mg/ml RNase A. Then we added 7.5 μL of lysis buffer containing SDS (to disrupt the cell membranes of the bacteria) and NaOH (to denature genomic and non-supercoiled bacterial DNA). Thereafter, we added neutralisation buffer to aggregate the denatured DNA together with protein into a precipitate. We centrifuged the mixture to separate the precipitate from the supernatant. We then added isopropanol to the solution and incubated at −20 °C for 15 min. We transferred the incubated solution to a spin column and centrifuged again, discarding the flow-through, and repeated with the addition of wash buffer. We then eluted the plasmid DNA with 50 μL of water. We used the restriction enzyme PvuI to cleave the CT83 DNA from the plasmid as it only has restriction sites on the plasmid and not the CT83 gene. The concentration of the obtained DNA template, measured using a nanodrop, was 2516 ng/μL.

2.5 In-vitro Transcription and Post-Transcriptional Modification I (Vazyme Biotech T7 High Yield RNA Transcription Kit TR101)

To 50 mg of the DNA template, we added NTPs, Pseudo-U, T7 reaction buffer, T7 polymerase and RNase free water to a total volume of 20 μL before incubating the mixture for 2 h at 37 °C. We then digested any residual DNA by adding 2 μL of DNaseI to the mixture and incubating it at 37 °C for 20 min.

2.6 RNA Purification

To purify the RNA from the mixture, we added absolute ethanol to half the volume of the filtrate and mixed thoroughly. We then transferred the mixture into a FastPure RNA column, centrifuged the mixture and discarded the filtrate. We washed the mixture and centrifuged it with 700 μL of RW1 and twice with RW2 (700 μL the first time and 500 μL the second time). To remove excess residual ethanol, we centrifuged one more time, before eluting with 30 μL of RNase-free ddH2O.

2.7 Cap and Tail Addition

To add a 5' methylguanosine cap to the RNA, we added 10 x capping buffer, GTP, RNase inhibitor, SAM, 2'O methyltransferase, Vaccinia capping enzyme and RNase free water and incubated the mixture at 37 °C in a thermocycler for 1 h. The concentration of mRNA measured using a nanodrop was 1.43 g/L (Fig. 3).

Fig. 3. Structure of vaccine. Cap = 5' methylguanosine cap, UTR = untranslated region, sig = extended signal sequence of the CT83 gene, CT83 = CT83 mRNA sequence, poly-A = polyadenylate signal tail.

2.8 Transfection of mRNA (Vazyme Biotech Lipomaster 2000 Transfection Reagent)

We used two petri dishes of HEK 293T cells at a density of about 106, one for the transfection of our mRNA and one as a negative control. We then prepared a diluted solution of Lipofectamine MessengerMAX Reagent using 125 μL of Opti-MEM Medium and 7.5 μL of Lipofectamine MessengerMAX Reagent. We incubated this solution at room temperature for 10 min. While waiting, we prepared a dilution of our modified CT83 mRNA, diluting 2.5 μg of our mRNA (12.3 μL of our purified RNA solution) in 125 μL of Opti-MEM Medium. Following this, we added 125 μL of the diluted Lipofectamine MessengerMAX Reagent Solution to our 125 μL of diluted mRNA.

After incubation for 5 min, we homogeneously spread the mRNA solution over one petri dish of HEK 293T cells. We then grew the cells in an incubator at 37 °C for 48 h before transferring them to –80 °C environment until we extracted the RNA, using the same procedure used for the MKN45 cells. The concentration of RNA from the negative control (-) cells measured using a nanodrop was 14.4 μg/mL and the concentration of RNA from the CT83-transfected (+) cells was 16 μg/mL.

2.9 Reverse Transcription to Form cDNA (Vazyme Biotech HiScript IV RT Supermix for qPCR)

To remove genomic DNA, we added 4 μL of 4x gDNA wiper mix to 1 μg of negative control RNA (6.94 μL of RNA solution), then topped it with 9.06 μL of RNase-free ddH2O. We also added 4 μL of 4x gDNA wiper mix to 1 μg of CT83-transfected cells (4.96 μL of RNA solution), then topped it with 9.06 μL of RNase-free ddH2O. Both mixtures were incubated at 42 °C for 2 min. To form cDNA, 4 μL of HiScript III qRT SuperMix was added to 16 μL of each mixture. Both mixtures were then incubated at 37 °C for 15 min, then 85 °C for 5 s.

2.10 Detecting Changes in Expression of CT83 Gene (Vazyme Biotech Taq Pro Universal SYBR QPCR Master Mix)

To detect any changes in the expression of CT83 compared to the negative control, we carried out a qPCR, using Actin (a component of the cytoskeleton that is always abundantly expressed in all eukaryotic cells) as a reference gene. We used 12 wells in total, 3 wells for each experiment. We added qPCR MasterMix to all of the wells. To the first 3 wells (A-), we added actin forward and reverse primers and cDNA from the – cells. To the next 3 wells (C–), we added CT83 forward and reverse primers and template cDNA from the – cells. To the third set of 3 wells (A+), we added forward and reverse primers for actin and the template cDNA from the + cells. To the last 3 wells (C+), we added CT83 forward and reverse primers and template cDNA from the + cells, as shown in the diagram below (Figs. 4 and 5).

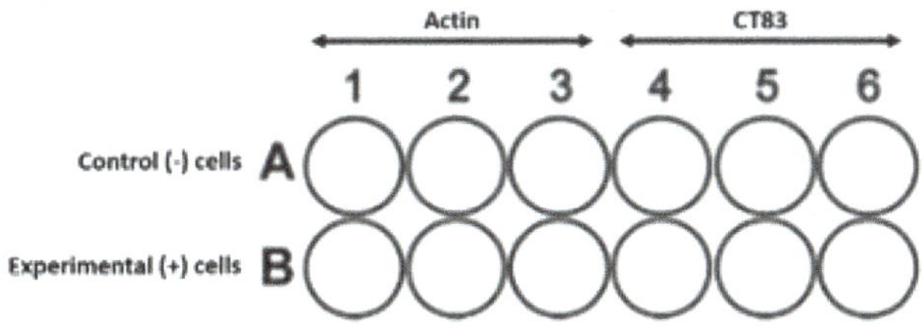

Fig. 4. Preparation of the 12 wells.

We then ran a qPCR with the settings as shown in the table below (Table 1).

Table 1. Mean number of cycles to reach threshold value of 1.000 (Ct).

Stage	Reps	Temperature/°C	Time/s
Initial Denaturation	1	95	30
Cycling Reaction	40	95	10
		60	30
Melting curve	1	95	15
		60	60
		95	15

3 Results and Discussion

The mean number of cycles taken to reach the threshold Ct value are shown in the table below (Table 2).

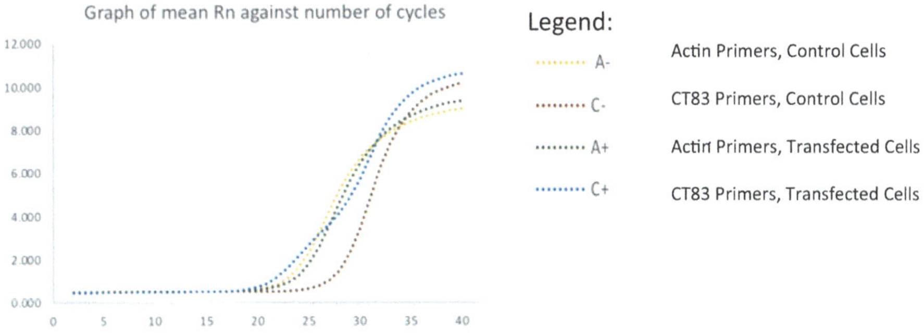

Fig. 5. Graph of Mean Rn against number of cycles.

Table 2. Mean number of cycles to reach threshold value of 1.000 (Ct).

Cells	Mean Ct
A-	23.019
C–	27.567
A+	23.812
C+	20.209

ΔCt for - cells = 4.548
ΔCt for + cells = –3.603
ΔΔCt = –8.151
Fold increase in expression caused by CT83 mRNA vaccine = 284.2

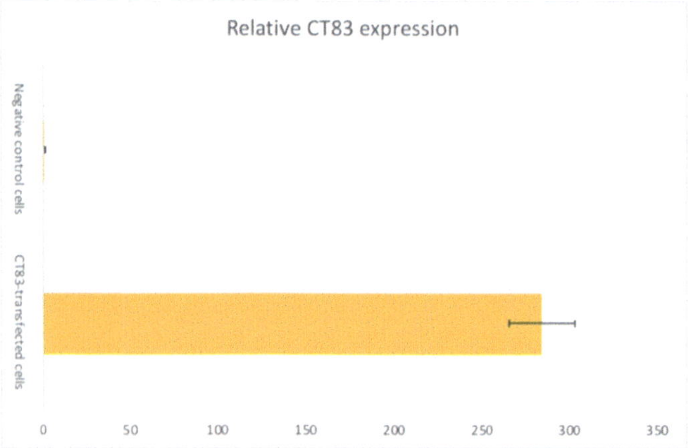

Fig. 6. Relative CT83 expression in CT83-transfected + cells and Negative Control – cells.

Our vaccine resulted in a 284-fold increase (calculated using the ΔΔCt value) in expression of CT83 in the HEK 293T epithelial cells (Fig. 6). This increase in expression, if applied to human cells, could be highly beneficial in creating immunity against CT83-related cancers. When the level of expression of CT83 increases in antigen-presenting cells (APCs), namely dendritic cells, macrophages and B lymphocytes, in somatic tissues, the APCs would recognise the CT83 proteins produced during translation as "foreign". The APCs would then present antigen fragments from the CT83 proteins on MHC II molecules. These MHC-antigen fragment complexes would then be recognised by both CD4+ helper T-cells and CD8 + cytotoxic T-cells, which would release cytokines, activating an appropriate clone of immunoglobulin-secreting B-lymphocytes, CD8 + cytotoxic T-cells, and more helper T-cells, thereby creating serum immunity. Memory B and T-cells are stored in the body and contribute to immunological memory. In the event a cancerous cell that expresses CT83 unnaturally is detected by the body, there are already antigen-specific B lymphocytes, natural killer cells, and cytotoxic T-cells available to attack and kill the cell before it divides uncontrollably, causing the cancer to progress.

4 Conclusions

Through this research, we have synthesised a vaccine that is highly effective in increasing the expression of the CT83 antigen in transfected cells. This means that our vaccine carries the potential to induce an immune response against CT83-related cancers, and with more research and development, be used in personalised preemptive medicine for those who are predisposed to CT83-related cancers, or as a therapeutic vaccine in combination with immunotherapy. Future research could confirm our findings through replication, or confirming an increase in CT83 protein synthesis by purifying intracellular proteins from the transfected cells and performing a western blot to detect the presence of the CT83 proteins. We also need to determine if the vaccine does indeed induce a response (antibody production and T-cell induction) by immune cells upon increased CT83 expression through co-culture of transfected HEK 293T epithelial cells with immune cells. After these hurdles are crossed and the vaccine is deemed successfully immunogenic, tests should be carried out in which the mRNA is delivered in vivo, to investigate the full effects of this vaccine on the body (Fig. 7).

Appendix

All quantities and methods used are as stated in the protocols linked in the bibliography (link numbers 5–12) unless stated otherwise either below, or under the materials and methods section (Tables 3, 4 and 5).

To obtain the standard curves from the qPCR, we used the raw data below that was obtained from the Vazyme Automatic Nucleic Acids Extraction Instrument VNP-32P qPCR machine (Table 6).

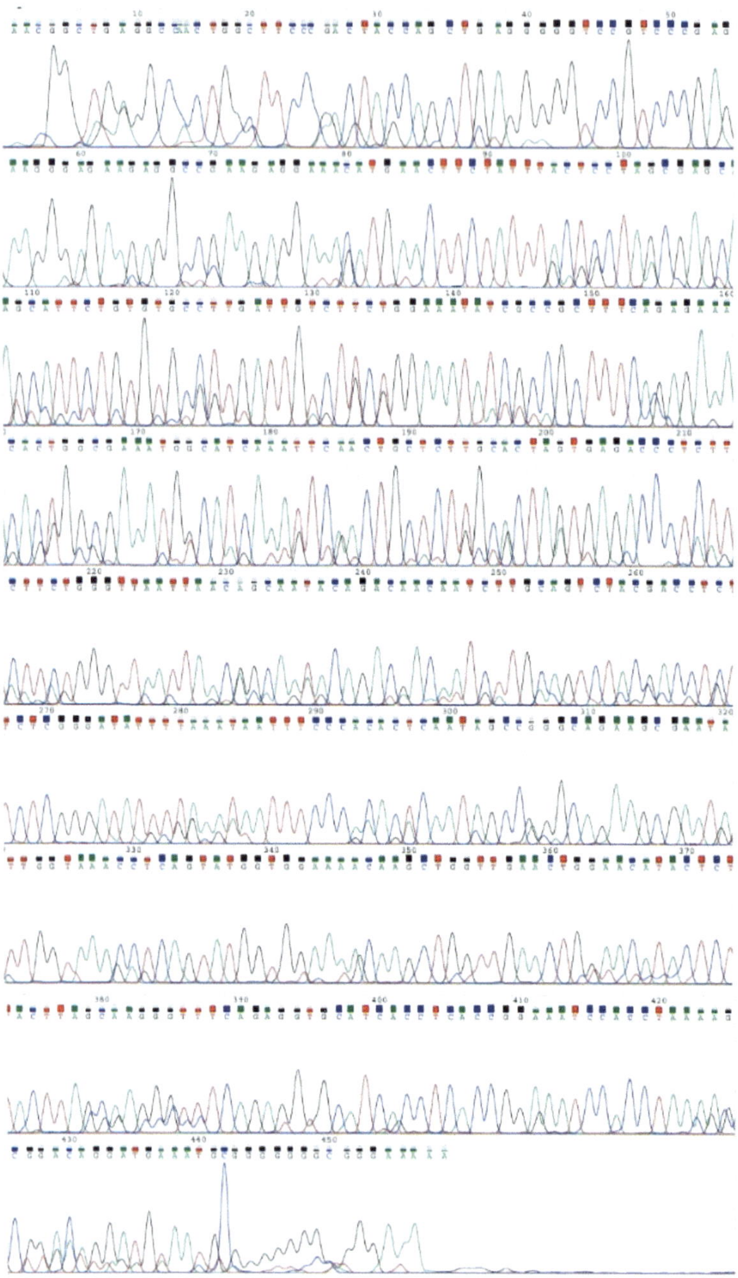

Fig. 7. Results of Sanger's sequencing of extracted mRNA.

Table 3. In-vitro Transcription.

Component	Mass/ng	Volume/μL
DNA template	500	5
ATP (20 mM)	-	2
CTP (20 mM)	-	2
GTP (20 mM)	-	2
Pseudo-U (20 mM)	-	2
T7 Reaction Buffer		2
T7 Polymerase		2
RNase-free ddH2O		3 (Top up to 20 μL)

Table 4. Cap addition.

Component	Mass/μg	Volume/μL
RNA	Up to 20	-
10 x capping buffer	-	4
GTP (10 mM)	-	2
RNase Inhibitor	-	1
SAM (4mM)	-	2
2'O methyltransferase		2
Vaccinia capping enzyme		2
RNase-free ddH2O		3 (Top up to 20 μL)

Table 5. qPCR Reaction Mix.

Reagent	Amount/μL
qPCR Master Mix	10
Forward primer	1
Reverse primer	1
Template cDNA	2
RNase-free ddH2O	6

Table 6. Amplification plot of qPCR, A– = actin primers, control cells; C– = CT83 primers, control cells; A+ = actin primers, transfected cells, C+ = CT83 primers, transfected cells

	Mean normalised reporter (Rn)			
Cycle	A–	C–	A+	C+
1	0.493	0.458	0.473	0.451
2	0.496	0.462	0.477	0.455
3	0.503	0.469	0.484	0.463
4	0.508	0.476	0.491	0.471
5	0.511	0.483	0.496	0.478
6	0.515	0.488	0.502	0.485
7	0.518	0.493	0.506	0.490
8	0.522	0.497	0.510	0.494
9	0.525	0.500	0.515	0.497
10	0.527	0.503	0.518	0.501
11	0.529	0.505	0.520	0.504
12	0.531	0.508	0.523	0.506
13	0.533	0.509	0.525	0.509
14	0.535	0.511	0.527	0.513
15	0.540	0.513	0.531	0.519
16	0.547	0.515	0.535	0.530
17	0.556	0.517	0.543	0.552
18	0.574	0.519	0.554	0.588
19	0.608	0.522	0.575	0.659
20	0.670	0.526	0.613	0.792
21	0.788	0.532	0.686	1.028
22	0.998	0.545	0.822	1.403
23	1.367	0.570	1.065	1.907
24	1.959	0.616	1.481	2.447
25	2.793	0.706	2.135	2.945
26	3.801	0.880	3.044	3.391
27	4.823	1.211	4.117	3.861
28	5.718	1.807	5.181	4.447
29	6.426	2.779	6.089	5.237
30	6.989	4.129	6.832	6.243
31	7.437	5.618	7.420	7.322

(continued)

Table 6. (*continued*)

	Mean normalised reporter (Rn)			
Cycle	A−	C−	A+	C+
32	7.794	6.904	7.879	8.256
33	8.082	7.885	8.243	8.957
34	8.308	8.609	8.542	9.484
35	8.491	9.137	8.771	9.866
36	8.657	9.518	8.959	10.142
37	8.783	9.788	9.108	10.336
38	8.884	9.984	9.223	10.478
39	8.978	10.136	9.331	10.588
40	9.046	10.235	9.406	10.660

References

1. Rausch, S., Schwentner, C., Stenzl, A., Bedke, J.: MRNA vaccine CV9103 and CV9104 for the treatment of prostate cancer. Hum. Vaccin. Immunother. **10**(11), 3146–3152 (2014). https://doi.org/10.4161/hv.29553
2. Chen, C., et al.: Multiomics analysis reveals CT83 is the most specific gene for triple negative breast cancer and its hypomethylation is oncogenic in breast cancer. Sci. Rep. **11**(1) (2021). https://doi.org/10.1038/s41598-021-91290-4
3. Yang, P., Meng, M., Zhou, Q.: Oncogenic cancer/testis antigens are a hallmarker of cancer and a sensible target for cancer immunotherapy. Biochimica et Biophysica Acta (BBA) – Rev. Cancer, **1876**(1), 188558 (2021). https://doi.org/10.1016/j.bbcan.2021.188558
4. Groeper, C., et al.: Cancer/testis antigen expression and specific cytotoxic T lymphocyte responses in non small cell lung cancer. Int. J. Cancer **120**(2), 337–343 (2006). https://doi.org/10.1002/ijc.22309

AI-Aided *De Novo* Design of Novel Irreversible Acetylcholinesterase Inhibitors

Solomon Lim(✉)

NUS High School of Mathematics and Science, Singapore 129957, Singapore
h2010077@nushigh.edu.sg

Abstract. Irreversible acetylcholinesterase inhibitors (IAIs) are a class of acutely cardinal biomolecules, with numerous political and industrial applications ranging from to biocidal deployment to infirmity ministration. However, multitudinous avenues of resistance against such compounds have been contrived, leading to a global necessitation for novel IAIs with augmented activity. The employment of artificial intelligence has also been evinced to be a highly efficacious tool for the *de novo* design of pharmaceutical compounds, enabling large compound libraries to be rapidly generated and screened. Hence, in this paper we report the effectual utilisation of AI based resources to design and screen large libraries of IAIs ($>10^5$·candidate conformers) to evaluate their inhibitory propensity and overall toxicity. Seed compound structures were employed to generate correlative compounds via an *in silico* bioisosteric replacement system (MolOpt) which utilises an AI generative model to elucidate candidate compounds. The compound library was then screened for binding affinity with human acetylcholinesterase using the screening webserver Pharmit to ascertain their effectiveness as irreversible acetylcholinesterase inhibitors. Finally, the overall acute toxicity of the candidate compounds was screened with the platform AD-METlab 2.0, which accorded the relative toxicity factors of each candidate. Analysis of the candidate compounds' properties indicated that five compounds were prognosticated that possessed a higher acetylcholinesterase binding affinity and overall acute toxicity than all existing IAIs, which alludes to a higher potency.

Keywords: AI generation · acetylcholinesterase · inhibition modelling · binding affinity

1 Introduction

First discovered in 1936, irreversible acetylcholinesterase inhibitors (IAIs) are a broad and diverse group of bioactive molecules that irreversibly bind to human acetylcholinesterase, inclusive of certain organophosphates [1] and carbamates [2]. They manifest this propensity from highly reactive functional groups, e.g. organophosphate or carbamoyl, that irreversibly effectuate the ordonnance of covalent bonds vis-à-vis serine residues within the acetylcholinesterase active site [3], thus indelibly binding to the enzyme and proscribing the subsequent breakdown of acetylcholine via suicide intervention (Fig. 1). Due to the resultant accumulation of the muscular neurotransmitter

H. Guo et al. (Eds.): IRC-SET 2024, PTT, pp. 424–432, 2025.
https://doi.org/10.1007/978-981-96-3770-6_43

acetylcholine between neuronal synapses and the sequent untrammelled engenderment of neurotransmission, rapid ictuses and paroxysms frequently ensue before the desired effects are reified [4].

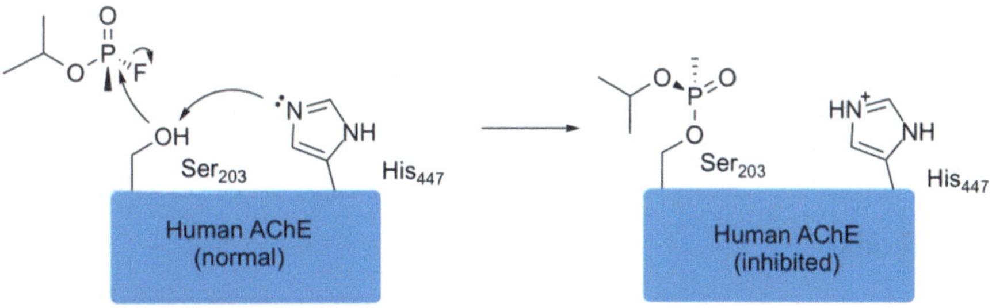

Fig. 1. Irreversible binding of isopropyl methylphosphonofluoridate to the Ser203 residue of human acetylcholinesterase

Naturally, IAIs have attained many applications due to their sui generis properties, such as insecticides, e.g. mercaptothion and propoxur [5] as the inhibition of acetylcholine breakdown serves as a potent mechanism for the rapid incapacitation and expiration of the targeted organisms [6]. Furthermore, IAIs are also highly suitable candidates for the treatment of several neuromuscular diseases such as myasthenia gravis, which utilises intravenous injections of the acetylcholinesterase inhibitor neostigmine in order to stimulate neuromuscular transmission [8]. Their employment in the treatment of the cognitive symptoms of certain neurodegenerative ailments, e.g. Alzheimer's disease and Parkinson's disease, has also been widely recognised, with a multitude of compounds being developed such as tacrine, donepzil, rivastigmine, and galantamine. This is primarily attributed to the induction of elevated levels of acetylcholine in neuronal synapses

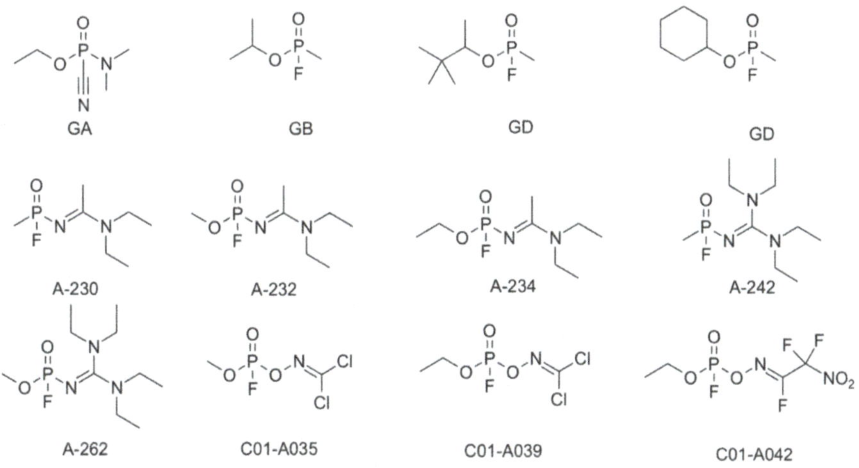

Fig. 2. Chemical structures of common IAIs along with their designation

in the central nervous system, thereby enhancing cholinergic brain neurotransmission [9] (Fig. 2).

Hence, this study utilises an open-source bioisosteric replacement system (MolOpt) [14] for the desideratum of aggregating derivative compound libraries from six primary base compounds, before being screened for human acetylcholinesterase binding affinity with the open-source webserver Pharmit [15] to ascertain functional suitability. Finally, the platform ADMETlab 2.0 [16] was employed to model the bioinhibitory capacity of the candidate compounds, and the candidate compounds were then juxtaposed with existing compounds to assess their overall efficacy.

2 Methods

2.1 Selection of Seed Compounds

Six organophosphate compounds were chosen as seed structures for the generation of the compound library, as shown in Table 1. They are subsequently referred to as seed structures A-F respectively. The seed structures were selected due to their intrinsically potent bioinhibitory properties observed experimentally [17], which indicates their suitability for subsequent modification via computational bioisosteric replacement. Furthermore, the well-defined pharmacophores present in the seed compounds, e.g. ethoxy and dialkylamine groups also enables the rapid and efficient elucidation of alternative candidate compounds using AI-aided deep generative machine learning [14], which compares the existing seed compound to potential fragment replacements in order to generate the compound libraries.

Utilising the AI-aided bioisosteric replacement platform MolOpt [14], the six seed compounds were used to generate a compound library containing $> 10^5$ candidate conformers. The bioisosteric replacement process proceeds via a sequence-based generative model to optimise molecular fragments within the seed compounds by mining bioisosteric groups from the CHEMBL database, enabling the generation of analogous compounds with similar bioinhibitory properties with differing binding affinities for human acetylcholinesterase.

2.2 Evaluation of Bioinhibitory Capacity

The bioinhibitory properties of the candidate compounds were also screened using the web-based platform ADMETlab 2.0 [16], which utilised a multi-task graph attention framework to process the regression tasks required to predict the ADMET (Absorption, Distribution, Metabolism, Excretion, and Toxicity) properties of the input molecules. The platform also builds upon an extensive collection of toxicological data collected from the CHEMBL, PubChem, and OCHEM databases, thus being able to provide a wide range of bioinhibitory results including AMES toxicity, LD_{50}, LC_{50}, IGC_{50}, Carcinogenicity, and Respiratory Toxicity. Out of these indicators, the LD_{50} value was selected as a key index of the bioinhibitory capacity of the candidate compounds, as this would serve to show how the compound holistically influences the target organism.

Table 1. Chemical name and structures of seed com

ID	Compound name	Compound structure	LD_{50} / mg kg^{-1}
A	Ethyl-N-2-diisopropylami-noethylmethyl Phosphono-thiolate	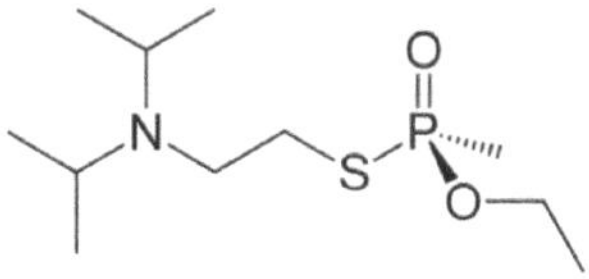	0.071 [17]
B	N-[(Ethoxy-phosphoro-fluoridoyl)oxy] carbonimidic dichloride	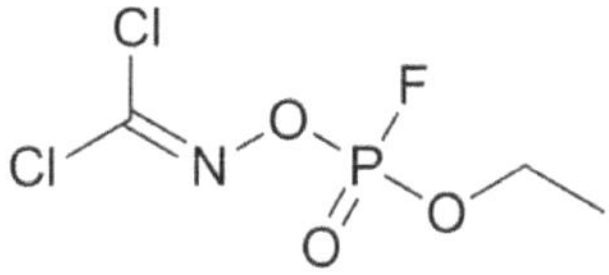	5.550 [18]
C	Propanyl-methylphospho nofluoridate	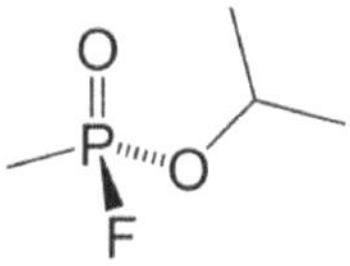	0.172 [19]
D	O-cyclopen-tyl-S-2-diethyla-minoethylme-thyl Phosphono-thioate	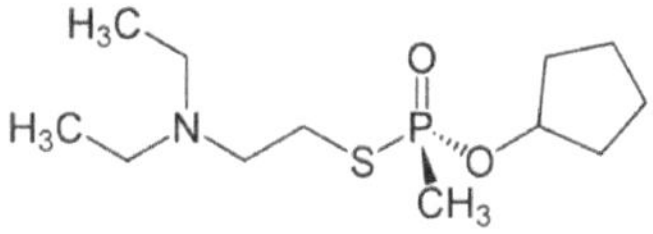	1.900 [20]
E	Methylfluoro-phosphonylcho-line	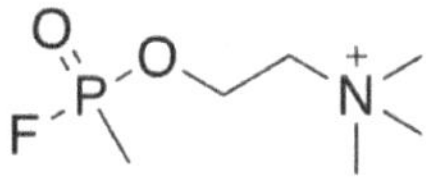	0.100 [21]
F	Dimethylami-noethyl-N,N-di-methylphospho-ramidofluori-date	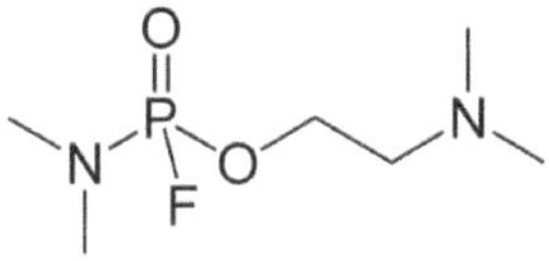	0.222 [22]

2.3 Comparison of Screened Compounds with Initial Dataset

The results of the two screening processes were used to evaluate the potently of the candidate compounds by comparing them with sixteen "baseline" compounds that have been experimentally shown to possess exceptionally high bioinhibitory capacities and binding affinities with human acetylcholinesterase [24]. Based on the results of the first round of screening, a subsequent screening process was also executed based on structural pharmacophore commonalities between the higher-scoring candidate compounds, in particular the *fluorophosphate* and *carbonimidic dichloride* moieties, and the second screened library was utilised to predict energy-minimised binding affinities with human acetylcholinesterase with the updated pharmacophore filters in place, so as to further optimise candidate screening.

3 Results and Discussion

3.1 Computational Screening Results

The bioisosteric replacement system proved to be a potent tool for the rapid design of candidate compounds (Fig. 4), with 98.8% of compounds in the second iteration found to be "toxic" (LD_{50} < 500 mg/kg) according to the Hodge-Sterner toxicity scale [25], in addition to 82.2% of compounds found to be "highly toxic" (LD_{50} < 50 mg/kg) and 19.3% of compounds found to be "extremely toxic" (LD_{50} < 1 mg/kg), the most severe toxicity class (Fig. 5). Furthermore, 9.3% of the candidate compounds were also found to outperform all reference target compounds, indicating a higher binding affinity and bioinhibitory capacity. In relation to the characterisation of the candidate compound libraries, it was observed that the subsequent screening round utilising streamlined pharmacophores filtering frameworks [26] was largely effective in improving candidate suitability with a higher mean binding affinity and a lower mean LD_{50}. This was attributed to the inherent constraints imposed by the correlation of binding affinity to the 6NTK human acetylcholinesterase model [27], as the docking of lower-RMSD candidates was further resolved (Fig. 3).

Five candidate compounds were also found to possess LD_{50} values below 0.1 mg/kg (herein referred to as optimised candidates), indicating extraordinarily high potency and bioinhibitory capacities representative of the optimised halophosphonyl and aminoalkyl moieties present, which contributes to the formation of secondary bioactive metabolites as well as increased reactivity due to the swifter formation of the phosphonium intermediate. An analysis of the compound library's performance with respect to the initial seed compounds (Fig. 6) also revealed that compounds generated from bioisosteric replacement of pharmacophores fragments of seed compound B led to the greatest enhancement in intrinsic potency, with 80.0% of the optimised candidates being derivatives of seed compound B. This is due to the carbonimidic dichloride functionality present in the derivatives, which not only functions as an analog of the nettle agent phosgene oxime which would present auxiliary mechanisms of action via respiratory, optical, gastrointestinal, and dermal routes, but also has been demonstrated to possess crosslinking capabilities which enables efficient binding to multiple sites in the active site of the human acetylcholinesterase enzyme and thus a higher efficacy.

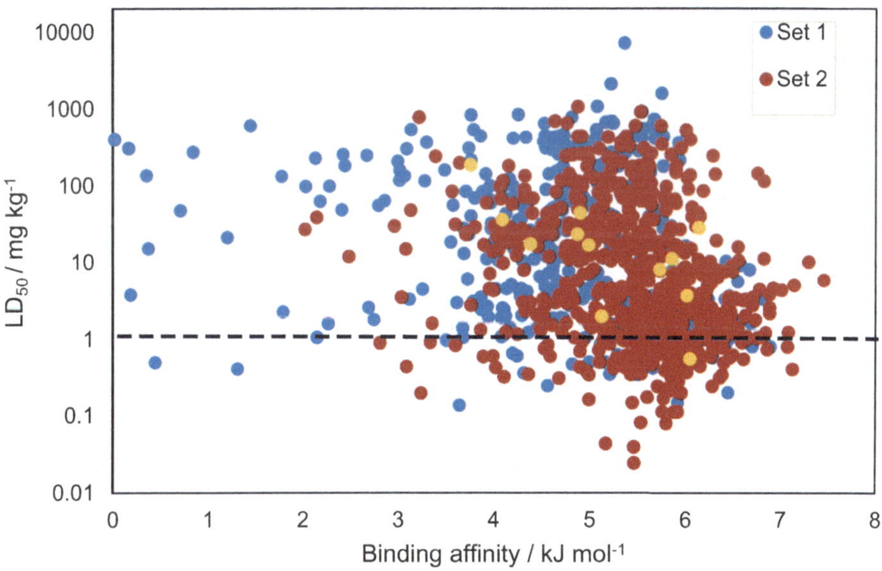

Fig. 3. Scatterplot of candidate compounds between screening iterations.

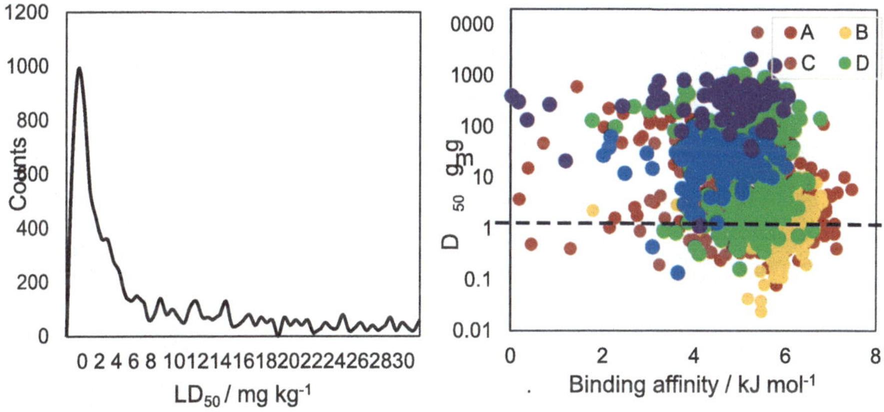

Fig. 4. Candidate count correlated with respective LD50 values (left) and toxicity scatterplot (right)

3.2 Analysis of Optimised Candidates

The structures of the five optimised candidates are shown below (Fig. 5), labelled in the format AI-XXXXY where XXXX represents the ID number of the candidate compound and Y represents the seed compound it was derived from. From the B-series optimised candidates AI-0215B, AI-0225B, and AI-0252B, it is readily observable that the sulfenyl ester moiety is highly favoured in regard to maximising binding affinity for human acetylcholinesterase (Fig. 8 - 10), as well as decreasing the LD_{50} value, due to the formation of reactive sulfur-mustard analogs upon hydrolysis at physiological conditions which not only serve as potent DNA alkylation agents [28], but are also able to cross-link multiple residues on the binding site of human acetylcholinesterase [29], thus increasing bioinhibitory capacity.

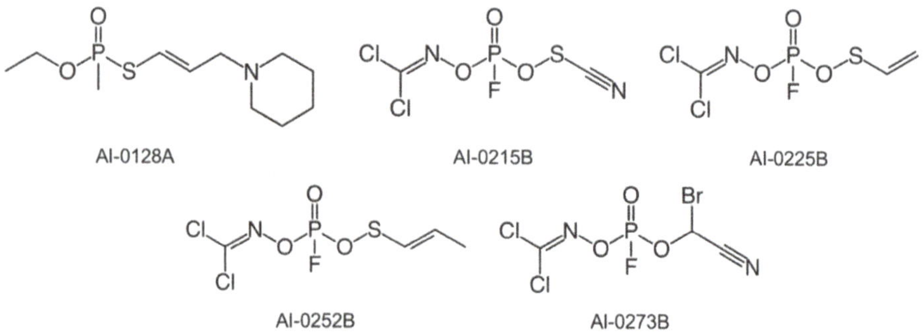

Fig. 5. Chemical structures and designations of the five optimised candidates

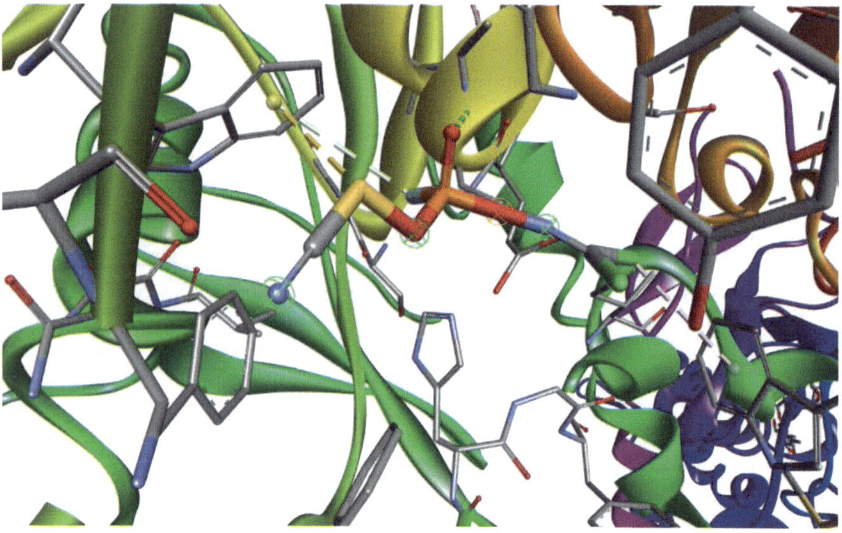

Fig. 6. Binding conformation of AI-0215

Furthermore, as observed from AI-0215B and AI-273B, the nitrile moiety also seems to be highly indicative of bioinhibitory capacity enhancement, due to the formation of cyanide *in vivo* which induces cellular asphyxiation. An analogous effect has also been observed with similar compounds, e.g. diphenylcyanoarsine and cyanogen chloride [30], which suggests that a multifaceted route of action with regards to a more holistic targeting mechanism would be especially effective in preventing early counteraction of the induced effects.

Similarly, the inclusion of a piperidine pharmacophore in AI-0128A also indicates that the complementarity of the IAI and the acetylcholinesterase is highly reminiscent of the structure of cyclosarin, which possesses a cyclohexyl moiety that enables more efficacious binding to the acetylcholinesterase binding site. Thus, it was also observed that the piperidine group enabled a much closer fit with the binding pocket. While the effect on binding affinity was relatively marginal, with a binding affinity of 5.79 kJ/mol, this however led to a marked increase in the bioinhibitory capacity of the compound, which was correlated with its similarity in pharmacophores features with the acetylcholine molecule while retaining the methylphosphonyl group that has been demonstrated to be highly instrumental in ensuring rapid delivery and localisation of the compound to the brain.

4 Conclusion

In conclusion, AI-driven resources have been successfully utilised to design organophosphate-based IAIs with a high binding affinity for human acetylcholinesterase and potent bioinhibitory capacities attributed to their exceptionally low LD_{50} values. Five optimised candidate compounds were also identified with extraordinarily high bioinhibitory capacities and enhanced binding capacities to human acetylcholinesterase due to the formation of secondary bioactive metabolites, as well as crosslinking and conformational characteristics that enable a tighter binding to the acetylcholinesterase binding site. The aforementioned results are highly indicative of the inherent potential of computational prediction for the rapid and accessible design of bioactive agents, which offers a less resource-intensive and convenient avenue for similar classes of compounds to be designed.

The candidate compounds that were screened also possess great potential in their applications in the biological and chemical fields of study, ranging from pesticides with a much higher efficacy due to their bioinhibitory properties to highly potent neuromuscular control agents due to their high binding affinity for human acetylcholinesterase. Furthermore, the optimised candidates are also predicted to be able to act as powerful bioinhibitory agents due to their relatively low molecular mass compared to existing IAIs, which suggests enhanced suitability for environmental dispersion.

5 Future Work

In the future, the synthetic routes to the optimised candidate compounds can also potentially be designed utilising existing computational tools, in order to demonstrate the feasibility of a wholly-automated process to design bioactive agents, from structural prediction to synthesis. Furthermore, this method can also be applied to similar classes of bioactive agents, e.g. sulfur-based DNA alkylation agents, which would enable a robust and versatile design route to be established for future use. Lastly, the data gleaned from the computational processes described above can also be experimentally validated in order to verify the accuracy of the resources.

References

1. Greathouse, B.: Acetylcholinesterase inhibitors toxicity. StatPearls-NCBI Bookshelf. (2023) https://www.ncbi.nlm.nih.gov/books/NBK535428/
2. Kovarik, Z., Radić, Z., Grgas, B., Škrinjarić-špoljar, M., Reiner, E., Simeon- Rudolf, V.: Amino acid residues involved in the interaction of acetylcholinesterase and butyrylcholinesterase with the carbamates Ro 02–0683 and bambuterol, and with terbutaline. Biochimica Et Biophysica Acta (BBA) - Proteins Proteomics, **1433**(1–2), 261–271 (1999) https://doi.org/10.1016/s0167-4838(99)00124-7
3. Hemmert, A., et al.: Nerve agent hydrolysis activity designed into a human drug metabolism enzyme. PLoS ONE **6**(3), e17441 (2011). https://doi.org/10.1371/journal.pone.0017441
4. Figueiredo, T.H., Apland, J.P., Braga, M.F.M., Marini, A.M.: Acute and long-term consequences of exposure to organophosphate nerve agents in humans. Epilepsia **59**(S2), 92–99 (2018). https://doi.org/10.1111/epi.14500
5. Krstić, D., et al.: Inhibition of AChE by malathion and some structurally similar compounds. J. Enzyme Inhib. Med. Chem. **23**(4), 562–573 (2008). https://doi.org/10.1080/14756360701632031
6. Čolović, M.B., Krstić, D., Lazarević-Pašti, T., Bondžić, A.M., Vasić, V.: Acetylcholinesterase inhibitors: pharmacology and toxicology. Curr. Neuropharmacol. **11**(3), 315–335 (2013). https://doi.org/10.2174/1570159x11311030006
7. Mladenović, M., et al.: The targeted pesticides as acetylcholinesterase inhibitors: comprehensive cross-organism molecular modelling studies performed to anticipate the pharmacology of harmfulness to humans in vitro. Molecules **23**(9), 2192 (2018). https://doi.org/10.3390/molecules23092192
8. Cridge, H., et al.: The clinical utility of neostigmine administration in the diagnosis of acquired myasthenia gravis. J. Vet. Emerg. Crit. Care **31**(5), 647–655 (2021). https://doi.org/10.1111/vec.13097
9. Marucci, G., Buccioni, M., Ben, D.D., Lambertucci, C., Volpini, R., Amenta, F.: Efficacy of acetylcholinesterase inhibitors in Alzheimer's disease. Neuropharmacology **190**, 108352 (2021). https://doi.org/10.1016/j.neuro-pharm.2020.108352
10. Nguyen, M., Alfonso, A., Johnson, C.D., Rand, J.B.: Caenorhabditis elegans mutants resistant to inhibitors of acetylcholinesterase. Genetics **140**(2), 527–535 (1995)
11. Fournier, D., Bride, J., Hoffmann, F., Karch, F.: Acetylcholinesterase. two types of modifications confer resistance to insecticide. J. Biol. Chem. **267**(20), 14270–14274 (1992) https://doi.org/10.1016/s00219258(19)49708-1

On the Computational Modelling of Chaotic Hamiltonian Systems

Mazereeuw Luc Martin Linus and Akshat Vijoy(✉)

Physics and Engineering, National University of Singapore High School of Mathematics and Science, 20 Clementi Avenue 1, Singapore 129957, Singapore
{h1910102,h1910157}@nushigh.edu.sg

Abstract. The present study aims to investigate the renowned Kicked Rotator, a classical system which exhibits regimes of either integrability or chaoticity under different system parameters. Our research will focus on the chaotic regime, and we will characterize the system by a 'diffusivity' constant. The aim of our project is to numerically investigate how this diffusivity constant depends on the system parameters i.e. the potential function and kicking strength, as well as to develop an analytic model to predict the diffusivity constant using appropriate series expansions. The findings have potential applications in the field of condensed matter physics such as with respect to the Anderson tight-binding model in cold atoms.

Keywords: Chaos · Kicked Rotator · Fixed Point Stability · Momentum Diffusion · Accelerator Modes

1 Preliminaries

1.1 The Kicked Rotator

The kicked rotator describes a free rotating stick in an inhomogeneous gravitational-like field that switches on periodically in extremely short pulses, which we call **kicks**. During each kick, some momentum is imparted into the stick instantaneously. This is modelled by a Dirac comb function, with the amplitude of each peak being dependent on the angle of the stick. Conventionally, the moment of inertia of the stick is denoted to be I and the field is modelled by **kicking strength** K and period T.

1.2 Chirikov Standard Map

We may derive the Hamiltonian of the system from its definition:

$$\mathcal{H} = \sum_{i=1}^{n} \dot{q}_i \frac{\partial \mathcal{L}}{\partial \dot{q}_i} - \mathcal{L} \tag{1}$$

H. Guo et al. (Eds.): IRC-SET 2024, PTT, pp. 433–440, 2025.
https://doi.org/10.1007/978-981-96-3770-6_44

where $\mathcal{L}$ represents the Lagrangian of the system. Given that the field **evolves over time**, our Hamiltonian must be defined in terms of t as well, obtaining:

$$\mathcal{H}(\theta, p_\theta, t) = \frac{p_\theta^2}{2I} + K\cos\theta \sum_{n=-\infty}^{\infty} \delta\left(\frac{t}{T} - n\right) \tag{2}$$

where $\theta \in [0, 2\pi]$ is the **angular position** of the stick and p_θ is its **conjugated momentum**. From this, we may write the equations of motion of the system:

$$\frac{d\theta}{dt} = \frac{\partial\mathcal{H}}{\partial p} = \frac{p}{I} \tag{3}$$

$$\frac{dp}{dt} = -\frac{\partial\mathcal{H}}{\partial\theta} = K\sin\theta \sum_{n=-\infty}^{\infty} \delta\left(\frac{t}{T} - n\right) \tag{4}$$

Hence, between two consecutive kicks i.e. when the field is "turned off", the **momentum is conserved** and the angular position grows linearly with time at the rate described by its conjugated momentum at that instant. On the other hand, given that the field is modelled by a Dirac comb, each kick can be seen to **impart a finite momentum** abruptly with the angular position of the stick remaining constant. Hence, we can model the dynamics using a discrete map which transforms the system. After de-dimensionalising this to reduce the system to 1 parameter, we obtain the Chirikov Standard Map:

$$p_{n+1} = p_n + K\sin\theta_n \tag{5}$$

$$\theta_{n+1} = \theta_n + p_{n+1} \tag{6}$$

Here, p_n is no longer periodic but for any two sticks with the same initial angular position and momenta differing by a multiple of 2π, it is clear that in the $\theta - p$ phase-space, these sticks would **exhibit the same appearance of motion** i.e. the same stroboscopic dynamics.

1.3 Modification of Kicking Function

We observe that the equations of motion of the kicked rotator can be generalised as:

$$p_{n+1} = p_n + Kf(\theta_n) \tag{7}$$

$$\theta_{n+1} = \theta_n + p_{n+1} \tag{8}$$

where $f(\theta_n)$ is the chosen "kicking" function dependent on the type of field. In this project, we will study $f(\theta_n) = |\sin(\theta_n)|^\alpha \cos\theta$ as it generalises the conventional kicked rotator to a system where its angular position has a quantifiable impact in terms of α, which we will denote as **kicking power**. The form of the equation is chosen such that it reflects a **potential function** of a form $V(\theta_n) = (\cos\theta)^\alpha$ while being continuous for all values of α. This allows us to contextualise our findings for real potentials studied in solid-state systems.

1.4 Stability Analysis

In general, we may adapt the above equations of motion to analyse the stability of fixed points in the region by performing a Taylor expansion in the neighbourhood of the fixed point and calculating the eigenvalues of the transformation matrix. The fixed points themselves can be found by solving:

$$p_n = p_n + Kf(\theta_n) \tag{9}$$

$$\theta_n = \theta_n + p_n \tag{10}$$

This implies that the fixed points are situated at the **roots of the kicking function** itself. Adjusting our transformation to expand about the fixed point θ_f, we obtain:

$$\begin{pmatrix} p_{n+1} \\ \theta_{n+1} \end{pmatrix} = \begin{pmatrix} 1 & \pm K\left(\sum\limits_{k=0}^{\infty} \frac{f^{(k)}(\theta_n)}{k!}(\theta_n - \theta_f)^{k-1}\right) \\ 1 & \pm K\left(\sum\limits_{k=0}^{\infty} \frac{f^{(k)}(\theta_n)}{k!}(\theta_n - \theta_f)^{k-1}\right) + 1 \end{pmatrix} \begin{pmatrix} p_n \\ \theta_n - \theta_f \end{pmatrix} \tag{11}$$

The $\pm$ corresponds to the direction of "travel" from the fixed point. Taking the trace of our transformation matrix, it is clear that the stability of this fixed point requires:

$$\left| 2 \pm K\left(\sum_{k=0}^{\infty} \frac{f^{(k)}(\theta_n)}{k!}(\theta_n - \theta_f)^{k-1}\right) \right| < 2 \tag{12}$$

This implies that fixed points can be categorised into one of the following:

1. **Elliptic**: Stable in both directions
2. **Hyperbolic**: Unstable in both directions
3. **Saddle**: Stable in one direction but unstable in the other

This property will be utilised subsequently to assist us in restricting the domain of the phase-space to only chaotic regions.

2 Numerical Simulations

2.1 Accelerator Modes

We begin by plotting the **2π-modded** phase-space of our modified kicked rotator (Fig. 1):

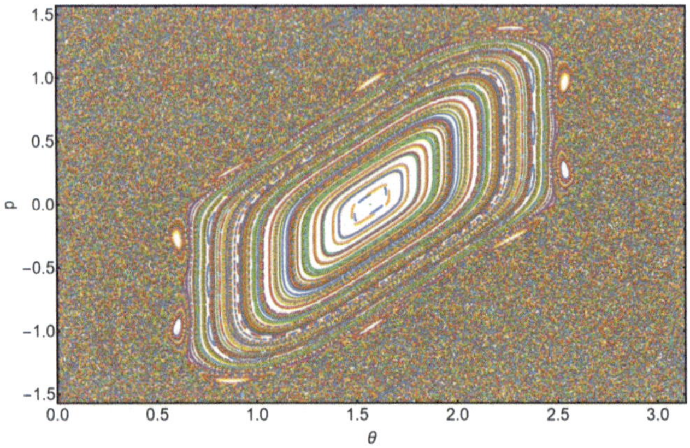

Fig. 1. Phase-Space of Kicked Rotator controlled by $f(\theta) = |\sin(\theta_n)|^\alpha \cos\theta$

Zooming out, we observe that the **toric** regions in the above phase-space correspond to **integrable** (i.e. non-chaotic) neighbourhoods wherein the $p - \theta$ phase-space exhibits **semi-periodicity** (Fig. 2).

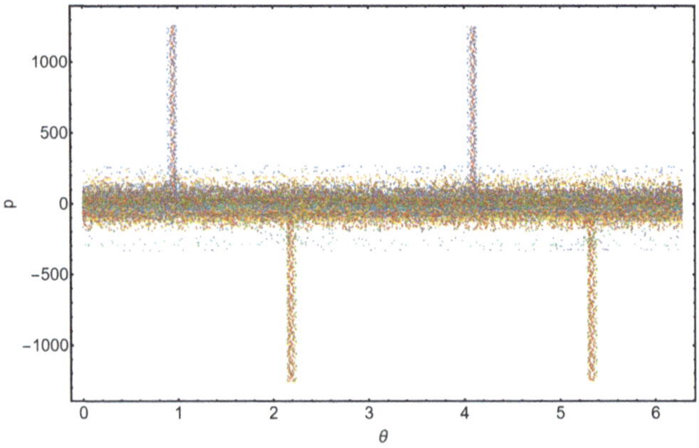

Fig. 2. Accelerator Modes

2.2 Momentum Diffusion

The above regions are known as **accelerator modes**, wherein the stick gets "trapped", resulting in p growing linearly with time. We wish to study the spatial **diffusion of momentum** across the phase-space, which only occurs in chaotic regions. This diffusion of momentum corresponds to the growth of p^2 against time. Hence, we will filter out these accelerator modes and study the chaotic region, which we will denote as Ω (Fig. 3).

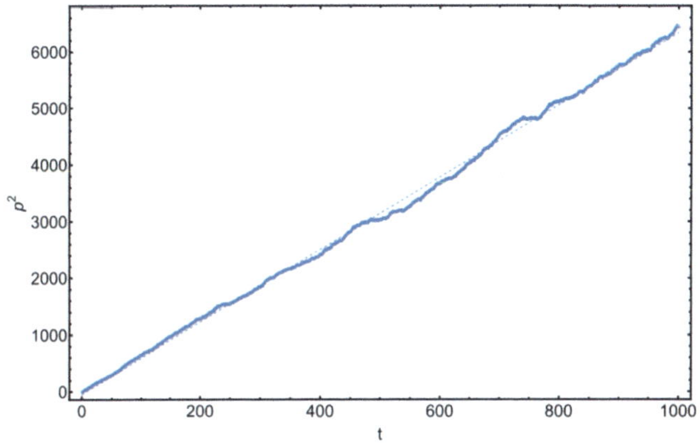

Fig. 3. Linear growth of p^2 against t

Plotting p^2 against t for a trajectory emanating from a randomised set of initial conditions, we observe that it always maintains a roughly-linear trend - as long as the set of initial conditions are contained within Ω. Denoting the constant gradient of this graph as the **diffusion constant** D, we will aim to analytically derive the effects of our system parameters on its value.

3 Analytic Derivation

3.1 Correlation

Since the system is deterministic, the evolution of a point in phase space **only depends on the initial conditions**, and any system parameter can be expressed as $f(x_0, p_0)$ for a chosen coordinate x and its conjugated momentum p. As stated previously, we are only interested in the region Ω of phase space whereby chaoticity is observed as momentum diffusion can occur here. The **spatial average** over the region an be defined as:

$$\langle f(x_0, p_0) \rangle = \left(\int_{\Omega} f(x_0, p_0)\, dx_0\, dp_o \right) \left(\int_{\Omega} dx_0\, dp_o \right)^{-1} \tag{13}$$

Since every point in this region maps to another point in this same region, the average is **time-invariant**. The evolution of this system can be reduced to the characteristic equation $x_{n+2} = 2x_{n+1} - x_n + kf(x_n)$, defining the characteristic functions:

$$\chi_n(m_0, m_1, m_2) = \left\langle \exp\left(i \sum_{j=0}^{n} m_j x_{n+j} \right) \right\rangle \tag{14}$$

Using the Taylor series to expand the function,

$$f(x) = \sum_{l=-\infty}^{\infty} a_l e^{ilx} \tag{15}$$

$$e^{ikf(x)} = \sum_{l=-\infty}^{\infty} g_l(k) e^{ilx} \tag{16}$$

we arrive at the recursion relation for χ_n:

$$\chi_n\left(\{m_i\}_{i=0}^{n}\right) = \sum_{l=-\infty}^{\infty} g_l(k)\chi_{n-1}(m_0, m_1, \ldots, m_{k-2} - m_k, m_{k-1} + 2m_k + l) \tag{17}$$

$$\text{with } \chi_0(m_0) = \delta(0, m_0) \;\&\; \chi_1(m_0, m_1) = \delta(0, m_0)\, \delta(0, m_1) \tag{18}$$

This allows us to obtain an **analytic expression** for the diffusion constant D, which can be utilised to study the effects of the kicking power α and strength K on the momentum diffusion occurring in the system:

$$D = \lim_{T\to\infty} \frac{\langle p_T \rangle^2}{2T} = \frac{k^2}{2} \sum_{m=-\infty}^{\infty} \sum_{n=-\infty}^{\infty} \left(\chi_0(m+n) + \sum_{i=1}^{\infty} 2\chi_i(m, 0, \ldots, 0, n) \right) a_m a_n \tag{19}$$

3.2 Results

We begin by numerically simulating the trajectories emanating from various sets of initial conditions and calculating the diffusion constant D in the region, followed by adjusting K and repeating the process, allowing us to obtain the following plots (Fig. 4):

The spikes in the graphs correspond to regions of integrability i.e. accelerator modes wherein $D = \lim_{T\to\infty} \frac{\langle p_T \rangle^2}{2T} = \infty$ owing to the linear growth of momentum in these regions. Hence, as mentioned above, we will filter out these modes when analysing the analytical model, the results of which are plotted below (Fig. 5): Zooming into a smaller region of $K \in (10, 30)$ with $\alpha = 0.5$, we observe excellent agreement between the analytic model and the numerical results. The analytical

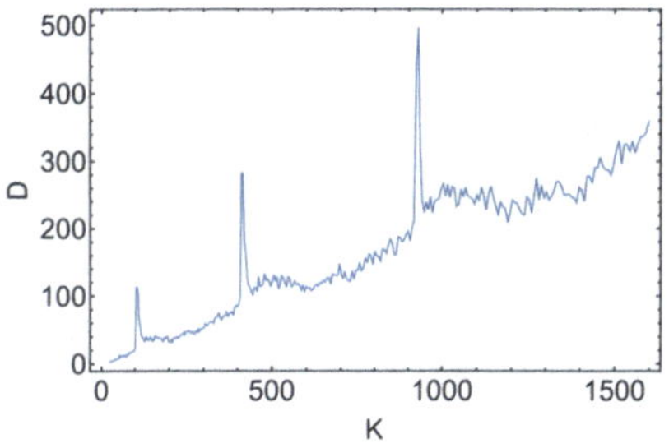

(a) α=0.5

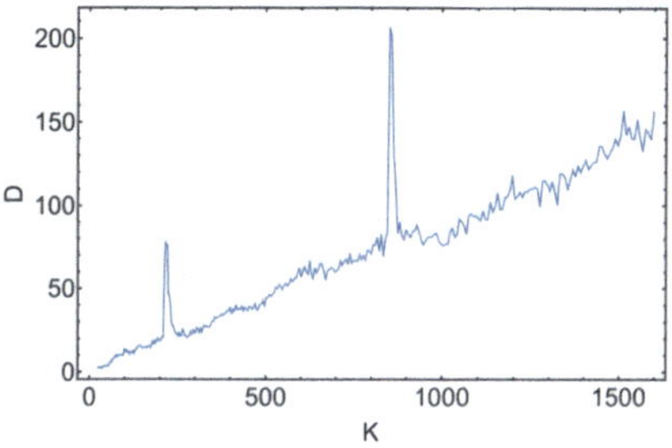

(b) α=1

Fig. 4. Numerical dependence of D on K for differing α

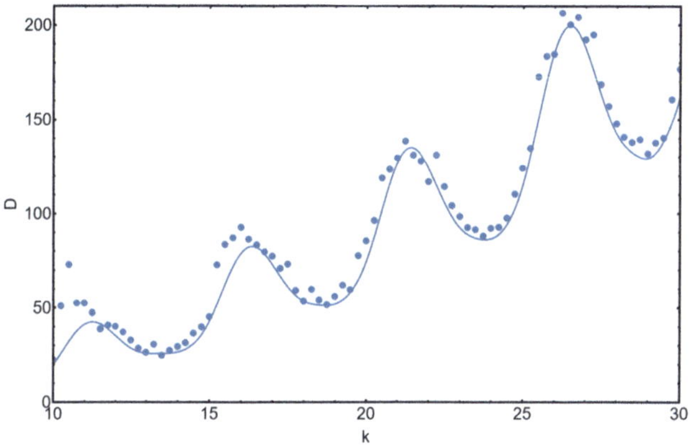

Fig. 5. Analytic dependence of D on K

model correctly predicts the general trend of $D \approx \frac{k^2}{2}$ as well as bumps which deviate, owing to the **series corrections**, as seen in Eq. 19. Lastly, we created a 3D plot to simultaneously show the numerical dependence of D with respect to both K and α. Given that the value of D is orders of magnitude higher than α, we plotted $\ln D$ to better show the trends observed. In general, we see the expected increasing trend with respect to K. However, there is **no clear roughly-monotonic dependence** of D on α. There is however, a regime of **relative stability** for $\alpha > 1$, while there is a much greater descent for $\alpha < 1$ (Fig. 6).

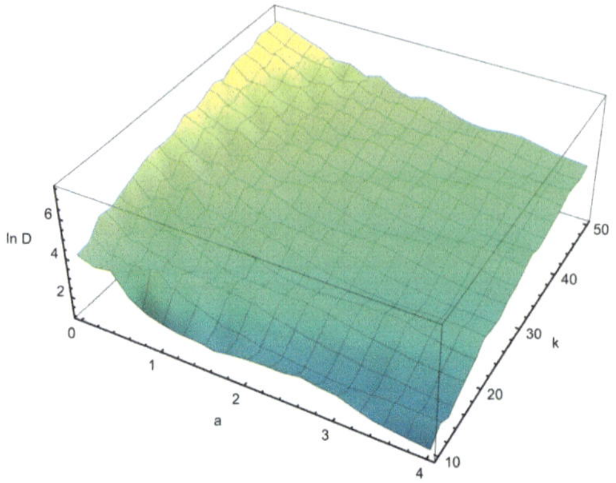

Fig. 6. 3D Plot of $\ln D$ against K and α

4 Discussion and Conclusion

Our work has elicited a rigorous quantitative model which analytically measures the dependence of D on the kicking power α and kicking strength K. The predictive power of this model was verified in accordance with numerical simulations generated by utilising the standard map equations. This model, owing to the usage of series expansions, can be easily modified to fit other types of kicking functions corresponding to other fields, allowing for the generalising of the theory to various types of potentials found in solid-state systems. Henceforth, the authors look to enhancing the robustness of our current model by accounting for and modelling relevant quantum effects - such as the suppression of the classical momentum diffusion (which was studied here) owing to dynamical localisation. Studying the potential applications of this model in the field of condensed matter physics - such as with respect to the Anderson tight-binding model - will require experimental realisation, which the authors look favourably towards achieving.

References

1. Cary, J.R., Meiss, J.D.: Rigorously diffusive deterministic map. Phys. Rev. A **24**(5), 2664–2668 (1981). https://doi.org/10.1103/physreva.24.2664
2. Cary, J.R., Meiss, J.D. Bhattacharjee, A.: Statistical characterization of periodic, area-preserving mappings. Phys. Rev. A **23**(5), 2744–2746 (1981). https://doi.org/10.1103/physreva.23.2744
3. Meiss, J.D., Cary, J.R., Grebogi, C., Crawford, J.D., Kaufman, A.N., Abarbanel, H.: Correlations of periodic, area-preserving maps. Physica D **6**(3), 375–384 (1983). https://doi.org/10.1016/0167-2789(83)90019-2
4. Rechester, A.B., White, R.B.: Calculation of turbulent diffusion for the chirikov-taylor model. Phys. Rev. Lett. **44**(24), 1586–1589 (1980). https://doi.org/10.1103/physrevlett.44.1586

On Ramsey Numbers and Monochromatic Cliques in a Complete Graph

Akshat Vijoy(✉), Lawrence Juin Yi Loke, Zheng Yu Choo, and Chan Lye Lee

Mathematics and Statistics, National University of Singapore High School of Mathematics and Science, 20 Clementi Avenue 1, Singapore 129957, Singapore
{h1910157,h1910075,h1910020}@nushigh.edu.sg, nhslcl@nus.edu.sg

Abstract. Variations of graphs in Ramsey theory are studied, in which not one, but multiple specified cliques are formed. Various walkthroughs of the construction and guaranteeing of such cliques, from the initial guaranteed conditions to the process of deduction of the other edge colourings are explained. We will not only be covering some simpler cases such as Ramsey's theorem for small cliques such as triangles and K_4, but also about graphs with more vertices, and the possibility of guaranteeing more than 1 of the specific monochromatic cliques.

Keywords: Ramsey's Theorem · Monochromatic Cliques · Graph Theory

1 Introduction

Ramsey's theorem, first introduced by mathematician Frank Plumpton Ramsey, states that one will find monochromatic cliques in any edge labelling (with colours) of a sufficiently large complete graph. [1] Consider a graph with k vertices where k is a positive integer, and every set of 2 vertices must be joined using an edge with n colours to choose from. Ramsey theory states that with enough vertices, at least of the given one monochromatic clique can be formed. The Ramsey number then states the minimum number of vertices needed to achieve this goal. For instance, common examples include R(3,3) = 6, R(3,4) = 9, R(4,4) = 18. In these cases, there are 2 colours to choose from for each edge, and the numbers in the brackets correspond to the number of vertices that a clique of one of the arbitrary colours must have for it to be counted. The notation K_n refers to a clique (complete graph) with n vertices. The function R(a, b, c, ...) states the minimum number of vertices such that any of a monochromatic K_a, K_b, K_c, ... in colours 1, 2, 3, ... respectively, can be formed using some colouring of a graph with R(a, b, c, ...) vertices.

H. Guo et al. (Eds.): IRC-SET 2024, PTT, pp. 441–453, 2025.
https://doi.org/10.1007/978-981-96-3770-6_45

2 Preliminary Results

2.1 Lemma 2.1: $R(3,3) = 6$

This is the well-known problem of a party with 6 people, and there having to be at least 3 mutual friends or 3 mutual strangers. It can also be interpreted as a 6-vertex graph with 2 line colours to choose (such as red and blue) for every edge, and there being at least either a red or blue triangle.

Let G be a complete graph with 6 vertices. We note that every vertex will be connected to 5 other vertices, each by an edge. Each of these 5 edges must be coloured using either blue or red. So, using Pigeon-hole principle, each vertex would be connected to at least 3 other vertices with the same colour. Without loss of generality, let the colour be blue. If any of the edges within these 3 vertices is coloured blue, then there would be a monochromatic triangle. If none of these edges are coloured blue, they must still be coloured, and then there would be a red monochromatic triangle. This shows that a monochromatic triangle in a G cannot be avoided, thus R(3,3)$\leq$ 6. To prove that R(3,3) is no less than 6, here is an example of a 5-vertex graph without any monochromatic triangles (Fig. 1):

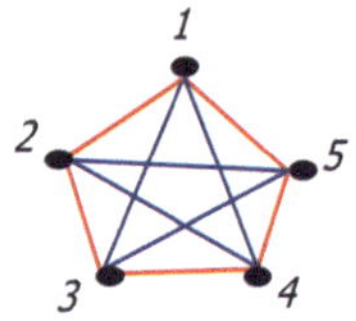

Fig. 1. An example of a 5-vertex graph without any monochromatic triangles

We know that there will be at least 1 monochromatic triangle. We also know that the maximum number of monochromatic triangles there can be in this 6-vertex graph is $\binom{6}{3} = 20$, which is when all the edges are the same colour. The natural question people may ask is: "Does a 2-colour 6-vertex graph with exactly 1 monochromatic triangle exist?" Surprisingly, the answer is no, which will be shown below.

3 Deeper Analysis

3.1 Lemma 3.1: Within Any 2 Colouring on a 6-Vertex Graph, the Minimal Number of Monochromatic Triangles is 2

(See Fig. 2).

We aim to construct a 6-vertex graph with only one monochromatic triangle or less to disprove the above statement. The goal is to arrive at a contradiction (Fig. 3).

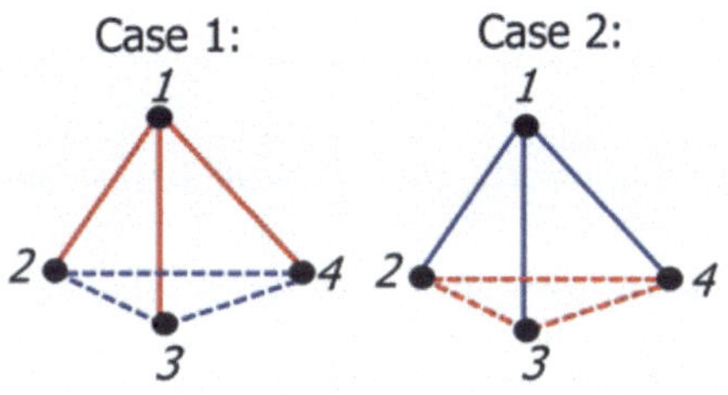

Fig. 2. The 2 cases show how a monochromatic triangle will definitely be formed with that set-up of the solid edges (dotted edges show how a monochromatic triangle may be formed)

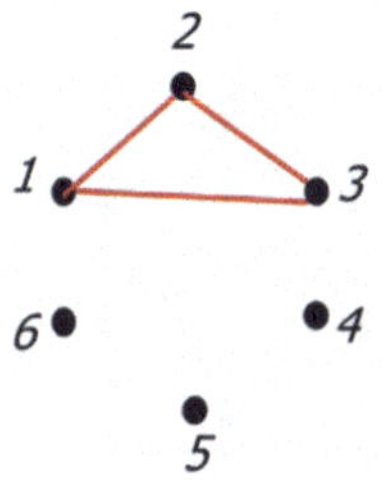

Fig. 3. Since we have established the existence of at least one monochromatic triangle in any 6-vertex graph with 2 edge colours, we can start the construction of the graph with it

Let numbered vertices 1,2,3 be the vertices of the triangle, which would be made of red edges. Since we have established the existence of at least one monochromatic triangle in any 6-vertex graph with 2 edge colours, we can start the construction of the graph with it. Let numbered vertices 1,2,3 be the vertices of the triangle, which would be made of red edges (Fig. 4).

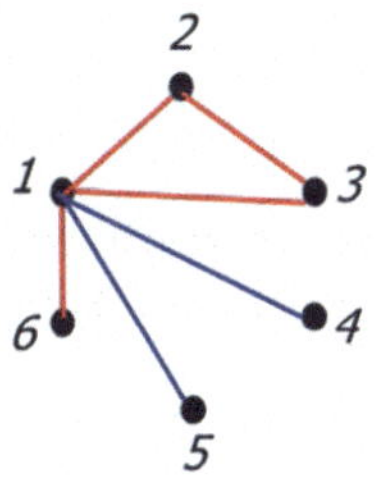

Fig. 4. In order to join vertex 1 to vertex 4,5,6, one red edge and two blue edges must be used. (Color figure online)

Connect vertex 1 to vertices 4,5 and 6. If two or more red edges are used, case 1 will be fulfilled, with vertex 1 connected to two of the vertices in the set of 4,5,6 and also vertex 2. This will force a monochromatic triangle within those vertices that is not the same as the first. If three blue edges are used, case 2 will be fulfilled with vertices 1,4,5,6. Therefore, vertex 1 is joined to vertex 6 with red and vertices 5,4 with blue, without loss of generality. With vertex 2, the same logic is used, while this time, it can no longer join to vertex 6 with red, as that will just make a 1-2-6 triangle. It is joined to vertex 5 with red and vertices 4,6 with blue, without loss of generality. Vertex 3 is joined to vertex 4 with a red edge and vertices 5,6 with blue, as that is the only option left. This creates the graph below (Fig. 5).

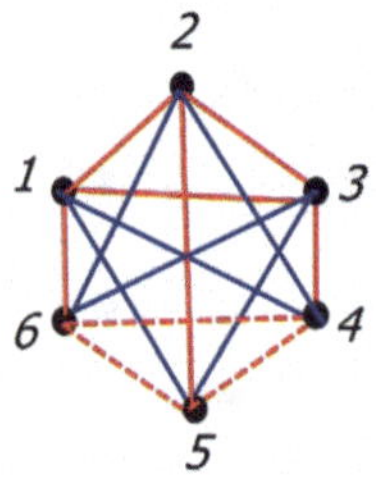

Fig. 5. For the remaining edges 4-6, 4-5, 5-6, if any of them are blue, there would be a blue triangle formed. If all of them are red, there would be a red triangle formed. (Color figure online)

This shows that there exists no 6-vertex graph with less than 2 monochromatic triangles.

4 A 7-Vertex Graph

We have investigated that R(3,3) = 6 provides 2 triangles of either colour, so what does it take to force a construction with 2 same-coloured monochromatic triangles? As it turns out, all it takes is one more vertex.

4.1 Proposition 1: A 7-Vertex Graph with 2 Edge Colours Will Have at Least 2 Monochromatic Triangles of the Same Colour

(See Fig. 6).

In case 1, there cannot be 2 or more blue edges among vertices 2,3,4,5, as that will simply form 2 blue triangles. However, if there is only 1 or less blue edges among them, the other edges must be red, resulting in 2 red triangles formed. Case 2 is similar to case 1 but red is used. For case 1 and case 2 to both be avoided, every single vertex in these 7 must have exactly 3 red edges joined to

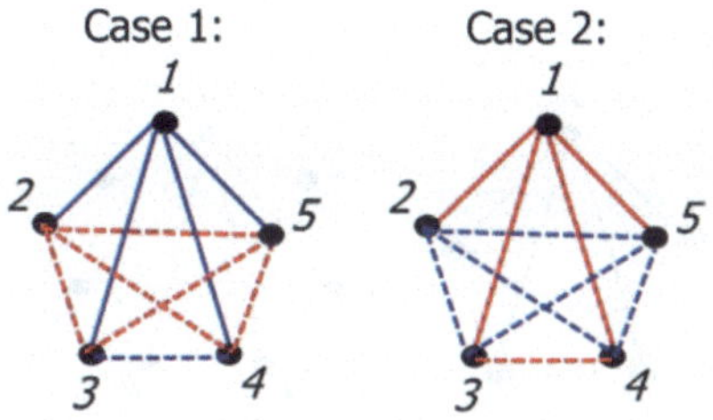

Fig. 6. These 2 cases will force 2 monochromatic triangles of the same colour to be formed

it. We show that this is not possible. Let the number of red edges joined to each vertex i be N_i. The sum of all N_1 to N_7 must be even, as every additional red edge in the graph adds 2 to the total (1 to each vertex of the edge). However, this sum would be 21 if we were to have 3 red edges to each vertex, which is odd, so this is not even possible. This proves that avoiding case 1 and case 2 in any 7-vertex graph with 2 edge colours is impossible, and therefore there will be at least 2 monochromatic triangles. [2] Before this, we first came up with a different proof, which will be in the **Appendix**.

5 R(3,4) and 10-Vertex Graphs

Now, what is $R(3,4)$, the minimum number of vertices such that any 2 colouring of the edges will have at least a triangle or a K_4 of the other colour?

5.1 Lemma 3.1: $R(3,4) = 9$

Firstly, we have to construct an 8-vertex graph without any blue triangle or red K_4. This disproves the statement that $R(3,4)<9$ (Fig. 7).

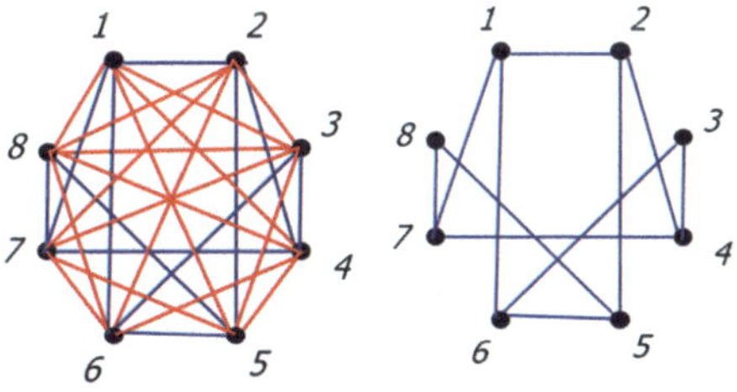

Fig. 7. Left is the full diagram; right is a simplified version without the red edges shown. (Color figure online)

Then, we try to construct a 9-vertex graph that avoids both a red K_4 and a blue triangle, and prove that it is not possible (Fig. 8).

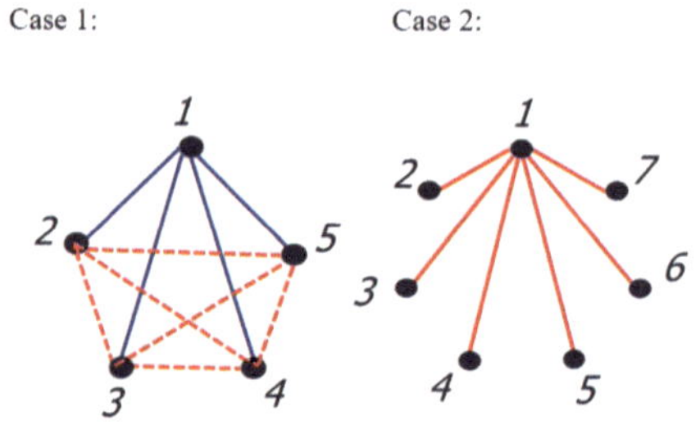

Fig. 8. These 2 cases must be avoided as case 1 will force a red K4 or blue triangle, and case 2 will do the same, as within vertices 2,3,4,5,6,7, there will be at least 1 (or 2 but this does not matter for our proof) monochromatic triangles of either colour. (Color figure online)

For case 1 and case 2 to both be avoided, every single 9 of these vertices must have exactly 5 red edges joined to it, if less, case 1 is fulfilled, if more, case 2 is fulfilled. We show that this is not possible.

Let the number of red edges joined to each vertex be $N_{vertex\ number}$. The sum of all N_1 to N_9 must be even as explained in Proposition 1. However, if we were to have 5 red edges to each vertex, the sum would be 45, which is odd, so trying to join exactly 5 red edges to every vertex would be impossible. This proves that either case 1 or case 2 will be fulfilled within any 9-vertex graph with 2 edge colours, and therefore a red K_4 or blue triangle is forced. This completes the proof that $R(3,4)=9$ [2] We also have a different method of proving that R(3,4) = 9, which will be in the **Appendix**.

5.2 Lemma 3.2

With 6 vertices, we have shown that there is not only at least 1, but 2 monochromatic triangles, while in 5 vertices we are able to avoid having any monochromatic triangle. But in 9 vertices, there are cases where there is only one of either the blue triangle or red K_4. Below are examples of these cases (Fig. 9).

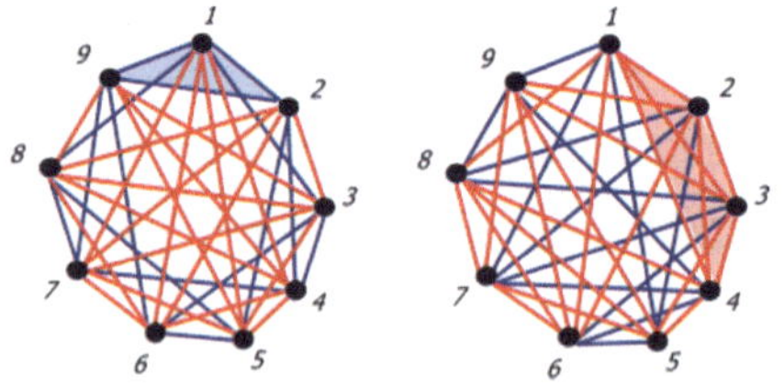

Fig. 9. Left: An example of a 9-vertex graph with only 1 blue K_3 and no red K_4. Right: An example of a 9-vertex graph with only 1 red K_4 and no blue K_3. (Color figure online)

On the left, vertices 1,2,3 form the blue K_3. Now we try to join 4,5,6,7,8,9 such that there will not be a red K_4. So we joined vertex 4 to 5, 6 to 7, and 8 to 9 using blue edges. Consider vertex 4. It has to be joined to the other vertices using at least 3 blue edges, if not 6 red edges are used and case 2 (see above) will be fulfilled. If 4 or more blue edges are used, case 1 will be fulfilled. So exactly 3 blue edges must be used. It cannot be joined to 1,2,3 with 2 or more blue edges, as that forms a second triangle. So vertex 4 will be additionally connected to vertex 7 with a blue edge. The same applies for 5,6,7,8,9, but we have to be careful not to form any more blue triangles. This drawing works but is not the only one that can exist.

On the right, vertices 1,2,3,4 are connected by red edges. This will be the (only) red K_4 here. Then, we try to draw edges between 5,6,7,8,9 such that we avoid any red K_4 with the minimal number of blue edges. Without loss of generality, edges 5-6 and 8-9 will be blue, while all the other edges among those 5 vertices are red. Now, try to connect vertex 5 with vertices 1,2,3,4. If we have 3 or more red edges used here, there will be a red K_4 as any set of 3 vertices in 1,2,3,4 form a red triangle already. However, if we have 1 or less red edge used here, we would have to use at least 3 blue edges. This creates a problem of trying to join vertex 6 with 1,2,3,4 with less than 3 red edges (meaning 2 or more blue edges), as a blue triangle will be formed, since 5-6 is already blue.

So, 2 red edges and 2 blue edges are used, and vertex 5 is joined to 1,2 with blue edges, and 3,4 with red edges. Vertex 6 will join to 1,2 with red edges, and 3,4 with blue edges, to avoid any blue triangles. This is also the reason we try to have as few as possible blue edges, as each vertex on these edges have to find a way to connect to 1,2,3,4 with at least 2 blue edges without connecting blue edges to the same vertex forming a blue triangle, and 1,2,3,4 are only 4 vertices.

Now, we consider vertex 8. Note that vertex 8 is connected to both vertices 5 and 6 with red edges, so it cannot connect to both 1,2 or both 3,4 with red edges, as 1-2 and 3-4 are also red, and this will form a second K_4. In this example, we connect vertex 8 to 1,4 with red edges, and 2,3 with blue edges. Vertex 9 now cannot connect to either of 2,3 with blue edges, so it will connect to 2,3 with red edges, and 1,4 with blue edges. To complete the diagram, vertex 7 is connected to all of 1,2,3,4 with blue edges.

5.3 Lemma 3.3: In 10 Vertices, at Least 1 Combination of the Two Structures, Red K_4 or Blue K_3 Will Be Formed

From the above, we saw how with 2 edge colours, 6 vertices must contain at least 2 monochromatic triangles. While in 9 vertices, there will be at least 1 red K4 or 1 blue triangle, and we have found cases where there is exactly one of these shapes. Some may wonder: "At what number of vertices can we guarantee at least 2 monochromatic shapes?" We will now prove that this lemma is always true. We will use 2 cases, 1 that starts with a red K_4 and 1 that starts with a blue triangle, as we already know that there must be at least one of these shapes in 10 vertices (Fig. 10).

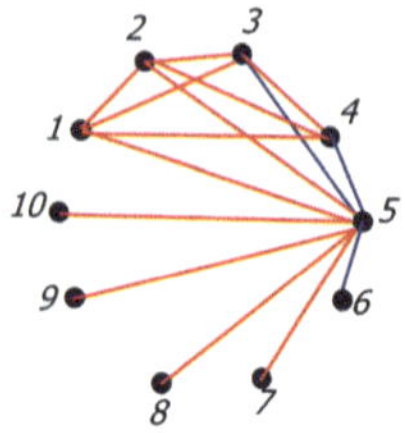

Fig. 10. In this case, we start from a red K_4, which will be 1234

Within the remaining 6 vertices, there must be a blue edge somewhere to try avoiding any more K_4. We assume this edge to connect vertices 5 and 6 without loss of generality. Vertex 5 must join to 1,2,3 and 4 using 2 or less red edges to avoid a second red K4. It also must use 2 or less blue edges in order to avoid case 1. So 2 red edges and 2 blue edges are used. To connect vertex 5 to vertices 7, 8, 9, 10, we cannot use any more blue edges as that fulfils case 1. If we use all red edges, case 2 will be fulfilled (Fig. 11).

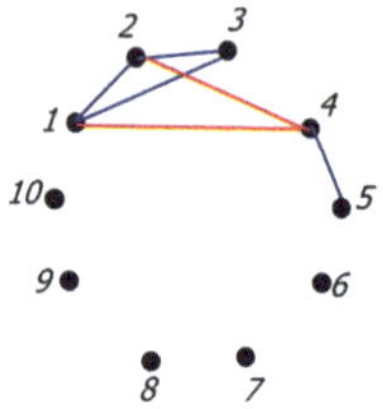

Fig. 11. In this case, we start with a blue triangle 123

Within the remaining 7 vertices, there must be at least 1 blue edge somewhere to try avoiding any red K4. We assume this blue edge joins vertex 4 and 5. Consider vertex 4. It still has to be joined to 8 other vertices (excluding 5) while trying to avoid both case 1 and case 2. After this, we have to use 2 more blue edges and 5 more red edges, which is not enough for 8 vertices, so this is not possible. If we fulfil case 2 and try to involve the triangle 123 to cause redundancy and avoid any more cliques, this is also not possible as case 2 will force not 1 but 2 cliques (as shown in Lemma 3.1). Therefore, there will be 2 blue triangles, 2 red K_4, or 1 of each forced in 10 vertices.

6 Conclusion

From the above, we have proved the Ramsey numbers for smaller cliques, such as $R(3,3) = 6$ and $R(3,4) = 9$. We have also discussed how multiple of these specified cliques can be guaranteed with enough vertices, such as 2 monochromatic triangles in a 6-vertex graph with 2 line colours, 2 same-coloured monochromatic triangles in a 7-vertex graph, and 2 of either a monochromatic triangle or a K_4 of the other colour in a 10-vertex graph. As further research, we believe more analysis could be done using graphs with more edge colours since we have only analysed those with 2 colours.

A Appendix

A.1 Alternative Method for Proposition 1

We will attempt to make a 7-vertex graph while trying to avoid 2 same coloured monochromatic triangle. Therefore, we will also try to avoid the 2 cases shown in the proposition. We start with a monochromatic triangle using the first 3 vertices. This triangle will be blue without loss of generality. For each of the remaining 4 vertices to connect to the first 3, we have a choice: Either connect using 1 blue edge (and 2 red edges) or connect using 3 red edges. Below shows what happens if you connect all of them using the first choice. Therefore, at least one of the vertices 4,5,6,7 has to connect to 1,2,3 using 3 red edges. Without loss of generality, we let that be vertex 4. Then, to connect vertex 4 to 5,6,7, it would have to use 3 blue edges, to avoid case 2. This will force 5,6,7 to form a red triangle, since none of the edges among them can be blue if we want to avoid a second blue triangle (Figs. 12, 13).

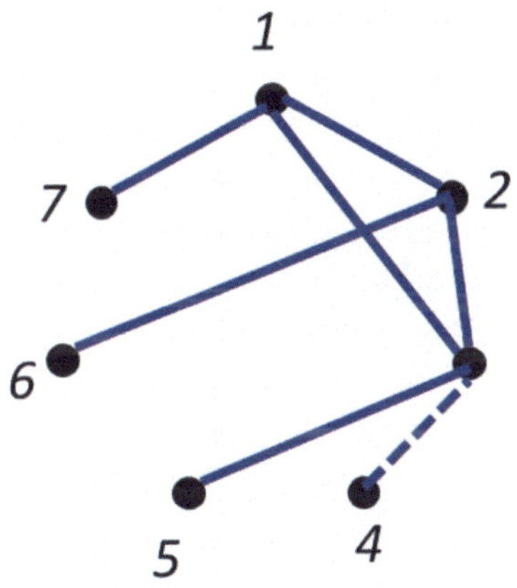

Fig. 12. In this case, vertex 4 joining to 1,2 or 3 with a blue edge would result in case 1 being fulfilled, which we do not want

Now, we try to connect vertex 7 back to 1,2,3. If it uses 2 or more red edges, we fulfil case 2, and if we use 2 or more blue edges, we get a second blue triangle. There is no way to avoid either scenario when there are 3 edges to be coloured. Since there is no way to construct a 7-vertex graph with 2 edge colours such that

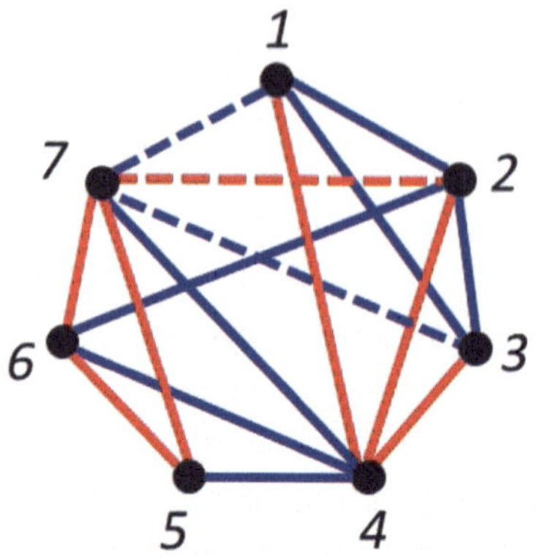

Fig. 13. This illustrates what might happen when we try to join vertex 7 back to vertices 1,2,3. In this case, since 7-2 is red and 7-3 is blue, there is no way to colour 7-1 such that case 2 and a second blue triangle are both avoided. (Color figure online)

2 monochromatic triangles of the same colour cannot be found, we have proven that a 7-vertex graph with 2 edge colours must have at least 2 same-coloured monochromatic triangles.

A.2 Alternative Initial Construction

(See Fig. 14).

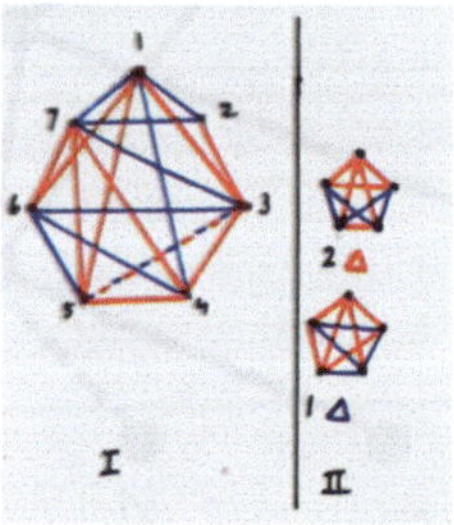

Fig. 14. Start with a blue K3 on vertices 1, 2, 7. To avoid either case in II, there must be 3 red and 3 blue edges branching off of vertex 7 and vertex 1. (Color figure online)

WLOG, the colours of the edges are as such in the figure (Look at all edges coming off vertices 1 and 7). However, vertex 7 and vertex 1 cannot connect to the same vertex in blue. Allow construction of red K_3 on vertices 4, 5 and 7. The rest can be worked out logically to attempt to avoid any K_3. However, that is not possible, and either coloured K_3 will be formed, giving 2 monochromatic coloured triangles of either colour.

A.3 Alternative Method for Lemma 3.1

The only way to avoid both cases shown in the lemma is by joining vertex 1 to the 8 other vertices with 5 red edges and 3 blue edges (Fig. 15).

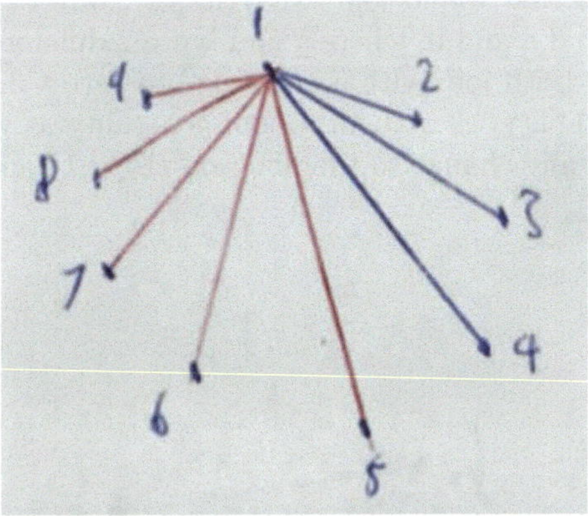

Fig. 15. Without loss of generality, vertex 1 is joined to vertices 2,3,4 with blue edges, and 5,6,7,8,9 with red edges. (Color figure online)

Vertices 2,3,4 will be forced to form a red triangle to avoid a blue one, and the vertices 5,6,7,8,9 have to connect like shown on the right to avoid any monochromatic triangle among them. This is also without loss of generality, as there is no other unique way to correctly join these 5 vertices other than essentially renumbering them (Fig. 16).

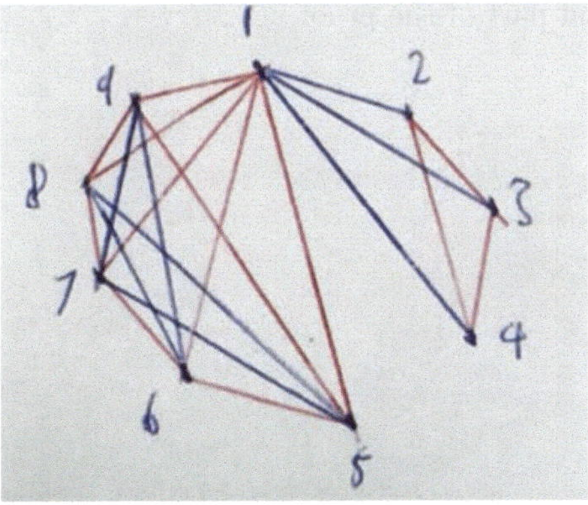

Fig. 16. .

We now consider how vertex 5 (or 6,7,8, or 9) can be joined to vertex 2,3,4. Joining all with 3 red edges will result in a red K_4. Joining with 2 blue edges or more fulfils case 1, as it is already connected to vertices 7,8 with blue edges. Therefore, it must use 2 red edges and 1 blue edge. There are only 3 possible ways to do this. As the same goes with vertices 6,7,8,9, there would be at least 2 vertices in 5,6,7,8,9 that would join to vertices 1,2,3 in the same way. Let the 2 vertices be numbered a and b, where a and b are 2 different integers from 6 to 9. Assume that they both join with a blue edge to vertex 3 and with red edges to vertices 2 and 4. This goes without loss of generality as vertices 2,3,4 can be renumbered without any change to the diagram (Figs. 17 and 18).

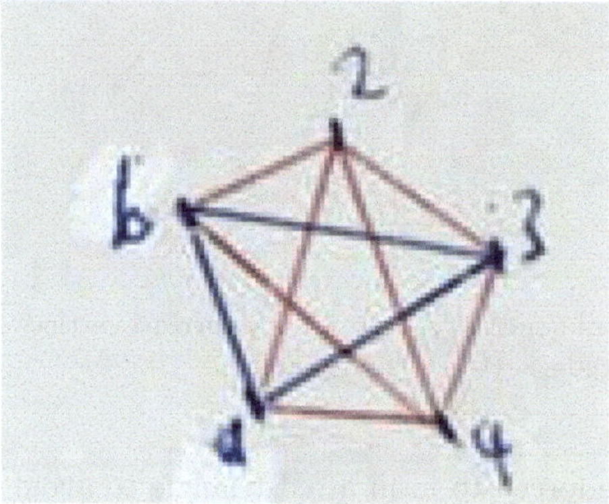

Fig. 17. If these 2 vertices are directly connected by a blue edge, there would be a blue triangle as they both also join to vertex 3 by blue edges

Therefore, it is not possible to have no red K_4 and no blue triangles in 9 vertices, completing the second part of the proof.

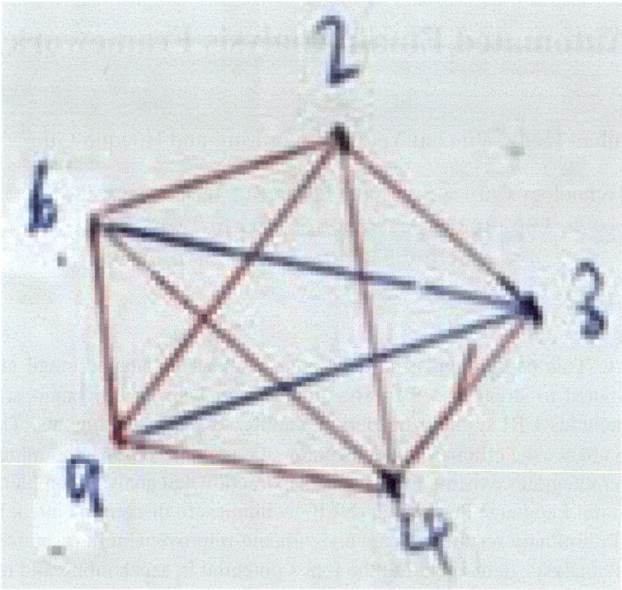

Fig. 18. If these 2 vertices are instead directly connected by a red edge, a red K_4 will be formed, as they each have another 2 red edges joining to vertices 2 and 4, and vertices 2 and 4 are also connected by a red edge. (Color figure online)

References

1. Plumpton, F.: On A Problem of Formal Logic (1930)

Automated Email Analysis Framework

Shibao Hong, Vincent Yeow Chong Lim, and Huaqun Guo(✉)

Infocomm Technology Cluster, Singapore Institute of Technology, Singapore, Singapore
1802963@sit.singaporetech.edu.sg,
huaqun.guo@singaporetech.edu.sg

Abstract. This paper details the development of an automated email analysis tool designed to streamline the extraction and analysis of key email artifacts, which includes URLs, email headers, IP addresses, and attachments. The tool aims to enhance the efficiency and accuracy of email analysis. Methodologically, this paper integrates existing tools for data extraction and analysis, supplemented with Natural Language Processing (NLP) techniques to discern the intent behind emails. Preliminary results indicate a significant improvement in response times for email analysis, demonstrating the tool's potential in accelerating and refining the process of email threat detection. This advancement not only contributes to the field of cybersecurity by providing a more efficient method for handling email-based threats but also sets a foundation for future enhancements in automated email analysis.

Keywords: automated email analysis · natural language processing · threat detection

1 Introduction

The industry is escalating reliance on digital communications, particularly email, as a cornerstone of business operations [1–4]. As such, the threat posed by email-based cyber-attacks [5–7] has dramatically increased over the years. There is a growing need for incident response team to efficiently manage the surge in reported emails, and the automated email analysis tool is designed to reduce the triage and analysis efforts required by these teams significantly.

This scope of our project is to design and develop an automated email analysis tool aimed at enhancing the efficiency of email triage and analysis procedures in an organization. As the complexity of modern cyber threats continues to rise, there is the challenge of extracting and accurately analysing intricate artefacts in emails such as URLs, email headers, IP addresses, domains, and attachments.

The goal of our project is to streamline the email analysis workflow by automating repetitive yet essential tasks. Examples of these include scanning attachments and URLs using reputable sources like VirusTotal [8], TrendMicro [9], Kaspersky [10], and Symantec [11] amongst many others, as well as analysing email headers and its authentication results to detect email spoofing. To enhance the interpretability and usefulness of the

H. Guo et al. (Eds.): IRC-SET 2024, PTT, pp. 454–472, 2025.
https://doi.org/10.1007/978-981-96-3770-6_46

analysis results, the integration of Natural Language Processing (NLP) methodologies will aid in transforming raw email contents into structured and meaningful information. With NLP, the tool will automatically identify and classify the intent of the email into categories such as phishing, extortion, threatening, and other form of malicious intent. This project also requires the need for the tool to operate in both online (internet-connected) and offline (non-internet connect) modes.

In the following Sect. 2, the architecture design is presented. Section 3 describes our planned tasks, and Sect. 4 reports the achieved outputs and deliverables. The comparisons and test cases are described in Sect. 5. Section 6 concludes the work.

2 Architecture Design

Figure 1 illustrates our architecture design.

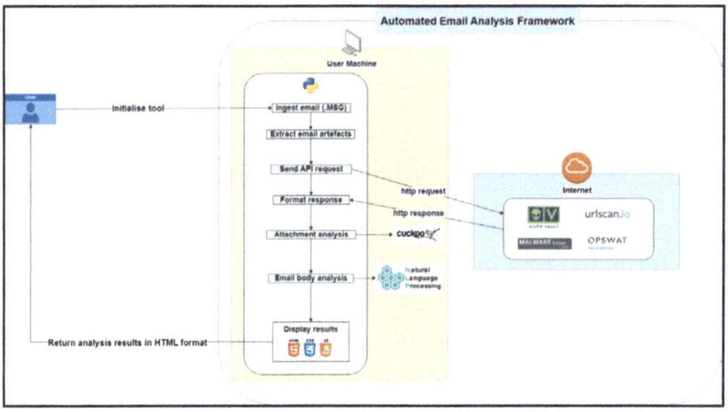

Fig. 1. Architecture design.

The architectural design for this the automated email analysis tool and framework was carefully chosen to ensure that the final product is lightweight, user-friendly, and easy to implement. Central to this design is the use of Python, a highly versatile and robust programming language, which serves as the foundation for programming various components of the tool.

This framework utilises Python libraries to perform email ingestion that is adept at efficiently processing incoming emails to prepare them for detailed analysis. Following this, the tool further uses these libraries to perform email artefact extraction. This process precisely extracts key artifacts from emails, such as headers, IP addresses, URLs, and attachments. These artifacts are crucial for a thorough and accurate analysis of the emails.

In addition to these analysis capabilities, the framework is designed to format and present the analysed data in an easily interpretable HTML format. This not only enhances the clarity and accessibility of the information but also allows for a more organised presentation of the analysis results.

Furthermore, the architecture uses API request to facilitate seamless interactions between the tool and various web-based analysis tools, as well as independent components like the Cuckoo sandbox for real-time attachment analysis and NLP algorithms for discerning the intent behind the email body content. This integration is key to providing a comprehensive analysis by leveraging external tools and resources effectively.

Overall, the architectural design decisions reflect a commitment to creating a tool that is not just functional and reliable but also intuitive and adaptable, catering to the dynamic needs of email analysis in today's digital landscape.

3 Planned Tasks

3.1 Developing Automated Email Analysis Tool

Creation of a robust, reliable, and functional automated email analysis tool. This tool will have the ability to extract and perform in-depth analysis both the email headers, body and attachments. This includes identifying URL links, IP addresses, domains in the email for potential red flags and conducting automated analysis of any attachments for harmful payloads. Once the analysis is complete, the tool will generate a report detailing the findings from the analysis and scanning and deliver a verdict based on the information gathered.

3.2 Developing the Tool to Be Capable of Function in Both Online and Offline Mode

In the most basic form, the tool in offline mode should provide users with the essential email analysis functions, such as extraction of email artefacts and evaluation of the email content. As a stretch goal, the online mode will potentially incorporate automated analysis with open-source cyber intelligence tools such as VirusTotal, TrendMicro, Kaspersky, and Symantec to elevate the level of scrutiny and provide real-time threat intelligence.

3.3 Preparation of Final Project Presentation Using Declassified Emails

This will highlight the specific methodologies employed, and the improvement in workflow efficiency such as the comparison of analysis time between manual and automated procedures from the result of implementing the automated email analysis tool.

4 Achieved Outputs and Deliverables

4.1 Conducted In-Dept Research into Email Analysis Procedures and Processes to Identify Malicious Intent

Conducted in-depth research into three comprehensive study revolving around email header analysis, identification of malicious email patterns and psychological tactics employed by threat actors.

4.2 Research on Structure and Semantics of Email Headers

Research into the structure and semantics of email headers explores the essential components of the header, such as the sender, recipient, return-path and more technical elements such as DKIM (DomainKeys Identified Mail), SPF (sender policy framework) and DMARC (Domain-based Message Authentication, Reporting & Conformance) which are crucial in preventing email spoofing. The aim of this research is to help pinpoint irregularities and signs of malicious activity.

4.3 Research on Malicious Email Patterns

Research into identifying malicious email patterns is critical into identifying tell-tale signs of phishing, extortion, and other forms of potentially malicious emails typically exhibits. These email categories were researched, and patterns were form based on the traits the certain emails exhibit, such as, spoofed emails, suspicious attachments, hidden or obfuscated URLs and its email body contents.

Research into psychological tactics employed by threat actors specifically aims at identifying the wording and sentence and scare tactics used in the email body contents. These tactics includes wording the content to create a sense of urgency or fear, impersonate trusted entities such as individuals or companies, and exploiting current events to deceive or trick recipients into replying or clicking on suspicious URLs and attachments.

4.4 Identified Viable Extraction Tools for Automated Email Analysis

Researched and identified a range of viable tools, plugin and libraries that could aid in the extraction of email artefacts. These artefacts include, sender, recipient, cc, bcc, IP addresses, domain names, URLs, and attachments.

This exploration took into the account the tool's extraction capabilities, such as extracting the relevant artefacts and its compatibility with the email format (.MSG). These tools were also selected based on their ease of integration into the overall system architecture of the automated email analysis workflow.

The tools tried are mail-parser [12], eml-parser [13], msg-parser [14], msg-extractor [15], and ioc-finder [16]. The tools that have been identified and used are msg-extractor, and ioc-finder.

4.5 Identified Viable Analysis Tools for Automated Email Analysis

Researched and identified a range of suitable services for the automated analysis portion of the tool. These include threat intelligence platforms that can analysis the extracted artefacts and was researched and utilised based on their reputability, efficiency and ease of use and integration.

The analysis tools that were researched and used are URLscan [17], MetaDefender Cloud [18], AlienVault Open Threat Exchange [19], MalwareBazaar [20], Cuckoo Sandbox [21]. The tools analysis tools that were research but not used due to rate limitations and incompatibility are VirusTotal, OpenPhish [22], Cymon [23], Symantec Sitereview [24], and AbuseIPDB [25].

4.6 Coded Segments of the Tool for Extraction of Email Artefacts

The main code segments for the extraction of email artefacts rely on the two chosen tools, msg-extractor and ioc-finder.

The code segment depicted in Fig. 2 leverages the capabilities of the 'msg-extractor' tool to extract the following artefacts from an email file (.MSG). The extracted artefacts are as follows - email header, sender's email address, recipient's email address, date, subject, cc, bcc, email body contents, email attachments. The function "extract_email_info" identifies all emails listed in a specified folder and extracts the artefacts as listed above. The extracted artefacts are stored in a nested dictionary and tied to their respective file name for ease of searching for use in the later segments of the code. The attachments are saved in a folder labelled with their respective filename.

```
def extract_email_info(EMAIL_FOLDER, EXTRACTED_EMAIL):
    """
    Locate all emails and extract its contents (headers, body, attachments)
    Attachment file naming convention stays the same
    """

    email_contents = {}
    """
    Initilise email contents dict
    """

    try:
        """
        Iterate through EMAIL_FOLDER path to find all emails to extract
        """
        list_email = os.listdir(EMAIL_FOLDER)
    except Exception as e:
        log.error("[ERR] - Unable to find emails in speicifed folder: {}".format(e))

    try:
        """
        Iterate through each found email and append EMAIL_FOLDER path to name of email (.msg) and extract contents
        Strips (.msg) extension for storing
        """
        for email in list_email:
            email_path = EMAIL_FOLDER + "\\" + email
            msg = extract_msg.Message(email_path)
            email = email[:len(email) - 4]
            email_contents[email] = {}
            attachment_list = []
            recipient_list = []
            recipient_email_list = []

            """
            Iterate through email recipients
            """
            for recipient in msg.recipients:
                recipient_list.append(recipient.formatted)
                recipient_email_list.append(recipient.email)
            """
            Iterate through found attachment in email and get filename
            """
            for attachment in msg.attachments:
                attachment_list.append(attachment.longFilename)

            """
            Extract email header and contents and attachments
            """
            email_contents[email]['Header'] = str(msg.header)
            email_contents[email]['Sender'] = str(msg.sender)
            email_contents[email]['Recipient'] = recipient_list
            email_contents[email]['RecipientEmail'] = recipient_email_list
            email_contents[email]['Date'] = str(msg.date)
            email_contents[email]['Subject'] = str(msg.subject)
            email_contents[email]['CC'] = str(msg.cc)
            email_contents[email]['BCC'] = str(msg.bcc)
            email_contents[email]['Body'] = str(msg.body)
            #email_contents[email]['Attachments'] = attachment_list

    except Exception as e:
        log.error("[ERR] - Unable to extract email contents: {}".format(e))

    """
    Return the data from email_contents dict
    """
    return email_contents
```

Fig. 2. Code snippet of the "extract_email_info" function.

The artefacts extracted are organised in a structured nested dictionary format for seamless data management and ease of reference. This nested dictionary design allows each artifact to be efficiently indexed with its corresponding email, optimising the search operations in subsequent stages of the code execution. The email attachments are extracted and stored in folders that are named in congruence with its original filename.

The code segment depicted in Fig. 3 leverages the capabilities of the 'ioc-finder' tool to extract potential Indicators of Compromise (IOC) from the previously identified and catalogued email artifacts. The various IOCs that are extracted are email addresses, IP addresses, domain names, URLs, and bitcoin addresses.

```
def extract_ioc(EMAIL_CONTENTS):
    """
    Placeholder
    """

    """
    Initilise iocs dict
    """
    iocs = {}

    try:
        """
        Identify IOC using extracted email contents from 'Header', 'Subject' and 'Body'
        - Email addresses
        - IP addresses
        - Domains
        - URLs
        - Bitcon addresses
        """
        for email, contents in EMAIL_CONTENTS.items():
            iocs[email] = {}
            buffer = ""

            """
            Iterate through email contents dict and identify IOCs using ioc_finder
            """
            for data in contents:
                if data == 'Header' or 'Subject' or 'Body':
                    buffer = buffer + str(contents[data])

            iocs[email]['email_address'] = ioc_finder.parse_email_addresses(buffer)
            iocs[email]['ipv4s'] = ioc_finder.parse_ipv4_addresses(buffer)
            iocs[email]['domains'] = ioc_finder.parse_domain_names(buffer)
            iocs[email]['urls'] = ioc_finder.parse_urls(buffer, parse_urls_without_scheme=False)
            iocs[email]['bitconAddress'] = ioc_finder.parse_bitcoin_addresses(buffer)

    except Exception as e:
        log.error("[ERR] - Extraction of IOCs has failed: {}".format(e))

    """
    Return identified ioc
    """
    return iocs
```

Fig. 3. Code snippet of the "extract_ioc" function.

The function "extract_ioc" will search through the nested dictionary from produced by the 'extract_email_info' function to identify any IOCs listed above and are similarly stored in the same nested dictionary format from 'extract_email_info' for ease of searching for later parts of the code segment. The function 'extract_ioc' has been coded to iterate through the nested dictionary generated by the 'extract_email_info' function. This function combs through each dictionary entry, identifying and capturing any potential IOCs as outlined above.

These identified IOCs are catalogued with the same nested dictionary structure established by the 'extract_email_info' function which will subsequently be used in later segments of the code to analysis the extracted IOCs.

4.7 Coded Segments of the Tool for Analysis of Email Artefacts

The analysis tools implemented are MetaDefender Cloud, AlienVault Open Threat Exchange, and URLscan.

1) *Implementation of MetaDefender Cloud*

MetaDefender Cloud, created by OPSWAT, is an all-encompassing security solution hosted in the cloud. Its standout feature is the multi-scanning technology, employing several antivirus engines to meticulously scan both files and URLs. This approach ensures a more comprehensive and dependable method for identifying malware and various other security threats.

Figure 4 displays a code snippet that utilizes the MetaDefender Cloud API to execute a GET request. This request encompasses sending the target URL for scanning and subsequently retrieving the analysis results.

```
def metadefender_api_scan(iocs_list):
    """
    MetaDefender Cloud Scanning API
    """

    """
    Initilise Metadefender iocs dict
    """
    metadefender_results = {}

    try:
        """
        Iterate through URLs
        """
        print("[JOB] {} - Starting MetaDefender...".format(datetime.datetime.now()))

        for email in iocs_list:
            metadefender_results[email] = {}
            """
            Initilise URL list for tracking and searching of results
            """
            url_list = []

            for url in iocs_list[email]["urls"]:

                if url.endswith(".jpg") or url.endswith(".png"):
                    """
                    Skip scanning urls with .jpg or .png
                    """

                else:
                    """
                    Submit scan request for url
                    """

                    meta_scan_url = "https://api.metadefender.com/v4/url/"
                    encoded_url = urllib.parse.quote(url, safe="")
                    url_to_scan = meta_scan_url + encoded_url
                    headers = {'apikey': os.getenv("METADEFENDER_API")}
                    response = requests.request("GET", url_to_scan, headers=headers)
                    results = response.json()

                    """
                    Extract sources with "status" of 1 and create a dictionary (1 = Malicious)
                    """
                    sources_dict = {}
                    for source in results["lookup_results"]["sources"]:
                        if source["status"] == 1:
                            sources_dict[source["provider"]] = source

                    metadefender_results[email][url] = sources_dict

            print("[INFO] {} - Urlscan completed successfully...".format(datetime.datetime.now()))

    except Exception as e:
        log.error("[ERR] {} - MetaDefender Scan API has failed: {}".format(datetime.datetime.now(),e))

    return metadefender_results
```

Fig. 4. Code snippet of the "metadefender_api_scan" function.

Figure 5 illustrates the data obtained from MetaDefender Cloud, highlighting the outcomes of the scans as determined by the multiple antivirus scanners.

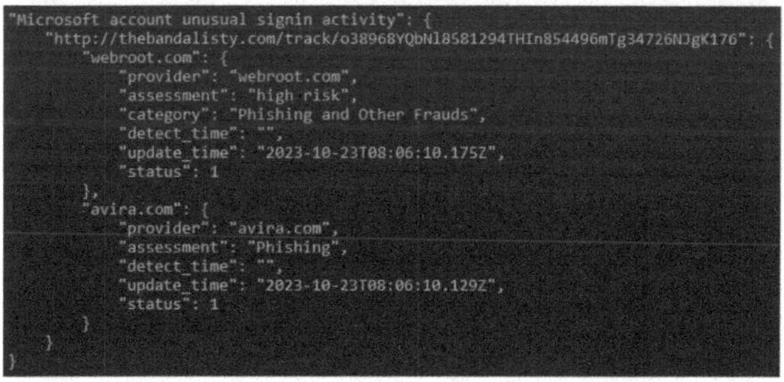

Fig. 5. Analysis results from MetaDefender Cloud.

2) *Implementation of AlienVault Open Threat Exchange*

AlienVault Open Threat Exchange (OTX) is a crowd-sourced threat intelligence platform that can be utilised for analysing URLs. Figure 6 displays the code that leverages the AlienVault API, designed for submitting a specific URL to be analysed. This code facilitates the search and retrieval of threat information associated with the input URL, utilizing AlienVault's extensive threat intelligence capabilities.

Figure 7 illustrates the results acquired from AlienVault following the submission of a URL for analysis. It highlights the 'pulse' information associated with the URL, a distinctive feature of AlienVault's Open Threat Exchange (OTX). This pulse data indicates that the URL is being monitored by specific user communities within OTX and has been identified as malicious. The pulse feature in AlienVault OTX is known for aggregating and sharing threat intelligence from a global network of contributors, offering insights into the reputation and threat level of URLs as perceived and reported by these communities.

3) *Implementation of URLscan*

URLscan is an online analysis tool tailored for comprehensive website analysis. Its core functionality centres around detecting potential malicious activities, which include phishing, malware and many others. The tool provides an extensive evaluation of a website, encompassing network request tracking and in-depth domain and IP analysis. This robust examination assists in identifying the potential threat of a website. Integration of URLscan is achieved through its dedicated API which can be implemented using Python.

```
def otx_api_scan(iocs_list):
    """
    AlienVault OTX Scanning API
    """

    """
    Initilise AlienVault OTX iocs dict
    """
    otx_results = {}

    try:
        """
        Iterate through URLs
        """
        print("[XXX] {} - Starting AlienVault OTX...".format(datetime.datetime.now()))

        """
        Connect to AlienVault OTX
        """
        OTX_SERVER = 'https://otx.alienvault.com/'
        otx = OTXv2(os.getenv("OTX_API"), server=OTX_SERVER)

        for email in iocs_list:
            otx_results[email] = {}
            """
            Initilise URL list for tracking and searching of results
            """
            url_list = []

            for url in iocs_list[email]["urls"]:

                if url.endswith(".jpg") or url.endswith(".png"):
                    """
                    Skip scanning urls with .jpg or .png
                    """

                else:
                    """
                    Submit scan request for url
                    """

                    alerts = []
                    result = otx.get_indicator_details_full(IndicatorTypes.URL, url)
                    pulses = getValue( result, ['general', 'pulse_info', 'count'])
                    if pulses > 0:
                        url_pulse = getValue( result, ['general', 'pulse_info', 'pulses'])
                        if url_pulse:
                            for pulse in url_pulse:
                                if 'name' in pulse:
                                    alerts.append('URL pulse info: ' + pulse['name'] + ' ' + 'Tags:' + ','.join(pulse['tags']))

                    google = getValue( result, ['url_list', 'url_list', 'result', 'safebrowsing'])
                    if google and 'response_code' in str(google):
                        alerts.append({'google_safebrowsing': google})
                        print(str(alerts))

                    clamav = getValue( result, ['url_list', 'url_list', 'result', 'multiav','matches','clamav'])
                    if clamav:
                            alerts.append({'clamav': clamav})
                            print(str(alerts))

                    avast = getValue( result, ['url_list', 'url_list', 'result', 'multiav','matches','avast'])
                    if avast:
                        alerts.append({'avast': avast})
                        print(str(alerts))

                    # Get the file analysis too, if it exists
                    has_analysis = getValue( result,  ['url_list','url_list', 'result', 'urlworker', 'has_file_analysis'])
                    if has_analysis:
                        hash = getValue( result,  ['url_list','url_list', 'result', 'urlworker', 'sha256'])
                        file_alerts = file(otx, hash)
                        if file_alerts:
                            for alert in file_alerts:
                                alerts.append(alert)


                    print(str(alerts))
                    otx_results[email][url] = alerts

            print("[INFO] {} - AlienVault OTX completed successfully...".format(datetime.datetime.now()))


    except Exception as e:
        log.error("[ERR] {} - AlienVault OTX Scan API has failed: {}".format(datetime.datetime.now(),e))

    return otx_results
```

Fig. 6. Code snippet of the "otx_api_scan" function.

```
"Microsoft account unusual signin activity": {
    "http://thebandalisty.com/track/o38968YQbNl8581294THIn854496mTg34726NJgK176": [
        "URL pulse info: SMBC Phishing Collection Tags:urls,phishing,scam,smbc"
    ]
}
```

Fig. 7. Results from AlienVault.

Figure 8 illustrates the code that utilises the API to dispatch a POST request to URLscan, targeting a specific URL for analysis.

```python
def urlscan_api_scan(iocs_list):
    """
    Urlscan.io Scanning API
    """

    """
    Initilise Urlscan iocs dict
    """
    urlscan_iocs = {}

    try:
        """
        Iterate through URLs
        """
        print("[JOB] {} - Starting Urlscan...".format(datetime.datetime.now()))

        for email in iocs_list:
            urlscan_iocs[email] = {}
            """
            Initilise URL list and UUID list for tracking and searching of results
            """
            url_list = []
            uuid_list = []

            for url in iocs_list[email]["urls"]:

                if url.endswith(".jpg") or url.endswith(".png"):
                    """
                    Skip scanning urls with .jpg or .png
                    """

                else:
                    """
                    Submit scan request for url
                    """

                    headers = {'API-Key': os.getenv("URLSCAN_API"), 'Content-Type':'application/json'}
                    data = {"url": url, "visibility": "unlisted"}
                    response = requests.post('https://urlscan.io/api/v1/scan/',headers=headers, data=json.dumps(data))
                    results = response.json()

                    if results["message"] == "Submission successful":
                        url_list.append(url)
                        uuid_list.append(results["uuid"])

            urlscan_iocs[email]["urls"] = url_list
            urlscan_iocs[email]["uuid"] = uuid_list
            print("[INFO] {} - Urlscan completed successfully...".format(datetime.datetime.now()))

    except Exception as e:
        log.error("[ERR] {} - Urlscan Scan API has failed: {}".format(datetime.datetime.now(),e))


    """
    Return identified ioc
    """
    return urlscan_iocs
```

Fig. 8. Code snippet of the "urlscan_api_scan" function.

Additionally, the code includes a verification process to confirm the successful execution of the request. The API also respond with a UUID that is tied to the URL sent for analysis so that the results can be retrieved when the analysis has completed.

Figure 9 presents the code segment responsible for fetching the analysis results of the specified URL. It further processes these results, extracting key information critical for determining whether the URL poses any malicious threats.

The key information from the URLScan analysis results is depicted in Fig. 10, such as – IP address, geolocation, server used, ASN information, domains contacted, URLs contacted, certificate information, verdict score, threat category, targeted brand, threat tag.

4) *Implementation of MalwareBazaar*

MalwareBazaar is a web-based analysis tool that houses a comprehensive repository of malware samples. In the context of our automated email analysis framework, MalwareBazaar is utilised for the analysis of email attachments, in a manner that prioritises security and confidentiality. Instead of submitting attachments directly, which might risk exposing sensitive information, the framework only submits hashes of the attachment to MalwareBazaar. This approach enables the identification of known malware through

```python
def urlscan_api_result(uuid_list):
    """
    Urlscan.io Results API
    """

    """
    Initilise urlscan results dict
    """
    urlscan_results = {}

    try:
        """
        Iterate through uuids
        """
        print("[JOB] {} - Retrieving scan results from Urlscan...".format(datetime.datetime.now()))

        for email in uuid_list:
            urlscan_results[email] = {}

            """
            Initilise URL list and UUID list for tracking and searching of results
            """
            detected = []

            for uuid in uuid_list[email]["uuid"]:

                ready_counter = 0

                """
                wait till scan finished
                """
                while ready_counter == 0:
                    headers = {
                        "Content-Type": "application/json",
                        "API-Key": os.getenv("URLSCAN_API")
                    }

                    base_url = f"https://urlscan.io/api/v1/result/{uuid}/"
                    response = requests.get(base_url, headers=headers)

                    if response.status_code == 200:
                        ready_counter = 1
                    else:
                        time.sleep(5)
                        ready_counter = 0

                if ready_counter:
                    result = response.json()

                    """
                    Check if URL was accessable
                    """
                    loaded = result["data"]["requests"][0]["response"]["dataLength"]
                    if loaded > 0:
                        urlscan_results[email]["successful_scan"] = 1
                    else:
                        urlscan_results[email]["successful_scan"] = 0
                    """
                    Gather scanned information: https://urlscan.io/docs/result/ (Description of each field)
                    """

                    """ Information on scanned URL"""
                    urlscan_results[email]["page_url"] = result["page"]["url"]
                    urlscan_results[email]["page_domain"] = result["page"]["domain"]

                    keys_to_check = ['ip', 'country', 'city', 'server', 'asn', 'asnname']

                    for key in keys_to_check:
                        if key in result["page"]:
                            urlscan_results[email]["page_" + key] = result["page"][key]
                        else:
                            urlscan_results[email]["page_" + key] = ""

                    """ Information on IP/Hostnames/URLs contacted, and Certs """
                    keys_to_check = ['ips', 'countries', 'domains', 'urls', 'linkDomains', 'certificates']

                    for key in keys_to_check:
                        if key in result["lists"]:
                            urlscan_results[email]["lists_" + key] = result["lists"][key]
                        else:
                            urlscan_results[email]["lists_" + key] = ""

                    """ Information on Urlscan verdict and score"""
                    urlscan_results[email]["urlscan.score"] = result["verdicts"]["overall"]["score"]
                    urlscan_results[email]["urlscan.categories"] = result["verdicts"]["overall"]["categories"]
                    urlscan_results[email]["urlscan.brands"] = result["verdicts"]["overall"]["brands"]
                    urlscan_results[email]["urlscan.tags"] = result["verdicts"]["overall"]["tags"]
                    urlscan_results[email]["urlscan.malicious"] = result["verdicts"]["overall"]["malicious"]
                    print("[INFO] {} - Urlscan results retrieved successfully...".format(datetime.datetime.now()))

    except Exception as e:
        log.error("[ERR] {} - Urlscan Results API has failed: {}".format(datetime.datetime.now(),e))

    return urlscan_results
```

Fig. 9. Code snippet of the "urlscan_api_result" function.

hash matching against MalwareBazaar's extensive database. The integration is achieved through MalwareBazaar's dedicated API.

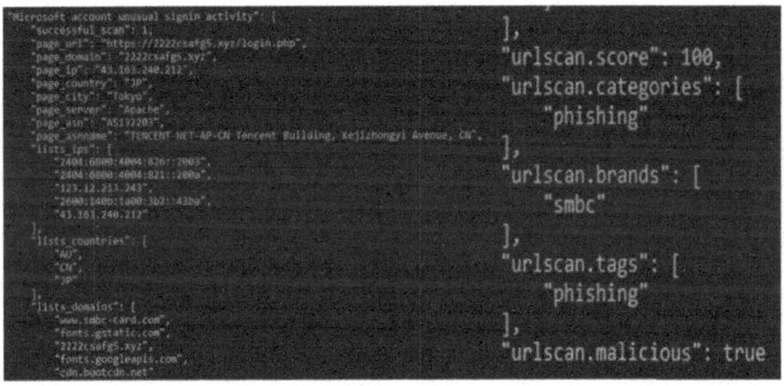

Fig. 10. Analysis results from URLscan

Figure 11 presents the segment of code responsible for interfacing with the MalwareBazaar API, the process for submitting hashes derived from the email attachments. This facilitates the hash matching operation against MalwareBazaar's extensive malware repository. The response from MalwareBazaar encompass critical metadata about the analysed attachment, including data such as file type, size and associated hashes. Significantly, it also provides intelligence from various security vendors alongside a definitive verdict regarding the hash's association with known malware entities.

Fig. 11. Code snippet of the "malwarebazaar_api_scan" function.

Figure 12 presents a portion of the analysis results from the various different security vendors that have conducted evaluations on the file correlated with the submitted hash. The diversity in the analytical perspectives offered by the multiple vendors attributes to a more comprehensive and definitive conclusion.

5) *Implementation of Text Summariser using NLP*

The utilisation of NLP and the Natural Language Toolkit (NLTK) is pivotal in the summarisation of email body content, offering an approach to distil and highlight essential information from extensive text content. NLP techniques enable the identification

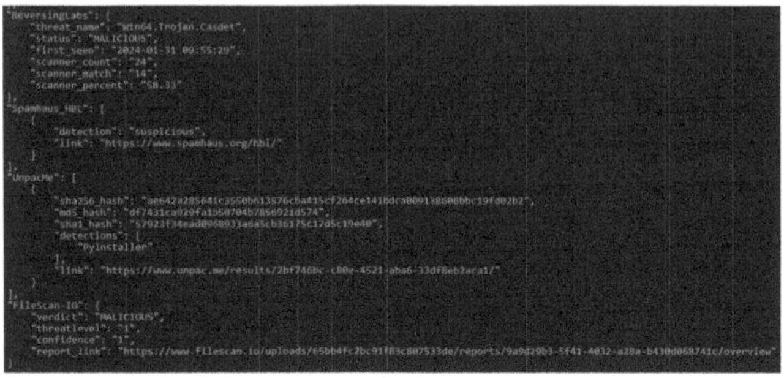

Fig. 12. Analysis results from MalwareBazaar.

and understanding of key themes, topics, and sentiments within the text, facilitating the extraction of relevant information. By leveraging NLTK, which is equipped with advanced NLP tools and algorithms, it will help aid in efficient summarisation processes.

The challenge lies not in the incorporation of NLP but doing so effectively, which involves careful selection of appropriate algorithms, rigorous training of the model, and optimisation to ensure its performance aligns with our project objectives. In the development of the tool, a key challenge has been ensuring the tool remains lightweight and efficient. This consideration is crucial for maintaining the tool's usability in a normal analyst operational environment, especially when computing power may be limited. Heavy models can lead to inefficiency, potentially hindering the tool's effectiveness in response times, as compared to manual analysis efforts.

To address this challenge while still utilising the benefits of NLP, NLTK has been strategically incorporated into the tool. NLTK has a comprehensive suite of libraries and tools designed for language processing, offering a balance between functionality and performance. It provides essential NLP capabilities such as tokenization, stemming, and classification, which are integral for analysing and summarising email contents, without the need for more complex models. The lightweight nature of NLTK, combined with its NLP features, makes it an ideal choice for our tool, striking the right balance between sophistication and resource efficiency.

A summarized email body content is produced using various algorithms and tools such as frequency mapping and sentences scoring. The function presented in Fig. 13 constructs a dictionary mapping to each significant word to its frequency of occurrence within a given text. This is done by tokenizing the input text into individual words and then applying stemming, which reduce words to their base or root forms. e.g. "running" and "runs" will be reduced to "run". The function filters out "stopwords" that are irrelevant to the meaning of the text such as "the", "is", and "in". This filtering process is crucial to ensure the frequency table reflects the prevalence of only meaningful words, which are more likely to contribute to the core themes and topics of the text.

The function presented in Fig. 14 assigns a numerical value to each sentence, reflecting its importance based on the words it contains. This scoring is rooted in the frequency table mention earlier. For each sentence, the function calculates a score by totaling the frequencies of its constituent words, after converting them to their stemmed forms to match the frequency table. The score is then normalized by the number of unique, meaningful words in the sentence to mitigate bias towards longer sentences. This normalization ensures that the significance of a sentence is judged by the relevance of its content rather than its length.

```python
def _create_frequency_table(text_string) -> dict:
    """
    we create a dictionary for the word frequency table.
    For this, we should only use the words that are not part of the stopWords array.

    Removing stop words and making frequency table
    Stemmer - an algorithm to bring words to its root word.
    :rtype: dict
    """
    stopWords = set(stopwords.words("english"))
    words = word_tokenize(text_string)
    ps = PorterStemmer()

    freqTable = dict()
    for word in words:
        word = ps.stem(word)
        if word in stopWords:
            continue
        if word in freqTable:
            freqTable[word] += 1
        else:
            freqTable[word] = 1

    return freqTable
```

Fig. 13. Frequency Table Function.

```python
def _score_sentences(sentences, freqTable) -> dict:
    """
    score a sentence by its words
    Basic algorithm: adding the frequency of every non-stop word in a sentence divided by total no of words in a sentence.
    :rtype: dict
    """

    sentenceValue = dict()

    for sentence in sentences:
        word_count_in_sentence = (len(word_tokenize(sentence)))
        word_count_in_sentence_except_stop_words = 0
        for wordValue in freqTable:
            if wordValue in sentence.lower():
                word_count_in_sentence_except_stop_words += 1
                if sentence[:10] in sentenceValue:
                    sentenceValue[sentence[:10]] += freqTable[wordValue]
                else:
                    sentenceValue[sentence[:10]] = freqTable[wordValue]

        if sentence[:10] in sentenceValue:
            sentenceValue[sentence[:10]] = sentenceValue[sentence[:10]] / word_count_in_sentence_except_stop_words
```

Fig. 14. Sentence Scoring Function.

The resultant score shown in Fig. 15 provides a quantitative basis for evaluating which sentences are the most critical in information in the text given. Using these scores, the summarization process can effectively distil the text down to its most substantive elements, thus generating a summary that captures the essence of the original content while discarding extraneous details.

6) *Implementation of Report Generation using HTML and CSS*

The utilisation of HTML and CSS to present the analysis reports of emails from the various web-based analysis platforms used offers a multitude of advantages, particularly when dealing with high volumes of analysed emails. HTML provides a structured format for embedding and organising the diverse types of contents such as text summaries, links

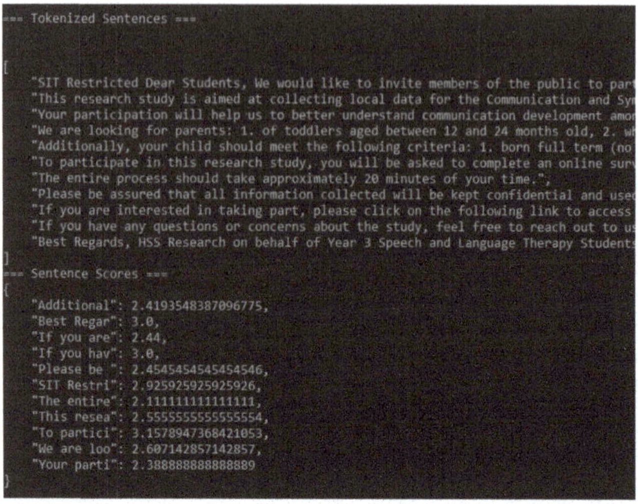

Fig. 15. Tokenized Sentence and Sentence Score Table.

and verdicts with the use of interactive elements such as buttons or dropdowns. Figure 16 is a screenshot of generated report.

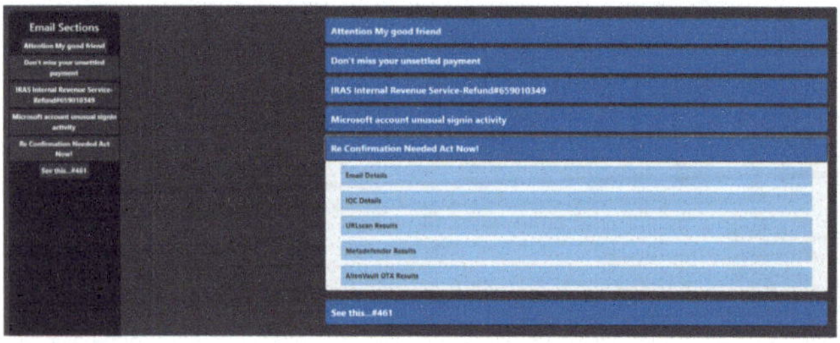

Fig. 16. Generated Report.

The report includes a side navigation panel, allow access to individual email analyses identified by their email title. Within each email section, interactive buttons are provided, encapsulating key information such as email specifics, IOCs, and results from the utilised web-based analysis platforms.

Upon clicking, each interactive button reveals the corresponding information as labelled, utilising boxed layouts to ensure clarity and ease of reading across the multiple segments (Fig. 17). This interactive capability is valuable in contexts where analysts need quickly sift through vast amounts of information to identify key insights, making HTML and CSS an ideal choice for presenting complex data in an accessible and user-friendly manner.

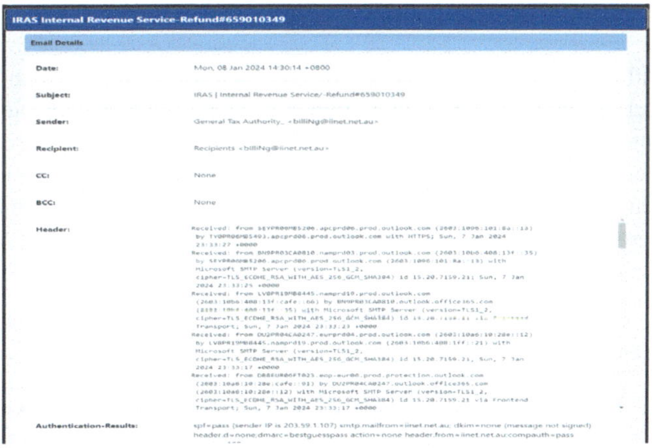

Fig. 17. Generated Report contents.

5 Comparisons and Test Cases

The goal of the automated email analysis framework for the organization is to streamline and enhance the email analysis process that is currently in place. While numerous tools and services exist for processes such as scanning, tracking email behaviour and identifying email threats such as phishing, malware and many others, they often operate as Software as a Service (SaaS) at the gateway level. The framework discussed in this report focuses specifically on email analysis. As such the tools that will be discussed in this section will be compared based on their ability to improve operational efficiency and its integration into the organization's existing workflow for handling reported emails. The tools that will be discussed in this comparison are Avanan [26], Barracuda Email Security Gateway [27], MailScanner [28] and ThePhish [29].

5.1 Operational Efficiency

Due to the nature of the workflow of handling reported emails, the tool should offer quick and effective functions to help aid analyst in conducting email analysis. Avanan, Barracuda Email Security Gateway and MailScanner are tools that are offered as SaaS which needs to be implemented on the email gateway. This constriction will prove difficult to implement and use by analyst. However, these tools do provide useful features that will help aid in the email analysis process, such as being able to detect malicious content in emails. The architecture of our tool has been crafted in a way that is easy to setup while offering similar capabilities to aid in the email analysis process while maintaining a minimal resource usage.

ThePhish on the other hand is an automated phishing email analysis tool. ThePhish is comprised of various architectural design elements and features that are similar to our automated email analysis framework, however, ThePhish is only designed to analyse phishing emails, and lack the capabilities to detect other types of malicious email that the automated email analysis framework addresses.

5.2 Integration

Integration with the current workflow is an important aspect that is focused on when creating the automated email analysis framework. The tool should remain lightweight in terms of integration and resource usage. Tools such as Avanan, Barracuda Email Security Gateway, MailScanner and ThePhish relies its integration on various resource intensive modules and complex setup procedures. The design of our automated email analysis framework relies on minimal modules and setup procedures while still keeping its efficiency in email analysis.

5.3 Test Cases

We will conduct 4 test cases, each involving a different set of emails, to evaluate how effectively the tool streamlines the email analysis process compared to traditional manual methods. These tests aim to demonstrate the tool's ability to reduce the time analysts spend on email analysis.

The first test case consists of 10 marketing emails, with a total of 45 URLs in total. The test shows that the manual email analysis process took 11 min and 31 s, while the automated email analysis tool took 1 min and 32 s (Fig. 18).

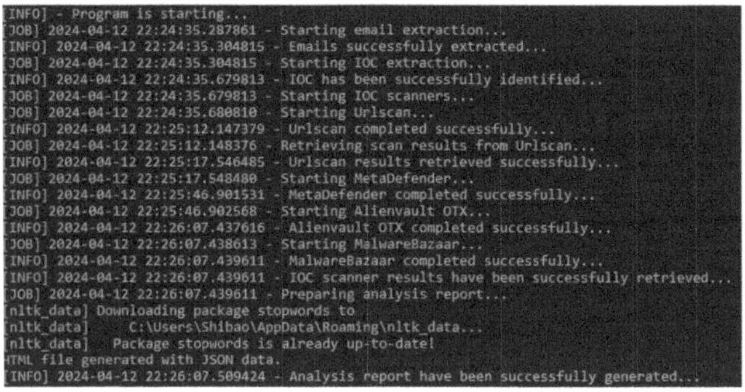

Fig. 18. Test Case 1 Results with automated email analysis tool.

The second test case consist of 10 phishing emails, with a total of 10 URLs in total. The third and fourth test case consist of a general pool of 10 emails such as phishing, fraudulent, marketing and malware emails. The results are listed in Table 1.

Table 1. Test Results

Dataset	Response Time by Manual email analysis	Response Time by Automated email analysis tool
1st Dataset	11 min and 31 s	1 min and 32 s
2nd Dataset	2 min and 13 s	56 s
3rd Dataset	3 min and 18 s	1 min and 9 s
4th Dataset	3 min and 9 s	1 min and 4 s

6 Conclusions

This paper presents the details of design and development of an automated email analysis tool. Our work integrates existing tools, supplemented with NLP techniques to streamline the extraction and analysis of key email artifacts, including URLs, email headers, IP addresses, and attachments. The comparisons and preliminary test results show that a significant improvement in response times for email analysis, demonstrating the tool's potential in accelerating and refining the process of email threat detection. This advancement not only contributes to the field of cybersecurity by providing a more efficient method for handling email-based threats but also sets a foundation for future enhancements in automated email analysis in the future work.

Acknowledgement. This research project is supported by an ignition grant with Grant No. R-IE2-A405-0002, Singapore Institute of Technology and Ministry of Education, Singapore.

References

1. Afridi, K., Ali Turi, J., Zaufishan, B.: Impact of digital communications on project efficiency through ease of use and top management support. Heliyon **9**(7) (2023)
2. DeFilippis, E., Michael Impink, S., Singell, M., Polzer, J.T., Sadun, R.: The impact of COVID-19 on digital communication patterns. Humanit. Soc. Sci. Commun. **9**(180) (2022)
3. Steinmaurer, A., Bajramovic, A., Pollhammer, D., Gütl, C.: Learning security awareness in email communication using a platform for digital skill teaching. In: 2022 IEEE International Conference on Teaching, Assessment and Learning for Engineering (TALE), Hong Kong, 2022
4. Mujtaba, G., Shuib, L., Raj, R.G., Majeed, N., Al-Garadi, M.A.: Email classification research trends: review and open issues. IEEE Access **5**, 9044–9064 (2017)
5. Ramprasath, J., Priyanka, S., Manudev, R., Goku, M.: Identification and mitigation of phishing email attacks using deep learning. In: 3rd International Conference on Advance Computing and Innovative Technologies in Engineering (ICACITE), Greater Noida, India, 2023
6. Zeng, Y.G.: Identifying email threats using predictive analysis. In: 2017 International Conference on Cyber Security and Protection of Digital Services (Cyber Security), London, UK, 2017
7. Yu, B., Li, P., Liu, J., Zhou, Z., Han, Y., Li, Z.: Advanced analysis of email sender spoofing attack and related security problems. In: 2022 IEEE 9th International Conference on Cyber Security and Cloud Computing (CSCloud)/2022 IEEE 8th International Conference on Edge Computing and Scalable Cloud (EdgeCom), Xi'an, China, 2022

8. VirusTotal, "VirusTotal". https://www.virustotal.com/. Accessed 15 Jan 2024
9. Trend Micro. https://www.trendmicro.com/. Accessed 10 Jan 2024
10. kaspersky. https://www.kaspersky.com/. Accessed 10 Jan 2024
11. Broadcom, "Symantec Enterprise Cloud". https://www.broadcom.com/products/cybersecurity. Accessed 15 Jan 2024
12. Mantuano, F.: "SpamScope/mail-parser," GitHub. https://github.com/SpamScope/mail-parser. Accessed 10 Jan 2024
13. Toth, G., et al.: "GOVCERT-LU / eml_parser," GitHub. cccs-rs. https://github.com/GOVCERT-LU/eml_parser. Accessed 10 Jan 2024
14. Puyp-bot, vikramarsid, A. R. Tahri and DayDotMe, "vikramarsid/msg_parser". https://github.com/vikramarsid/msg_parser. Accessed 10 Jan 2024
15. Peterson, D., et al.: "TeamMsgExtractor/msg-extractor," GitHub. https://github.com/TeamMsgExtractor/msg-extractor. Accessed 10 Jan 2024
16. f. Floyd, nor3th, wesinator, Frikky and P. McGuire, "fhightower/ioc-finder," GitHub. https://github.com/fhightower/ioc-finder. Accessed 10 Jan 2024
17. "URLScan". https://urlscan.io/. Accessed 15 Jan 2024
18. Opswat, "MetaDefender Cloud". https://metadefender.opswat.com/. Accessed 16 Jan 2024
19. AlienVault, "AlienVault - Open Threat Exchange". https://otx.alienvault.com/. Accessed 2 Jan 2024
20. abuse.ch, "MalwareBazaar". https://bazaar.abuse.ch/. Accessed 2 Jan 2024
21. Cuckoo, "What is Cuckoo?" https://cuckoo.readthedocs.io/en/latest/introduction/what/. Accessed 10 Jan 2024
22. OpenPhish. https://openphish.com/. Accessed 15 Jan 2024
23. cyjax, "CYMON: Tailored Cyber Threat Intelligence for Real-Time Protection,". https://www.cyjax.com/cymon/. Accessed 1 Feb 2024
24. Broadcom, "Symantec Sitereview,". https://sitereview.symantec.com/. Accessed 1 Feb 2024
25. Abuseipdb, "AbuseIPDB making the internet safer, one IP at a time,". https://www.abuseipdb.com/. Accessed 1 Feb 2024
26. Avanan, "The Best Way to Protect Enterprise Email & Collaboration from ...,". https://www.avanan.com/. Accessed 2 Jan 2024
27. barracuda, "Barracuda Email Security Gateway,". https://www.barracuda.com/products/email-protection/email-security-gateway. Accessed 2 Jan 2024
28. MailScanner, "The #1 Open Source Email Filter". https://www.mailscanner.info/. Accessed 2 Jan 2024
29. Galdi, E.: ThePhish: an automated phishing email analysis tool, GitHub. https://github.com/emalderson/ThePhish. Accessed 2 Jan 2024

An Effective Mitigation Scheme to Defend Against Ransomware Attacks

Anurag Prabhakar, Gladys Ling Hui Chua, James Jun Heng Ong, Yun Zhen Toh, Huaqun Guo(✉), and Liming Lu

Information and Communications Technology (Information Security), Singapore Institute of Technology, Singapore, 1 Punggol Coast Road, Singapore 828608, Singapore
{2200562,2201906,2203208,2202044}@sit.singaporetech.edu.sg, {huaqun.guo,liming.lu}@singaporetech.edu.sg

Abstract. Ransomware has evolved from its inception with PC Cyborg, which encrypted files and demanded a ransom, to sophisticated threats employing advanced encryption and evasion techniques. This evolution has shifted the target from individual users to larger organisations, creating a multi-billion dollar industry. The exploitation of system vulnerabilities by ransomware highlights the importance of maintaining an up-to-date Common Vulnerabilities and Exposures (CVE) database. The project focuses on identifying ransomware infection factors, evaluating attack vectors, and implementing preventive measures. It also involves developing a local CVE database to filter for ransomware-related CVEs and an automated checklist to assess the presence of specific antiviruses, firewalls, and installed software versions. These efforts are designed to enhance the organisation's ability to detect and mitigate ransomware threats effectively.

Keywords: Ransomware · Mitigation · Common Vulnerabilities and Exposures (CVEs)

1 Introduction

Ransomware has become an evolving multi-billion-dollar industry, predicted to exceed $265 billion (USD) by 2031 [1]. This lucrative industry preys on victims who are unprotected and weak with little to no cybersecurity knowledge [2]. In today's world, technology serves as a bridge to connect organisations and individuals, creating an interconnected environment that offers both convenience and heightened risk. Cybercriminals have found sophisticated ways to get their fair share in this lucrative industry by exploiting vulnerabilities in systems. This leads to significant financial losses and operational disruptions. For example, the notorious WannaCry ransomware attack that happened back in 2017, has affected more than 200,000 computers worldwide and cost an estimated $8 billion in a span of four days [3].

Common Vulnerabilities and Exposures (CVEs) play a critical role in understanding and mitigating ransomware threats. CVEs are standardised identifiers for known vulnerabilities in software and hardware, providing essential information for organisations

H. Guo et al. (Eds.): IRC-SET 2024, PTT, pp. 473–480, 2025.
https://doi.org/10.1007/978-981-96-3770-6_47

and professionals to learn and address these potential risks. Ransomware can leverage on vulnerabilities that are in the wild, by taking advantage of unpatched systems and outdated security measures to infiltrate the systems.

Organisations require a systematic approach to effectively identify and mitigate vulnerabilities that are commonly exploited by ransomware. The lack of accessible, up-to-date, and comprehensive CVE information, especially in environments with restricted internet access, hampers their ability to proactively defend against these evolving threats (Fig. 1).

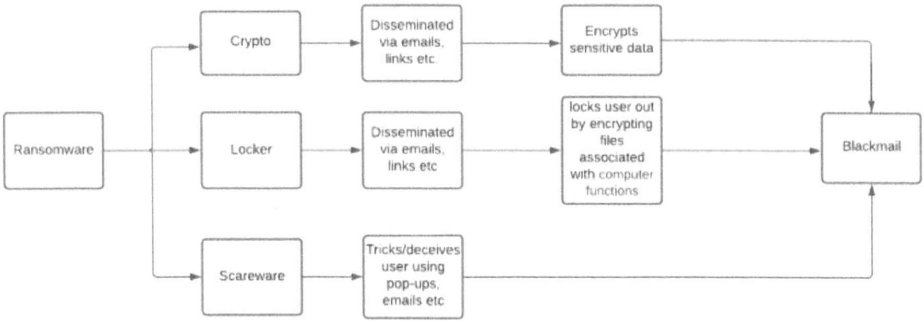

Fig. 1. The various types of ransomware

2 Literature Review

Ransomware continues to pose a significant threat to organisations, exploiting vulnerabilities to infiltrate and damage systems. A comprehensive understanding of these vulnerabilities, particularly those catalogued as Common Vulnerabilities and Exposures (CVEs), is essential for effective mitigation strategies.

Lanza et al. [4] conducted an analysis of the CVEs exploited by ransomware, which is crucial for understanding which vulnerabilities are most commonly targeted. This enables more focused and effective mitigation strategies. Knowing which vulnerabilities are frequently exploited allows the project to prioritise these in data collection and mitigation display implementations.

Brewer [5] underscores the importance of awareness of historical vulnerabilities. Early ransomware relied on exploiting specific vulnerabilities that are well-documented. Including older CVEs in the database instead of just recent ones increases the likelihood of finding attack vectors, especially for legacy systems. Additionally, mapping these CVEs to specific phases of ransomware attacks, such as delivery and execution or file encryption, enables the ability to address vulnerabilities at each stage of the attack lifecycle.

Adamov and Carlsson [6] discuss the nature of ransomware and how keeping defences updated with threat intelligence is important. Their paper discusses the need for constant monitoring and analysis of CVEs to keep up with ransomware trends.

Connolly et al. [7] emphasise the critical role of comprehensive security measures, including technical controls and security education, in mitigating the impact of ransomware attacks. They highlight the need for a holistic approach to addressing vulnerabilities by prioritising CVEs most relevant to ransomware, particularly those affecting an organisation's security posture.

Prioritising frequently exploited vulnerabilities, including older CVEs, and mapping them to specific ransomware attack phases can significantly enhance an organisation's security posture.

3 Solution

Building upon the insights gathered from the literature review, which emphasised the critical role of Common Vulnerabilities and Exposures (CVEs) in identifying recurring vulnerabilities often exploited by ransomware, our proposed solution focuses on two key components: a local CVE database that filters ransomware related CVEs and an automated checklist program. These components are designed to proactively aid organisations in identifying and addressing vulnerabilities that could be exploited by ransomware, thereby reducing the risk of such attacks.Compared to existing projects like cve-search [8] that display and parse CVE data, this project offers the additional capability of frequent database updates and specialised filtering specifically tailored to ransomware-related vulnerabilities.

3.1 Local CVE Database

To create a comprehensive list of CVEs, the team has scraped data from four primary sources as shown in Table 1. Multiple sources were chosen to ensure its comprehensiveness and accuracy of the information we retrieved.

After identifying the sources, the team retrieved the essential API keys and began scraping data using web scraping techniques. The collected data were then combined into a single dataset. Duplicate entries and rejected CVEs were removed, resulting in approximately 240,000 unique CVEs, ensuring the dataset's accuracy and comprehensiveness.

The local CVE database is stored in a MySQL database and displayed in a django web application. The proposed solution allows efficient and rapid retrieval of CVE data. The web application allows users to search by keyword or CVE ID to find specific vulnerabilities (Fig. 2).

3.2 Ransomware Related CVEs

In the web application, there exists a separate table for CVEs related to ransomware campaigns. Each entry in this table includes information on ransomware groups and comprehensive guidance on how to mitigate the vulnerabiltiy. Mitigations were sourced from various industry-recognised sources, such as NIST and Cloud Security Alliance (CSA). For solutions that require the installation of software, the links to the respective developers' website are included.

Table 1. Sources chosen

Sources	Description	Reason for Selection
NVD	Comprehensive database of CVEs maintained by NIST	Highly reputable and widely used in the cybersecurity community; provides detailed and standardised information on vulnerabilities
MITRE CVE	Detailed CVE information including impact scores and related exploits	Offers in-depth information useful for risk assessment and analysis; includes impact metrics and exploit data
OpenCVE	The official CVE database managed by MITRE Corporation	Centralised and authoritative source for CVE entries; widely recognized and trusted in the cybersecurity field
CirCL	Provides vulnerability information and CVE data	Valuable for understanding regional cybersecurity threats and trends

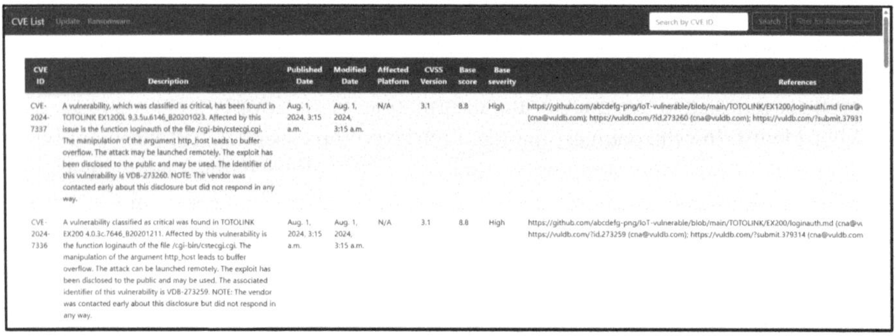

Fig. 2. CVE entries in django web app

In order to identify vulnerabilities that have been exploited in ransomware attacks, multiple data sources and methodologies have been used to ensure wider coverage and accuracy. Information was gathered from CISA's Known Exploited Vulnerabilities Catalog, NVD's detailed CVE data, and scraped reports and white papers. These sources were instrumental in identifying the ransomware families associated with the CVEs. This approach ensures comprehensive identification of relevant CVEs and provides critical information, making the database more effective for ransomware prevention (Fig. 3).

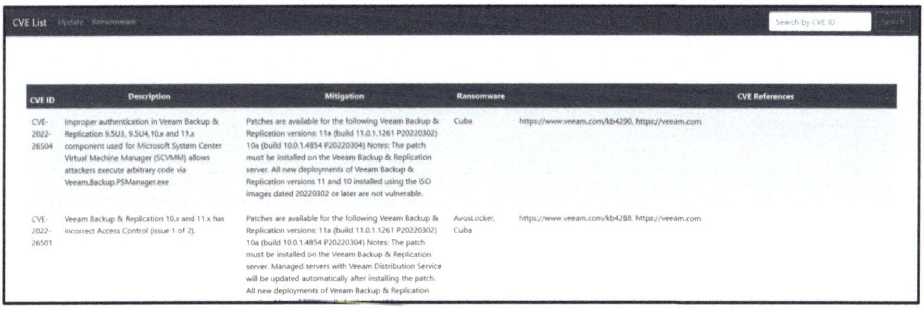

Fig. 3. GUI page for showing only ransomware related CVEs

3.3 CVE Database Update

To ensure that the CVE database remains current and updated, the web app can be used to manually update the local database by extracting the latest entries from NVD APO. Furthermore, it can be automatically executed on a daily/weekly basis via a cron job on linux systems. On a windows system, the script can be executed weekly with the use of scheduled tasks (Fig. 4).

Fig. 4. Updating database using web app

3.4 Automated Checklist Program

Using the Django framework and Python, another web application is created to assess the user's environment. This web application leverages Python scripts that utilise the `os` module and `subprocess` module to directly check for the presence of specific antivirus software, firewalls, agent-based log collection, and installed software versions. The environment check can be initiated through the application, typically completing in approximately 30 s, with the results automatically saved to a text file upon completion (Fig. 5).

Program	Status
Firewall	Windows Firewall: ON, Version: Windows-11-10.0.22621-SP0
Antivirus	Windows Defender, Version: Unknown version
Firewall logging	domainprofile Logging: ON, Version: Unknown version, privateprofile Logging: ON, Version: Unknown version, publicprofile Logging: ON, Version: Unknown version
Open Ports	Listening Ports: , 49664, 49665, 49666, 49667, 49668, 49669, 49670, 4097, 4096, 49673, 49674, 49675, 49676, 51726, 51727, 1042, 1043, 51730, 49686, 49687, 49688, 17945, 4125, 10782, 49697, 5159, 10795, 10801, 53, 10810, 60483, 60484, 80, 65113, 22112, 54368, 10853, 10854, 10862, 10864, 10865, 10866, 10867, 10868, 10870, 10871, 10872, 10873, 9848, 62587, 62588, 17532, 62589, 62590, 10361, 2177, 52358, 135, 52360, 52361, 52359, 139, 4098, 54925, 137, 138, 4608, 4767, 54950, 9382, 54955, 3702, 10436, 10439, 27339, 13010, 5851, 10459, 50911, 50912, 9956, 13030, 13031, 13032, 5353, 3306, 6379, 5355, 10482, 4362, 49419, 4364, 51979, 51980, 52503, 52504, 10010, 33060, 50473, 51500, 4403, 9012, 9013, 52023, 6463, 8000, 52544, 59722, 60747, 59723, 59724, 59726, 64847, 50000, 10752, 56657, 56658, 59729, 51541, 59733, 59728, 59732, 59731, 59730, 56659, 53084, 10755, 1900, 63342, 10616, 902, 10632, 912, 8080, 60307, 4500, 1434, 55207, 5040, 63415, 3000, 5050, 443, 445, 62403, 59725, 4564, 59727, 10714, 10717, 10718, 2020, 2021, 2022, 10726, 65001, 10733, 500, 51708
Syslog	Not detected
Agent-based Log Collection	Not detected
	Python 3.11.8 Development Libraries (64-bit): 3.11.8150.0, Python 3.11.8 pip Bootstrap (64-bit): 3.11.8150.0, Python 3.8.10 Tcl/Tk Support (64-bit): 3.8.10150.0, Python 3.8.10 Test Suite (64-bit): 3.8.10150.0, Python 3.11.8 Utility Scripts (64-bit): 3.11.8150.0, Python 3.8.10 Add to Path (64-bit): 3.8.10150.0, Python 3.8.10 Standard Library (64-bit): 3.8.10150.0, Python 3.8.10 Utility Scripts (64-bit): 3.8.10150.0, Python 3.11.8 Tcl/Tk Support (64-bit): 3.11.8150.0, Python 3.8.10 Development Libraries (64-bit): 3.8.10150.0, Python 3.8.10 Core Interpreter (64-bit): 3.8.10150.0, Python 3.11.8 Test Suite (64-bit): 3.11.8150.0

Fig. 5. Environment check results on GUI

4 Implementation

The local CVE database becomes more effective when integrated with Security Information and Event Management (SIEM) tools by providing additional information on CVEs identified in SIEM alerts, which enhances the accuracy and relevance of threat detection. Additionally, it can provide support during incident response by offering immediate access to detailed vulnerability information, allowing security teams to quickly apply the necessary mitigations.

Furthermore, the database's offline accessibility is a significant advantage for organisations operating in environments with restricted or no internet access, such as air-gapped networks commonly found in organisations. It ensures that critical vulnerability data remains available even without an external connection.

The environment checks provide a quick way to obtain information on a system, and checking if programs found are on a version with known vulnerabilities. For instance, software and version numbers obtained from the check can be checked against the CVE database to find out what vulnerabilities exist and how users can patch them.

5 Limitations

The primary challenge in obtaining detailed mitigation or patch information for ransomware CVEs lies in the automation process. This information is often vendor-specific and subject to frequent format changes, which makes automated extraction difficult. As a result, our current approach also involves manually entering this data into the database after consulting each vendor's site for the necessary details.

Although Feedly is one of the few platforms used to identify CVEs associated with ransomware and their corresponding ransomware strains, it does not consistently provide this information in every feed or article. As a result, the team must also manually review CVEs to determine whether it has been associated with ransomware attacks. Additionally, for the data scraping process, the Feedly API used during this project was on a trial period. Production builds would require an API key.

With regard to scanning user systems and displaying installed programs and linking them to vulnerabilities, this was deemed too difficult to achieve as the current way CVEs display vulnerable versions is too different from how Windows and Ubuntu Operating systems display their program versions.

6 Future Works

The CVE scraping program could be expanded to cross-reference more sources, and perhaps have the database include more detailed instructions on which steps need to be taken to mitigate or patch a given CVE. A solution using The CVE database with this information would be useful for industries that require their computers to be physically isolated and do not have access to the internet, or automatic update solutions. Given that some malwares, and most ransomware needs to communicate with a server, the data collated could highlight known IP addresses or domains associated with a CVE or CWE that has been previously exploited, so that said domains and IP addresses can be blocked to prevent the malware from activating if it were to be dropped onto a system.

The checklist for the user's environment could also be expanded by correlating installed programs with Known CVEs or CWEs within the database. Currently, the checks account for Windows 11 and Ubuntu 20.04 Operating Systems, more Operating systems can be accounted for in these checks in future. The checklist could be expanded to check for software on other devices on the network, with methods such as SSH for linux systems or WMIC for windows systems.

7 Conclusion

This project addresses the critical need for organisations to manage and mitigate ransomware threats by leveraging an updated, comprehensive CVE database. By focusing on vulnerabilities commonly exploited by ransomware, the project provides a targeted approach to identifying and addressing security risks. Additionally, the automated checklist program offers a practical tool for assessing system security, identifying potential vulnerabilities based on installed software versions, and providing actionable insights for remediation. Future enhancements, such as expanding the database's sources, incorporating more detailed mitigation instructions, and improving the checklist program's coverage, will further strengthen the system's effectiveness in safeguarding organisational security. While there are limitations, such as the challenges of obtaining detailed mitigation information and the constraints of the current system's scope, the foundation laid by this project offers a meaningful contribution to the ongoing effort to defend against ransomware attacks.

Acknowledgement. The authors deeply acknowledge the Professors from Singapore Institute of Technology (SIT) for their help and support for this project research. The authors would also like to thank the industry supervisor and the organisation for their continued support and guidance. The authors would like to declare that there is no conflict of interest.

This research project is supported by an ignition grant with Grant No. R-IE2-A405-0002, Singapore Institute of Technology and Ministry of Education, Singapore.

References

1. Morgan, S.: Global ransomware damage costs predicted to exceed $265 billion by 2031, Cybercrime Magazine (2023). https://cybersecurityventures.com/global-ransomware-damage-costs-predicted-to-reach-250-billion-usd-by-2031/. Accessed 24 July 2024

2. Parkinson, S.: Use of access control to minimise ransomware impact, Network Security (2017). https://www.sciencedirect.com/science/article/abs/pii/S1353485817300697. Accessed 24 July 2024
3. Chin, K.: The impact of cybercrime on the economy: Upguard, Upguard (2023). https://www.upguard.com/blog/the-impact-of-cybercrime-on-the-economy. Accessed 24 July 2024
4. Lanza, C., Lahmadi, A., Osmond, F.: An empirical study of ransomware vulnerabilities descriptions. In: 10th International Conference on Information Systems Security and Privacy, 2024, pp. 146–153 (2024)
5. Brewer, R.: Ransomware attacks: detection, prevention and cure. Netw. Secur. **9**, 5–9 (2016). https://doi.org/10.1016/s1353-4858(16)30086-1
6. Adamov, A., Carlsson, A.: The state of ransomware. trends and mitigation techniques. In: 2017 IEEE East-West Design; Test Symposium (EWDTS) [Preprint] (2017). https://doi.org/10.1109/ewdts.2017.8110056
7. Connolly, L.Y., Wall, D.S., Lang, M., Oddson, B.E.: An empirical study of ransomware attacks on organisations: an assessment of severity and salient factors affecting vulnerability. J. Cybersecur. **6**(1) (2020). tyaa023. https://doi.org/10.13140/RG.2.2.25264.92169
8. CVE-search- a tool to perform local searches for known vulnerabilities Github. Available at: https://github.com/cve-search/cve-search. Accessed 9 Aug 2024

The Feasibility of OpenSearch's Security Analytics with Wazuh

Cheryl Ching Tay, Darshan Kumar, Rachel Ying Xuan Poh, Randy Yi Ping Cheang, Villarico Gabriel Babiera, Huaqun Guo(✉), and Liming Lu

Infocomm Technology Cluster, Singapore Institute of Technology, Singapore, 1 Punggol Coast Road, Singapore 828608, Singapore
{2203483,2202089,2202272,2201708, 2200479}@sit.singaporetech.edu.sg, {huaqun.guo, liming.lu}@singaporetech.edu.sg

Abstract. In the context of rapidly advancing cyber threats, ransomware stands out as a particularly destructive challenge that exploits vulnerabilities to significantly disrupt operations. This project explores the feasibility of using OpenSearch's Security Analytics in conjunction with Wazuh to detect cyberattacks through log data correlation analysis. Various integration methods, including server integration, indexer integration, and direct plugin integration, were tested for its ability to effectively correlate alerts and identify potential ransomware threats in simulated settings. The findings provide essential insights into optimal practices and configurations for using OpenSearch's Security Analytics to enhance Wazuh's functionality.

Keywords: Wazuh · OpenSearch · Security Analytics · Security Information and Event Management · Ransomware · Correlation Analysis

1 Introduction

In recent years, ransomware has emerged as a significant cyber threat, affecting organisations across various industries. This form of malware encrypts victims' files and demands a ransom for their release, often causing severe operational and financial disruptions [1]. Notable incidents such as WannaCry attack in 2017 [2] and the more recent Colonial Pipeline attack in 2021 [3] have highlighted the devastating impact of ransomware, emphasising the need for robust detection and prevention mechanisms. Hence, this paper will explore the feasibility of using OpenSearch's Security Analytics with Wazuh to perform correlation to detect potential ransomware attacks.

2 Background and Motivation

Ransomware attacks are a significant threat that could endanger any business. These attacks can compromise computer systems, encrypt files that are important to the company and steal other sensitive information to demand a ransom amount in exchange for

H. Guo et al. (Eds.): IRC-SET 2024, PTT, pp. 481–489, 2025.
https://doi.org/10.1007/978-981-96-3770-6_48

the safe return of these compromised assets. The consequence of losing these important items may cause the business to have major disruptions and financial losses. Therefore, it is critical to detect and prevent ransomware by identifying infection sources, recognising attack activities, and reducing their impact.

This paper will focus on the correlation of Wazuh alerts to detect potential ransomware attacks. Specifically, it will evaluate the feasibility of using OpenSearch's Security Analytics plugin with Wazuh to correlate Wazuh alerts.

3 Literature Review

Ransomware is one of the most common types of cybersecurity threats that holds victims' data hostage, through the use of encryption techniques, in exchange for ransom [4]. Ransomware can be categorised into two main categories, crypto and locker [5].

Crypto ransomware makes sensitive files and data inaccessible by encrypting them, but leaves the computer functionalities intact. After the files are encrypted, victims will be notified and provided with instructions on how to pay the ransom for the decryption key. Typically, there is also a countdown and warning to victims that their files will no longer be accessible after the time has expired [6, 7].

Locker ransomware, on the other hand, locks the entire device and only allows access to a screen showing the ransom note, preventing victims from accessing their files and applications. In this variant, only the device is locked while the data stored within the device is left untouched [7, 8].

To achieve a scalable and adaptable system for ransomware detection, the use of expert systems can be used to enhance security. Expert systems can play a crucial role in combating ransomware by leveraging predefined rules and the expertise of cybersecurity professionals to identify patterns typical of ransomware behaviour, such as sudden encryption of numerous files or blocking system functionalities. They use heuristic analysis to detect new ransomware variants by analysing program behaviours and spotting anomalies, flagging suspicious activities like unauthorised file encryption [9]. With real-time monitoring and response capabilities, expert systems continuously analyse system activities and network traffic, providing immediate detection and isolation of ransomware to minimise its impact. These systems automate decision-making processes, enabling them to disconnect infected devices, initiate system restores from safe backups, or notify IT teams with detailed incident reports upon detecting ransomware. Advanced expert systems also incorporate machine learning algorithms, allowing them to adapt to new threats by learning from past incidents and updating their knowledge base. This continuous learning and adaptation improve detection accuracy over time, ensuring effectiveness against evolving ransomware [10]. Integrating expert systems into cybersecurity frameworks can enhance organisations' abilities to detect, respond to, and mitigate ransomware attacks, protecting sensitive data and maintaining operational integrity.

4 Research and Development

4.1 Wazuh

Wazuh is a free and open-source security platform designed to safeguard digital assets from cybersecurity threats. It integrates both Security Information and Event Management (SIEM) and Extended Detection and Response (XDR) capabilities, therefore providing a unified platform for comprehensive security monitoring, threat detection and incident response [11, 12].

The Wazuh architecture consists of agents, which are deployed on monitored endpoints to gather and send data, as well as three central components, the server, indexer and dashboard, which helps with the collection, analysis and visualisation of data [13]. Figure 1 shows a simplified Wazuh architecture.

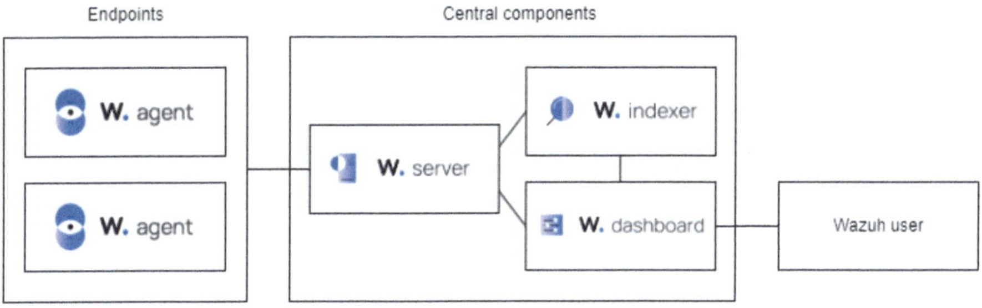

Fig. 1. Simplified Wazuh architecture adapted from [14].

4.2 OpenSearch

OpenSearch is an open-source search and analytics suite which is usually used for application monitoring on a broad range. It allows for high scalability on the system and is also able to provide fast access to large amounts of data with an easy to explore dashboard enabling further insight [15].

Presently, some functionalities of OpenSearch are already integrated into Wazuh. However, the current integration does not fully utilise OpenSearch's advanced analytics features, such as Security Analytics, that can help to detect and analyse threats.

4.3 OpenSearch Security Analytics

Security Analytics is an OpenSearch Security Information and Event Management (SIEM) solution to mitigate security threats. It is capable of correlating events, defined by the correlation rules, across systems of different types. This plugin uses detectors, which must be defined by the user, to evaluate logs to generate findings. The findings can then be used to generate the correlation graph to provide a visualisation of the relationship between different findings [16].

4.4 OpenSearch Security Analytics with Wazuh

To utilise both Security Analytics and Wazuh simultaneously, there are several ways to integrate them together, namely three.

Server Integration. The Wazuh server integration with OpenSearch works by reading the */var/ossec/logs/alerts/alerts.json* file and forwarding the alerts to the OpenSearch platform using a data forwarder. To implement this integration, one needs to install a data forwarder on the same system as the Wazuh server. In multi-node configurations, install the data forwarder on each Wazuh server node. Ensure that the Wazuh server configuration file (*/var/ossec/etc./ossec.conf*) has the jsonout_output option set to yes for logging alerts in JSON format. This setup is recommended if one lacks the resources to host both the OpenSearch indexer and the Wazuh indexer, especially for large-scale Wazuh deployments generating numerous alerts.

Indexer Integration. To set up the Wazuh indexer integration with OpenSearch, one needs a running Wazuh indexer and Logstash as the data forwarder. Logstash queries the Wazuh indexer for alerts and forwards them to the OpenSearch indexer. This requires installing Logstash either on a dedicated server or on the server hosting the OpenSearch indexer. This integration is ideal if one wants to keep analysing security data within Wazuh while archiving it in a data lake for storage or out-of-band analysis. Both the Wazuh indexer and the OpenSearch indexer will create indexes for the same alerts, which can lead to redundant data storage and increased resource costs.

Direct Plugin Integration. This method involves directly installing the OpenSearch Security Analytics plugin onto the Wazuh server, which operates within Wazuh's existing data collection and analysis infrastructure.

5 Alternatives

5.1 Composite Rules

Composite rules in Wazuh are designed to detect complex attack patterns by correlating multiple events over a specified period. These rules are particularly useful for identifying advanced persistent threats (APTs) that often employ a sequence of actions to penetrate and exploit systems. Composite rules function by collecting and normalising log data from a variety of courses, including network devices, servers, applications, and cloud services (Event Collection). These rules then correlate the gathered events according to predefined criteria to detect patterns indicative of security events (Event Correlation). For instance, a composite rule might identify a security threat by detecting a sequence of a failed login attempt followed by a successful one within a brief period. Once these specified criteria are met, an alert is generated, offering valuable insights into potential security incidents and enabling proactive response measures (Alert Generation).

Composite rules offer several benefits that enhance cybersecurity measures. Firstly, by correlating multiple events, these rules are capable of detecting sophisticated attacks that single-event rules might overlook, improving overall detection capabilities. Secondly, they provide greater contextual awareness, enabling security teams to comprehend the sequence of actions and assess potential impacts more effectively. Lastly, by focusing on identifying patterns rather than isolated events, composite rules significantly reduce the occurrence of false positives. This targeted approach not only streamlines the detection process but also enhances the accuracy of threat identification (Fig. 2).

```
<group name="ossec,syslog,sshd,">
<rule id="100001" level="3" overwrite="yes">
    <if_sid>500,5762</if_sid>
    <match>Manager started|Connection reset|Failed password|Failed keyboard|authentication error</match>
    <description>Wazuh server started,sshd connection reset,aut5hentication failed</description>
    <group>pci_dss_10.6.1,gpg13_10.1,gdpr_IV_35.7.d,hipaa_164.312.b,nist_800_53_AU.6,tsc_CC7.2,tsc_CC7.3
        ,</group>
  </rule>

</group>
```

Fig. 2. Example of a Composite rule.

5.2 Custom Rules

Custom rules in Wazuh allow organisations to tailor the detection capabilities to meet their specific needs. These rules can be created to monitor unique behaviours or to comply with specific compliance requirements. Custom rules in Wazuh are defined by security teams to target specific threats or meet specific requirements. These rules are integrated into Wazuh's event analysis engine, which facilitates real-time monitoring and alerting. When the conditions of these custom rules are met, alerts are generated and configured to notify the relevant personnel promptly. This structured approach ensures that security responses are both timely and tailored to the specific security landscape of the organisation.

Custom rules offer several distinct advantages. They provide flexibility, allowing organisations to address specific security concerns that are unique to their operational environment. This adaptability not only enhances the overall security posture but also ensures that the organisation can meet industry-specific compliance requirements. By tailoring rules to fit precise needs, organisations can ensure that their security measures are both effective and compliant with relevant regulations and standards.

6 Limitations

6.1 Interaction of Wazuh with OpenSearch Security Analytics

Wazuh uses its own XML-based ruleset to analyse logs and generate alerts. These alerts are stored in *alerts.json* and indexed as *wazuh-alerts-** indices. On the other hand, the OpenSearch Security Analytics plugin uses SIGMA rules to analyse logs and events. The correlation engine in the Security Analytics plugin operates based on these SIGMA rules.

The primary challenge in integrating Wazuh with OpenSearch Security Analytics lies in the difference between the rulesets used by each system. Wazuh's XML rules are not directly compatible with SIGMA rules used by OpenSearch Security Analytics. To utilise the Security Analytics plugin's correlation engine, Wazuh alerts need to be fed into OpenSearch and then processed using the SIGMA rules defined within the Security Analytics plugin. Filebeat forwards these alerts to OpenSearch, creating the *wazuh-alerts-** indices. Since the Security Analytics plugin is installed directly onto Wazuh, it automatically utilises these indices for further analysis and correlation.

The *wazuh-alerts-** indices, populated by the alerts.json file, are ready for analysis within OpenSearch Security Analytics. The plugin uses these indices to perform correlation using SIGMA rules. This integration requires mapping Wazuh alert fields to the corresponding fields expected by the SIGMA rules to ensure accurate correlation.

6.2 Direct Plugin Integration

Despite successfully installing the plugin directly into the all-in-one Wazuh server, we faced issues getting the Security Analytics detector to pick up Wazuh-generated alerts. The detector could not pick up any alerts other than the two that are shown in Fig. 3.

Attempts to create custom detector rules, such as the ones shown in Fig. 4, also did not help in detecting the alerts that were generated from Wazuh as nothing was showing up as seen in Fig. 5, despite there being alerts generated in the Wazuh server as seen in Fig. 6.

While searching for possible solutions on the Security Analytics Github repository [17], we came across a bug report [18] stating that false findings were generated by the detector for the rule "Publicly Accessible RDP Services." This led us to suspect that the findings for the "Raw Disk Access Using Illegitimate Tools" rule could also be a bug, however, searching through the repository did not yield any findings.

Fig. 3. The two rules picked up by the Security Analytics detector.

Fig. 4. Custom detector rules created to trigger the plugin.

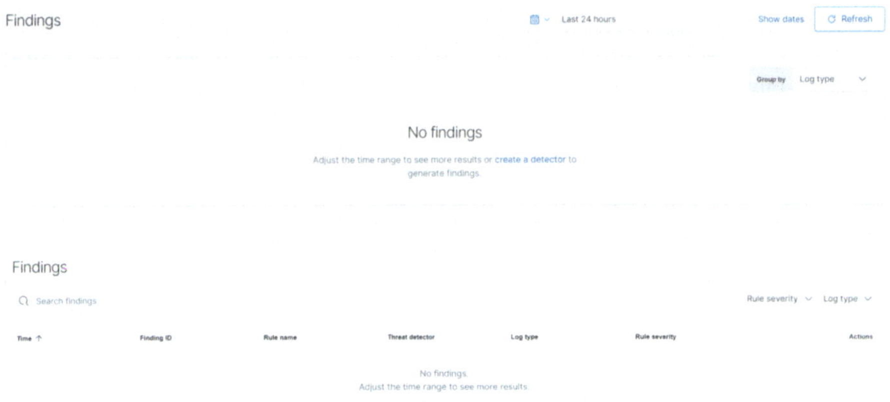

Fig. 5. No findings in the Security Analytics plugin.

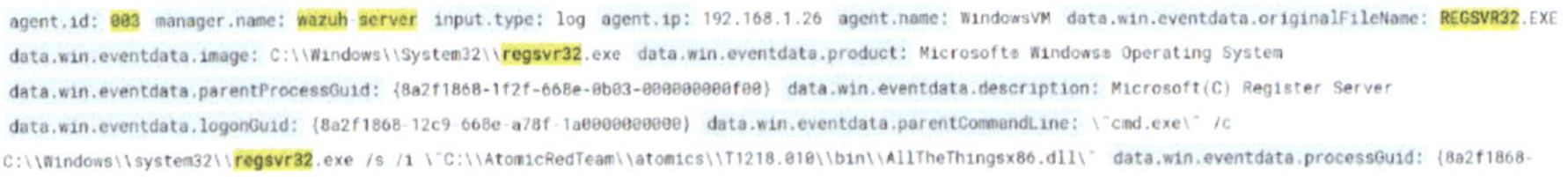

Fig. 6. One of the alerts generated seen on the Wazuh server.

6.3 Custom and Composite Rules

While composite and custom rules in Wazuh offer significant advantages, they also come with several limitations. Firstly, the manual effort required to implement, maintain, and continuously update a large number of these rules is considerable. This process involves constantly monitoring and fine-tuning to maintain effectiveness. Moreover, the creation of effective composite and custom rules demands an in-depth understanding of both the operational environment and potential threats, adding complexity and consuming significant time.

Scalability is another limitation of implementing composite and custom rules. There's a practical limit to how many custom rules can be implemented and managed. Although Wazuh does not have a strict limit, the system's performance and manageability might suffer if too many rules are in place. Additionally, despite their advanced nature, these rules are still susceptible to generating false positives or missing threats, especially if they are not meticulously defined and regularly updated. Furthermore, Wazuh's limit of handling up to 9,000 logs may prove inadequate for organisations that generate a high volume of logs, potentially hindering comprehensive monitoring and analysis.

7 Conclusion

Through research and hands-on trial and error, we have found that it is very difficult to use the Security Analytics plugin by integrating Wazuh and OpenSearch, whether through indexer or server integration, as it is rather tedious and there is insufficient documentation and support on the integration. Furthermore, integrating both Wazuh and OpenSearch, regardless through indexer or server integration, will also consume more resources compared to just using Wazuh as a standalone. We have also discovered that it is infeasible to use the Security Analytics plugin with Wazuh at this stage with version 4.7.5, as Wazuh uses its own set of XML rules while the plugin uses SIGMA rules, which are incompatible. However, Wazuh is currently working on integrating the plugin for its version 5.0 release. We have managed to use composite rules for correlation, and we also found out that maintaining a correlation database will be very tedious, as each correlation rule needs to be manually added. This is a part of our future work of further research for correlation analysis.

Acknowledgement. We would like to extend their sincere gratitude to the Professors from the Singapore Institute of Technology for their invaluable guidance and support for this research project. We would also like to thank the industry supervisors and the organization for their continued support and guidance.

This research project is supported by an ignition grant with Grant No. R-IE2-A405-0002, Singapore Institute of Technology and Ministry of Education, Singapore.

References

1. Morgan, S., Sausalito, C.: 'Global Ransomware Damage Costs Predicted To Exceed $265 Billion By 2031', Cybercrime Magazine, 2023. https://cybersecurityventures.com/global-ransomware-damage-costs-predicted-to-reach-250-billion-usd-by-2031/. Accessed 20 July 2024
2. 'What is WannaCry ransomware?', Kaspersky. https://www.kaspersky.com/resource-center/threats/ransomware-wannacry. Accessed 20 July 2024
3. Easterly, J., Fanning, T.: The Attack on Colonial Pipeline: What We've Learned & What We've Done Over the Past Two Years, Cybersecurity & Infrastructure Security Agency. https://www.cisa.gov/news-events/news/attack-colonial-pipeline-what-weve-learned-what-weve-done-over-past-two-years. Accessed 20 July 2024
4. Kosinski, M.: What is ransomware? IBM. https://www.ibm.com/topics/ransomware. Accessed 20 July 2024
5. Razaulla, S., et al.: The age of ransomware: a survey on the evolution, taxonomy, and research directions. IEEE Access **11**, 40698–40723 (2023). https://doi.org/10.1109/ACCESS.2023.3268535
6. Gonzalez, D., Hayajneh, T.: Detection and prevention of crypto-ransomware. In: 2017 IEEE 8th Annual Ubiquitous Computing, Electronics and Mobile Communication Conference (UEMCON), IEEE, Oct. 2017, pp. 472–478 (2017). https://doi.org/10.1109/UEMCON.2017.8249052
7. Richardson, R., North, M.M.: Ransomware: evolution, mitigation and prevention. Int. Manag. Rev. **13**(1), 10 (2017)
8. Wickramasinghe, S.: Most Common Types of Ransomware Today, Splunk. https://www.splunk.com/en_us/blog/learn/ransomware-attack-types.html. Accessed 20 July 2024
9. Aslan, O., Samet, R.: A comprehensive review on malware detection approaches. IEEE Access **8**, 6249–6271 (2020). https://doi.org/10.1109/ACCESS.2019.2963724
10. Mahdavifar, S., Ghorbani, A.A.: DeNNeS: deep embedded neural network expert system for detecting cyber attacks. Neural Comput. Appl. **32**(18), 14753–14780 (2020). https://doi.org/10.1007/s00521-020-04830-w

11. 'Overview', Wazuh. https://wazuh.com/platform/overview/. Accessed 20 July 2024
12. 'Quickstart', Wazuh. https://documentation.wazuh.com/current/quickstart.html. Accessed 20 July 2024
13. 'Components - Getting started with Wazuh', Wazuh. https://documentation.wazuh.com/current/getting-started/components/index.html. Accessed 20 July 2024
14. 'Architecture - Getting started with Wazuh', Wazuh. https://documentation.wazuh.com/current/getting-started/architecture.html. Accessed 20 July 2024
15. 'About OpenSearch', OpenSearch. https://opensearch.org/about.html. Accessed 20 July 2024
16. 'About Security Analytics', OpenSearch. https://opensearch.org/docs/latest/security-analytics/. Accessed 20 July 2024
17. 'security-analytics'. https://github.com/opensearch-project/security-analytics. Accessed 25 July 2024
18. '[BUG] Detector Generated False Findings For Rule "Publicly Accessible RDP Service". https://github.com/opensearch-project/security-analytics/issues/701. Accessed 25 July 2024

Author Index

H. Guo et al. (Eds.): IRC-SET 2024, PTT, pp. 491–493, 2025.
https://doi.org/10.1007/978-981-96-3770-6

The manufacturer's authorised representative in the EU is Springer Nature Customer Service Centre GmbH, Europaplatz 3, 69115 Heidelberg, Germany. If you have any concerns regarding our products, please contact ProductSafety@springernature.com

Printed and bound by CPI Group (UK) Ltd, Croydon, CR0 4YY
01/10/2026
02225911-0013